Molecular Biology Intelligence Unit

Calreticulin

Second Edition

Paul Eggleton, Ph.D.
Peninsula Medical School, Devon, U.K.
MRC Immunochemistry Unit
Department of Biochemistry
University of Oxford
Oxford, U.K.

Marek Michalak, Ph.D.
CIHR Membrane Protein Research Group
Department of Biochemistry
University of Alberta
Edmonton, Alberta, Canada

Landes Bioscience / Eurekah.com
Georgetown, Texas
U.S.A.

Kluwer Academic / Plenum Publishers
New York, New York
U.S.A.

Calreticulin
Second Edition
Molecular Biology Intelligence Unit

Eurekah.com / Landes Bioscience
Kluwer Academic / Plenum Publishers

Designed by Celeste Carlton

Printed in the U.S.A.

Kluwer Academic / Plenum Publishers, 233 Spring Street, New York, New York, U.S.A. 10013
http://www.wkap.nl/

Please address all inquiries to the Publishers:
Eurekah.com / Landes Bioscience, 810 South Church Street
Georgetown, Texas, U.S.A. 78626
Phone: 512/ 863 7762; FAX: 512/ 863 0081
www.Eurekah.com
www.landesbioscience.com

Calreticulin, 2nd Edition, edited by Paul Eggleton and Marek Michalak, Landes / Kluwer dual imprint / Landes series: Molecular Biology Intelligence Unit

ISBN: 0-306-47845-5

Library of Congress Cataloging-in-Publication Data

Calreticulin / [edited by] Paul Eggleton, Marek Michalak.-- 2nd ed.
p. ; cm. -- (Molecular biology intelligence unit)
Includes bibliographical references and index.
ISBN 0-306-47845-5
1. Calreticulin.
[DNLM: 1. Calreticulin. QU 55 C165 2003] I. Eggleton, Paul, Ph.D.
II. Michalak, Marek. III. Molecular biology intelligence unit
(Unnumbered)
QP552.C29C35 2003
572'.69--dc21

2003012624

CONTENTS

EDITORS

Paul Eggleton, Ph.D.
Peninsula Medical School, Devon, U.K.
MRC Immunochemistry Unit
Department of Biochemistry
University of Oxford
Oxford, U.K.
Chapters 1, 17, 21

Marek Michalak, Ph.D.
CIHR Membrane Protein Research Group
Department of Biochemistry
University of Alberta
Edmonton, Alberta, Canada
Chapter 1

CONTRIBUTORS

Raju Adhikari
Everest Biotech Ltd.
Oxford Biobusiness Centre
Littlemore, Oxford, U.K.
Chapter 9

Joohong Ahnn
Department of Life Science
Kwangju Institute of Science
and Technology
Pukgu, Kwangju, Republic of Korea
Chapter 22

Serge Arnaudeau
Department of Physiology
University of Geneva
Geneva, Switzerland
Chapter 13

John J.M. Bergeron
Department of Biochemistry
Department of Anatomy
and Cell Biology
McGill University
Montreal, Quebec, Canada
Chapter 4

Ben E. Black
Center for Cell Signaling
Department of Biochemistry
and Molecular Genetics
Cell and Molecular Biology Program
University of Virginia
Charlottesville, Virginia, U.S.A.
Chapter 8

Edward R. Block
Division of Pulmonary Medicine
University of Florida and Research
Service
VA Medical Center
Gainesville, Florida, U.S.A.
Chapter 19

A. Brown
Boots Science Institute
School of Pharmaceutical Sciences
University of Nottingham
Nottingham, U.K.
Chapter 21

R. Caddick
Boots Science Institute
School of Pharmaceutical Sciences
University of Nottingham
Nottingham, U.K.
Chapter 21

Patricia Camacho
Department of Physiology
University of Texas Health Science Center at San Antonio
San Antonio, Texas, U.S.A.
Chapter 12

Mounia Chami
Department of Experimental and Diagnostic Medicine
Section of General Pathology
Center for the Study of Inflammatory Diseases (CESMI)
University of Ferrara
Ferrara, Italy
Chapter 11

Alain Debrabant
Division of Emerging and Transfusion Transmitted Diseases
CBER, FDA
Bethesda, Maryland, U.S.A.
Chapter 20

Nicolas Demaurex
Department of Physiology
University of Geneva
Geneva, Switzerland
Chapter 13

Suzanne Donnelly
Department of Rheumatology
St. George's Hospital and Medical School
London, U.K.
Chapter 17

Dennis M. Dwyer
Laboratory of Parasitic Diseases
NIAID, NIH
Bethesda, Maryland, U.S.A.
Chapter 20

Lars Ellgaard
Institute of Biochemistry
ETH Zürich
Hoenggerberg, Zürich, Switzerland
Chapter 3

Tim Elliott
Cancer Sciences Division
University of Southampton School of Medicine
Southampton General Hospital
Southampton, U.K.
Chapter 9

Maud Frieden
Department of Physiology
University of Geneva
Geneva, Switzerland
Chapter 13

N. Girod
Boots Science Institute
School of Pharmaceutical Sciences
University of Nottingham
Nottingham, U.K.
Chapter 21

Silvia M. Goicoechea
Department of Cell and Molecular Physiology
University of North Carolina
Chapel Hill, North Carolina U.S.A.
Chapter 18

Lei Guo
CIHR Membrane Protein Research Group
Department of Biochemistry
University of Alberta
Edmonton, Alberta, Canada
Chapter 23

Kjell O. Håkansson
August Krogh Institute
Laboratory of Cellular and Molecular Physiology
Universitetsparken Copenhagen, Denmark
Chapter 2

Ari Helenius
Institute of Biochemistry
ETH Zürich
Hoenggerberg, Zürich, Switzerland
Chapter 3

Peter M. Henson
Program in Cell Biology
Department of Pediatrics
National Jewish Medical and Research Center
Denver, Colorada, U.S.A.
Chapter 15

D.S.W. Hooi
Boots Science Institute
School of Pharmaceutical Sciences
University of Nottingham
Nottingham, U.K.
Chapter 21

Dorothy Hudig
Cell and Molecular Biology Program
School of Medicine
University of Nevada
Reno, Nevada, U.S.A.
Chapter 14

Linu John
Genentech Inc.
San Francisco, California, U.S.A.
Chapter 12

Steven J. Johnson
Sir William Dunn School of Pathology
University of Oxford,
Oxford, U.K.
Chapters 2, 21

Reza Karimi
Cell and Molecular Biology Program
School of Medicine
University of Nevada
Reno, Nevada, U.S.A.
Chapter 14

Michael R. Leach
Department of Biochemistry
University of Toronto
Toronto, Ontario, Canada
Chapter 6

Jin Il Lee
Department of Life Science
Kwangju Institute of Science and Technology
Pukgu, Kwangju, Republic of Korea
Chapter 22

Nancy Lee
Division of Emerging and Transfusion Transmitted Diseases
CBER, FDA
Bethesda, Maryland, U.S.A.
Chapter 20

Yong D. Li
Division of Pulmonary Medicine
University of Florida
Gainesville, Florida, U.S.A.
Chapter 19

Yun Li
Department of Physiology
University of Texas Health Science Center at San Antonio
San Antonio, Texas, U.S.A.
Chapter 12

Paola Mariani
Department of Biology
University of Padova
Padova, Italy
Chapter 10

J.E. Murphy-Ullrich
Cell Adhesion and Matrix Research Center
University of Alabama at Birmingham
Birmingham, Alabama, U.S.A.
Chapter 18

Hira L. Nakhasi
Division of Emerging and Transfusion Transmitted Diseases
CBER, FDA
Bethesda, Maryland, U.S.A.
Chapter 20

William M. Nauseef
Inflammation Program and Department of Medicine
Roy J. and Lucille A. Carver College of Medicine
University of Iowa and Veterans Affairs Medical Center
Iowa City, Iowa, U.S.A..
Chapter 7

Lorella Navazio
Department of Biology
University of Padova
Padova, Italy
Chapter 10

Michal Opas
Department of Anatomy and Cell Biology
University of Toronto
Toronto, Ontario, Canada
Chapter 5

Sylvia Papp
Department of Anatomy and Cell Biology
University of Toronto
Toronto, Ontario, Canada
Chapter 5

R. Madelaine Paredes
Department of Physiology
University of Texas Health Science Center at San Antonio
San Antonio, Texas, U.S.A.
Chapter 12

Byung-Jae Park
Department of Life Science
Kwangju Institute of Science and Technology
Pukgu, Kwangju, Republic of Korea
Chapter 22

Bryce M. Paschal
Center for Cell Signaling
Department of Biochemistry and Molecular Genetics
Cell and Molecular Biology Program
University of Virginia
Charlottesville, Virginia, U.S.A.
Chapter 8

Jawaharlal M. Patel
Division of Pulmonary Medicine
University of Florida and Research Service
VA Medical Center
Gainesville, Florida, U.S.A.
Chapter 19

Sandra E. Pike
Experimental Transplantation and Immunology Branch
Center for Cancer Research
National Cancer Institute
National Institutes of Health
Bethesda, Maryland, U.S.A.
Chapter 16

Paolo Pinton
Department of Experimental and Diagnostic Medicine
Section of General Pathology
Center for the Study of Inflammatory Diseases (CESMI)
University of Ferrara
Ferrera, Italy
Chapter 11

D.I. Pritchard
Boots Science Institute
School of Pharmaceutical Sciences
University of Nottingham
Nottingham, U.K.
Chapter 21

R.J. Quinnell
School of Biology
University of Leeds
Leeds, U.K.
Chapter 21

Doina Racila
Department of Dermatology
University of Iowa College of Medicine
Iowa City, Iowa, U.S.A.
Chapter 17

Emil Racila
Holden Cancer Center
University of Iowa College of Medicine
Iowa City, Iowa, U.S.A.
Chapter 17

Rosario Rizzuto
Department of Experimental
and Diagnostic Medicine
Section of General Pathology
Center for the Study of Inflammatory
Diseases (CESMI)
University of Ferrara
Ferrara, Italy
Chapter 11

H. Llewelyn Roderick
Laboratory of Molecular Signalling
The Babraham Institute
Babraham, Cambridge, U.K.
Chapter 12

Richard D. Sontheimer
Department of Dermatology
University of Iowa College of Medicine
Iowa City, Iowa, U.S.A.
Chapter 17

György Szabadkai
Department of Experimental
and Diagnostic Medicine
Section of General Pathology
Center for the Study of Inflammatory
Diseases (CESMI)
University of Ferrara
Ferrara, Italy
Chapter 11

David Y. Thomas
Department of Biochemistry
Department of Anatomy and Cell
Biology
McGill University
Montreal, Quebec, Canada
Chapter 4

Giovanna Tosato
Experimental Transplantation
and Immunology Branch
Center for Cancer Research
National Cancer Institute
National Institutes of Health
Bethesda, Maryland, U.S.A.
Chapter 16

David B. Williams
Department of Biochemistry
University of Toronto
Toronto, Ontario, Canada
Chapter 6

Lei Yao
Experimental Transplantation
and Immunology Branch
Center for Cancer Research
National Cancer Institute
National Institutes of Health
Bethesda, Maryland, U.S.A.
Chapter 16

Jianliang Zhang
Division of Pulmonary Medicine
University of Florida
Gainesville, Florida, U.S.A.
Chapter 19

Anna Zuppini
Department of Biology
University of Padova
Padova, Italy
Chapter 10

ABBREVIATIONS

Å	Ångstrom
AAB	autoantibody(s)
ACE	angiotensin converting enzyme
ALLM	acetyl-leu-leu-norleucinal
ALLnL	acetyl-leu-leu-methional
Ang	angiotensin
ATP	adenosine triphosphate
β2m	β2-microglobulin
BAE	bovine aortic endothelial
BAPTA-AM	1,2-bis(2-aminophenoxy) ethane-N,N,N,N'-tetraacetic acid-AM
BiP	binding protein
BiP	immunoglobulin-heavy-chain-binding protein
BSA	bovine serum albumin
C1q	first subcomponent of complement 1
C319A	mutant form of myeloperoxidase with the cysteine at codon 319 replaced by alanine
$[Ca^{2+}]_{cyt}$	cytosolic Ca^{2+} concentration
$[Ca^{2+}]_{ER}$	endoplasmic reticulum luminal Ca^{2+} concentration
Ca	calcium
Ca^{2+}	calcium
cADPR	cyclic ADP-ribose
CAS	castanospermine
cC1qR	collagenous tail C1q receptor
CD	circular dichroism
$CD8^+$	cluster of differentiation marker 8-positive cells (T lymphocytes which recognize MHC class I antigens)
cGMP	guanosine 3',5'-cyclic monophosphate
CHO	chinese hamster ovary
CICR	Ca^{2+} induced Ca^{2+} release
CK2	casein kinase 2
CMG	calmegin
CN	calcineurin
CNX	calnexin
CON A	concanavalin A
COOH	carboxyl
CRT	calreticulin
CRT-1	*C. elegans* calreticulin protein
crt-1	*C. elegans* calreticulin gene
CS	citrate synthase
CTL	cytotoxic T cell
DBD	DNA binding domain
Dex	dexamethasone
DHPR	dihydropyridine receptor
DNJ	deoxynojirimycin
EBV	Epstein-Barr virus
ECM	extracellular matrix
EDTA	ethylene diamine tetra-acetic acid (disodium salt)
EGTA	ethylene glycol-bis (?-aminoethyl ether)-N,N', N',N' –tetraacetic acid
Endo H	endoglycosidase H
eNOS	endothelial nitric oxide synthase
ER	endoplasmic reticulum
ERAD	ER-associated degradation
ERK	extracellular signal-regulated kinase
ERO	endoplasmic reticulum over load
ES	excretory-secretory
ETE	electron transfer control element
FACS	fluorescence-activated cell sorting
FAD	flavin adenine dinucleotide / familial Alzheimer's disease
FBHE	fetal bovine heart endothelial cells
FGF	fibroblast growth factor
FMN	flavin mononucleotide
gC1qR	globular head C1q receptor
GFP	green fluorescent protein

Glc	glucose
GlcNAc	N-acetyl glucosamine
GPCRs	G-protein coupled receptors
GR	glucocorticoid receptor
Gr	granzyme
GRP	glucose-regulated proteins
GST	glutathione S-transferase
GT	UDP-glc:glycoprotein glucosyltransferase
H chain	heavy chain of class I histocompatibility molecule
HA	hemagglutinin
HACBP	high affinity calcium-binding protein
HBD	heparin binding domain
HBV	hepatitis B virus
HC	class I heavy chain
HOCl	hypochlorous acid
Hsp	heat shock protein
HUVEC	human umbilical vein endothelial cells
IFN-γ	interferon
$InsP_3$	inositol 1,4,5-trisphosphate
IP	immunoprecipitation
IP-10	IFN-γ inducible protein-10
IP_3	inositol 1,4,5 trisphosphate
IP_3R	inositol 1,4,5 trisphosphate receptor
IRM	interference reflection microscopy
KO	knockout
LdCR	*Leishmania donovani* calreticulin
LE	lupus erythematosus
LMP	latency membrane protein
LRP	LDL receptor related protein
LTP	long term potentiation
MBL	mannose binding lectin (also sometimes called MBP for mannose binding protein)
MBP	maltose binding protein
MEC-4	Na^{2+} degenerin channel
MEF	mouse embryonic fibroblasts
MHC	major histocompatibility complex
Mig	monokine induced by IFN-γ
MMP	mitochondrial membrane permeabilization
MPO	myeloperoxidase
mPTP	mitochondrial permeability transition pore
Mr	relative molecular mass
mu	map unit
MW	molecular weight
NAADP	nicotinic acid adenine dinucleotide phosphate
NES	nuclear export signal
NFAT	nuclear factor of activated T-cells
NIDDM	non-insulin-dependent diabetes mellitus
NK	natural killer (lymphocyte)
NLS	nuclear localization signal
NMR	nuclear magnetic resonance
NMR	nuclear magnetic resonance spectroscopy
NO	nitric oxide
NOS	nitric oxide synthase
NPC	nuclear pore complex
NR	nuclear receptor
PAGE	polyacrylamide gel electro-phoresis
PCL/PLD	phosphlipase C/D

PCR	polymerase chain reaction
PDI	protein disulfide isomerase
PDK	proline directed kinase
PI3K	phosphoinositide 3-kinase
PIP3	phosphatidylinositol 3,4,5-triphosphate
PKC	protein kinase C
PKG	cyclic GMP-dependent protein kinase
PKI	protein kinase inhibitor
PLC	phospholipase C
PMNs	polymorphonuclear neutrophils
PS	phosphatidylserine
PTX	pertussis toxin
QC	quality control
R569W	mutant form of myeloperoxidase with the arginine at codon 569 replaced by tryptophan
rbc=s	red blood cells
RNAse	ribonuclease
RyR	ryanodine receptor
SAcP	secretory acid phosphatase
SCLE	subacute cutaneous lupus erythematosus
SERCA	sacroplasmic/endoplasmic reticulum calcium ATPase; $[Ca^{2+}]_{cyt}$, and $[Ca^{2+}]_{ER}$, cytosolic and ER free Ca^{2+} concentration, respectively
SjS	Sjogren's syndrome
SLE	systemic lupus erythematosus
SP	mammalian semiperm-eabilized cell system
SP-A	surfactant protein A
SP-D	surfactant protein D
SPR	surface plasmon resonance
SR	sarcoplasmic reticulum
T	thymically processed lymphocyte
T134K	threonine to lysine class I heavy chain point mutant at position 134
TAP	transporter associated with Antigen Processing
TCR	T cell receptor
TM	transmembrane
TROSY	transverse relaxation-optimized spectroscopy
TSP	thrombospondin
UGGT	UDP-glucose glycoprotein: glucosyltransferase
UPR	unfolded protein response
UV	ultraviolet
VDAC	voltage dependent anion channel
VEGF	vascular endothelial growth factor
VSV	vesicular stomatitis virus
VSVG	vesicular stomatitis virus G protein
WT	wild type
Y173C	mutant form of myeloperoxidase with the tyrosine at codon 173 replaced by cysteine.
$\Delta\Psi_m$	mitochondrial membrane potential

PREFACE

Calreticulin has been first identified and characterized over 30 years ago as a soluble calcium-binding protein of skeletal muscle sarcoplasmic reticulum. It took over 20 years before it was realized that the protein is in fact a key calcium-binding chaperone of endoplasmic reticulum, a major calcium storage organelle in non-muscle cells. Today calreticulin is considered one of the best markers for the endoplasmic reticulum. The cDNA encoding calreticulin was isolated in 1989, and it was then recognized that the protein plays an important role in virtually every aspect of cell biology. The first edition of calreticulin book was published in 1996. This new edition focuses on the latest discoveries on calreticulin, calnexin (an integral membrane protein similar to calreticulin) and other endoplasmic reticulum proteins. Findings described in the book identify calreticulin and other ER proteins as important molecules involved in many diseases, including protein folding disorders, cardiac pathologies, cancer and autoimmunity. The effects of calreticulin in the modulation of cellular calcium homeostasis have profound effects on many cellular functions. Cell surface calreticulin becomes an important player in modulation of many different pathologies. Gene knockout studies on different animal models point out the critical role of calreticulin in organogenesis and other developmental pathways. Lastly, the structural studies on calreticulin and calnexin revealed a highly unusual three-dimensional arrangement for these chaperones. These observations will undoubtedly have a profound impact on the future studies of other endoplasmic reticulum proteins. The book raises many intriguing questions about calreticulin, calnexin and the endoplasmic reticulum, and gives a unique opportunity to realize the significance of these calcium-binding chaperones.

Paul Eggleton, Ph.D.
Oxford, U.K.
Marek Michalak, Ph.D.
Edmonton, Alberta, Canada

Acknowledgments

I would like to thank the most important people in my life, my wife Hanna and my daughter Karolina for their love, understanding and continuing support. I thank Michel Puceat (Montpellier, France) for his hospitality during my sabbatical work in his laboratory and for his support during preparation of this book. Research in our laboratory is supported by the Canadian Institutes of Health Research and the Alberta Heritage Foundation for Medical Research.

Marek Michalak, Ph.D.
Edmonton, Alberta, Canada

During the production of this book my wife, Lucy give birth to our lovely new daughter – Nicole Marie ('Buzzy') and I know I should have spent more time with them both. So I thank them for their understanding and look forward to spending more time with them now that this book is complete. I also thank Ms. Alison Marsland for her wonderful secretarial and organizational skills and for pestering the contributors to complete their manuscripts on time. Both Marek and I are very grateful that some of the world's experts in the calreticulin field have agreed to contribute to this book and hope they are proud of their book. Finally I thank the Arthritis Research Campaign and Medical Research Council of Great Britain for their generous support over the years.

Paul Eggleton, Ph.D.
Oxford, U.K.

Chapter 1

Introduction to Calreticulin

Paul Eggleton and Marek Michalak

Introduction

Over 30 years ago calreticulin, then known as the high affinity calcium binding protein (HACBP), was identified and purified from isolated skeletal muscle sarcoplasmic reticulum vesicles.[1,2] Surprisingly, it took almost 20 years to realize that the protein is a major component of the endoplasmic reticulum (ER) in non-muscle cells[3]. However, today, calreticulin is considered one of the best markers for the ER. In 1989 isolation of cDNA encoding calreticulin was reported[3,4] and provided a useful tool to carry out biochemical, molecular biological and cell biological studies of the protein. This led to a number of advances on the structure and function of calreticulin. The recent application of calreticulin gene deletion in mice,[5,6] *C. elegans*[7,8] and in *Dictyostelium*[9] have led to exciting discoveries of the role of calreticulin in organogenesis and several pathologies. Moreover, long awaited structural studies on calreticulin[10] and calnexin[11] provided the first insights into 3D structure of ER luminal proteins and their domains. This will have a tremendous impact on the future studies on these and other ER chaperones.

The first edition of calreticulin book was published in 1996 and encompassed a series of diverse articles introducing this peculiar protein to the World.[12] In the last 6 years, chaperone and Ca^{2+} binding functions of calreticulin have been well described and are now wildly accepted. New and exciting areas of research have emerged focusing now on the role of calreticulin and other ER protein in diseases including protein folding disorders, cardiac pathologies, cancer and autoimmunity. The protein has attracted a lot of attention in many diverse areas of basic and now clinical research. The popularity in this protein has led to the organization of a number of international workshops specifically on the function of calreticulin. The first International Workshop on calreticulin was held in Banff, Alberta, Canada in 1994. This was followed by 4 further workshops now alternating between European and North American locations on a biannual basis: in 1996 in Como, Italy;[13] in 1998 in Banff, Alberta,[14] in 2000 in Oxford, UK[15] and in 2002 in San Antonio, Texas, USA. In addition, an exciting research session organized by Michal Opas (University of Toronto) dedicated to calreticulin was included at the meeting of the American Cell Biology Society in San Francisco in 1996.[16]

Structure and Function of Calreticulin

In the late 1980's, Waisman's group in Calgary, contributed enormous information on structure and many biophysical properties of the protein (then referred to as calregulin).[17-21] This set the foundation for later studies that have revealed many new properties of the protein. In particular attention has focussed on its role in protein folding and modulation of protein-protein interaction in the ER lumen.[22-29] An overview of the biochemical and structural aspects of the protein is reviewed by Johnson in Chapter 2. Recent structural data available for the central region or P-domain of calreticulin[10] and the crystallography studies on the soluble, ER luminal domain of calnexin[11] reveal a highly unusual structure (Chapters 3 and 4). The P-domain of calreticulin and calnexin form an extended "arm" connected to globular regions.[10] These

Calreticulin, 2nd ed., edited by Paul Eggleton and Marek Michalak.

structural studies enable us to now speculate how chaperones interact and help fold other glycoproteins (Chapters 3, 4 and 6).

Protein Folding and Quality Control

Calreticulin is a Ca^{2+}-binding chaperone and a component of the calreticulin/calnexin pathway. There have been an extraordinary number of studies on the role of these chaperones in protein folding.[30] Recently, chaperone function of calreticulin has been investigated in calreticulin deficient cells. These studies show a critical role for the protein in the protein quality control and they point out that ER chaperones are not redundant. Latest immunolocalization studies indicate that quality control is not restricted to the whole ER but may also be present in the pre-Golgi intermediates.[31] Interestingly, calreticulin along with the UDP-glucose-glycoprotein transferase have been localized to this most distal region of the ER forming an exit "gate" to the Golgi suggestive that the two proteins may be critical for the protein quality control and secretion.[31] A fascinating finding is that calreticulin is essential for MHC class I assembly and cell surface targeting (Chapter 9). It has been generally accepted that calnexin must be the most important chaperone involved in MHC Class I folding. Surprisingly, there is no problem with MHC class I assembly in calnexin deficient cells.[32]

Early evidence for calreticulin function as a protein-protein chaperone came from studies by Nauseef's group[33] in the mid 1990's. In these studies, the importance of calreticulin as a chaperone and it's selective ability to act as quality controller of pro-inflammatory enzymes in neutrophils was demonstrated in a series of mutational experiments (Chapter 7). In 1993 Bergeron and Thomas group reported that calnexin functions as chaperone for glycoproteins.[34] Helenius' group showed in elegant studies that calnexin and calreticulin are both lectin-like chaperones and they can recognize monoglucosylated carbohydrate on proteins.[35] One important and still unanswered question remains: do calreticulin and calnexin interact with their substrates via monoglucosylated carbohydrate only or do they utilize both sugar and polypeptide? Latest findings indicate that the proteins may indeed function as molecular chaperones and interact with misfolded polypeptides (Chapter 6). Likely the two proteins utilize both carbohydrate (Chapter 3) and specific amino acid sequences to recognize their substrates and assist in protein folding and quality control. Perhaps understanding of the 3D structure of both proteins will help to identify, at the molecular level, mechanisms of interaction with their substrates. The importance of calreticulin as a chaperone and as a Ca^{2+} binding protein is emphasized by the highly conserved regions of its DNA that encode for the amino acid sequences associated with these functions in both complex and simple animal systems. The fact that calreticulin functions in a similar regard in the plant kingdom as described by Mariani et al (Chapter 10) serves to illustrate the importance of this protein in the both the animal and plant kingdoms.

Ca^{2+} Binding and Ca^{2+} Homeostasis

It is well documented now that calreticulin is a Ca^{2+} binding protein responsible for Ca^{2+} storage in the ER lumen (Chapter Demaurex 11, 12, and 13). Although Ca^{2+} binding to calnexin has been documented in vitro there is no information available on the role of this protein in Ca^{2+} storage in the ER. In contrast calreticulin is a major Ca^{2+} buffer in the lumen of the ER. Changes in the level of expression of calreticulin have profound effect on Ca^{2+} capacity of the ER.[36-40] An intriguing finding is that over-expression of calreticulin in fibroblasts results in significant changes in the free ER luminal Ca^{2+} concentration (Chapter 13). This may have a tremendous impact on Ca^{2+}-dependent protein folding, modification and trafficking within the ER lumen. It seems as calreticulin may affect both the Ca^{2+} capacity and free Ca^{2+} concentration in the ER and, therefore, many Ca^{2+}-dependent pathways in the ER and other cellular compartments. It is not surprising, therefore, that several laboratories showed that modulation of expression of calreticulin might influence the function of the store-operated Ca^{2+} influx.[36-39] However, the essential role of calreticulin, if any, in the regulation of the store-operated Ca^{2+}

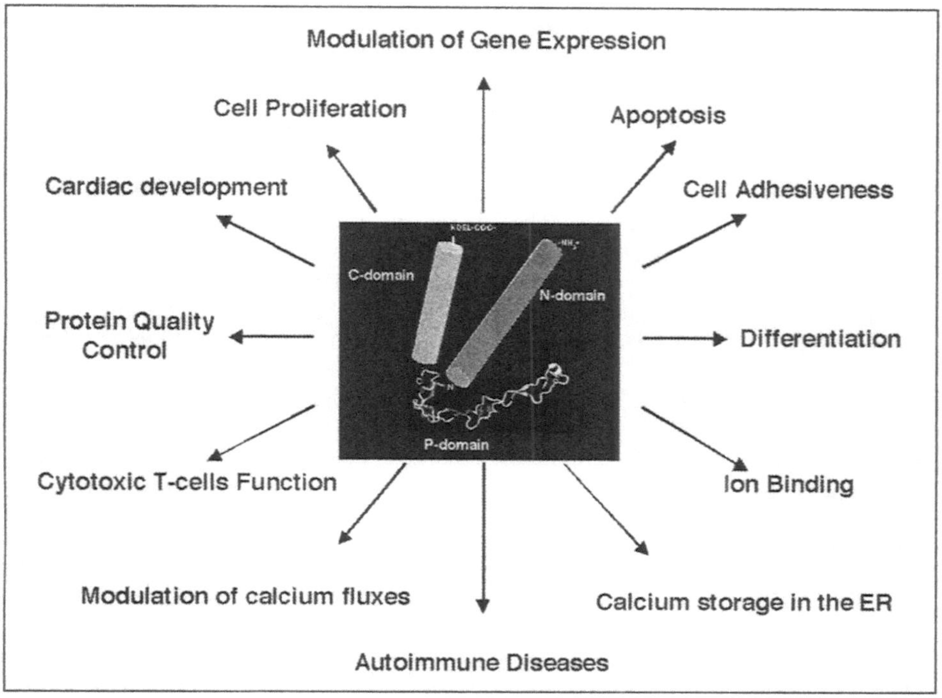

Figure 1.1. Summary of the functions associated with calreticulin.

influx remains to be determined. Camacho's group showed in a series of elegant studies that calreticulin and calnexin play very important role in modulation of the function of SERCA Ca^{2+}-ATPase in the ER.[41-43] There may be very specific and regulated dynamic interactions between calreticulin, calnexin and SERCA2b affecting function of the Ca^{2+} pump, but the molecular mechanisms for this regulation remains to be defined.

Immunological Functions of Calreticulin

Over the past fifteen years or so various functions for calreticulin in the regulation of immune function have been cited, both physiological and pathological. Initial interest in calreticulin as an immunological molecule was raised when McCauliffe and co workers suggested calreticulin might be part of a cluster of intracellular autoantigens known as the Ro/SS-A complex recognized by autoantibodies typically found in sera from patients with primary Sjögren's syndrome and in subsets of patients with systemic lupus erythematosus (SLE).[44] The fact that autoantibodies were generated against clusters of intracellular proteins and other molecules, suggested they were released from the cell during cell death. Calreticulin was subsequently found not to be a specific protein of the Ro/SS-A complex,[45] but an autoantigen in its own right which bore high homology with a number of calreticulin homologues found on the surface and possibly secreted from various human parasites, including nematodes, trypanosomes malarial protozoa and ticks—reviewed by Nakhasi[46] among others. The role of calreticulin in parasite physiology has been speculated upon and may be important in avoiding innate immune responses of the host.[47] These parasitic forms are also immunogenic and humans often develop antibodies to these various forms of parasitic calreticulin (Chapter 21). Serological cross-reactivity between the various forms of calreticulin then became of interest to auto-immunologists, since this raised the possibility that infectious organisms might play a triggering or exacerbating role in the human autoimmune response.[48-51] It then became clear from the work of Sontheimer and co-workers (Chapter 17) that calreticulin was not just an autoantigen, but could under specific inflammatory conditions alter the response of the immune system to various diseases (reviewed by Eggleton et al in refs. 52,53). Work in Eggleton's laboratory in Oxford has shown that calreticulin binds to the globular heads of the first component of complement system, C1q. An important patho-physiological consequence of this is that such interaction can inhibit classical complement activation. This has important implications in some autoimmune diseases, as C1q knockout studies in mice[54] have emphasized the importance of C1q in both the recognition and clearance of apoptotic cells (Chapter 15) and immune complexes.[55]

The recognition of self and non-self antigens on the surface of antigen presenting cells occurs with the binding of peptides to major histocompatibility complex (MHC) class I and II molecules occurs in the ER. Peptides that bind to class I MHC molecules are derived from viruses that have infected the host cells. Efficient peptide binding requires a number of components in addition to the MHC class I- beta 2 microglobulin or MHC class II molecules. These include the two subunits of the transporter associated with antigen presentation (TAP1 and TAP2), which aid peptide movement into the ER from the cytosol, and tapasin, an MHC-encoded membrane protein. In addition the chaperone properties of calreticulin and the thiol oxidoreductase activity of ERp57, are also essential components of this complex. Tim Elliott and co-workers have through a series of mutational studies (Chapter 9) shown that calreticulin is an important component of the peptide loading complex and that mutated MHC class I molecules fail to bind TAP and prevents association of calreticulin with the MHC class I. The immunological implication of this is that antigen presenting cells do not present viral antigens to T cells.[56]

Over the past five years there has been confirmation that calreticulin is released from cells either by secretion from T cell cytotoxic (CTL) granules or via cell death in the form of necrosis. The implications of extracellular calreticulin are only just being addressed. However a number of groups have observed calreticulin on the surface of cells and have proposed a number of intriguing functions for the protein at the cell surface. One of the most recent inflammatory functions has been proposed by Joanne Murphy-Ullrich and co-workers[57] in which surface calreticulin on endothelial cells interacts with the N-terminal portion of soluble thrombospondin. Thrombospondin/hep I stimulate focal adhesion disassembly through a mechanism involving phosphoinositide 3-kinase activation. This group describes their most recent findings in Chapter 18 and propose that cell surface calreticulin mediates focal adhesion disassembly.

What Have We Learned from Calreticulin Gene Knockout?

The calreticulin deficient mouse was recently created by homologous recombination.[5] This work showed that the protein is essential for cardiac development.[5] Initially it was surprising finding but we know today that calreticulin is a new cardiac embryonic gene[5] control by many cardiac specific transcription factors[58] (Chapter 23). Importantly, calreticulin expression is tightly control during cardiac development and high levels of calreticulin postnataly leads to sever cardiac pathology, complete heart block and sudden death.[59] Calreticulin may become a future target for control of congenital heart block and perhaps other congenital heart diseases. Tissue specific gene knockout should help to sort out function of this unique protein in different tissues. The role of calreticulin in adult heart needs further investigations. Calnexin gene knockout has not been reported as yet. It will be interesting to investigate functional consequences of calnexin and calreticulin double knockout.

Calreticulin gene has also been deleted in *C. elegans* affecting cell necrosis.[7,8] In *Dictyostelium* a double knockout of calreticulin and calnexin has severe effects on phygocytosis whereas the deletion of one gene only has no significant phenotype.[9]

Calreticulin and Disease

Do calreticulin and calnexin play a role in any disease states and can their functions be exploited to disrupt normal physiological processing in pathogens? The proteins clearly must play a role in many protein conformational disorders. Further studies await clarification of a specific association of these chaperones with a specific protein folding problems and pathologies. Differential expression of calreticulin has been seen in prostate cancer and in the future it may be useful as a marker of prostate and likely other cancers.[60,61] A serendipitous observation in the 1990's has led to a whole new field of research concerning the possible therapeutic uses of calreticulin as a treatment for cancers. Culture supernatants from a number of Epstein-Barr immortal cell lines were observed to inhibit to varying degrees the proliferation of cultured endothelial cells. As Tosato, Yao and Pike describe in Chapter 16. The active inhibitor of cell proliferation and ultimately angiogenesis was found to be the NH_2 terminal fragment of calreticulin. This group has gone on to show that calreticulin can prevent the growth of sub cutaneous tumors in mice. The future molecular mechanisms of these observations will surely contribute to a better understanding of spread of tumors via angiogenesis. In an elegant study described by Debrabant et al (Chapter 20) show that the secretion of acid phosphatase and possibly other proteins trafficking through the secretory pathway of the parasite were affected as a result of overexpression of *Leishmamia* calreticulin P- or C-domain. In addition, parasites expressing either the N- or P-domain showed significant decrease in survival inside macrophages *in vitro*. These results suggest that disruption of the functions of calreticulin in *Leishmania* may result in an alteration of the parasite secretory pathway and also reduce its virulence *in vitro*.

Calreticulin and Apoptosis

Calreticulin has been implicated in several regulatory pathways of regulated cell death or apoptosis. For example, calreticulin deficient cells are relatively resistant to apoptosis.[64] In contrast, increased expression of calreticulin makes cells more sensitive to apoptotic stimuli.[64] Although, the precise mechanism how calreticulin affects cell sensitivity to apoptosis is not clear at present, it indicates that ER membrane is an important player in pathways leading to apoptosis. Interestingly, calreticulin deficient cells have a relatively low level of intra-ER Ca^{2+}, similar to Bcl-2 over-expressers,[62,63] and this may be responsible for their relatively resistant to apoptosis.[64]

Recent studies suggest several roles for calreticulin in apoptosis, perhaps as a cell surface receptor "marking" apoptotic cells (Chapter 15). *In vitro*, calreticulin inhibits target gene transcription by interacting with steroid hormone receptors, thereby masking their DNA-binding sites. It was suggested that if such a process occurs *in vivo*, blocking of DNA binding sites could trigger the onset of the apoptotic process.[65] Whether this general process occurs or not remains to be resolved. More specific studies have been performed on cytotoxic T lymphocytes. These cells possess secretory granules that contain granzyme B a serine proteinase that has been implicated in CTL-induced apoptosis and the lytic molecule perforin. Bleackley and Tschopp's research groups independently identify calreticulin as a new component of the CTL granules.[66] Calreticulin within granules co-associates with perforin in a calcium dependent manner, suggesting that calreticulin may act as Ca^{2+}-regulated chaperone for perforin. This action could serve to protect the CTL during biogenesis of granules and may also serve to regulate perforin lytic action after release. Hudig and Karimi describe in Chapter 14, the role of calreticulin in regulating granule mediated lysis of target cells by natural killer and cytotoxic lymphocytes and its ability to prevent autolysis of the effector cell. Quite a different apoptotic role for extracellur calreticulin has been suggest by Henson et al (Chapter 15) who have observed that calreticulin acts as a linker molecule allowing the recognition of a cells earmarked for cell death by phagocytes. They propose the first component of complement, C1q and the collectins, a family of proteins that share similar ultrastructure to C1q bind to cells and cell debris undergoing apoptosis. These protein-coated cells are then recognized by CD91 expressing phagocytes via calreticulin.[67] If this work is confirmed it will illustrate yet another important physiological role of calreticulin. It would appear from the independent work of a number of laboratories that extracellular calreticulin might play a dual role in anti-inflammatory mechanisms by binding to C1q to reduce over activation of complement while acting as a bridging molecule allowing C1q-coated cell debris to be removed by professional phagocytes via CD91 preventing the accumulation of cell debris in vital organs.

Conclusions

In summary since protein folding is important, Ca^{2+} homeostasis is important and they seem to be tightly link and it not surprising that modulation of calreticulin expression and/or function may have a profound effects on many cellular function. Moreover, it is now becoming clear that the human genome encodes far fewer proteins than first thought and therefore it is not unreasonable to imagine highly ubiquitous proteins such as calreticulin are required to be multifunctional. What is intriguing about this is that as the protein is observed in various locations within the cell, cell surface or in the extra-cellular environment, its function changes as it interacts with different molecules. This is more apparent when looking at the whole organism when cell-cell communication and recognition are involved (adhesion).

Acknowledgments

Research in the authors' laboratories is supported by the Canadian Institutes of Health Research (to M.M.) and the Arthritis Research Campaign (to P.E.). M.M. is a Medical Scientist of the Alberta Heritage Foundation for Medical Research and a Senior Investigator of the Canadian Institutes of Health Research. P.E. is a senior investigator at the MRC Immunochemistry Unit, Oxford University and senior lecturer in Biomedical Sciences at Peninsula Medical School, Devon, U.K..

References

1. Ostwald TJ, MacLennan DH. Isolation of a high affinity calcium-binding protein from sarcoplasmic reticulum. J Biol Chem. 1974; 249:974-79.
2. Ostwald TJ, MacLennan DH, Dorrington KJ. Effects of cation binding on the conformation of calsequestrin and the high affinity calcium-binding protein of sarcoplasmic reticulum. J Biol Chem 1974; 249:5867-71.
3. Fliegel L, Burns K, MacLennan DH et al. Molecular cloning of the high affinity calcium-binding protein (calreticulin) of skeletal muscle sarcoplasmic reticulum. J Biol Chem 1989; 264:21522-28.
4. Smith MJ, Koch GLE. Multiple zones in the sequence of calreticulin (CRP55, calregulin, HACBP), a major calcium binding ER/SR protein. EMBO J 1989; 8:3581-86.
5. Mesaeli N, Nakamura K, Zvaritch E et al. Calreticulin is essential for cardiac development. J Cell Biol 1999; 144:857-68.
6. Rauch F, Prud'homme J, Arabian A et al. Heart, brain, and body wall defects in mice lacking calreticulin. Exp Cell Res 2000; 256:105-11.
7. Park BJ, Lee DG, Yu JR et al. Calreticulin, a calcium-binding molecular chaperone, is required for stress response and fertility in Caenorhabditis elegans. Mol Biol Cell 2001; 12:2835-45.
8. Xu K, Tavernarakis N, Driscoll M. Necrotic cell death in C. elegans requires the function of calreticulin and regulators of Ca^{2+} release from the endoplasmic reticulum. Neuron 2001; 31:957-71.
9. Muller-Taubenberger A, Lupas AN, Li H et al. Calreticulin and calnexin in the endoplasmic reticulum are important for phagocytosis. EMBO J 2001; 20:6772-82.
10. Ellgaard L, Riek R, Herrmann T et al. NMR structure of the calreticulin P-domain. Proc Natl Acad Sci USA 2001; 98:3133-38.
11. Schrag JD, Bergeron JJM, Li Y et al. The structure of calnexin, an ER chaperone involved in quality control of protein folding. Mol Cell 2001; 8:633-44.
12. Michalak M. Calreticulin. Austin: Landes Bioscience, 1996.
13. Meldolesi J, Krause K-H, Michalak M. Calreticulin: how many functions in how many cellular compartments? Como, April 1996. Cell Calcium 1996; 20:83-86.
14. Michalak M, Mariani P, Opas M. Calreticulin, a multifunctional Ca^{2+} binding chaperone of the endoplasmic reticulum. Biochem Cell Biol 1998; 76:779-85.
15. Llewellyn DH, Johnson S, Eggleton P. Calreticulin comes of age. Trends Cell Biol 2000; 10:399-402.
16. Krause K-H, Michalak M. Calreticulin. Cell 1997; 88:439-43.
17. Khanna NC, Waisman DM. Development of a radioimmunoassay for quantitation of calregulin in bovine tissues. Biochemistry 1986; 25:1078-82.
18. Waisman DM, Khanna NC, Tokuda M. Identification of a major bovine heart Ca^{2+} binding protein. Biochem. Biophys Res Commun 1986; 139:596-603.

19. Tokuda M, Khanna NC, Waisman DM. Identification of bovine brain calcium binding proteins. Cell Calcium 1987; 8:229-39.
20. Khanna NC, Tokuda M, Waisman DM. Comparison of calregulins from vertebrate livers. Biochem J 1987; 242:245-51.
21. Khanna NC, Tokuda M, Waisman DM. Calregulin: purification, cellular localization, and tissue distribution. Meth Enzymol 1987; 139:36-50.
22. Baksh S, Michalak M. Expression of calreticulin in Escherichia coli and identification of its Ca^{2+} binding domains. J Biol Chem 1991; 266:21458-65.
23. Heilmann C, Spamer C, Leberer E et al. Human liver calreticulin: characterization and Zn^{2+}-dependent interaction with phenyl-sepharose. Biochem Biophys Res Commun 1993; 193:611-16.
24. Vassilakos A, Michalak M, Lehrman MA et al. Oligosaccharide binding characteristics of the molecular chaperones calnexin and calreticulin. Biochemistry 1998; 37:3480-90.
25. Corbett EF, Oikawa K, Francois P et al. Ca^{2+} regulation of interactions between endoplasmic reticulum chaperones. J Biol Chem 1999; 274:6203-11.
26. Saito Y, Ihara Y, Leach MR et al. Calreticulin functions in vitro as a molecular chaperone for both glycosylated and non-glycosylated proteins. EMBO J 1999; 18:6718-29.
27. Corbett EF, Michalak KM, Oikawa K et al. The conformation of calreticulin is influenced by the endoplasmic reticulum lumenal environment. J Biol Chem 2000; 275:27177-85.
28. Bouvier M, Stafford WF. Probing the three-dimensional structure of human calreticulin. Biochemistry 2000; 39:14950-59.
29. Li Z, Stafford WF, Bouvier M. The metal ion binding properties of calreticulin modulate its conformational flexibility and thermal stability. Biochemistry 2001; 40:11193-201.
30. Michalak M, Corbett EF, Mesaeli N et al. Calreticulin: one protein, one gene, many functions. Biochem J 1999; 344:281-92.
31. Zuber C, Fan JY, Guhl B et al. Immunolocalization of UDP-glucose:glycoprotein glucosyltransferase indicates involvement of pre-Golgi intermediates in protein quality control. Proc Natl Acad Sci USA 2001; 98:10710-15.
32. Scott JE, Dawson JR. MHC class I expression and transport in a calnexin-deficient cell line. J Immunol 1995; 155:143-48.
33. Nauseef WM, McCormick SJ, Clark RA. Calreticulin functions as a molecular chaperone in the biosynthesis of myeloperoxidase. J Biol Chem 1995; 270:4741-47.
34. Ou WJ, Cameron PH, Thomas DY et al. Association of folding intermediates of glycoproteins with calnexin during protein maturation. Nature 1993; 364:771-76.
35. Ellgaard L, Molinari M, Helenius A. Setting the standards: quality control in the secretory pathway. Science 1999; 286:1882-88.
36. Bastianutto C, Clementi E, Codazzi F et al. Overexpression of calreticulin increases the Ca^{2+} capacity of rapidly exchanging Ca^{2+} stores and reveals aspects of their lumenal microenvironment and function. J Cell Biol 1995; 130:847-55.
37. Mery L, Mesaeli N, Michalak M et al. Overexpression of calreticulin increases intracellular Ca^{2+} storage and decreases store-operated Ca^{2+} influx. J Biol Chem 1996; 271:9332-39.
38. Fasolato C, Pizzo P, Pozzan T. Delayed activation of the store-operated calcium current induced by calreticulin overexpression in RBL-1 cells. Mol Biol Cell 1998; 9:1513-22.
39. Xu W, Longo FJ, Wintermantel MR et al. Calreticulin modulates capacitative Ca^{2+} influx by controlling the extent of inositol 1,4,5-trisphophate-induced Ca^{2+} store depletion. J Biol Chem 2000; 275:36676-82.
40. Nakamura K, Zuppini A, Arnaudeau S et al. Functional specialization of calreticulin domains. J Cell Biol 2001; 154:961-72.
41. Camacho P, Lechleiter JD. Calreticulin inhibits repetitive intracellular Ca^{2+} waves. Cell 1995; 82(5):765-71.
42. John LM, Lechleiter JD, Camacho P. Differential modulation of SERCA2 isoforms by calreticulin. J Cell Biol 1998; 142:963-73.
43. Roderick HL, Lechleiter JD, Camacho P. Cytosolic phosphorylation of calnexin controls intracellular Ca^{2+} oscillations via an interaction with SERCA2b. J Cell Biol 2000; 149:1235-48.
44. McCauliffe DP, Lux FA, Lieu TS et al. Molecular cloning, expression, and chromosome 19 localization of a human Ro/SS-A autoantigen. J Clin Invest 1990; 85:1379-91.
45. Rokeach LA, Haselby JA, Meilof JF et al. Characterization of the autoantigen calreticulin. J Immunol 1991; 147:3031-39.
46. Nakhasi HL, Pogue GP, Duncan RC et al. Implications of calreticulin function in parasite biology. Parasitol Today 1998; 14:157-60.

47. Kasper G, Brown A, Eberl M et al. A calreticulin-like molecule from the human hookworm Necator americanus interacts with C1q and the cytoplasmic signalling domains of some integrins. Parasite Immunol 2001; 23:141-52.
48. Lux FA, McCauliffe DP, Buttner DW et al. Serological cross-reactivity between a human Ro/SS-A autoantigen (calreticulin) and the lambda Ral-1 antigen of Onchocerca volvulus. J Clin Invest 1992; 89:1945-51.
49. Meilof JF, Van der Lelij A, Rokeach LA et al. Autoimmunity and filariasis. Autoantibodies against cytoplasmic cellular proteins in sera of patients with onchocerciasis. J Immunol 1993; 151:5800-09.
50. Pritchard DI, Brown A, Kasper G et al. A hookworm allergen which strongly resembles calreticulin. Parasite Immunol 1999; 21:439-50.
51. Sanders ML, Glass GE, Nadelman RB et al. Antibody levels to recombinant tick calreticulin increase in humans after exposure to Ixodes scapularis (Say) and are correlated with tick engorgement indices. Am J Epidemiol 1999; 149:777-84.
52. Eggleton P, Reid KB, Kishore U et al. Clinical relevance of calreticulin in systemic lupus erythematosus. Lupus 1997; 6:564-71.
53. Eggleton P, Llewellyn DH. Pathophysiological roles of calreticulin in autoimmune disease. Scand J Immunol 1999; 49:466-73.
54. Botto M. C1q knock-out mice for the study of complement deficiency in autoimmune disease. Exp Clin Immunogenet 1998; 15:231-34.
55. Nash JT, Taylor PR, Botto M et al. Immune complex processing in C1q-deficient mice. Clin Exp Immunol 2001; 123:196-202.
56. Lewis JW, Elliott T. Evidence for successive peptide binding and quality control stages during MHC class I assembly. Curr Biol 1998; 8:717-20.
57. Goicoechea S, Orr AW, Pallero MA et al. Thrombospondin mediates focal adhesion disassembly through interactions with cell surface calreticulin. J Biol Chem 2000; 275:36358-68.
58. Guo L, Lynch J, Nakamura K et al. COUP-TF1 antagonizes Nkx2.5-mediated activation of the calreticulin gene during cardiac development. J Biol Chem 2001; 276:2797-801.
59. Nakamura K, Robertson M, Liu G et al. Complete heart block and sudden death in mouse over-expressing calreticulin. J Clin Invest 2001; 107:1245-53.
60. Zhu N, Pewitt EB, Cai XY et al. Calreticulin: An intracellular Ca^{++}-binding protein abundantly expressed and regulated by androgen in prostatic epithelial cells. Endocrinology 1998; 139:4337-44.
61. Zhu N, Wang Z. Calreticulin expression is associated with androgen regulation of the sensitivity to calcium ionophore-induced apoptosis in LNCaP prostate cancer cells. Cancer Res 1999; 59:1896-902.
62. Foyouzi-Youseffi R, Arnaudeau S, Borner C et al. Bcl-2 decreases the free Ca^{2+} concentration within the endoplasmic reticulum. Proc Natl Acad Sci USA 2000; 97:5723-28.
63. Pinton P, Ferrari D, Magalhaes P et al. Reduced loading of intracellular Ca^{2+} stores and downregulation of capacitative Ca^{2+} influx in Bcl-2-overexpressing cells. J Cell Biol 2000; 148:857-62.
64. Nakamura K, Bossy-Wetzel E, Burns K et al. Changes in endoplasmic reticulum luminal environment affect cell sensitivity to apoptosis. J Cell Biol 2000; 150:731-40.
65. Bruchovsky N, Snoek R, Rennie PS et al. Control of tumor progression by maintenance of apoptosis. Prostate 1996; 6:13-21.
66. Bleackley RC, Atkinson EA, Burns K et al. Calreticulin: a granule-protein by default or design? Curr Topics Microbiol Immunol 1995; 198:145-59.
67. Ogden CA, deCathelineau A, Hoffmann PR et al. C1q and mannose binding lectin engagement of cell surface calreticulin and CD91 initiates macropinocytosis and uptake of apoptotic cells. J Exp Med 2001;194:781-95.

Chapter 2

Biochemical and Molecular Properties of Calreticulin

Steven J. Johnson and Kjell O. Håkansson

Abstract

Calreticulin is a highly abundant Ca^{2+}-storage protein found in all cells of higher organisms, with the exception of erythrocytes. It is predominantly located in the endoplasmic reticulum where, in tandem with the homologue calnexin, it performs an important role in glycoprotein folding, such as in the assembly of MHC Class I complexes. Under conditions of cellular stress, calreticulin may be released into the extracellular environment, and autoantibodies against the protein have been detected in the sera of a number of autoimmune conditions including patients with systemic lupus erythematosus (SLE) and Sjögren's syndrome. Although there is currently no crystal structure of calreticulin, the structure of the protein has recently been extensively studied by protein chemistry, circular dichroism (CD) and nuclear magnetic resonance (NMR) spectroscopy. These studies have revealed that the P-domain is an extended, flexible hairpin loop and contains regions of localized secondary structure. Investigation of full length calreticulin has revealed that calreticulin can be classed as an $\alpha + \beta$ protein, and that it is a highly extended prolate ellipsoid. Cation binding to calreticulin has been demonstrated to modulate both the structure and function of the protein, with Ca^{2+} and Zn^{2+} binding increasing the lectin activity and polypeptide binding capacity of calreticulin respectively. Proteolytic digestion of full-length calreticulin has demonstrated a Ca^{2+}-dependent protease-resistant fragment which encompasses the N-terminal half of the molecule. This fragment shows homology to the legume lectin family. The crystal structure of calnexin has now been solved and this has illustrated a more complex domain organization than was originally envisaged for calreticulin and calnexin.

Introduction

Since the discovery in 1974[1] and molecular cloning in 1989[2,3] of calreticulin, the structure and function of the protein have been intensively studied. Predominantly located in the endoplasmic reticulum (ER), calreticulin is now known to perform an important role in glycoprotein folding (Chapter 3,6) and Ca^{2+}-homeostasis and signalling (Chapter 11,12). Calreticulin may also be released into the extracellular environment, and autoantibodies against the protein have been detected in the sera of patients with a number of autoimmune conditions (Chapter 17). Importantly, in the past couple of years new information regarding the structure of calreticulin has been forthcoming. This chapter will summarize the most recent advances in the relationship between structure and function of calreticulin.

Domain Organization of Calreticulin

Based on the primary amino acid sequence of calreticulin, a domain organization was proposed for the protein[2,3] (Fig. 2.1), and the protein was subsequently divided into these

Calreticulin, 2nd ed., edited by Paul Eggleton and Marek Michalak.

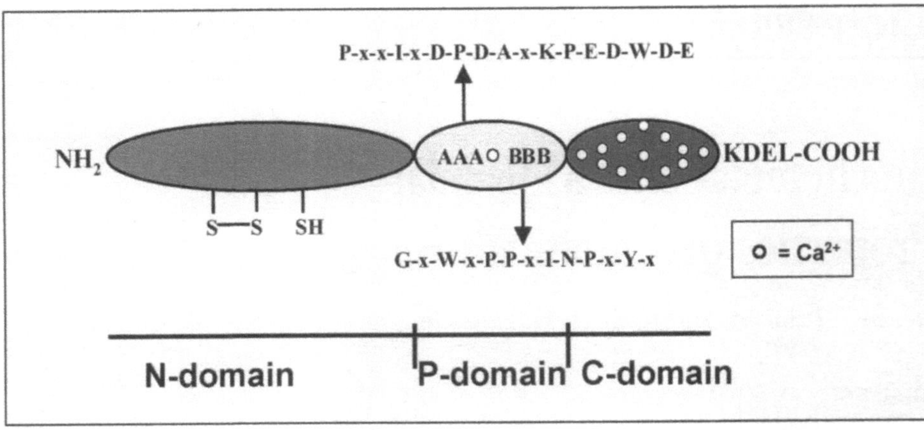

Figure 2.1. Historical domain organization of calreticulin. Original domain organization assigned to calreticulin based on the zonal nature of its primary sequence. A compact globular domain (N-domain) is immediately followed by a proline-rich sequence (P-domain) and a highly acidic C-terminus (C-domain). The P-domain contains 2 sequence repeats (Repeats A and B) which are highly conserved in the calreticulin family.

domains for functional studies. The N-terminal domain (residues 1-180) was predicted to form a globular structure consisting of 8 anti-parallel β-strands, with a short helical segment at residues 91-96. Two sequences in the N-domain are highly conserved throughout the calreticulin family: residues 77-96 (Q-[FY]-x-[LIVM]-[KRN]-x-[DEQN]-[DEQNK]-x-x-x-C-G-G-[AG]-[FY]-[LIVM]-K-[LIVMFY]-[LIVMFY]) and residues 113-121 ([LIVM]-[LIVM]-F-G-P-D-x-C-[AG]). However, no particular activity has been assigned to these residues, and therefore the reason for their high level of conservation is currently unknown.

Immediately following the N-domain is a proline-rich segment termed the P-domain (residues 181-290). The P-domain contains 17 mol% proline residues, which are regularly spaced throughout this region of the protein. The domain also contains 7 of the 11 tryptophan residues of calreticulin and 31 mol% acidic residues. A striking feature of the P-domain is the presence of 2 sets of amino acid repeats which are unique to calreticulin and its molecular homologues calnexin and calmegin. The first such repeat, termed Repeat A, consists of the amino acid sequence P-x-x-I-x-D-P-D-A-x-K-P-E-D-W-D-E and exists as 3 copies in calreticulin. Repeat A is followed by 3 copies of the amino acid sequence G-x-W-x-P-P-x-I-N-P-x-Y-x, termed Repeat B. These repeats were originally proposed to contribute to both the high-affinity Ca^{2+}-binding[4] and the carbohydrate-binding sites[5] of the protein. However, as yet the presence of either Ca^{2+} or carbohydrate binding sites in the NMR structure of the P-domain has not been reported.[6,7] The P-domain also contains a nuclear localization signal, P-P-K-K-I-K-D-P-D, at residues 187-195. This nuclear localization signal is functional in vitro (Burns and Michalak, unpublished) and the protein has recently been implicated to play a role in nuclear export of the glucocorticoid receptor.[8]

The C-terminal domain of calreticulin (residues 291-400) is responsible for up to 50% of the Ca^{2+} capacity of the endoplasmic reticulum (ER)[9] and binds Ca^{2+} with low affinity but high capacity (Kd = 2mM, Bmax = 20-30 mole Ca^{2+}/mole protein)[4]. Thirty seven of the last 57 residues of this region are the acidic residues aspartic and glutamic acid. These residues are arranged in acidic stretches, which are interrupted by the basic residues lysine and arginine. Intriguingly, disruption of these basic residues, by chemical modification, reduces the Ca^{2+}-binding capacity of calreticulin.[10] One important feature of the C-domain of calreticulin is that it terminates with the ER-retrieval amino acid sequence, Lys-Asp-Glu-Leu-COOH (KDEL) which is, at least in part, responsible for ER lumenal retention of the protein.[11]

Glycosylation

The glycosylation state of calreticulin appears to be species and tissue specific.[12-21,23] Human calreticulin contains a consensus site for N-linked glycosylation at Asn 327, a site which is conserved in mammalian and *Xenopus* and *Aplysia* calreticulins, but not in *Drosophila*, tick, plant and nematode calreticulins. An alternative site at amino acid residue 162 exists in bovine brain and *Schistosoma mansoni* calreticulins, while tobacco calreticulin contains an N-linked glycosylation site at amino acid residue 131. As yet, no carbohydrate has been detected in calreticulin from human lymphocytes or placenta, mouse, rat sperm, dog or chicken liver sources.[15] However, calreticulin from human myeloid cells,[16] bovine brain,[14] rat liver,[17] Chinese hamster ovary (CHO) cells,[18] *L. donovani*[19] and spinach cells[20] has been shown to be glycosylated. For example bovine brain calreticulin is glycosylated at amino acid residue 162 and contains a high mannose sugar structure ((GlcNAc)2Man5) consistent with its ER localization.[14] Rat liver calreticulin, on the other hand, contains a complex hybrid oligosaccharide with a terminal galactose.[21] This suggests that the calreticulin has passed through the trans-Golgi rather than simply being retained in the ER, a fact which is supported by evidence that this terminal galactosylation is blocked when vesicular transport from the intermediate- to the trans-Golgi is blocked.

Disulphide Bridge

Full-length calreticulin contains a disulphide bond in the N-terminal half of the molecule, which has been demonstrated to be important for the chaperone function of calreticulin, and hence it must play an important role in the folding of the protein and its ability to interact with specific substrates. Two independent studies have now demonstrated that human calreticulin contains a disulphide bond between Cys88 and Cys120.[22,23] This is inconsistent with data published on bovine brain material[14] but is in agreement with the sequence homologies with calnexin and calmegin as there is no equivalent of calreticulin's Cys146 in these proteins. Such a discrepancy between mammalian forms of the protein seems unusual. However, bovine brain calreticulin is unusual in other ways, compared with other mammalian calreticulins, as it is glycosylated in the N-domain and is the only form of calreticulin to have documented isoforms.

Phosphorylation

Calreticulin contains several consensus amino acid sequences that are potential recognition sites for a variety of protein kinases, including protein kinase C, casein kinase II and tyrosine kinase.[15] However, the phosphorylation state of the molecule has been shown to vary between species.[24,25] One phospho-form of calreticulin has been identified with phosphate non-covalently associated with the sequence 367L-K-E-E-E-E-D-K-K in the C-terminus of the protein.[26] Intriguingly, calreticulin contains amino acid sequence with low similarity/identity to the amino acid sequence of the active site of protein kinase C (residues 215-224; K-P-E-D-W-D-K-P-E-H). Calreticulin binds ATP but it does not contain the consensus sequence (G-x-x-G-x-G) found in the ATP-binding motif of all protein kinases.[27] However, calreticulin undergoes autophosphorylation in Vero 76 cells in response to rubella virus infection, at serine and threonine residues in the N-domain.[25]

Recent Structural Studies on Calreticulin

Despite 15 years of intense study, there is still no X-ray crystallographic structure of full-length calreticulin. A number of alternative studies, however, have shed some light on structural elements of the protein.

Circular dichroism spectroscopy has been carried out on native full-length calreticulin in order to investigate the secondary structure of the protein.[22,27-29] The far-UV spectrum demonstrated a strong minimum at 208 nm and a shallow minimum at 226 nm. This is consistent with calreticulin being an $\alpha + \beta$ protein, i.e., containing separate domains of α-helix and β-sheet, rather than an α/β structure.[30] This result is in agreement with the predictions of calreticulin's secondary structure with, the N-domain predicted to be predominantly β-sheet and the C-domain consisting of α-helices. Provencher-Glöckner secondary structure calculations based on the CD spectra indicate that calreticulin contains 10 % α-helix, 37 % β-sheet and 29 % β-turn.

Gel filtration chromatography analysis of calreticulin has surprisingly revealed that the protein elutes at a position corresponding to a significantly larger size than its calculated Mw (46,500). Calreticulin consistently elutes as a single sharp peak at a molecular weight corresponding to approximately 150,000, i.e., more than three times greater than its calculated Mw.[22,31] However, cross-linking experiments and native-PAGE both suggest that native calreticulin is monomeric. Given that the elution position from a gel filtration column actually correlates most closely with the Stokes radius of a molecule,[32] this unusual gel filtration profile may arise from calreticulin having a highly elongated and/or flexible structure. This is supported by results from analytical ultracentrifugation experiments, which demonstrated that calreticulin is a monomeric, prolate ellipsoid with an apparent length of 29.8 nm and a diameter of 2.44 nm.[31]

Structure of the P-Domain

Thus far the only region of calreticulin for which 3-dimensional structural information is available is the P-domain. Ellgaard *et al* solved the solution structure of residues 189-288 of rat calreticulin by nuclear magnetic resonance (NMR) spectroscopy; a construct which encompasses all of Repeats A and B. The structure has revealed several key points about this unusual stretch of polypeptide.[6,7]

Firstly, the sequence contains large stretches of polypeptide with very little structure due to the regular spacing of the proline residues. Secondly, the identifiable secondary structure consists of a short 3_{10} helix and 6 short β-strands, arranged as 3 anti-parallel sheets, and therefore places the P-domain in the β-II class of secondary structures.[33] The overall topology of the P-domain highlights the repetitive nature of the sequence. The first 3 β-strands (encompassing the type A repeats) are 4 residues long and separated by 17 residues. Residues 238-241 then form a hairpin, and the remaining 3 β-strands are formed by the type B repeats, again being 4 residues long but separated by 14 residues. Proline residues clustered at the ends of the β-strands prevent the formation of longer sheets. The hairpin nature of the domain leads to the pairing of β1 with β6, β2 with β5 and β3 with β4. Furthermore, it also leads to the N- and C-termini of the domain being in close proximity. This suggests that the P-domain could form an extension from calreticulin. Thirdly, 3 hydrophobic clusters appear to be important in holding the hairpin together. Each hydrophobic cluster involves 2 highly conserved Trp residues (1 from a K-P-E-D-W-D and 1 from a G-x-W sequence) packing against the aliphatic region of conserved Pro and Lys sidechains. Extension of the P-domain away from the bulk of the polypeptide may play an important role in chaperone function of calreticulin. It is well established that calreticulin forms functional complexes with ERp57 a thiol-disulfide oxidoreductase that promotes the formation of disulfide bonds in glycoproteins bound by calreticulin.[34] Recent NMR studies revealed that interactions with ERp57 occur at the tip of the hairpin structure of the P-domain.[35]

Structure/Function Relationships—Role of Cations

Calreticulin is a Ca^{2+}-binding protein with both high affinity/low capacity and low affinity/high capacity Ca^{2+}-binding sites [4]. In addition, calreticulin has also been demonstrated to bind Zn^{2+} with a concomitant increase in the hydrophobicity of the protein.[36,37] A detailed study into the chaperoning functions of calreticulin has implicated roles for Ca^{2+}, Zn^{2+} and Mg^{2+}-ATP in controlling the ability of calreticulin to interact with different substrates.[37] While Ca^{2+} increased calreticulin's capacity to prevent aggregation of unfolded glycosylated proteins, Zn^{2+} and Mg^{2+}-ATP both augmented calreticulin's ability to solubilize aggregated non-glycosylated proteins. This is consistent with earlier findings indicating that Zn^{2+}-binding to calreticulin increase the hydrophobicity of the molecule.[36] These results demonstrate the importance of the environmental conditions in the lumen of the ER on both the structure and function of calreticulin. In response to external stimuli, the concentrations of small molecules, such as the ions studied here, fluctuate significantly in the ER lumen.[28] A consequence of these fluctuations may be to alter the structure of calreticulin, and therefore its ability to interact with different proteins, be they unfolded protein substrates, or other chaperones such as ERp57 (Fig. 2.2). Indeed, it has also been demonstrated that Ca^{2+} and Zn^{2+} regulate the interactions between ER lumenal chaperones, including calreticulin.[28]

Limited proteolysis, using a variety of proteolytic enzymes, has confirmed that the binding of small molecules to calreticulin does induce conformational changes. Calreticulin is much more resistant to digestion in the presence of Ca^{2+} compared with EDTA, suggesting a general stabilization of the whole molecule.[22,27] Even more strikingly, 4 of the 5 proteolytic enzymes used produced a Ca^{2+}-dependent, ~25-kDa fragment encompassing the N-terminal half of the molecule. Mapping of a tryptic digest pinpointed residue 205 as a potential C-terminus of this fragment.[22] Current literature highlights the P-domain of calreticulin, and in particular the first set of repeats, as containing the high affinity binding site for Ca^{2+}, while the C-domain contains the high-capacity Ca^{2+}-binding site.[4] These proteolysis results suggest a role for the N-domain in Ca^{2+}-binding, as the fragment stabilized by the Ca^{2+} cannot contain all of the first set of repeats. Ca^{2+} appears to stabilize the complete N-terminal half of the molecule. These studies also demonstrated that Zn^{2+}-binding to CRT protected a central region of the protein spanning a portion of the N-domain and the whole of the P-domain, while Mg^{2+}-ATP protects the full-length protein.[27] The same study also showed that although calreticulin does bind ATP, it does not hydrolyse Mg^{2+}-ATP, unlike several other ER chaperones.

More recently, other groups have also used limited proteolysis for structural analysis of calreticulin. Bouvier and Stafford[31] and Højrup et al[23] demonstrated that limited proteolysis of human calreticulin produced a stable fragment with only the C-terminal acidic residues (340-400) cleaved off. Significantly, a fragment consisting of residues 1-337 of rat calreticulin retains the chaperoning ability of full-length CRT.[38] Højrup et al found that the addition of any of $CaCl_2$, $MgCl_2$, $ZnCl_2$, EDTA or EGTA to the reaction increased the rate of proteolysis, which they ascribed to the effects of these molecules on the proteinase K used in the digests. Although Bouvier and Stafford did not include any divalent cations in their digest reaction, they did note the production of distinct fragments, one of which corresponded almost exactly to the P-domain of calreticulin.

Calreticulin Shows Sequence Homology to the Legume Lectins

The results from the proteolytic digests in the presence of Ca^{2+} are made more interesting by the identification of the legume lectin fold as a potential tertiary structure for the N-terminal half of calreticulin. The legume lectins are a family of proteins which bind simple sugars in a Ca^{2+}-dependent manner.[39] Most are split into 2 domains or even 2 separate chains; an N-terminal ~180 residues which have a β-sheet fold and provide the Ca^{2+} binding site as well as half the sugar binding residues, and a shorter ~50 residue section which provides the other half of the sugar binding site. Despite low overall similarity between legume lectins and calreticulin (Fig. 2.3) there are a number of striking comparisons which can be made:-

1. The size of the legume lectin Ca^{2+} binding domain and the fragment of calreticulin stabilized by Ca^{2+} are very similar.
2. Legume lectins have a β-sheet structure and the calreticulin N-domain is predicted to have a β-sheet structure. These β-strands align well.
3. A high level of homology in the Ca^{2+}/sugar binding residues of the legume lectin, specifically residues known to be in contact with the ligand from the crystal structure. Incidentally this is the region of the calreticulin N-domain which shares highest homology with calnexin and calmegin.

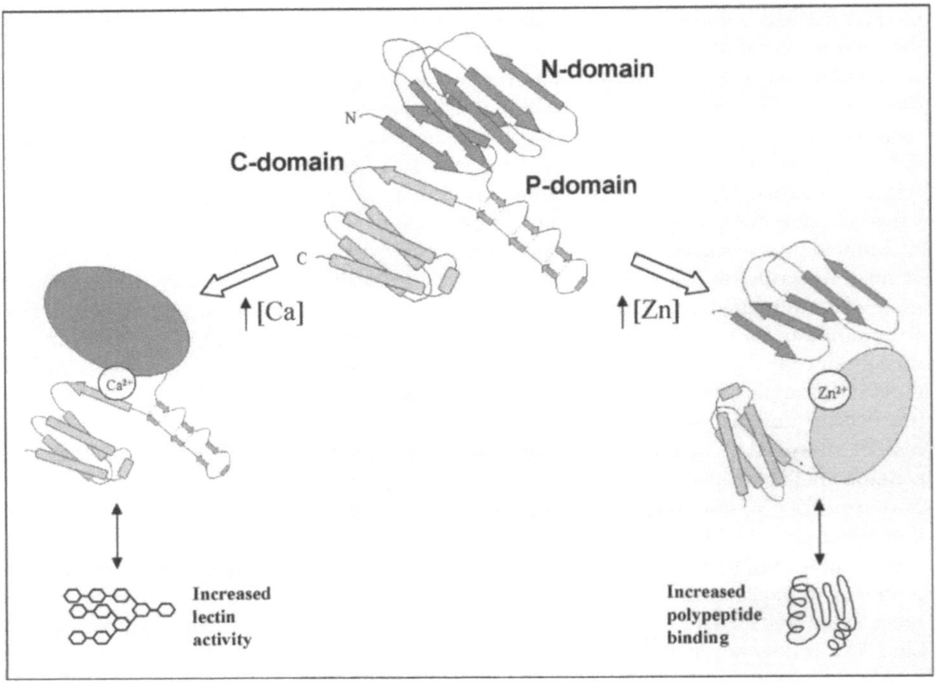

Figure 2.2. Modulation of calreticulin structure/function by small molecules. Cation binding to calreticulin modulates the chaperone function of the protein. Ca^{2+}-binding to calreticulin stabilizes the N-domain and enhances its lectin activity. Zn^{2+}-binding to calreticulin stabilizes the P-domain and promotes the solubilization of unfolded, non-glycosylated polypeptides.

```
           1                                                   38
CRTC_HUMAN -EPAVYFKEQ F--------L DGDGWTSRW- --IESKHKSD FGKFVLSSGK
LEC_PEA.   TETTSFLITK FSPDQQNLIF QGDGYTTKEK LTLTKAVKNT VGRALYSSPI
                                                    *

           39                                                  88
CRTC_HUMAN FYGDEEKDKG LQTSQDARFY ALSASFEPFS NKGQTLVVQF TVKHEQNIDC
LEC_PEA.   HIWDRETGN- VANFVTSFTF VINAPNSYNV ADGFTFFIAP VDTKPQ---T
                                             *

           89                                                 133
CRTC_HUMAN GGGYVKLF-P NSLDQTDMHG ----DSEYNI MFGPDICGPG TKKVHVIFNY
LEC_PEA.   GGGYLGVFNS AEYDKTTQTV AVEFDTFYNA AWDPS----- NRDRHIGIDV
           *****                     * * *   **            *

          134                                                 183
CRTC_HUMAN KGKNVLINKD IRCKDDEFTH LYTLIVRPDN TYEVKIDNSQ VESGSLEDDW
LEC_PEA.   NSIKSVNTKS WKLQNGEEAN VVIAFNAATN VLTVSL---- TYPNSLEEEN
```

Figure 2.3. Sequence homology between calreticulin and the legume lectins. Sequence alignment of calreticulin and pea lectin generated using ClustalW. Identical residues are shown in red, similar residues are shown in green. Residues involved in binding to Ca^{2+} and sugar in the pea lectin crystal structure are highlighted by an asterisk (*). Predicted β-strands (___) and α-helices (___) in calreticulin align well with known β-strands and α-helices from the pea lectin.

Interestingly, other mammalian proteins have been demonstrated to have a homologous structure to the legume lectins, despite having no sequence homology. The pentraxins, which include C-reactive protein and serum amyloid protein, are constructed from multimers of the legume lectin fold.[40] However, in addition to lectin activity these molecules are also capable of binding lipids and DNA in a Ca^{2+}-dependent manner, indicating that the structure is a useful framework on which more specific functions can be built. Recent studies on the mammalian lectins VIP36 and ERGIC53 have identified them as being potential members of the legume lectin family, although they too have fairly low homology.[41] Physical characterization of these proteins has demonstrated Ca^{2+}-dependent lectin activity with a Ca^{2+} affinity similar to that measured for calreticulin.[42] As these are also proteins involved in trafficking of glycoproteins from the ER to the cell surface, it is of great interest to note that they have independently been linked to the legume lectins.

Model of Calreticulin—Implications of the Calnexin Structure

In the last 2 years the entire field of calreticulin and calnexin research has progressed with the publication of the three-dimensional structure of the lumenal domain of calnexin, as solved using X-ray crystallography.[43] Given the high level of sequence identity between calreticulin and calnexin, the structure of calreticulin will almost certainly look very similar (Fig.2.4– model of CRT). The structure is split into two distinct regions, a compact globular domain and a highly extended hairpin loop.

The globular domain is a β-sandwich of two antiparallel β-sheets, and is composed of residues corresponding to residues 1-189 and 284-358 of calreticulin, i.e., the N-domain plus the first 74 residues of the C-domain. Closer examination of the structure reveals that the β-strands from the C-domain form an integral part of the core of the domain. Significantly, the fold of the globular domain is homologous to that of the legume lectins. Calnexin crystals soaked in α-D-glucose demonstrated a putative carbohydrate binding site on the concave β-sheet near strands β7-β9. The residues involved in contacting the glucose molecule are highly conserved residues in the N-domain, and correspond to those suggested as potential Ca^{2+} and carbohydrate binding sites.[22] The researchers also identified a putative Ca^{2+} binding site in the calnexin structure between Asp118 and Asp437 (Glu44 and Asp311 in CRT), i.e., between the N- and C-domains. As this site is distant from the carbohydrate binding site, Ca^{2+} binding to the globular domain was suggested to serve a structural, rather than functional, role.

The extended arm of the calnexin structure corresponds to the P-domain, and agrees very well with the NMR structure reported for the calreticulin P-domain,[7] forming an extended, flexible hairpin loop. This region in calnexin extends 140 Å away from the globular domain, but in calreticulin would be expected to be shorter due the presence of only 3 copies of each repeat compared to calnexins 4. Based on the putative carbohydrate-binding site in the globular domain, the researchers modelled a complete N-linked high-mannose oligosaccharide onto the structure and found that, even in a fully extended structure, the N-linked Asn residue remained 20 Å away from the nearest P-domain repeat motif. This is consistent with the P-domain playing a role in binding polypeptide substrates.

Conclusions

The three-dimensional structure of calnexin should lead to major advances in the study of both calnexin and calreticulin. Combining this structure with the results of studies with cations highlights the importance of the domain organization of calreticulin and calnexin; the compact globular domain being responsible for binding carbohydrates, while the extended P-domain is almost certainly involved in protein-protein interactions as recently demonstrated for ERp57.[35] It is imperative to recognize, however, that the domain structure is more complex than the original linear model depicted in Figure 2.1, as illustrated in Figure 2.4B. The contribution of β-strands formed by the C-domain to the globular domain appears to be essential in the calnexin structure. Interpretation of existing data and future functional studies carried out using calreticulin domains should take this into account.

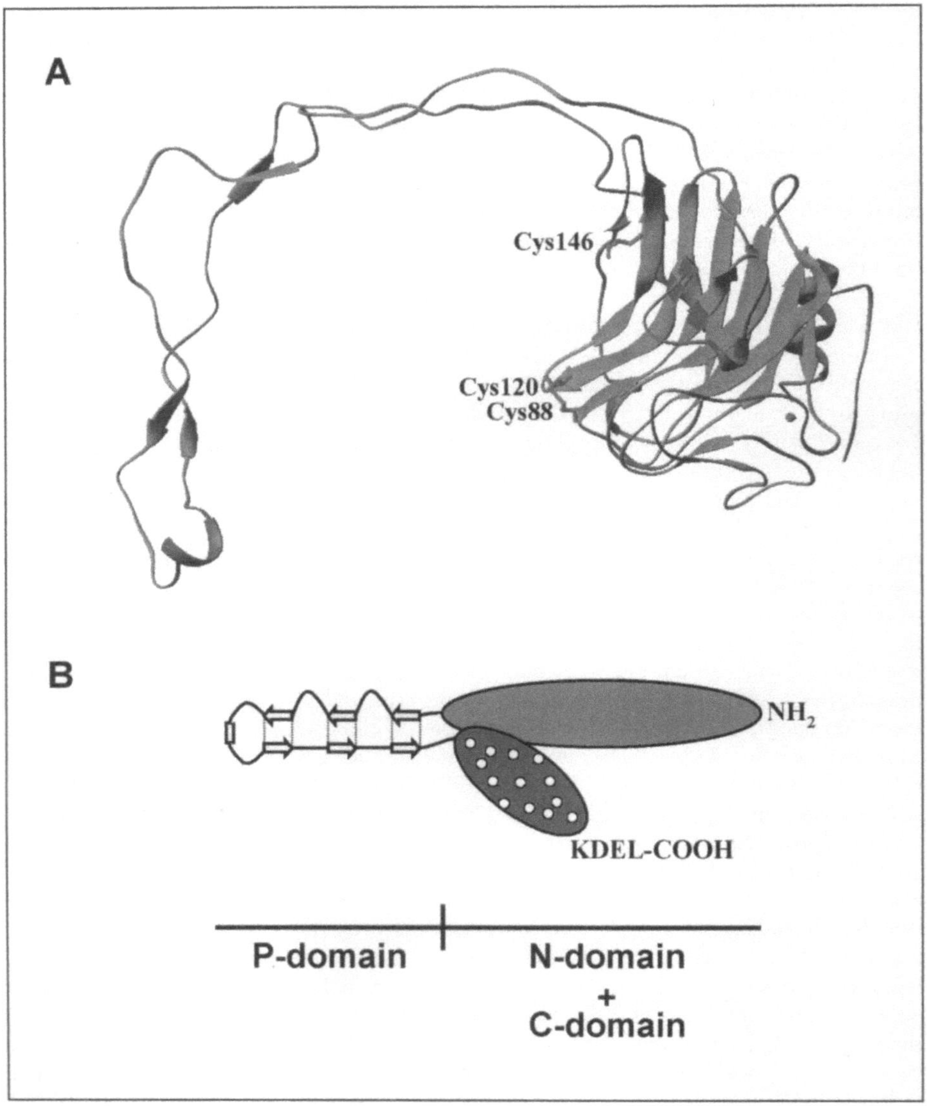

Figure 2.4. A) Model of calreticulin made by fusing the NMR structure of calreticulin P-domain (pdb entry 1HHN) with the crystal structure of calnexin lumenal domain (pdb entry 1JHN). The structure of the compact domain is composed of two antiparallel β-sheets stacked on each other. Helices are shown in red, β-strands in blue and residues interacting with carbohydrate are highlighted in magenta. The side chains of cysteine residues and a calcium ion are shown in green. The P-domain (to the left) is made up of a long hairpin loop, which in parts is held together by β-strands, and contains a short helix at the end. A gap present in the calnexin structure prior to the P-domain has been mended for clarity. B) Revised domain organization of calreticulin, presented in the same style as Figure 2.1.

References

1. Ostwald TJ, MacLennan DH. Isolation of a high affinity calcium binding protein from sarcoplasmic reticulum. J Biol Chem 1974; 249:974-979.
2. Fliegel L, Burns K, MacLennan DH et al. Molecular cloning of the high affinity calcium-binding protein (calreticulin) of skeletal muscle sarcoplasmic reticulum. J Biol Chem 1989; 264:21522-8.
3. Smith MJ, Koch GL. Multiple zones in the sequence of calreticulin (CRP55, calregulin, HACBP), a major calcium binding ER/SR protein. Embo J 1989; 8:3581-6.
4. Baksh S, Michalak M. Expression of calreticulin in Escherichia coli and identification of its Ca^{2+} binding domains. J Biol Chem 1991; 266:21458-65.
5. Vassilakos A, Michalak M, Lehrman MA et al. Oligosaccharide binding characteristics of the molecular chaperones calnexin and calreticulin. Biochemistry 1998; 37:3480-90.
6. Ellgaard L, Riek R, Braun D et al. Three-dimensional structure topology of the calreticulin P-domain based on NMR assignment. FEBS Lett 2001; 488:69-73.
7. Ellgaard L, Riek R, Herrmann T et al. NMR structure of the calreticulin P-domain. Proc Natl Acad Sci USA 2001; 98:3133-8.
8. Holaska JM, Black BE, Love DC et al. Calreticulin Is a receptor for nuclear export. J Cell Biol 2001; 152:127-40.
9. Nakamura K, Zuppini A, Arnaudeau S et al. Functional specialization of calreticulin domains. J Cell Biol 2001; 154:961-72.
10. Breier A, Michalak M. 2,4,6-Trinitrobenzenesulfonic acid modification of the carboxyl- terminal region (C-domain) of calreticulin. Mol Cell Biochem 1994; 130:19-28.
11. Pelham HRB. Control of exit from the endoplasmic reticulum. Annu Rev Cell Biol 1989; 5:1-23.
12. Waisman DM, Salimath BP, Anderson MJ. Isolation and characterization of CAB-63, a novel calcium-binding protein. J Biol Chem 1985; 260:1652-60.
13. Van PN, Peter F, Soling HD. Four intracisternal calcium-binding glycoproteins from rat liver microsomes with high affinity for calcium. No indication for calsequestrin-like proteins in inositol 1,4,5-trisphosphate-sensitive calcium sequestering rat liver vesicles. J Biol Chem 1989; 264:17494-501.
14. Matsuoka K, Seta K, Yamakawa Y et al. Covalent structure of bovine brain calreticulin. Biochem J 1994; 298:435-42.
15. Michalak M, Milner RE, Burns K et al. Calreticulin. Biochem J 1992; 285:681-92.
16. Denning GM, Leidal KG, Holst VA et al. Calreticulin biosynthesis and processing in human myeloid cells: demonstration of signal peptide cleavage and N-glycosylation. Blood 1997; 90:372-81.
17. Zuber C, Spiro MJ, Guhl B et al. Golgi Apparatus Immunolocalization of Endomannosidase Suggests Post- Endoplasmic reticulum Glucose Trimming: Implications for Quality control. Mol Biol Cell 2000; 11:4227-4240.
18. Jethmalani SM, Henle KJ, Kaushal GP. Heat shock-induced prompt glycosylation. Identification of P-SG67 as calreticulin. J Biol Chem 1994; 269:23603-9.
19. Joshi M, Pogue GP, Duncan RC et al. Isolation and characterization of Leishmania donovani calreticulin gene and its conservation of the RNA binding activity. Mol Biochem Parasitol 1996; 81:53-64.
20. Navazio L, Baldan B, Mariani P et al. Primary structure of the N-linked carbohydrate chains of Calreticulin from spinach leaves. Glycoconj J 1996; 13:977-83.
21. Peter F, Nguyen Van P, Soling HD. Different sorting of Lys-Asp-Glu-Leu proteins in rat liver. J Biol Chem 1992; 267:10631-7.
22. Johnson SJ. Characterization of the Structure and Pathophysiological Roles of Human Calreticulin. D. Phil. Thesis 2001.
23. Hojrup P, Roepstorff P, Houen G. Human placental calreticulin characterization of domain structure and post-translational modifications. Eur J Biochem 2001; 268:2558-65.
24. Houen G, Koch C. Human placental calreticulin: purification, characterization and association with other proteins. Acta Chem Scand 1994; 48:905-11.
25. Singh NK, Atreya CD, Nakhasi HL. Identification of calreticulin as a rubella virus RNA binding protein. Proc Natl Acad Sci USA 1994; 91:12770-4.
26. Cala SE. Determination of a Putative Phosphate-Containing Peptide in Calreticulin. Biochem Biophys Res Commun 1999; 259:233-238.
27. Corbett EF, Michalak KM, Oikawa K et al. The conformation of calreticulin is influenced by the endoplasmic reticulum luminal environment. J Biol Chem 2000; 275:27177-85.
28. Corbett EF, Oikawa K, Francois P et al. Ca^{2+} regulation of interactions between endoplasmic reticulum chaperones. J Biol Chem 1999; 274:6203-11.

29. Li Z, Stafford WF, Bouvier M. The metal ion binding properties of calreticulin modulate its conformational flexibility and thermal stability. Biochemistry 2001; 40:11193-201.
30. Manavalan P, Johnson WC Jr. Nature 1983; 305:831-32.
31. Bouvier M, Stafford WF. Probing the three-dimensional structure of human calreticulin. Biochemistry 2000; 39:14950-9.
32. Uversky VN. Use of fast protein size-exclusion liquid chromatography to study the unfolding of proteins which denature through the molten globule. Biochemistry 1993; 32:13288-98.
33. Wu J, Yang JT, Wu C-SC. β-II conformation of all-β proteins can be distinguished from unordered form by circular dichroism. Anal Biochem 1992; 200:359-64.
34. High S, Lecomte FJ, Russell SJ et al. Glycoprotein folding in the endoplasmic reticulum: a tale of three chaperones? FEBS Lett 2000; 476:38-41.
35. Frickel EM, Riek R, Jelesarov I et al. TROSY-NMR reveals interaction between ERp57 and the tip of the calreticulin P-domain. Proc Natl Acad Sci USA 2002; 99:1954-9.
36. Khanna NC, Tokuda M, Waisman DM. Conformational changes induced by binding of divalent cations to calregulin [published erratum appears in J Biol Chem 1986 Dec 5; 261(34):16279]. J Biol Chem 1986; 261:8883-7.
37. Saito Y, Ihara Y, Leach MR et al. Calreticulin functions in vitro as a molecular chaperone for both glycosylated and non-glycosylated proteins. Embo J 1999; 18:6718-29.
38. Peterson JR, Helenius A. In vitro reconstitution of calreticulin-substrate interactions. J Cell Sci 1999; 112:2775-84.
39. Einspahr H, Parks EH, Sugana K et al. The crystal structure of pea lectin at 3.0-A resolution. J Biol Chem 1986; 261:16518-27.
40. Emsley J, White HE, O'Hara BP et al. Structure of pentameric human serum amyloid P component. Nature 1994; 367:338-45.
41. Fiedler K, Simons K. A putative novel class of animal lectins in the secretory pathway homologous to legume lectins (letter). Cell 1994; 77:625-6.
42. Fiedler K, Simons K. Characterization of VIP36, an animal lectin homologous to leguminous lectins. J Cell Sci 1995; 109:271-6.
43. Schrag JD, Bergeron JJ, Li Y et al. The Structure of calnexin, an ER chaperone involved in quality control of protein folding. Mol Cell 2001; 8:633-44.

Chapter 3

A Chaperone System for Glycoprotein Folding: The Calnexin/Calreticulin Cycle

Lars Ellgaard and Ari Helenius

Abstract

The endoplasmic reticulum (ER) contains a particularly wide range of molecular chaperones and other proteins that assist the folding and quality control of newly synthesized protein. Some, like BiP/GRP78 and GRP94, belong to classical chaperone families. Others, like protein disulfide isomerase, ERp57, and ERp72, belong to the family of thiol-disulfide oxidoreductases especially well represented in the ER. The ER lectin chaperones calnexin (CNX) and calreticulin (CRT) have unique features that distinguish them from other known molecular chaperones. They interact with proteins that carry N-linked glycans, and cooperate with a number of accessory enzymes during the folding process. Here we review work on calnexin/calreticulin-assisted glycoprotein folding in the ER, with an emphasis on recent molecular and structural studies.

The ER As a Compartment for Protein Folding and Quality Control

The majority of secretory proteins, plasma membrane proteins, and proteins of the secretory and endocytic pathways in eukariotic cells are synthesized on membrane-bound ribosomes, and undergo maturation in the endoplasmic reticulum (ER). The ER is responsible for translocation, modification, folding, oligomeric assembly, and quality control. It contains a wide variety of molecular chaperones, folding enzymes, escort proteins, folding sensors, and enzymes involved in co- and post-translational modification. The milieu in the lumen is, moreover, strictly controlled with respect to redox state and ionic environment.

Among the most common covalent changes that newly synthesized proteins undergo are signal peptide cleavage, disulfide bond formation, N-linked glycosylation, and the addition of glycosylphosphatidyl inositol (GPI) anchors. These modifications frequently influence the folding process, and are necessary for subsequent transport of proteins out of the ER. Correct timing of certain modifications can also be important because they are part of a stepwise folding and assembly program.

Although conditions inside the ER are optimized for protein folding, the folding process is frequently far from 100% efficient. Aggregated and misfolded proteins as well as unassembled subunits of oligomers are often observed as side products. To overcome the problems created by the generation of such aberrant, nonfunctional proteins, the ER employs a rigorous quality control (QC) system that allows sorting according to conformation.[1,2] Typically, nonnative conformers are recognized and retained in the ER by chaperones and folding enzymes. Proteins that have acquired their native conformation are free to leave. QC ensures that the proteins deployed by the cell are functional and correctly folded. It also keeps immature proteins in the folding environment of the ER thereby improving their chances for folding.

Calreticulin, 2nd ed., edited by Paul Eggleton and Marek Michalak.

If persistently misfolded, proteins are retro-translocated to the cytosol and degraded by the proteasome. This process is referred to as ER-associated degradation (ERAD) (for reviews see refs. 3-5). Overall, the existence of the QC system is needed to prevent escape from the ER of malfunctioning, nonnative conformers that may pose a danger to the cell.

The Calnexin/Calreticulin Cycle

The majority of soluble and membrane-bound proteins synthesized in the ER are glycoproteins, i.e., they contain one or more N-linked oligosaccharides.[6] For these, the ER has evolved a dedicated system for folding assistance and QC centered around two abundant, homologous, resident ER lectins, calnexin (CNX) and calreticulin (CRT).

CNX is a type 1 transmembrane protein, and CRT a soluble, lumenal protein. Both are lectins and share the specificity for binding the di-, tri-, and tetra-saccharides Glcα1-3Man, Glcα1-3Manα1-2Man and Glcα1-3Manα1-2Manα1-2Man with increasing avidity.[7,8] This ensures that they bind selectively to glycoproteins that carry monoglucysylated glycans ($Glc_1Man_{7-9}GlcNAc_2$).[9-11] Situations are also known where they can bind to nonglycosylated proteins but the functional significance of this remains to be determined (see Discussion).[10,12-16] Monoglucosylated glycans are generated either when two glucoses are removed from the orginal core oligosaccharide ($Glc_3Man_9GlcNAc_2$) by ER glucosidases I and II, or when fully deglucosylated glycans are reglucosylated by a ubiquitous ER enzyme called UDP-glc:glycoprotein glucosyltransferase (GT). During normal maturation of glycoproteins, the monoglucosylated glycans occur as transient trimming intermediates.

Like other molecular chaperones, CNX and CRT assist the folding of proteins by interacting with folding intermediates in a cycle of binding and release. This cycle—the so-called calnexin/calreticulin cycle—is illustrated in Figure 3.1. Substrate binding and release is driven by the addition and removal of glucose residues through the action of three independently acting enzymes, glucosidases I and II, and GT. First, glucosidases I and II are responsible for generating the monoglucosylated glycan, thus initiating the interaction with CNX and CRT. By removing the last of the glucose residues, glucosidase II later serves as a dissociation factor. GT on the other hand is an association promoting enzyme because it reglucosylates high mannose glycans. Since it only reglucosylates incompletely folded glycoproteins, it is also the folding sensor in the cycle.

By binding to the substrate glycoproteins, CNX and CRT provide a protected environment for folding. They also help to inhibit aggregation, and prevent premature ER exit. In addition, they mediate the interaction between substrate glycoproteins and ERp57, a glycoprotein specific thiol-disulfide oxidoreductase with which they can form complexes. ERp57 is similar to protein disulfide isomerase (PDI), the best studied of the ER oxidoreductases. Both ERp57 and PDI have been shown to form transient, mixed disulfide bonds with newly synthesized proteins in the ER.[17]

Below, we will discuss the processess of the CNX/CRT cycle in more detail. In addition to the two lectins, we will review recent work on the three key enzymes GT, glucosidase II, and ERp57.

The Structure of Calnexin and Calreticulin

CNX and CRT have several regions in their amino acid sequence that are highly similar. Both have a proline-rich region in the middle, the so-called P-domain. It contains two types of short sequence repeats, termed type 1 and type 2. CRT contains three copies of each organized as '111222', whereas CNX has four of each arranged as '11112222'. Other more N- and C-terminal regions also show high sequence similarity. The main differences are seen close to the C-terminus where CNX has a single transmembrane domain and a cytosolic domain of about 70 residues.

Recently, structural studies have brought important, new insight into the function of CNX and CRT. The structure of the CNX ectodomain was solved by X-ray crystallography,[18] and

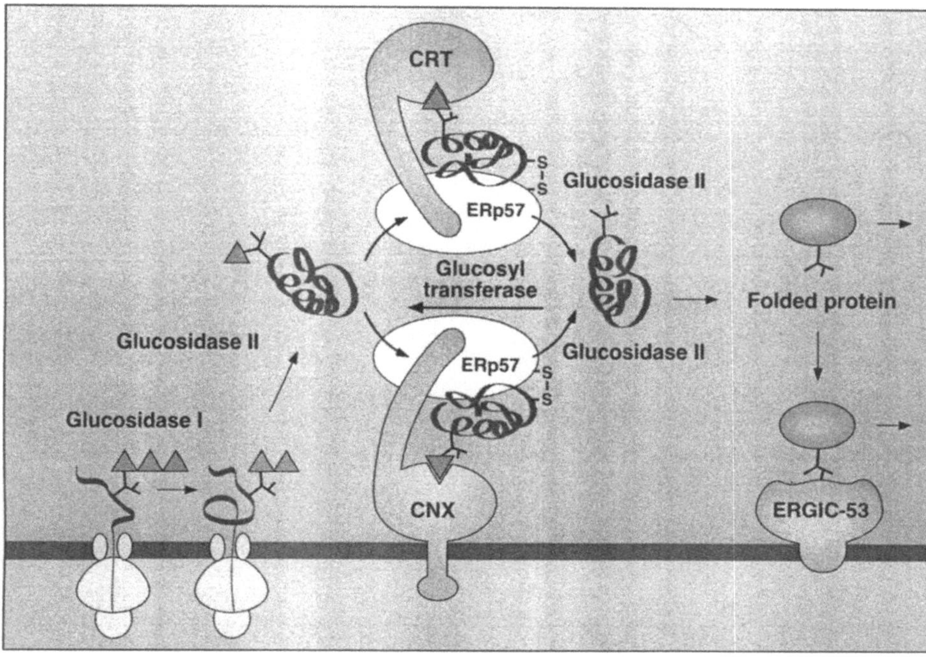

Figure 3.1. The calnexin/calreticulin cycle. The core oligosaccharide ($Glc_3Man_9GlcNAc_2$) is transferred to the growing nascent chain as it enters the ER lumen. Trimming by glycosidases I and II to the monoglucosylated form of the glycan ($Glc_1Man_9GlcNAc_2$) sets up the interaction with CNX and CRT. These proteins bind the thiol-disulfide oxidoreductase ERp57 at the tip of the P-domain, which forms an unusual extended hairpin structure protruding from the more globular lectin domain. While bound by either lectin, disulfide bond formation in the substrate glycoprotein is promoted by ERp57 through a series of transient intermolecular disulfide bonds. Release from CNX and CRT is ensured by glucosidase II trimming of the remaining glucose. If, at this point, the glycoprotein is correctly folded and thereby no longer retained in the ER by the QC system, it is free to leave the ER. In certain cases, the forward transport to the Golgi complex involves another lectin, ERGIC-53. If, on the other hand, the glycoprotein has not yet reached its native conformation, it is recognized by the folding sensor in the CNX/CRT cycle, UDP-glc:glycoprotein glucosyltransferase (GT). Using UDP-glucose as the glucose donor, this enzyme is capable of readding a glucose unit to the α1-3 branch of the oligosaccharide. This recreates the monoglucosylated form of the oligosaccharide and thereby allows renewed interaction of the glycoprotein with CNX and CRT. If persistently misfolded, the substrate glycoprotein is subjected to retro-translocation to the cytosol where it is degraded by the proteasome. This figure was adapted from ref. 60.

the structure of the CRT P-domain by NMR spectroscopy.[19] The crystal structure reveals two domains. One comprises a compact globular β-sandwich fold with structural homology to leguminous lectins as well as to so-called LNS domains.[20] LNS domains are found in neuronal cell surface proteins known as neurexins.[21] The globular domain is formed by the sequences flanking the P-domain. By soaking glucose into the crystal, a single oligosaccharide binding site was identified within this domain. In addition, a Ca^{2+} binding site was observed at some distance from the oligosaccharide binding site.

The second domain, corresponding to the P-domain, has an unusual structure. In CNX it forms an 'arm' approximately 140 Å long, protruding from the lectin domain as an extension of a loop between two of the β-strands. In CRT, the arm is similar, but somewhat shorter (110 Å). In both proteins, the P-domain shows a hairpin fold stabilized by short anti-parallel β-sheets, and small hydrophobic clusters. Each of the clusters result from the packing of two tryptophan side chains against the aliphatic regions of a lysine and a proline. The four residues are highly conserved within the sequences of the CNX and CRT P-domains. The number of β-sheets as well as the number of hydrophobic clusters mirror the number of type 1 and type 2 repeat sequences.

The P-domain structure also shows that each type 1 repeat pairs up with a type 2 repeat by forming interactions across the hairpin. All such '12' units are structurally homologous. Recently, we have found that the outermost of these '12' units in CRT, i.e., residues 221-256, constitutes an autonomous folding unit (L. Ellgaard, P. Bettendorff, D. Braun, T. Herrmann, F. Fiorito, P. Güntert, A. Helenius and K. Wüthrich, in preparation). It is conceivable, that the P-domain sequence has evolved from one such '12' unit—capable of folding independently—by insertion of new units into the middle of a preexisting '12' unit.

NMR studies of the CRT P-domain suggest that a central region of the P-domain arm exhibits slow conformational exchange.[19] Thus, this poorly ordered linker region is flanked by two ordered subdomains for which the relative orientation could not be determined. The conformational instability of the P-domain is a plausible cause for the difficulties in obtaining crystals of CNX and CRT that diffract to a resolution beyond 3 Å. At the same time, the potential mobility in the P-domain could well be of functional importance in the process of glycoprotein folding, as discussed below.

GT

The role of GT as a sensor of nonnative glycoproteins in the ER is intriguing (for a recent review on the role of GT in glycoprotein folding see ref. 22). The enzyme comprises approximately 1500 amino acids of which the ~300 C-terminal residues show clear sequence similarity to members of glycosyltransferase family 8.[23-26] Potential active site residues have been identified within this region.[24,25] The remaining ~1200 residues of the enzyme, which display no obvious homology with known proteins, are presumably involved in the ability of the protein to distinguish between folded and incompletely folded glycoproteins.

Two homologous human GT genes have been identified.[25] Only one of the variants seems to be active and stress-induced. In S. cerevisae, which lacks a functional calnexin cycle, the GT homologue, Kre5p, does not serve as a reglucosylating folding sensor but rather as an enzyme involved in glucan synthesis for the cell wall.[27,28]

The features by which GT senses the folding status of its substrates have been studied in vitro using isolated GT. As the rate of glucosylation rapidly decreases with decreasing mannose content of the core glycan, activity is clearly dependent on the presence of the polymannose core structure.[29] The glycan must, moreover, be linked to a protein; free glycans or small glycopeptides are not substrates.[29] If the protein to which the glycan is bound is folded to its native conformation, the glycan is not reglucosylated. If it is in a random coil conformation, the glycan also fails to be reglucosylated indicating that GT preferentially recognizes partially folded polypeptides.[30,31] This is in agreement with in vivo experiments demonstrating that GT reglucosylates glycoproteins in late stages of folding.[32,33]

Studies using different conformers of bovine pancreatic RNase B, a small, well characterized glycoprotein, as a substrate in vitro, have demonstrated that GT reglucosylates those N-linked glycans that are directly attached to misfolded regions of a protein ignoring nearby glycans in folded regions (C. Ritter, K. Quirin and A. Helenius, in preparation).[31] Substrate recognition is thus 'local' rather than 'global'. This allows the folding machinery to concentrate on those regions of glycoproteins that remain incompletely folded. For example in a multi-domain protein, GT only reglucosylated glycans in the misfolded domains.[31] Also, when a limited region of the RNase B (a single domain protein) is misfolded by point mutations, GT only reglycosylates glycans directly within the affected region (C. Ritter, K. Quirin and A. Helenius, in preparation).

Generalizing from these observations one can hypothesize that the precise location of individual N-linked oligosaccharides on the surface of a glycoprotein is quite important in regulating the stringency by which the protein exposes itself to quality control. It remains to be determined whether recognition of the incompletely folded protein is based on the increased dynamics of incorrectly folded regions such as the 'breathing' motions in the peptide backbone, or whether it occurs via recognition of exposed hydrophobic residues.

An interesting new finding relating to GT is its recently published co-purification with the 15 kDa selenoprotein, Sep15.[34] Sep15 was found to localize to the ER, most likely as a result of its interaction with GT since it lacks a C-terminal KDEL localization signal.[34] Selenoproteins often contain selenocysteine (Sec) residues in redox active "Cys-Xaa-Xaa-Sec" sequence motifs, equivalent to the "CXXC" motifs present in a number of ER resident thiol-disulfide oxidoreductases (see below). In the case of Sep15, a conserved Cys-Gly-Sec-Lys motif is found.[34] The functional significance is not yet clear, but it is conceivable that Sep15 participates as a thiol-disulfide oxidoreductase in the reshuffling or recognition of cysteines or incorrect disulfides in GT substrates.

Glucosidase II

Glucosidase II performs a central function in the CNX/CRT cycle by trimming the two innnermost glucoses of the core oligosaccharide of potential substrate glycoproteins (Fig. 3.1). Trimming of the first of these two glucoses thus generates the monoglucosylated form of the glycan—the form capable of interacting with CNX and CRT. Trimming of the remaining glucose ensures substrate release from, or prevents rebinding to, CNX and CRT.

Glucosidase II is a heterodimer composed of an α-subunit (104 kDa) and a HDEL-containing β-subunit (58 kDa).[35] Both chains occur in differentially spliced forms.[36-39] The sequence similarity with known glycosidases indicates that the C-terminal region of the α-subunit corresponds to the catalytic domain. Indeed, it was recently shown that a 70 kDa fragment from this domain is catalytically active in vitro using *p*-nitrophenyl-α-D-pyranoside as a substrate.[40] Genetic evidence obtained in *S. pombe* confirms the catalytic role of the α-subunit.[41]

The function of the β-subunit is not well understood. The sequence shows the presence of an EF hand motif for calcium binding, and a long stretch of negatively charged residues also likely to interact with calcium.[35] Recently, Sean Munro found that the C-terminal segment contains a sequence similar to the glycan binding domain of the cation-dependent mannose 6-phosphate receptor suggesting a lectin function.[42] The β-subunit seems to interact with the α-subunit through two distinct regions.[43] In vivo, the β-subunit is essential for the formation of an active enzyme.[37,38,41]

An important question concerning the function of glucosidase II is whether the protein is capable of cleaving the innermost glucose while the substrate glycoprotein is bound to CNX and CRT, or whether it acts on the unbound form of the glycoprotein and thereby prevents renewed association with the lectin. The partial protection of a glycoprotein from glucosidase II digestion by CNX argues in favor of the latter.[44] In the crystal structure of CNX, the glycan binding site is, in fact, located in a shallow groove where acces by a glycosidase would seem difficult.[18]

ERp57

Most of the proteins synthesized in the ER acquire disulfide bonds. As mentioned above, the ER maintains an oxidizing environment and is rich in enzymes capable of catalyzing the formation of correct disulfide bonds. PDI is the most extensively studied among ER thiol-disulfide oxidoreductases. While it is known to assist disulfide bond formation in certain glycoproteins, it is ERp57 that cooperates directly with CNX and CRT.[17,45] This enzyme is the closest known homologue of PDI. Like PDI it contains four thioredoxin-like domains, of which the N- and C-terminal ones harbor redox-active cysteines within signature "CXXC" sequence motifs.

Whether ERp57 functions as a thiol oxidase and/or a disulfide isomerase in vivo is at present unclear. If it is an oxidase, the active site cysteines become reduced and have to be reoxidized. For PDI, the FAD-binding protein Ero1 provides the oxidizing equivalents needed for reoxidation.[46,47] However, this membrane-associated enzyme does not seem to reoxidize ERp57.[48,49]

The role of ERp57 as a co-factor for CNX and CRT is evident from studies in vivo and in vitro. In live cells, complexes of ERp57 and glycoproteins have been isolated in combination with CNX or CRT.[17,50-52] Such complexes are not present when the interaction of glycoprotein with CNX and CRT is inhibited using glucosidase inhibitors.[17,50-52] It has further been demonstrated that ERp57 can associate with either lectin directly without the presence of a glycoprotein substrate.[45] In vivo, the ERp57-catalyzed disulfide bond formation in glycoproteins bound by CNX and CRT has been shown to proceed through a series of transient intermolecular disulfide bonds.[17] Furthermore, an elegant in vitro study demonstrated that ERp57-enhanced kinetics and efficiency of disulfide bond formation in a model glycoprotein substrate was observed only in the presence of CNX or CRT.[53]

ERp57 Binds the P-Domain of CRT

Our recent work shows that ERp57 forms a 1:1 complex with the CRT P-domain.[54] TROSY-NMR spectroscopy allowed the mapping of the binding site to residues 225-251, a region which forms the tip of the P-domain. This region also closely corresponds to the independently folding subdomain comprising the outermost of the '12' units in CRT, as mentioned above.

Using isothermal titration calorimetry, a dissociation constant of $9 \pm 3 \times 10^{-6}$ M for the complex was determined, and NMR spectrospic measurements allowed an estimate of the off-rate for the complex of $k_{off} > 1000\ s^{-1}$.[54] Recent biochemical studies show that the affinity of monoglucosylated IgG for CRT is 1.9×10^{-6} M, with an off-rate of $k_{off} = 0.1\ s^{-1}$.[55] Taken together, these results imply that the CRT-glycoprotein interaction is more long-live than the CRT-ERp57 interaction.

The main conclusion drawn from the above results regarding the mechanism of chaperone action, is that the lectin domain, the P-domain, and ERp57 together form a partially closed space in which a substrate glycoprotein can be sequestered and partially protected from the outside (see Fig. 3.2). Within this space the substrate is bound to the oligosaccharide binding site. That the substrate molecule is in fact partially shielded is shown by the observation that CNX and CRT generally prevents the simultaneous binding of BiP/GRP78.[56]

We propose that binding to the distal end of the P-domain allows ERp57 easy access to cysteines and/or cystines in the substrate glycoprotein, which itself is positioned at a distance from the lectin domain by the presence of the oligosaccharide (Fig. 3.2). The extended P-domain forms a protective arm around the lectin-bound substrate polypeptide chain. The observed plasticity of the CRT P-domain might allow the P-domain arm to adapt itself to substrate proteins of varying sizes and shapes. Thus, as pointed out earlier,[57] a certain degree of flexibilty might be important for the function of CNX and CRT as molecular chaperones.

A similar ternary complex with the same characteristics as described above can be imagined for CNX, ERp57 and substrate glycoprotein. However, the fact that CNX is membrane bound clearly influences substrate selection as compared to CRT.[58] With respect to substrate binding by CNX, it has been demonstrated that CNX can interact with an N-linked glycan placed only 13 residues from the membrane.[59] Since the distance from the transmembrane anchor to the lectin domain of CNX is approximately 24 amino acids,[18] this linker sequence is probably rather flexible. This feature most likely allows the protein to interact with oligosaccharides at various positions, and thus to adapt itself to a variety of different glycoprotein substrates.

Discussion

N-linked glycans can play multiple roles during the life cycle of a glycoprotein.[60] Many glycans are essential already during co- and post-translational folding and assembly; some affect the folding process directly, others serve as an 'entrance ticket' into the CNX/CRT cycle. Moreover, in view of what we now know about GT, it seems likely that some are used primarily

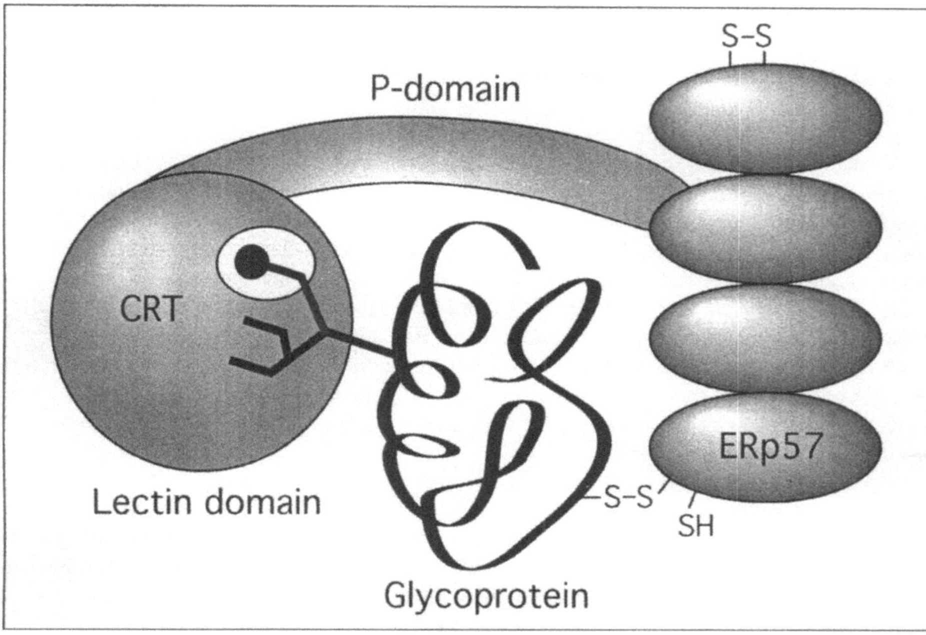

Figure 3.2. Schematic representation of the ternary complex of CRT, ERp57 and substrate glycoprotein. The branched, monoglucosylated oligosaccharide of the substrate glycoprotein interacts with the lectin domain of CRT. In the figure, the oligosaccharide binding site is oval shaped and shown to interact with the α1-3 branch, including the glucose (circle). ERp57 is placed at the distal end of the protruding P-domain arm. This positions the protein favorably for the formation of transient intermolecular disulfide bonds with the substrate glycoprotein. At the same time, the structural arrangement of the lectin domain, the P-domain and ERp57 creates a partially solvent-shielded space in which the folding glycoprotein is likely to be protected from interaction with other ER folding intermediates and chaperones. The four thioredoxin-like domains in ERp57 are drawn as ovals and active site cysteines are represented as SH-groups or disulfide bonds (S-S).

for quality control purposes. They may be strategically localized to regions of the protein where folding needs to be more tightly controlled.

After completed folding, the oligosaccharides can be used as sorting signals for targeting and intracellular transport.[61-63] Finally, once modified in the Golgi complex, they fulfill a variety of different functions as part of the mature protein.[64]

In many of the functions, the N-linked glycan is used as a 'tag', i.e., as a structurally independent, information-carrying element that is used in a wide variety of proteins. Modification through trimming and terminal glycosylation allows the information contents of the tag to be varied. After these changes the tags can be selectively recognized by lectins and modifying enzymes. In this way, whole classes of proteins, and—when extended to the cell surface—whole cell populations, can be tagged for specific recognition. Usually, the message transmitted by the tag depends on the configuration of the most terminal saccharide residues. One property of such a glycan tag is that, in contrast to for example phosphate groups added by kinases, it can be recognized independently of the protein.

It is increasingly apparent that the compartments of the secretory pathway contain lectins for recognition and sorting of glycan tagged proteins. The first such lectins recognized were the mannose-6-phosphate receptors.[61] CNX and CRT represent the second generation. Research on these lectins has provided a starting point for understanding the role of glycan trimming and modification, and led to the concept of glycan tags. In these cases, the lectins work in collaboration with specific glycosidases and glycosyltransferases needed for generating the specific tags.

Additional lectins such as ERGIC-53,[65,66] VIP36,[67] Htm1p/EDEM,[68-70] and Mrl1p[42,71] have recently been identified in the ER and the Golgi complex. Some of these seem to play a role in forward transport of glycoproteins.[67,71-73] Others have a central function in ERAD.[68-70]

In their mode of action, CNX and CRT show similarities and differences compared with other well-studied molecular chaperones. Like members of the Hsp70 and GroEL families they interact with substrate proteins in an on-off cycle, and they prevent aggregation of newly synthesized proteins. In the 'classical' chaperones, the on-off cycle generally depends on conformational changes caused by binding and hydrolysis of ATP (see for instance ref. 74). The ATPase is the chaperone itself. In the case of CNX and CRT, the on-off cycle is driven by covalent modifications performed by two enzymes that act independently of the chaperones.

Another difference is that instead of binding to the polypeptide chain, the main interaction of CNX and CRT is with the tags, i.e., the oligosaccharide appendices. Tethering in this way is likely to provide the polypeptide chain with extensive freedom to adopt different conformations during the folding process.

In addition to the well documented oligosaccharide binding observed for CNX and CRT, it is conceivable that protein-protein interactions could contribute to substrate binding. Although the surface of the P-domain does not seem to contain any obvious hydrophobic polypeptide binding sites, it is possible that protein-protein contacts can occur with a bound substrate. It is also clear that CNX, and under some experimental conditions also CRT, can interact with proteins and protein aggregates that do not contain monoglucosylated glycans.[10,12-15] A series of papers from David Williams' lab have presented data, primarily obtained in vitro, that suggest a function of CNX and CRT in protein folding similar to that of classical chaperones.[16,75-77] At present, it remains unclear how important such a function might be for the folding process in vivo.

Acknowledgments

The authors are thankful for the financial support obtained from the Swiss National Science Foundation.

References

1. Hurtley SM, Helenius A. Protein oligomerization in the endoplasmic reticulum. Annu Rev Cell Biol 1989; 5:277-307.
2. Ellgaard L, Molinari M, Helenius A. Setting the standards: quality control in the secretory pathway. Science 1999; 286(5446):1882-1888.
3. Plemper RK, Wolf DH. Retrograde protein translocation: ERADication of secretory proteins in health and disease. Trends Biochem Sci 1999; 24(7):266-270.
4. Bonifacino JS, Weissman AM. Ubiquitin and the control of protein fate in the secretory and endocytic pathways. Annu Rev Cell Dev Biol 1998; 14:19-57.
5. Fewell SW, Travers KJ, Weissman JS et al. The action of molecular chaperones in the early secretory pathway. Annu Rev Genet 2001; 35:149-191.
6. Gahmberg CG, Tolvanen M. Why mammalian cell surface proteins are glycoproteins. Trends Biochem Sci 1996; 21(8):308-311.
7. Spiro RG, Zhu Q, Bhoyroo V et al. Definition of the lectin-like properties of the molecular chaperone, calreticulin, and demonstration of its copurification with endomannosidase from rat liver Golgi J Biol Chem 1996; 271(19):11588-11594.
8. Vassilakos A, Michalak M, Lehrman MA et al. Oligosaccharide binding characteristics of the molecular chaperones calnexin and calreticulin. Biochemistry 1998; 37(10):3480-3490.
9. Hammond C, Braakman I, Helenius A. Role of N-linked oligosaccharide recognition, glucose trimming, and calnexin in glycoprotein folding and quality control. Proc Natl Acad Sci USA 1994; 91(3):913-917.
10. Ware FE, Vassilakos A, Peterson PA et al. The molecular chaperone calnexin binds Glc1Man9GlcNAc2 oligosaccharide as an initial step in recognizing unfolded glycoproteins. J Biol Chem 1995; 270(9):4697-4704.

11. Hebert DN, Foellmer B, Helenius A. Glucose trimming and reglucosylation determine glycoprotein association with calnexin in the endoplasmic reticulum. Cell 1995; 81(3):425-433.
12. Zhang Q, Tector M, Salter RD. Calnexin recognizes carbohydrate and protein determinants of class I major histocompatibility complex molecules. J Biol Chem 1995; 270(8):3944-3948.
13. Arunachalam B, Cresswell P. Molecular requirements for the interaction of class II major histocompatibility complex molecules and invariant chain with calnexin. J Biol Chem 1995; 270(6):2784-2790.
14. Cannon KS, Hebert DN, Helenius A. Glycan-dependent and -independent association of vesicular stomatitis virus G protein with calnexin. J Biol Chem 1996; 271(24):14280-14284.
15. Bennett MJ, Van Leeuwen JE, Kearse KP. Calnexin association is not sufficient to protect T cell receptor alpha proteins from rapid degradation in CD4+CD8+ thymocytes. J Biol Chem 1998; 273(37):23674-23680.
16. Danilczyk UG, Williams DB. The lectin chaperone calnexin utilizes polypeptide-based interactions to associate with many of its substrates in vivo. J Biol Chem 2001; 276(27):25532-25540.
17. Molinari M, Helenius A. Glycoproteins form mixed disulphides with oxidoreductases during folding in living cells. Nature 1999; 402(6757):90-93.
18. Schrag JD, Bergeron JJ, Li Y et al. The structure of calnexin, an ER chaperone involved in quality control of protein folding. Mol Cell 2001; 8(3):633-644.
19. Ellgaard L, Riek R, Herrmann T et al. NMR structure of the calreticulin P-domain. Proc Natl Acad Sci USA 2001; 98(6):3133-3138.
20. Rudenko G, Hohenester E, Muller YA. LG/LNS domains: multiple functions — one business end? Trends Biochem Sci 2001; 26(6):363-368.
21. Rudenko G, Nguyen T, Chelliah Y et al. The structure of the ligand-binding domain of neurexin Ibeta: regulation of LNS domain function by alternative splicing. Cell 1999; 99(1):93-101.
22. Parodi AJ. Protein glucosylation and its role in protein folding. Annu. Rev. Biochem. 2000; 69:69-93.
23. Parker CG, Fessler LI, Nelson RE et al. Drosophila UDP-glucose:glycoprotein glucosyltransferase: sequence and characterization of an enzyme that distinguishes between denatured and native proteins. EMBO J 1995; 14(7):1294-1303.
24. Tessier DC, Dignard D, Zapun A et al. Cloning and characterization of mammalian UDP-glucose glycoprotein: glucosyltransferase and the development of a specific substrate for this enzyme. Glycobiology 2000; 10(4):403-412.
25. Arnold SM, Fessler LI, Fessler JH et al. Two homologues encoding human UDP-glucose:glycoprotein glucosyltransferase differ in mRNA expression and enzymatic activity. Biochemistry 2000; 39(9):2149-2163.
26. Campbell JA, Davies GJ, Bulone V et al. A classification of nucleotide-diphospho-sugar glycosyltransferases based on amino acid sequence similarities. Biochem J 1997; 326(Pt 3):929-939.
27. Meaden P, Hill K, Wagner J et al. The yeast KRE5 gene encodes a probable endoplasmic reticulum protein required for (1----6)-beta-D-glucan synthesis and normal cell growth. Mol Cell Biol 1990; 10(6):3013-3019.
28. Simons JF, Ebersold M, Helenius A. Cell wall 1,6-beta-glucan synthesis in Saccharomyces cerevisiae depends on ER glucosidases I and II, and the molecular chaperone BiP/Kar2p. EMBO J 1998; 17(2):396-405.
29. Sousa MC, Ferrero-Garcia MA, Parodi AJ. Recognition of the oligosaccharide and protein moieties of glycoproteins by the UDP-Glc:glycoprotein glucosyltransferase. Biochemistry 1992; 31(1):97-105.
30. Trombetta ES, Helenius A. Conformational requirements for glycoprotein reglucosylation in the endoplasmic reticulum. J Cell Biol 2000; 148(6):1123-1129.
31. Ritter C, Helenius A. Recognition of local glycoprotein misfolding by the ER folding sensor UDP-glucose:glycoprotein glucosyltransferase. Nature Struct Biol 2000; 7(4):278-280.
32. Kearse KP. Calnexin associates with monomeric and oligomeric (disulfide-linked) CD3delta proteins in murine T lymphocytes. J Biol Chem 1998; 273(23):14152-14157.
33. Labriola C, Cazzulo JJ, Parodi AJ. Trypanosoma cruzi calreticulin is a lectin that binds monoglucosylated oligosaccharides but not protein moieties of glycoproteins. Mol Biol Cell 1999; 10(5):1381-1394.
34. Korotkov KV, Kumaraswamy E, Zhou Y et al. Association between the 15-kDa selenoprotein and UDP- glucose:glycoprotein glucosyltransferase in the endoplasmic reticulum of mammalian cells. J Biol Chem 2001; 276(18):15330-15336.
35. Trombetta ES, Simons JF, Helenius A. Endoplasmic reticulum glucosidase II is composed of a catalytic subunit, conserved from yeast to mammals, and a tightly bound noncatalytic HDEL-containing subunit. J Biol Chem 1996; 271(44):27509-27516.

36. Arendt CW, Dawicki W, Ostergaard HL. Alternative splicing of transcripts encoding the alpha- and beta- subunits of mouse glucosidase II in T lymphocytes. Glycobiology 1999; 9(3):277-283.
37. Treml K, Meimaroglou D, Hentges A et al. The alpha- and beta-subunits are required for expression of catalytic activity in the hetero-dimeric glucosidase II complex from human liver. Glycobiology 2000; 10(5):493-502.
38. Pelletier MF, Marcil A, Sevigny G et al. The heterodimeric structure of glucosidase II is required for its activity, solubility, and localization in vivo. Glycobiology 2000; 10(8):815-827.
39. Ziak M, Meier M, Etter KS et al. Two isoforms of trimming glucosidase II exist in mammalian tissues and cell lines but not in yeast and insect cells. Biochem Biophys Res Commun 2001; 280(1):363-367.
40. Trombetta ES, Fleming KG, Helenius A. Quaternary and domain structure of glycoprotein processing glucosidase II. Biochemistry 2001; 40(35):10717-10122.
41. D'Alessio C, Fernandez F, Trombetta ES et al. Genetic evidence for the heterodimeric structure of glucosidase II. The effect of disrupting the subunit-encoding genes on glycoprotein folding. J Biol Chem 1999; 274(36):25899-25905.
42. Munro S. The MRH domain suggests a shared ancestry for the mannose 6-phosphate receptors and other N-glycan-recognising proteins. Curr Biol 2001; 11(13):R499-501.
43. Arendt CW, Ostergaard HL. Two distinct domains of the beta-subunit of glucosidase II interact with the catalytic alpha-subunit. Glycobiology 2000; 10(5):487-492.
44. Zapun A, Petrescu SM, Rudd PM et al. Conformation-independent binding of monoglucosylated ribonuclease B to calnexin. Cell 1997; 88(1):29-38.
45. Oliver JD, Roderick HL, Llewellyn DH et al. ERp57 functions as a subunit of specific complexes formed with the ER lectins calreticulin and calnexin. Mol Biol Cell 1999; 10(8):2573-2582.
46. Frand AR, Kaiser CA. Ero1p oxidizes protein disulfide isomerase in a pathway for disulfide bond formation in the endoplasmic reticulum. Mol Cell 1999; 4(4):469-477.
47. Tu BP, Ho-Schleyer SC, Travers KJ et al. Biochemical basis of oxidative protein folding in the endoplasmic reticulum. Science 2000; 290(5496):1571-1574.
48. Benham AM, Cabibbo A, Fassio A et al. The CXXCXXC motif determines the folding, structure and stability of human Ero1-Lalpha. EMBO J 2000; 19(17):4493-4502.
49. Mezghrani A, Fassio A, Benham A et al. Manipulation of oxidative protein folding and PDI redox state in mammalian cells. EMBO J 2001; 20(22):6288-6296.
50. Oliver JD, van der Wal FJ, Bulleid NJ et al. Interaction of the thiol-dependent reductase ERp57 with nascent glycoproteins. Science 1997; 275(5296):86-88.
51. Elliott JG, Oliver JD, High S. The thiol-dependent reductase ERp57 interacts specifically with N-glycosylated integral membrane proteins. J Biol Chem 1997; 272(21):13849-13855.
52. Van der Wal FJ, Oliver JD, High S. The transient association of ERp57 with N-glycosylated proteins is regulated by glucose trimming. Eur J Biochem 1998; 256(1):51-59.
53. Zapun A, Darby NJ, Tessier DC et al. Enhanced catalysis of ribonuclease B folding by the interaction of calnexin or calreticulin with ERp57. J Biol Chem 1998; 273(11):6009-6012.
54. Frickel E-M, Riek R, Jelesarov I et al. TROSY-NMR reveals interaction between ERp57 and the tip of the calreticulin P-domain. Proc Natl Acad Sci USA 2002; 99(4):1954-1959.
55. Patil AR, Thomas CJ, Surolia A. Kinetics and the mechanism of interaction of the endoplasmic reticulum chaperone, calreticulin, with monoglucosylated (Glc1Man9GlcNAc2) substrate. J Biol Chem 2000; 275(32):24348-24356.
56. Molinari M, Helenius A. Chaperone selection during glycoprotein translocation into the endoplasmic reticulum. Science 2000; 288(5464):331-333.
57. Bouvier M, Stafford WF. Probing the three-dimensional structure of human calreticulin. Biochemistry 2000; 39(48):14950-14959.
58. Danilczyk UG, Cohen-Doyle MF, Williams DB. Functional relationship between calreticulin, calnexin, and the endoplasmic reticulum luminal domain of calnexin. J Biol Chem 2000; 275(17):13089-13097.
59. Andersson H, Nilsson I, von Heijne G. Calnexin can interact with N-linked glycans located close to the endoplasmic reticulum membrane. FEBS Lett 1996; 397(2-3):321-324.
60. Helenius A, Aebi M. Intracellular functions of N-linked glycans. Science 2001; 291(5512):2364-2369.
61. Kornfeld S, Mellman I. The biogenesis of lysosomes. Annu Rev Cell Biol 1989; 5:483-525.
62. Fiedler K, Simons K. The role of N-glycans in the secretory pathway. Cell 1995; 81(3):309-312.
63. Hauri H, Appenzeller C, Kuhn F et al. Lectins and traffic in the secretory pathway. FEBS Lett 2000; 476(1-2):32-37.
64. Varki A. Biological roles of oligosaccharides: all of the theories are correct. Glycobiology 1993; 3(2):97-130.

65. Schindler R, Itin C, Zerial M et al. ERGIC-53, a membrane protein of the ER-Golgi intermediate compartment, carries an ER retention motif. Eur J Cell Biol 1993; 61(1):1-9.
66. Itin C, Roche AC, Monsigny M et al. ERGIC-53 is a functional mannose-selective and calcium-dependent human homologue of leguminous lectins. Mol Biol Cell 1996; 7(3):483-493.
67. Fiedler K, Parton RG, Kellner R et al. VIP36, a novel component of glycolipid rafts and exocytic carrier vesicles in epithelial cells. EMBO J 1994; 13(7):1729-1740.
68. Jakob CA, Bodmer D, Spirig U et al. Htm1p, a mannosidase-like protein, is involved in glycoprotein degradation in yeast. EMBO Rep 2001; 2(5):423-430.
69. Hosokawa N, Wada I, Hasegawa K et al. A novel ER alpha-mannosidase-like protein accelerates ER-associated degradation. EMBO Rep 2001; 2(5):415-422.
70. Nakatsukasa K, Nishikawa S, Hosokawa N et al. Mnl1p, an alpha -mannosidase-like protein in yeast Saccharomyces cerevisiae, is required for endoplasmic reticulum-associated degradation of glycoproteins. J Biol Chem 2001; 276(12):8635-8638.
71. Whyte JR, Munro S. A yeast homolog of the mammalian mannose 6-phosphate receptors contributes to the sorting of vacuolar hydrolases. Curr Biol 2001; 11(13):1074-1078.
72. Fiedler K, Simons K. Characterization of VIP36, an animal lectin homologous to leguminous lectins. J Cell Sci 1996; 109(Pt 1):271-276.
73. Appenzeller C, Andersson H, Kappeler F et al. The lectin ERGIC-53 is a cargo transport receptor for glycoproteins. Nat Cell Biol 1999; 1(6):330-334.
74. Grantcharova V, Alm EJ, Baker D et al. Mechanisms of protein folding. Curr Opin Struct Biol 2001; 11(1):70-82.
75. Ihara Y, Cohen-Doyle MF, Saito Y et al. Calnexin discriminates between protein conformational states and functions as a molecular chaperone in vitro. Mol Cell 1999; 4(3):331-341.
76. Saito Y, Ihara Y, Leach MR et al. Calreticulin functions in vitro as a molecular chaperone for both glycosylated and non-glycosylated proteins. EMBO J 1999; 18(23):6718-6729.
77. Stronge VS, Saito Y, Ihara Y et al. Relationship between calnexin and BiP in suppressing aggregation and promoting refolding of protein and glycoprotein substrates. J Biol Chem 2001; 276(43):39779-39787.

CHAPTER 4

Calnexin, an ER Integral Membrane Chaperone in Health and Disease

John J.M. Bergeron and David Y. Thomas

Abstract

This review discusses the ER protein calnexin that is related in structure and function to calreticulin. In vivo and in vitro experiments from many laboratories have provided evidence that calnexin and calreticulin interact transiently with glycoproteins while they are folding in the ER a that this interaction is via a specific $Glc_1Man_9GlcNAc_2$ glycoform. The structure of calnexin has recently been determined to 2.9Å resolution by X-ray crystallography and has a unique and remarkable structure a globular domain and an extended 140Å arm termed the P domain. The P domain recruits a member of the protein disulfide isomerase family, ERp57, that specifically catalyzes disulfide bond exchange on glycoproteins bound to calnexin. Calnexin links N-glycosylation and protein folding and forms the quality control system for glycoproteins. Mutant glycoproteins are the basis of many human protein trafficking diseases and the ER quality system is responsible for their retention in the ER and their proteolytic degradation in the cytosol.

Introduction

The quality of proteins is of central importance for cells. They have evolved molecular chaperone mechanisms that enhance the probability of correct protein folding and also quality control mechanisms to assess their folded state and to remove incorrectly folded proteins. Molecular chaperones bind to or sequester unfolded proteins to prevent their aggregation and thus to promote their folding. Quality control systems are composed of sensor molecules that may also be molecular chaperones that discriminate the folding status of proteins and associated proteolytic systems that remove unfolded proteins. These quality control systems survey newly synthesized proteins for their folding and also operate at later stages to remove incorrectly folded proteins. The biological consequences of incorrectly folded proteins that escape the action of molecular chaperones and their associated quality control mechanisms can be the presence of non-functional proteins in the cell, or their failure to be transported to their correct cellular location, or the formation of protein aggregates that may be toxic. Cells employ many resources to ensure the fidelity of protein folding and quality control and there are mechanisms to ensure this in all cellular compartments.

Eukaryotic cells, for secreted and membrane proteins, use a molecular chaperone system and quality control system that links N-glycosylation and protein folding. The initial step in N-glycosylation in the Endoplasmic reticulum (ER) is the transfer to polypeptide chains translocating into the ER of a $Glc_3Man_9GlcNAc_2$ carbohydrate to asparagine residues in N-X-Ser/Thr sequence motifs and then there is extensive processing of this glycan. The molecule that links protein folding and glycosylation is the ER membrane protein calnexin and it shares many or most of its functions with its ER lumenal homolog calreticulin. They are components of a highly effective quality control apparatus that together with other proteins constitute the

Calreticulin, 2nd ed., edited by Paul Eggleton and Marek Michalak.

"Calnexin/calreticulin Cycle" or more simply the "Calnexin cycle". This molecular machine ensures the correct folding and oligomeric assembly of many secretory and membrane glycoproteins. Some of the other constituents are known and include enzymes that act directly on proteins such as the protein disulfide isomerase family member ERp57, but also includes the enzymes involved in modification of carbohydrates such as the glucosidases I and II and UDP-glucose:glycoprotein glucosyltransferase (UGGT). The calnexin cycle links protein N-glycosylation and protein folding, but the linkage to other protein folding machines in the ER such as the molecular chaperones BiP, protein prolyl isomerases and GRp94 is not clear. However, it is clear that in other mechanisms in the ER that the glycan moiety is used to direct the trafficking and fate of glycoproteins, (see refs. 1-4 for recent reviews). The calnexin cycle is shown in Figure 4.1.

There are many human hereditary diseases that are due to mutations in secreted and membrane proteins. Mutant secretory proteins are recognized by the calnexin and associated quality control system and retained in the ER and thus prevented from reaching their correct cellular location. The mutant proteins are then subjected to the ER associated degradation. Alternatively some mutant proteins form aggregates that are resistant to the quality control pathway and lead to inclusion body formation in the ER. A large number of human diseases are thus due to mutations in secretory glycoproteins and the underlying cause of these diseases is the action of the protein folding and quality control machinery in the ER (see refs. 5-6 for recent reviews).

The Structure of Calnexin

The properties of calnexin and its interaction with glycoproteins have been the subject of intensive investigation since its discovery in 1991. Calnexin was first identified as an integral membrane phosphoprotein in canine microsomes.[7] Subsequent physiological experiments identified calnexin both as an ER protein involved in the assembly of MHC1 and as a protein that was transiently associated with glycoproteins.[8-9] The basis for this interaction has extensively studied using many cell types in experiments in which secretory proteins are labeled followed by immunoprecipitation of calnexin and the analysis of the associated proteins. In these experiments cells are treated with a variety of conditions to modify glycosylation and to inhibit protein folding. Results showed that calnexin interacts with many glycoproteins and not with nonglycoproteins in the ER such as serum albumen.[9] Also for glycoproteins for which the folding state could be assessed, it was shown that calnexin apparently interacts with unfolded proteins and not with folded proteins. Based upon these studies and the use of specific inhibitors of glycan processing enzymes it was proposed that calnexin specifically interacts with the $Glc_1Man_9GlcNAc_2$ glycoform.[10]

Direct evidence for this interaction with the glycan came from in vitro experiments with the purified lumenal domain of calnexin and different glycoforms and conformers of RnaseB.[11] This domain of calnexin has been shown to be functional in cells. For example, with the calnexin cnx1 in *Schizosaccharomyces pombe*,[12] which in this yeast calnexin is essential. It was shown that deletions of the C-terminal cytosolic and the transmembrane domain creating a lumenal domain could support growth. The equivalent lumenal domain of mammalian calnexin was produced in insect cells and extensively characterized and tested for its ability to bind the folding intermediates of the glycoprotein RNaseB that results from disulfide bond rearrangement.[13] These experiments confirmed that calnexin binds glycoproteins bearing $Glc_1Man_9GlcNAc_2$ glycans and they showed that it binds to glycoproteins irrespective of their conformation.[13] However, a series of in vitro experiments with classical nonglycoprotein substrates of cytosolic chaperones have shown that calnexin has a profound effect on their folding and effectively functions as a molecular chaperone.[14] Of particular interest with these latter experiments were that they showed an effect of ATP folding of nonglycoproteins. ATP was previously identified bound to calnexin and to have profound effects on its conformation.[13] Thus effects of calnexin on the folding of both glycoproteins and nonglycoproteins has been found in in vitro experiments.

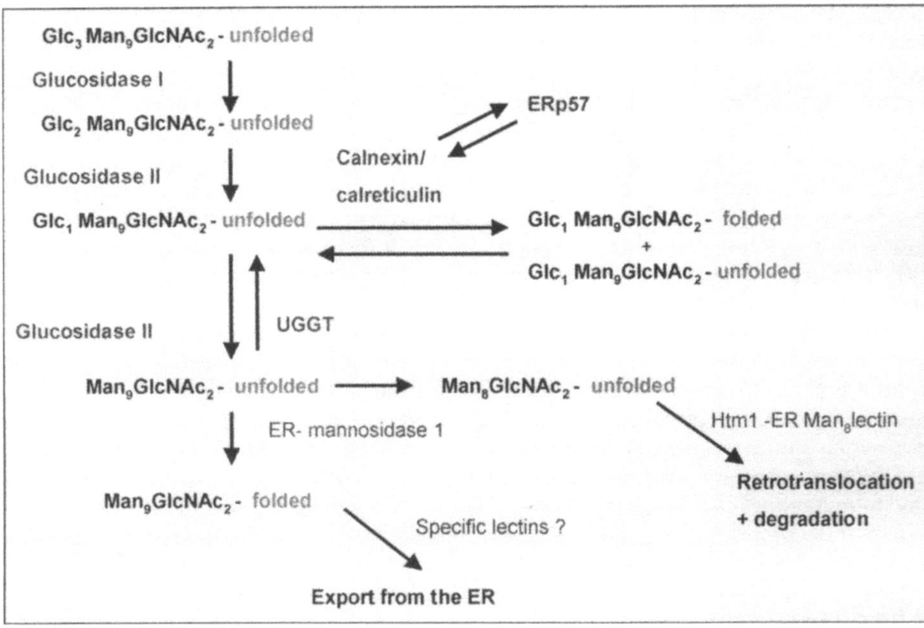

Figure 4.1. This presents a schematic of the calnexin/calreticulin cycle. By the action of oligosaccharyl transferase, glycoproteins acquire a $Glc_3Man_9GlcNAc_2$ glycan at N-X-S/T sequence motifs soon after translocation into the ER. The terminal glucose residues are trimmed by the consecutive actions of glucosidase I and glucosidase II. Glycoproteins bearing the intermediate $Glc_1Man_9GlcNAc_2$ are specifically recognized and bound by calnexin. and calreticulin. The protein disulfide isomerase ERp57 catalyzes the exchange of disulfides of calnexin bound $Glc_1Man_9GlcNAc_2$ glycoproteins. The terminal glucose-mannose bond is then cleaved by glucosidase II. Then the $Man_9GlcNAc_2$ glycoprotein if it is folded is subjected to further glycan modification and transported from the ER to other compartments of the secretory pathway. Unfolded glycoproteins are recognized by UGGT and reglucosylated and rebind with calnexin or calreticulin for surther cycles of folding. Glycoproteins that cannot fold are substrates for ER mannosidsae I and the $Man_8GlcNAc_2$ glycoprotein is recognized by an ER Man_8 specific lectin Htm1p and introduced into the Sec61/AAA protease dependent-retrotranslocation pathway and then degraded by proteasomes in the cytosol.[44]

There are two general models of how calnexin functions. The first is that it functions solely as a lectin binding monoglucosylated glycoproteins irrespective of their conformation. The second model is that it can also recognize the conformation of proteins. The structure of calnexin provides strong evidence that the first model is correct.

The structure of the lumenal domain of calnexin was determined at 2.9Å resolution. The overall structure is remarkable, it is highly asymmetric and composed of two very distinct regions. A compact domain comprised of a β-sandwich of two antiparallel β-sheets, and a long (140Å) proline-rich "P-domain" (see Fig. 4.2).[15] Attempts at cocrystallization of calnexin with disaccharides and oligosaccharides were not successful but with glucose crystals were obtained and a structure derived at 3.2Å. The carbohydrate binding site is located in a shallow groove in the compact β-sheet domain. This result is consistent with some of the experimental evidence of the recognition elements in the $Glc_1Man_9GlcNAc_2$ oligosaccharide residing in the Glc_1Man_3 moiety and on the inaccessibility of the terminal glucose mannose bond to the processing enzyme glucosidase II.[11,16] This domain also shares a fold that is found in the plant lectins, galectins, and pentraxins and thus it was concluded that this domain has a lectin function. Apart from a small region of hydrophobicity at the base of the P loop there is no apparent

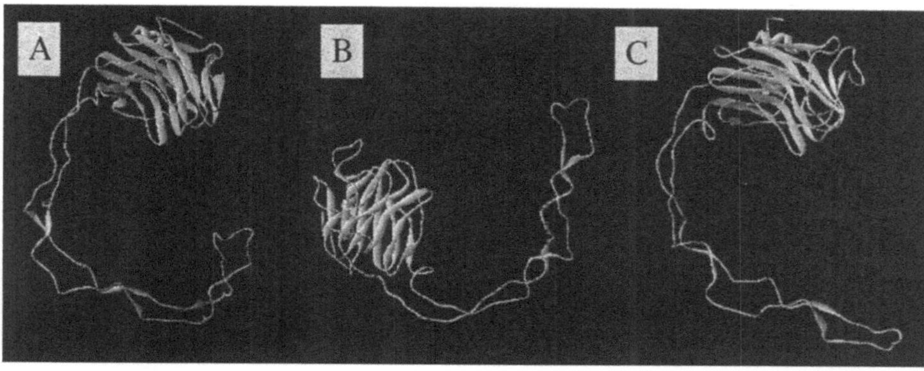

Figure 4.2. The structure of Calnexin (A) is derived from the coordinates determined from the crystal structure.[15] The β-sheet globular lectin domain is in gray, and the extended P domain is shown with the individual repeats in blue, yellow, purple, and the terminal domain in green. Panel B shows a model of calreticulin with the lectin domain in the same orientation. Panel C shows the structure of the one repeat shorter calreticulin P domain in a similar orientation to the P domain of calnexin in Panel A.

extensive hydrophobic domain in the whole structure determined that might bind unfolded polypetides and thus provide a basis for the effects of calnexin on the folding of nonglycoproteins in vitro. Although this result from the structure may seem to be conclusive and support the lectin-only model, it must be emphasized that the structure determined is only of residues 61-458 of the luminal domain and that parts of the molecule are missing in this structure. In addition the molecule is highly flexible and a structure with a known ligand ATP that promotes major conformational changes or with adenine were not obtained. Thus there is a possibility that another conformation of calnexin induced by ATP may expose hydrophobic surfaces for interaction with unfolded polypeptides. Molecular modelling of the sequence of calreticulin onto the structure of calnexin yields a very similar structure with many of the structural features and predicted lectin binding site conserved (see Fig. 4.2).

A remarkable feature of the structure of calnexin is the extended P-domain comprised of four copies each of two different proline-rich sequence motifs in a linear sequence pattern of 11112222. These repeats fold into a large hairpin which is formed by antiparallel interactions between each of the four tandem repeats of the two different sequence motifs. Each repeat of motif 1 interacts with a repeat of sequence motif 2, forming a small modular subdomain of the whole P domain. Ring stacking interactions between conserved tryptophan residues and hydrogen bonds between the backbone atoms of conserved isoleucine residues stabilize the folding of these small domains. There is strong sequence conservation of this region with calreticulin and the structure of this domain has been determined by NMR for calreticulin[17] and for calnexin (Ekiel et al, in preparation). The difference being that for calreticulin the number of repeat units is fewer in a 111222 pattern. The functional consequences of this shorter repeat are difficult to interpret as calnexin and calreticulin so far show very similar properties in vivo and in in vitro in their interaction with glycoproteins. The structures determined for the P domains of calnexin and calretivculin repeats and terminal structure repeats by crystallography and by NMR are superimposable (Fig. 4.2).[15] The potential function of the P domain is the subject of continuing investigation. Molecular modelling of the $Glc_1Man_9GlcNAc_2$ glycoprotein β-glucanase shows that a protein of 42 kDa can comfortable fit within the arm generated by the P domain. One speculation is that the arm provides a barrier against free diffusion of glycoproteins increasing the weak μmolar range affinity of the oligosaccharide for calnexin. It is not certain whether calnexin acts as a monomer or oligomer in the ER and there is a possibility that there are protein interactions mediated by the P domain or other parts of the molecule.

Thus at present the only known member of the calnexin cycle which has unequivocally been shown to recognize unfolded proteins is UGGT. This ER lumenal protein has the unique property of recognizing unfolded proteins and if they carry a $Man_9GlcNAc_2$ glycan, glucosylating them and enabling them to rebind calnexin and calreticulin (Fig. 4.1).[3] Sequencing and mutagenesis studies have shown that UGGT is a >180kDa protein with the catalytic residues restricted to the C-terminal 38kDa of the protein and that the protein preferentially interacts with and glucosylates glycoproteins in vitro.[18]

Specific Interaction with ERp57

The strongest evidence for a function of the P domain comes from studies of calnexin interaction with ERp57 a member of the protein disulfide isomerase family. Initially, experiments with the import of nascent glycoproteins into microsomes indicated that ERp57 as well as calnexin and calreticulin could be cross-linked to glycoproteins.[19] Subsequently this was shown to be due to the cross-linking of ERp57 to calnexin and calreticulin and that this effect was specific and that PDI was not cross-linked.[20] The functional significance of this interaction was demonstrated by in vitro experiments with purified calnexin and calreticulin that showed that ERp57 accelerated the rate of folding specifically of $Glc_1Man_9GlcNAc_2$-RNaseB bound to calnexin and calreticulin but not of unbound RNaseB.[11] Subsequently it was shown that ERp57 and calnexin and also calreticulin associate specifically in a calcium dependent fashion and that PDI does not interact with these chaperones.[21] These results show that the CNX/ERp57 and CRT/ERp57 complexes act as folding scaffolds specific for monoglucosylated secretory and membrane glycoproteins. Various approaches to mapping of the site of interaction indicate that this is specific for the P-domain (Pelletier et al, in preparation) and NMR studies of the interaction of the P domain and ERp57 have shown that the site of interaction is at the tip of the P domain of calreticulin and calnexin (Ekiel et al, in preparation).[22] From results with other molecular chaperone systems the physical and functional interaction of components of the calnexin cycle is to be expected and a search for further interactors will help elucidate the mechanism. In particular the functional relation of the calnexin cycle with the other major molecular chaperones in the ER remains to be elucidated.

Functions of Calnexin, Calreticulin and Calmegin

There are three members of this family in mammalian cells. Calnexin and calreticulin are found in all cells but expression of calmegin is restricted to the testis.[23] They all have highly conserved structural features in their presumed lectin domain and in their P domains.[15] Most of the evidence for equivalence of their function relies on experiments with calnexin and calreticulin. In pulse-chase experiments the majority of glycoproteins interact both with calnexin and with calreticulin but there are distinct preferences for some glycoproteins. Thus there is an overlapping spectrum of glycoproteins that associates with these chaperones. However, this type of result is difficult to interpret in terms of specificity as it could be due differences location abundance and affinity for substrates. Experiments where calreticulin was converted into a membrane tethered molecule using the cytosolic tail of calnexin showed that it could recognize calnexin preferred substrates.[24] Studies with specific viral glycoproteins have shown that calnexin and calreticulin do differ in their ability to bind different glycans on the same molecule and this result is interpreted as resulting from the accessibility of different glycans to calnexin and calreticulin.[25]

Calnexin and calreticulin occur in all eukaryotic species so far examined, although unicellular eukaryotes appear to have only calnexin, but the high degree of conservation of these paralogs attests to their key roles. The results of a knock-out mutation of the calreticulin gene in mouse shows that the apparent redundancy of calnexin and calreticulin from the results of in vitro experiments and from cell studies, is incorrect. Mice lacking the calreticulin gene die on embryonic day 18 with defects in heart development and umbilical hernia.[26] Overexpression of the calreticulin gene in mice is also lethal with the effected animals also showing defects in heart function.[27] The fact that homozygous ES cells and mouse embryonic fibroblasts could be derived from these animals shows that calreticulin is not essential for cell viability but that calreticulin has a specific function in development. The evidence for the role of calnexin is, as yet, less complete. A calnexin deficient NK resistant lymphocyte cell line is available and shows variation in T cell receptor assembly and apoptosis.[28,29] Full characterization of calnexin function in vivo awaits the publication of the results of a calnexin knock out mouse strain. Homozygous knock outs of the testis specific calmegin in mouse results in mice that live but are sterile.[30] In the nematode *C. elegans* both calnexin and calreticulin are found, and the knock out of the calreticulin gene is not lethal but is pleiotropic with effects on many cell and organ functions.[31,32]

Despite the large number of human protein trafficking diseases caused by mutations in secretory proteins no naturally occurring mutations of calnexin, calreticulin or calmegin have been found. Mutations in these genes might be expected to lead reduced efficiencies in protein folding and assembly and thereby to subtle alterations in phenotypes but they have not been found so far in humans. Mutations have been found in the other components of the calnexin cycle that are involved in glycosylation. There has been a recent report of a human infant with glucosidase I deficiency with a severe and progressive disease state characterized by hypotnia, dysmorphic features, hypoventilation, and death at 74 days.[33] A redundancy of some functions and perhaps effective shunt mechanisms for the processing of oligosaccharides may make the results of mutations in the other calnexin cycle components less pronounced and they may only be found with slight effects on phenotypes. There is, however, a human null mutation of the high-mannose specific lectin ERGIC53 involved in the trafficking of glycoproteins in the intermediate compartment. Patients with this mutation have a combined reduced level of the blood coagulation factors V and VIII and the levels of other secretory proteins are not disturbed although compensation by the homolog VIP36 may account for this.[34]

Experiments with inhibitors such as deoxynojirimycin of glucosidase II that trims oligosaccharides and mediates their binding to calnexin indicate that this pathway may have useful therapeutic targets.[35] Viral glycoproteins are synthesized in the ER and the virus assembled there. In the case of HIV1 the gp160 precursor of the viral gp120 glycoprotein shows prolonged interaction with calnexin mediated by the unusual signal sequence and perhaps reflecting a viral host-evasion strategy.[36,37] For HBV it has been experimentally shown that more cell-permeable derivatives of deoxynojirimycin inhibit viral assembly and deplete viable WHV virus in an animal model.[38,39] The therapeutic window and opportunity in this pathway for the development of antivirals may result from the increased molecular chaperone requirements for viral glycoprotein folding in comparison with endogenous glycoproteins of similar complexity. Anti-viral therapies based upon host ER targets would avoid the problems of the development of viral resistance encountered in other antiviral therapies.

Information on the functions of calnexin and calreticulin has also been obtained from their study in model organisms. *Dictyostelium* is the only microorganism in which calnexin and calreticulin occur and knock outs of both these genes results in cells that are severely reduced in phagocytosis.[40] Calnexin is present in the yeast *Schizosaccharomyces pombe* and its function appears to be similar to that in mammalian cells.[12] In this yeast calnexin is essential and there is also a functional UGGT.[41] However, in the yeast *Saccharomyces cerevisiae* the calnexin gene is not essential and the UGGT apparent homolog encoded by the KRE5 gene is essential but does not have detectable enzymatic activity. In this yeast the deletion of the gene for calnexin has an effect on β-glucan synthesis and is epistatic with known genes in this pathway.[42] Thus the results from model organisms give us some clues that calnexin and its partners may have functions other than directly on glycoprotein folding.

Conclusions

The critical role of correct protein folding in the ER for human health is readily apparent from the many pathologies arising from misfolded secretory proteins. [1,4,43] The calnexin cycle plays a key role in the folding and quality control of secretory glycoproteins and apparently the major components of this molecular machine have been found and their functions are being established. Present evidence is that calnexin and calreticulin function as lectins that specifically recognize monoglucosylated N-glycans and that the role of sensing unfolded proteins in the calnexin cycle is played by UGGT. It is clear that the calnexin cycle does not operate in isolation from the other molecular machines and chaperone systems in the ER and the challenge will be to identify them and establish how they are integrated.

Acknowledgments

We thank Leo Lin and Dr. Joe Schrag for help with Figure 4.2. We apologize to those authors whose work we have been unable to quote. Work in the authors' laboratories is supported by operating grants from the Canadian Institutes for Health Research.

References

1. Chevet E, Jakob CA, Thomas DY et al. Calnexin family members as modulators of genetic diseases. Semin Cell Dev Biol 1999; 10:473-80.
2. Zapun A, Jakob CA, Thomas DY et al. Protein folding in a specialized compartment: the endoplasmic reticulum. Structure Fold Des 1999;7: R173-82.
3. Parodi AJ. Protein glucosylation and its role in protein folding. Annu Rev Biochem 2000; 69:69-93.
4. Pelletier MF, Bergeron JJM, Thomas DY. Molecular chaperone systems in the endoplasmic reticulum. Chapter 8. Oxford: Oxford University Press, 2001.
5. Cabral CM, Choudhury P, Liu Y et al. Processing by endoplasmic reticulum mannosidases partitions a secretion- impaired glycoprotein into distinct disposal pathways. J Biol Chem 2000; 275:25015-22.
6. Molinari M, Helenius A. Chaperone selection during glycoprotein translocation into the endoplasmic reticulum. Science 2000; 288:331-3.
7. Wada I, Rindress D, Cameron PH et al. SSR alpha and associated calnexin are major calcium binding proteins of the endoplasmic reticulum membrane. J Biol Chem 1991; 266:19599-610.
8. Ahluwalia N, Bergeron JJ, Wada I et al. The p88 molecular chaperone is identical to the endoplasmic reticulum membrane protein, calnexin. J Biol Chem 1992; 267:10914-8.
9. Ou WJ, Cameron PH, Thomas DY et al. Association of folding intermediates of glycoproteins with calnexin during protein maturation. Nature 1993; 364:771-6.
10. Hammond C, Braakman I, Helenius A. Role of N-linked oligosaccharide recognition, glucose trimming, and calnexin in glycoprotein folding and quality control. Proc Natl Acad Sci USA 1994; 91:913-7.
11. Zapun A, Petrescu SM, Rudd PM et al. Conformation-independent binding of monoglucosylated ribonuclease B to calnexin. Cell 1997; 88:29-38.
12. Parlati F, Dignard D, Bergeron JJ et al. The calnexin homologue cnx1+ in Schizosaccharomyces pombe, is an essential gene which can be complemented by its soluble ER domain. Embo J 1995; 14:3064-72.
13. Ou WJ, Bergeron JJ, Li Y et al. Conformational changes induced in the endoplasmic reticulum luminal domain of calnexin by Mg-ATP and Ca2+. J Biol Chem 1995; 270:18051-9.
14. Ihara Y, Cohen-Doyle MF, Saito Y et al. Calnexin discriminates between protein conformational states and functions as a molecular chaperone in vitro. Mol Cell 1999; 4:331-41.
15. Schrag JD, Bergeron JJ, Li Y et al. The Structure of calnexin, an ER chaperone involved in quality control of protein folding. Mol Cell 2001; 8:633-44.
16. Vassilakos A, Michalak M, Lehrman MA et al. Oligosaccharide binding characteristics of the molecular chaperones calnexin and calreticulin. Biochemistry 1998; 37:3480-90.
17. Ellgaard L, Riek R, Herrmann T et al. NMR structure of the calreticulin P-domain. Proc Natl Acad Sci USA 2001; 98:3133-8.
18. Tessier DC, Dignard D, Zapun A et al. Cloning and characterization of mammalian UDP-glucose glycoprotein: glucosyltransferase and the development of a specific substrate for this enzyme. Glycobiology 2000; 10:403-12.
19. Oliver JD, van der Wal FJ, Bulleid NJ et al. Interaction of the thiol-dependent reductase ERp57 with nascent glycoproteins. Science 1997; 275:86-8.

20. Oliver JD, Roderick HL, Llewellyn DH et al. ERp57 functions as a subunit of specific complexes formed with the ER lectins calreticulin and calnexin. Mol Biol Cell 1999; 10:2573-82.
21. Corbett EF, Michalak KM, Oikawa K et al. The conformation of calreticulin is influenced by the endoplasmic reticulum luminal environment. J Biol Chem 2000; 275:27177-85.
22. Frickel EM, Riek R, Jelesarov I et al. TROSY-NMR reveals interaction between ERp57 and the tip of the calreticulin P-domain. Proc Natl Acad Sci USA 2002; 99:1954-9.
23. Watanabe D, Yamada K, Nishina Y et al. Molecular cloning of a novel Ca(2+)-binding protein (calmegin) specifically expressed during male meiotic germ cell development. J Biol Chem 1994; 269:7744-9.
24. Ho SC, Rajagopalan S, Chaudhuri S. Membrane anchoring of calnexin facilitates its interaction with its targets. Mol Immunol 1999; 36:1-12.
25. Hebert DN, Foellmer B, Helenius A. Calnexin and calreticulin promote folding, delay oligomerization and suppress degradation of influenza hemagglutinin in microsomes. Embo J 1996;15 :2961-8.
26. Mesaeli N, Nakamura K, Zvaritch E et al. Calreticulin is essential for cardiac development. J Cell Biol 1999; 144:857-68.
27. Nakamura K, Robertson M, Liu G et al. Complete heart block and sudden death in mice overexpressing calreticulin. J Clin Invest 2001; 107:1245-53.
28. Malyguine AM, Scott JE, Dawson JR. The role of calnexin in NK-target cell interaction. Immunol Lett 1998; 61:67-71.
29. Zuppini A, Groenendyk J, Cormack LA et al. Calnexin deficiency and endoplasmic reticulum stress-induced apoptosis. Biochemistry 2002; 41:2850-8.
30. Ikawa M, Wada I, Kominami K et al. The putative chaperone calmegin is required for sperm fertility. Nature 1997; 387:607-11.
31. Park BJ, Lee DG, Yu JR et al. Calreticulin, a Calcium-binding Molecular chaperone, Is Required for Stress Response and Fertility in Caenorhabditis elegans. Mol Biol Cell 2001; 12:2835-45.
32. Xu K, Tavernarakis N, Driscoll M. Necrotic Cell Death in C. elegans Requires the Function of Calreticulin and Regulators of Ca(2+) Release from the Endoplasmic reticulum. Neuron 2001; 31:957-71.
33. De Praeter CM, Gerwig GJ, Bause E et al. A novel disorder caused by defective biosynthesis of N-linked oligosaccharides due to glucosidase I deficiency. Am J Hum Genet 2000; 66:1744-56.
34. Nichols WC, Seligsohn U, Zivelin A et al. Mutations in the ER-Golgi intermediate compartment protein ERGIC-53 cause combined deficiency of coagulation factors V and VIII. Cell 1998; 93:61-70.
35. Dwek RA, Butters TD, Platt FM et al. Targeting glycosylation as a therapeutic approachA. Nature Reviews Drug Discovery 2002; 1:65-75.
36. Li Y, Bergeron JJ, Luo L et al. Effects of inefficient cleavage of the signal sequence of HIV-1 gp 120 on its association with calnexin, folding, and intracellular transport. Proc Natl Acad Sci USA 1996; 93:9606-11.
37. Li Y, Luo L, Thomas DY et al. The HIV-1 Env protein signal sequence retards its cleavage and down- regulates the glycoprotein folding. Virology 2000; 272:417-28.
38. Block TM, Lu X, Mehta AS et al. Treatment of chronic hepadnavirus infection in a woodchuck animal model with an inhibitor of protein folding and trafficking. Nat Med 1998; 4:610-4.
39. Block TM, Lu X, Mehta A et al. Role of glycan processing in hepatitis B virus envelope protein trafficking. Adv Exp Med Biol 1998; 435:207-16.
40. Muller-Taubenberger A, Lupas AN, Li H et al. Calreticulin and calnexin in the endoplasmic reticulum are important for phagocytosis. Embo J 2001; 20:6772-6782.
41. Fanchiotti S, Fernandez F, D'Alessio C et al. The UDP-Glc:Glycoprotein glucosyltransferase is essential for Schizosaccharomyces pombe viability under conditions of extreme endoplasmic reticulum stress. J Cell Biol 1998; 143:625-35.
42. Shahinian S, Dijkgraaf GJ, Sdicu AM et al. Involvement of protein N-glycosyl chain glucosylation and processing in the biosynthesis of cell wall beta-1,6-glucan of Saccharomyces cerevisiae. Genetics 1998; 149:843-56.
43. Ellis RJ, Pinheiro TJ. Medicine: danger—misfolding proteins. Nature 2002; 416:483-4.
44. Jakob CA, Bodmer D, Spirig U et al. Htm1p, a mannosidase-like protein, is involved in glycoprotein degradation in yeast. EMBO Rep 2001; 2:423-30.

Chapter 5

Sub-Cellular Distribution of Calreticulin

Sylvia Papp and Michal Opas

Abstract

Calreticulin is a KDEL-containing protein, yet in many cell types and under variable conditions, it has been found outside of its major residence, the endoplasmic reticulum. While the mechanism(s) behind this ER evasion remain elusive, the importance of calreticulin in certain extra-ER locations cannot be disputed. It plays pivotal roles at the cell surface and in the extracellular milieu, under both pathological and normal cellular processes. A cytosolic/nuclear calreticulin localization is still disputable but this view has, of late, been challenged by studies implicating it in nuclear export. Despite the salient roles undertaken by calreticulin at these extra-ER sites, its most well-characterized actions are carried out from within the lumen of the ER, affecting both intra and extra-ER processes. For example, molecular chaperoning and Ca storage are roles carried out by calreticulin within the ER lumen, but these actions may impact on the extra-ER roles attributed to calreticulin such as cell adhesion, steroid-sensitive gene expression and cellular Ca homeostasis. Cytosolically targeted counterparts have no effects on these latter processes. Calreticulin has transpired as a multifunctional protein, with some of these functions being carried out by calreticulin outside its classical ER residence. It will be important to elucidate the mechanism(s) behind the multi-compartmentalization of calreticulin, which will also shed light on the as yet incompletely understood ER-retention and retrieval machinery.

Introduction

Many studies have been performed to elucidate calreticulin's intracellular distribution. Calreticulin carries an N-terminal amino acid signal sequence, which targets it to the endoplasmic reticulum (ER), as well as the KDEL ER-retrieval sequence,[1] making it predominantly an ER-resident protein. However, a variety of studies have shown calreticulin to be localized not only to the ER and its related compartments such as the sarcoplasmic reticulum (SR) or the lumen of the nuclear envelope,[2-6] but also to the nucleus and cytosol,[7,8] the cell surface,[9-14] extracellular space,[15-19] and in specialized compartments such as the lytic granules of cytotoxic T-lymphocytes,[20] the cortical granules of oocytes,[21] and the cytoplasmic droplets and acrosomal matrix of spermatozoa.[22] Before exploring the possible residences of calreticulin, we must first examine how this product of a single gene may be directed to such diverse cellular compartments.

Various possibilities exist for the multi-compartmentalization of single gene products, such as calreticulin. To summarize, the gene product may be directed to alternate compartments through the creation of multiple transcripts, alternative translation, RNA splicing, inefficient protein targeting and/or translocation, or proteolytic modification.[23] Different transcripts of the same gene have the potential to encode different targeting information. Alternate transcription initiation sites lead to transcripts which differ only at their 5' ends, ultimately leading to polypeptides differing in their N-termini. This means that certain N-terminal targeting sequences may be potentially left out of some transcripts via alternate transcription. Thus,

Calreticulin, 2nd ed., edited by Paul Eggleton and Marek Michalak.

without its N-terminal signal sequence, calreticulin may remain in the cytosol, from which it may also be transported into the nucleus via the nuclear pore complexes.[24] Because calreticulin expression is driven by only one promoter,[25] it may be argued that it is unlikely to have multiple transcripts. In fact, Krause and Michalak reported that in a variety of tissues, there is only one mRNA transcript encoding calreticulin,[26] and RT-PCR and Northern Blot analysis from previous studies in spermatozoa have shown that there is only a single transcript for human testicular calreticulin.[22] Interestingly, however, alternate promoters are not always necessary to achieve multiple transcripts. There may be trans-acting factors on a gene which influence its transcription, thereby altering the amino acid sequence and subsequent targeting of the protein.[23] For example, fibroblast growth factor 2 is localized to the cytosol upon activation of the protein kinase C signalling pathway, but it is targeted to the nucleus when the cAMP pathway is initiated.[23] Evidence for such trans-acting factors has yet to be elucidated for calreticulin. Another method to achieve extra-ER calreticulin distribution is alternate translation from a single transcript,[23] which also results in polypeptides varying only in their N-termini. Alternate RNA splicing, on the other hand, leads to polypeptides containing variable regions anywhere across the length of the peptide.[23] Calreticulin mRNA contains a putative nuclear localization sequence in the middle P-domain,[25,27] in which variations caused by alternative splicing may act to redirect calreticulin to alternate compartments. However, to date, there is no evidence for RNA splicing of calreticulin transcripts.[26] Protein retargeting may also occur if the translocation process (into the ER) is not completed by a nascent polypeptide.[23] Additionally, proteolytic modification may also play a role in the multi-compartmentalization of proteins. An interesting study by Naaby-Hansen and colleagues described four calreticulin isoforms in spermatozoa, each containing the KDEL signal, and suggested proteolytic modification of the N-terminus to be responsible for the differences in molecular weight between the isoforms and the subsequent variability in their distribution.[22] Finally, alternate protein targeting may be inadvertently achieved by the addition of tagging sequences designed to elucidate intracellular protein distribution. For example, flagging p32, a small, acidic protein that has been implicated in transport processes between cellular compartments and the cell surface, with an N-terminal epitope tag, results in the retargeting of the protein.[28] It is diverted from its usual mitochondrial location to cytoplasmic vesicles, some of which are continuous with the ER, as well as to the cell surface. This tag is thought to block the mitochondrial targeting signal which is adjacent to it, resulting in the retargeting of p32 away from the mitochondria. In addition, the retargeted p32 binds calreticulin and relocates it from its normal reticular like ER pattern to a punctate/vesicular pattern.[28] Therefore, one must be cautious when using tagging sequences for localization studies, to ensure that the tag does not interfere with normal protein localization. Although numerous possibilities exist for the multi-compartmentalization of single gene products, as yet there is no conclusive evidence showing the precise mechanism by which such gene products, and in particular calreticulin, is localized to multiple cellular compartments. Future studies will need to address the pathways by which such multi-compartmentalization may be achieved. Before attempting to describe the extra-ER distributions of calreticulin, it is best to first examine its localization within its major residence, the ER.

Endoplasmic Reticulum

Despite its localization to a variety of cellular compartments, calreticulin is viewed primarily as an ER-resident protein, as it possesses the well-characterized KDEL ER-retrieval sequence. Immunocytochemistry of numerous cell types has revealed an exclusively ER distribution of calreticulin.[2,3,29] Figure 5.1 illustrates such an ER distribution in wild type mouse embryonic fibroblasts, and contrasts it to its calreticulin-null counterparts.

From its home in the ER, calreticulin is able to carry out both intra and extra-ER functions, making it an interesting molecule in terms of its multi-functionality and involvement in signal transduction. It is well established that calreticulin is the major Ca-binding protein within the lumen of the ER,[27] where it acts as a molecular chaperone[26,30] and is a crucial mediator of Ca

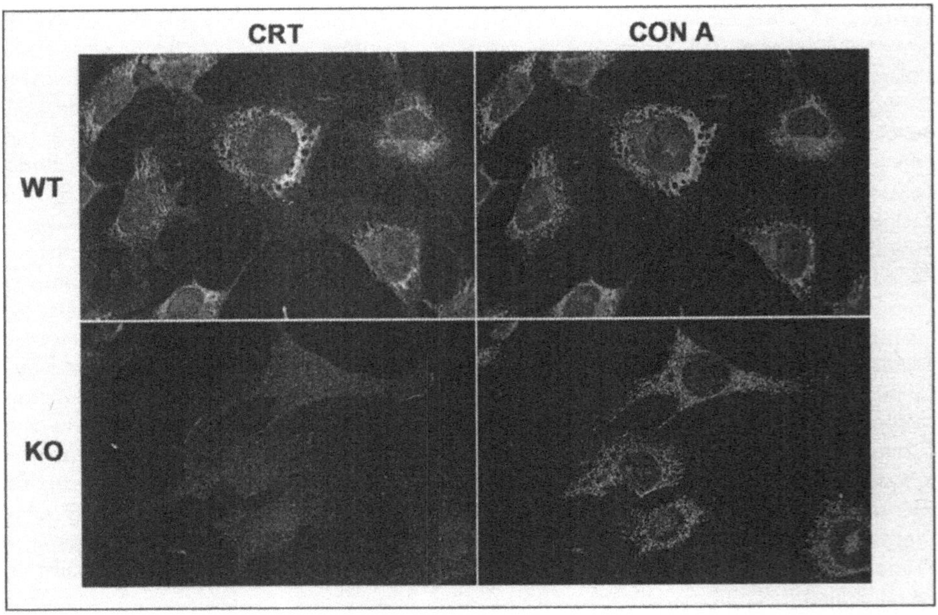

CRT
CON A
WT
KO

to be due to the lack of protein synthesis at these sites. For instance, ER exit sites represent a dangerous location for newly synthesized polypeptides as the possibility of escape of unfolded proteins is high here. In the case of pancreatic acinar cells, the ER in the apical region is extremely small and thin, and may not be able to support efficient protein synthesis. Thus, although calreticulin and other ER chaperones have been used as general markers of the ER, they may not always represent the whole of this endomembrane system due to the heterogeneity in their distribution.[24]

The non-uniform distribution of calreticulin in the ER also impacts on the distribution of ER calcium stores. Ca homeostasis is carefully maintained by calreticulin, which not only binds Ca with low affinity and high capacity,[1] therefore making it an ideal candidate for Ca storage and buffering, but it also modulates the activity of the Ca uptake and release channels of the ER, namely the SERCAs and IP3Rs.[39,42] It is not surprising, then, that calreticulin has been shown to co-localize with these channels, which are also non-uniformly distributed across the ER network.[24,43-45] Together, calreticulin and these channels may create local Ca gradients where they are enriched within the ER, leading to spatially distinct Ca stores, as described by several studies.[44,46-48] The sites of rapid Ca uptake and release may therefore differ spatially from the housekeeping areas of the ER, involved in protein turnover, which may be enriched in calreticulin but not necessarily in the Ca channels.

Interestingly, in striated muscle, calreticulin distribution takes on another level of complexity. Calreticulin was originally identified in the SR of skeletal muscle,[49] where it was later shown to be only a minor component. Calreticulin is also a minor and obscure component of adult cardiac muscle, as calreticulin-null cardiomyocytes have been shown to develop a functional SR and are capable of contraction.[50] Even though calreticulin may not be a pertinent component of adult cardiac muscle SR, new evidence indicates that it may play a crucial role within the ER of these same muscle cells. Accordingly, calreticulin-null cardiomyocytes have been shown to be impaired in their housekeeping functions, such as protein synthesis and turnover, which are the hallmarks of ER activity.[31] Thus, the importance of calreticulin within striated muscle cells should be re-examined, as calreticulin has also been shown to be critical during cardiac development[51] and may also be involved in pathological processes such as cardiac hypertrophy.

Nucleus and Cytosol

Calreticulin localization to the cytosol and nucleus has been very controversial. The presence of calreticulin in the nucleus was previously shown by immunofluorescence,[7,41] but was later believed to be merely an artefact of immunostaining.[52] A putative cytosolic isoform of calreticulin, named mobilferrin, has been identified and is implicated in intestinal iron absorption and transport.[53] The amino terminus of mobilferrin shares 100% sequence homology with rat calreticulin, and the two proteins have a similar molecular weight (56kD) and isoelectric point (4.7).[53,54] Antibodies raised against the individual proteins cross-react with one another.[54] Iron is essential to cells, but at high concentrations it induces the production of free radicals, causing cell toxicity. Such iron-mediated oxidative stress may be combated by calreticulin, a known stress protein, whose expression is inducible by heat shock, amino acid starvation and heavy metals.[55-57] Iron-induced oxidative stress causes increased calreticulin expression,[54] and conversely, the overexpression of calreticulin protects cells against oxidant-induced cell death.[37] Thus, calreticulin is emerging as an important molecule in the fight against oxidative cell stress. However, the main cellular location of calreticulin mediating these events is postulated to be the ER, which houses the bulk of calreticulin. Therefore, although mobilferrin may be essential for iron transport, it may be its ER-counterpart which is responsible for the protective effects against iron-induced damage.

There is yet another area of controversy surrounding ER versus cytosolic calreticulin forms and their methods of action. It has been shown that in vitro, calreticulin is able to bind to the KXFF(K/R)R motif of the cytosolic tails of α-integrins,[58,59] thereby modulating cell adhesion.

In vitro, calreticulin also binds the same motif in the DNA binding domain of steroid receptors, preventing receptor binding to its respective DNA-response elements,[34] thus implicating calreticulin in the modulation of steroid-sensitive gene expression. However, the binding of calreticulin to integrins has not been shown in vivo, and recent evidence indicates that calreticulin from the ER lumen modulates glucocorticoid -sensitive gene expression. A cytosolic form of calreticulin, created using an expression vector encoding calreticulin minus the leader peptide, does not significantly inhibit glucocorticoid receptor-mediated gene expression.[34]

Evidently, past studies have been unable to conclusively localize calreticulin to the cytosol/nucleus. Recently, however, a study using digitonin-based subcellular fractionation has localized calreticulin to the cytosol.[8] This cytosolic calreticulin was shown to act as a nuclear export receptor in a manner similar to Crm1, which is a general nuclear export receptor. Both receptors bind cargo along with Ran GTP to form a trimeric complex, which is exported out of the nucleus. However, calreticulin and Crm1 are proposed to have unique nuclear export functions as calreticulin, but not Crm1, mediates nuclear export of glucocorticoid receptors.[8] Previous studies have shown that calreticulin levels increase in the nucleus upon glucocorticoid receptor addition,[7] which may antagonize transcriptional activation by glucocorticoid receptors by binding to and blocking the DNA-binding domain of these receptors. Alternately, in light of this new export study, calreticulin may antagonize glucocorticoid receptor-mediated transcription by exporting the glucocorticoid receptor out of the nucleus. However, this is in direct conflict with a previous study showing that calreticulin in the cytosol is unable to inhibit glucocorticoid receptor-mediated gene expression, but that this task is performed by an ER form of calreticulin.[34] Apparently, much conflicting data has yet to be resolved concerning the nuclear/cytosolic localization of calreticulin and it will be interesting to unravel the functions attributable to cytosolic calreticulin.

Cell Surface

Contrary to the contestable localization of calreticulin to the cytosol/nucleus, the protein has been definitively localized to the cell surface. There are several suggested mechanisms by which a protein such as calreticulin, with a KDEL ER-retrieval sequence, may escape the ER and reach the cell surface. In the case of NG108-15 cells, a widely used neuronal cell model, the synthesis of KDEL proteins is very efficient whereas the synthesis of KDEL receptors is not.[10] Newly synthesized proteins, therefore, are able to escape from the ER as their guardian receptors are occupied. Alternatively, the KDEL sequence of calreticulin may be masked by proteins with which it interacts,[10] thus increasing the chance for ER escape. Calreticulin may also undergo proteolytic processing by ER lumenal proteases.[60] KDEL containing amino acid sequences may be specifically removed by ER or Golgi proteases, and this event may be calcium-dependent. Thus, the KDEL-containing C-domain of calreticulin is very sensitive to proteolysis and in vitro, it has been shown that only ATP protects it from complete proteolysis.[60] It is not surprising then, that calreticulin and other KDEL-containing proteins such as protein disulfide isomerase and glucose regulated protein 78 have been localized to the cell surface.[61] Additionally, different isoforms of calreticulin have been shown to exist in the central nervous system, and these isoforms may account for the complex cellular distribution observed for calreticulin.[62] Whatever the reason behind ER escape, a probable route by which calreticulin reaches the cell surface is the secretory pathway.[10,61] Evidence for calreticulin in the secretory pathway has stemmed from studies using brefeldin A, an inhibitor of ER-Golgi traffic, which was shown to inhibit calreticulin targeting to the cell surface.[61] Finally, under cell stress, calreticulin becomes glycosylated and may be redistributed from the ER to various parts of the cell, including the cell surface.[63]

Cell surface localization of calreticulin has been shown by several methods, including immunofluorescence visualized by confocal microscopy, FACS, cell surface biotinylation followed by affinity chromatography, radiolabelling, and functional inhibition studies using anti-calreticulin antibodies. Calreticulin has been detected on the surface of many cell types, including activated lymphocytes, Chinese Hamster Ovary cells, Jurkat T cells and COS (simian ovary) cells. In addition, calreticulin on the surface of BAE cells has been shown to interact with the hep1 peptide in thrombospondin, which is responsible for mediating focal adhesion disassembly. Binding of thrombospondin to cell surface calreticulin induces reorganization of stress fibres and the redistribution of vinculin and α-actinin, while leaving αvβ3 integrins undisturbed.[9] Antibodies against calreticulin block this focal adhesion disassembly.[9] Human monocyte-derived macrophages, which are involved in the phagocytosis of apoptotic and/or necrotic cells and debris, also possess calreticulin on their cell surface.[64,65] On the surface of these cells, calreticulin, via its N-domain, functions as a receptor for C1q, which opsonizes apoptotic cells in preparation for phagocytosis.[65] Since calreticulin does not contain a hydrophobic membrane-spanning segment, it most likely conveys intracellular signals via a transmembrane protein, such as CD91 in human monocyte-derived macrophages. Calreticulin on NG108-15 cells is found in surface patches and may be involved in neuronal precursor cell migration.[10] Anti-calreticulin antibodies added to unpermeabilized NG108-15 cells were shown to disturb neuronal adhesion and process (neurite) outgrowth.[10] Calreticulin was also found to be a receptor for anti-double stranded DNA antibodies,[11] which are the pathogenic hallmarks of systemic lupus erythromatosus (SLE). These auto-antibodies are able to penetrate the plasma membrane of living cells, and may potentially do so by utilizing cell surface calreticulin. B16 mouse melanoma cells also possess calreticulin on their surface, where the protein triggers cell spreading following engagement of the β1 integrin.[12] On the surface of human fetal lung fibroblasts, calreticulin binds the βb chain of fibrinogen, and thus stimulates cell replication.[13] Anti-calreticulin antibodies incubated with live cells inhibited the mitogenic effect of calreticulin, supporting the localization of calreticulin to the cell surface and its mitogenic function there.[13] It would be interesting to test whether calreticulin-deficient fibroblasts exhibit a decrease in rate of proliferation as compared to wild type cells when cultured in medium containing fibrinogen.

Calreticulin may reach the cell surface from the inside of the cell, or it may associate with it from the cell's external milieu. In an interesting study, calreticulin was intravenously injected into mice, and shown by immunohistochemistry to associate with the walls of blood vessels.[14] To elucidate the function of calreticulin on the surface of vessel walls, calreticulin was administered into canine coronary arteries, which were partially occluded by a thrombus. It was found that calreticulin prevented thrombosis, as calreticulin treated animals did not develop this pathological condition, whereas saline treated animals all developed occlusive thrombosis. The mechanism by which cell surface calreticulin prevents thrombosis is presently unclear, although it was shown that calreticulin on the surface of endothelial cells stimulated the production of nitric oxide, a known potent anti-coagulant.[14] Clearly, calreticulin may be implicated in many critical functions occurring at the cell surface yet the mechanisms underlying most of these actions remain obscure.

Extracellular

Extracellular calreticulin has emerged as a fascinating molecule in a number of important phenomena such as autoimmune disease, cancer and reproduction. The presence of calreticulin in human plasma was detected over a decade ago, when its origins and functions there remained elusive.[16] A mechanism for the secretion of KDEL-containing proteins was suggested earlier by Booth and Koch, stemming from the observation that ER lumenal proteins are secreted following Ca perturbation of the ER.[66] Disruption of normal ER function by chronic depletion of its Ca stores using calcium ionophores or thapsigargin, or inhibition of glycosylation by tunicamycin, lead to increased expression of stress proteins such as immunoglobulin heavy-chain binding protein, glucose-regulated protein 94 and calreticulin.[66,67] Concomitant with the overexpression is the increased secretion of these proteins. The simplest explanation for increased calreticulin secretion is that of saturation of the KDEL receptors, as receptor levels have been found to remain constant during ER stress.[67] However, the mechanism for secretion is likely more complex, as exogenous overexpression of calreticulin does not lead to its increased secretion, nor does the removal of its KDEL sequence,[66] suggesting that stress may alter the KDEL retrieval mechanism. In addition, the mechanisms may differ in plant and animal cells, as perturbation of ER function in *Arabidopsis thaliana* does result in concomitant increase of KDEL- receptors along with KDEL containing proteins.[66]

Calreticulin release has also been reported in activated neutrophils,[18] lymphocytes,[68] as well as necrotic[15] and apoptotic cells.[69] It was recently shown that necrotic, but not apoptotic cell death leads to the release of heat shock proteins, including calreticulin.[15] However, apoptotic surface blebs have previously been shown to contain calreticulin.[69] Many studies have since implicated calreticulin as an autoantigen, both on its own and in complexes, such as the Ro/SS-A antigenic complex.[70-73] Auto-antibodies against calreticulin have been found in several autoimmune diseases such as SLE and its associated Sjorgen's syndrome, rheumatoid arthritis, congenital heart block, coeliac disease and mixed connective tissue disease.[17,17,70,72,73] The major antigenic epitopes have been mapped and localized to the N-domain of calreticulin.[73] Excellent reviews regarding the role of extracellular calreticulin in autoimmunity have recently been published.[60,70] Much of the knowledge concerning the functions of calreticulin in the extracellular milieu have come form studies on systemic lupus erythromatosus. This disease is associated with a high concentration of immune complexes (antigen-antibody complexes) which are inefficiently cleared from the body, thus leading to their deposition, the elicitation of inflammation and eventual tissue damage and disease. The contribution of calreticulin to this process is two-fold. Firstly, as an autoantigen, it contributes to the formation of immune complexes, but additionally, it interferes with the ability of C1q, part of the first component of the classical complement pathway, to associate with immune complexes and thereby stimulate their clearance.[71] Extracellular calreticulin might therefore contribute to the pathogenesis of autoimmune diseases by preventing the clearance of immune complexes.[60] Several parasites have found a way to defend against host immune attack by mimicking calreticulin. *Schisostoma, Onchocerca,* and *Necator* secrete calreticulin isoforms, which, by competing with C1q binding for antibodies, likely inhibit complement mediated action against the parasites.[60,70]

Whereas calreticulin may be a contributing factor to the pathology of autoimmune diseases, it may instead play a protective role in cancer. The N-terminal fragment of calreticulin was recently isolated from the supernatant of an Epstein-Barr immortalized cell line, and was shown to inhibit the proliferation of cells exclusively from the endothelial lineage, in addition to suppressing angiogenesis in vivo.[19] Although the mechanism by which this fragment, renamed vasostatin, inhibits endothelial cell growth remained unknown, it did not seem to involve the production of nitric oxide by the endothelial cells.[19] Nitric oxide production was previously shown to be responsible for the inhibition of thrombosis by cell surface calreticulin.[14] New data has recently emerged, however, implicating laminin on the surface of the endothelial cells as the target of extracellular calreticulin.[74] Vasostatin was shown to block the interaction of endothelial cells with laminin, thus reducing their ability for extracellular matrix attachment and subsequent growth.[74] Since solid tumors are dependent on an adequate blood supply, angiogenesis is vital in maintaining their growth. Inhibiting angiogenesis has been key in minimizing tumor growth and effective antiangiogenic agents have been widely sought. Vasostatin has emerged as an ideal angiogenesis inhibitor, as it suppresses endothelial cell proliferation and it is small, soluble, stable and easy to produce and deliver.[19] It may also be used in combination with other anti-angiogenic agents, such as interleukin-12, to take advantage of the different modes of action of these agents, resulting in increased suppression of tumor growth.[75]

In addition to playing a role in pathological conditions, extracellular calreticulin has also been implicated in normal processes such as fertilization. During the cortical reaction in oocytes, which is stimulated by the binding of gametes at fertilization, cortical granules are secreted into the perivitelline space. Cortical granules are important in the block to polyspermy, and calreticulin was shown to be a component of these cortical granules.[21] Thus, the release of calreticulin into the perivitelline space during the cortical reaction may directly contribute to the block to polyspermy. The exact mechanisms responsible for this activity of calreticulin, as well as the majority of its other extracellular functions, remain elusive and warrant further investigation as they underlie critical cellular processes, both normal and pathological.

Concluding Remarks

Since its initial discovery as a Ca-binding protein of the SR, calreticulin has emerged as a protein with a vast array of functions, both in normal cellular processes and under pathological conditions. The bulk of calreticulin is found within the ER of non-muscle cells and the SR of smooth muscle cells, where it is heterogeneously distributed. Although the reasons underlying this heterogeneity are incompletely understood, it may act to "compartmentalize" the ER into spatially and functionally distinct units, while at the same time allowing the lumen of the ER to remain physically continuous. In cardiac muscle cells, which possess both an ER and SR, calreticulin is the major component of the ER only, where it most likely carries out housekeeping functions. Calreticulin remains a minor component of the SR, where its role is less clear, but may act as a backup for the major calcium binding protein of the SR, calsequestrin. Current evidence indicates that calreticulin carries out the majority of its functions from the lumen of the ER. Despite this, calreticulin has also been definitively localized to the cell surface, where it has been implicated in a wide array of cellular processes, such as cell adhesion, spreading, replication, and in systemic events such as thrombosis. Calreticulin may also be found secreted into the extracellular space, but its localization to the cytosol and nucleus remains controversial. The multi-compartmentalization and concurrent multi-functionality of calreticulin remain intriguing as they are the manifestations of a single gene with a single transcript. The mechanisms by which this KDEL-containing protein is targeted to such diverse cellular compartments remain elusive, demonstrating that despite the wealth of present knowledge on the subject, much information is still lacking regarding the complexity of ER-retrieval and escape mechanisms.

Acknowledgements

We would like to thank Marc Fadel for critical reading of this manuscript and his valuable comments. The MEFs and anti-calreticulin antibodies were a kind gift from Dr. Michalak. This work was supported by grants from the CIHR and the Heart and Stroke Foundations of Ontario.

References

1. Fliegel L, Burns K, MacLennan DH et al. Molecular cloning of the high affinity calcium-binding protein (calreticulin) of skeletal muscle sarcoplasmic reticulum. J Biol Chem 1989; 264:21522-21528.
2. Tharin S, Dziak E, Michalak M et al. Widespread tissue distribution of rabbit calreticulin, a non-muscle functional analogue of calsequestrin. Cell Tissue Res 1992; 269:29-37.
3. Milner RE, Baksh S, Shemanko C et al. Calreticulin, and not calsequestrin, is the major calcium binding protein of smooth muscle sarcoplasmic reticulum and liver endoplasmic reticulum. J Biol Chem 1991; 266:7155-7165.
4. Imanaka-Yoshida K, Amitani A, Ioshii SO et al. Alterations of expression and distribution of the Ca^{2+}-storing proteins in endo/sarcoplasmic reticulum during differentiation of rat cardiomyocytes. J Mol Cell Cardiol 1996; 28:553-562.
5. Allen BG, Katz S. Calreticulin and calsequestrin are differentially distributed in canine heart. J Mol Cell Cardiol 2000; 32:2379-2384.
6. Tharin S, Hamel PA, Conway EM et al. Regulation of expression and distribution of calreticulin and calsequestrin during L6 skeletal muscle differentiation. J Cell Physiol 1995; 166:547-650.
7. Roderick HL, Campbell AK, Llewellyn DH. Nuclear localisation of calreticulin in vivo is enhanced by its interaction with glucocorticoid receptors. FEBS Lett 1997; 405:181-185.
8. Holaska JM, Black BE, Love DC et al. Calreticulin is a receptor for nuclear export. J Cell Biol 2001; 152:127-140.
9. Goicoechea S, Orr AW, Pallero MA et al. Thrombospondin mediates focal adhesion disassembly through interactions with cell surface calreticulin. J Biol Chem 2000; 275:36358-36368.

10. Xiao GQ, Chung TF, Fine RE et al. Calreticulin is transported to the surface of NG108-15 cells where it forms surface patches and is partially degraded in an acidic compartment. J Neurosci Res 1999; 58:652-662.
11. Seddiki N, Nato F, Lafaye P et al. Calreticulin, a potential cell surface receptor involved in cell penetration of anti-dna antibodies. J Immunol 2001; 166:6423-6429.
12. White TK, Zhu Q, Tanzer ML. Cell surface calreticulin is a putative mannoside lectin which triggers mouse melanoma cell spreading. J Biol Chem 1995; 270:15926-15929.
13. Gray AJ, Park PW, Broekelmann TJ et al. The mitogenic effects of the Bβ chain of fibrinogen are mediated through cell surface calreticulin. J Biol Chem 1995; 270:26602-26606.
14. Kuwabara K, Pinsky DJ, Schmidt AM et al. Calreticulin, an antithrombotic agent which binds to vitamin K-dependent coagulation factors, stimulates endothelial nitric oxide production, and limits thrombosis in canine coronary arteries. J Biol Chem 1995; 270:8179-8187.
15. Basu S, Binder RJ, Suto R et al. Necrotic but not apoptotic cell death releases heat shock proteins, which deliver a partial maturation signal to dendritic cells and activate the NF-kappa B pathway. Int Immunol 2000; 12:1539-1546.
16. Sueyoshi T, McMullen BA, Marnell LL et al. A new procedure for the separation of protein Z, prothrombin fragment 1.2 and calreticulin from human plasma. Thromb Res 1991; 63:569-575.
17. Verreck FA, Elferink D, Vermeulen CJ et al. DR4Dw4/DR53 molecules contain a peptide from the autoantigen calreticulin. Tissue Antigens 1995; 45:270-275.
18. Eggleton P, Lieu TS, Zappi EG et al. Calreticulin is released from activated neutrophils and binds to C1q and mannan-binding protein. Clin Immunol Immunopathol 1994; 72:405-409.
19. Pike SE, Yao L, Jones KD et al. Vasostatin, a calreticulin fragment, inhibits angiogenesis and suppresses tumor growth. J Exp Med 1998; 188:2349-2356.
20. Dupuis M, Schaerer E, Krause K-H et al. The calcium-binding protein calreticulin is a major constituent of lytic granules in cytolytic T lymphocytes. J Exp Med 1993; 177:1-7.
21. Munoz-Gotera RJ, Hernandez-Gonzalez EO, Mendoza-Hernandez G et al. Exocytosis of a 60 kDa protein (Calreticulin) from activated hamster oocytes. Mol Reprod Dev 2001; 60:405-413.
22. Naaby-Hansen S, Wolkowicz MJ, Klotz K et al. Co-localization of the inositol 1,4,5-trisphosphate receptor and calreticulin in the equatorial segment and in membrane bounded vesicles in the cytoplasmic droplet of human spermatozoa. Mol Hum Reprod 2001; 7:923-933.
23. Danpure CJ. How can the products of a single gene be localized to more than one intracellular compartment. Trends Cell Biol 1995; 5:230-238.
24. Baumann O, Walz B. Endoplasmic reticulum of animal cells and its organization into structural and functional domains. Int Rev Cytol 2001; 205:149-214.
25. Michalak M, Corbett EF, Mesaeli N et al. Calreticulin: one protein, one gene and many functions. Biochem J 1999; 344:281-292.
26. Krause K-H, Michalak M. Calreticulin. Cell 1997; 88:439-443.
27. Michalak M, Milner RE, Burns K et al. Calreticulin. Biochem J 1992; 285:681-692.
28. Van Leeuwen HC, O'Hare P. Retargeting of the mitochondrial protein p32/gC1Qr to a cytoplasmic compartment and the cell surface. J Cell Sci 2001; 114:2115-2123.
29. Fliegel L, Burns K, Opas M et al. The high-affinity calcium binding protein of sarcoplasmic reticulum. Tissue distribution, and homology with calregulin. Biochim Biophys Acta 1989; 982:1-8.
30. Saito Y, Ihara Y, Leach MR et al. Calreticulin functions in vitro as a molecular chaperone for both glycosylated and nonglycosylated proteins. EMBO J 1999; 18:6718-6729.
31. Michalak M, Nakamura K, Papp S et al. Calreticulin and dynamics of the endoplasmic reticulum environment. In: Pochet R, Donato R, Haiech J, eds. The Molecular Basis of Calcium Action in Biology and Medicine. Kluwer Academic Publishers, 2000:245-258.
32. Opas M, Szewczenko-Pawlikowski M, Jass GK et al. Calreticulin modulates cell adhesiveness via regulation of vinculin expression. J Cell Biol 1996; 135:1913-1923.
33. Fadel MP, Dziak E, Lo CM et al. Calreticulin affects focal contact-dependent but not close contact-dependent cell-substratum adhesion. J Biol Chem 1999; 274:15085-15094.
34. Michalak M, Burns K, Mesaeli N et al. Endoplasmic reticulum form of calreticulin modulates glucocorticoid-sensitive gene expression. J Biol Chem 1996; 271:29436-29445.
35. Muller-Taubenberger A, Lupas AN, Li H et al. Calreticulin and calnexin in the endoplasmic reticulum are important for phagocytosis. EMBO J 2001; 20:6772-6782.
36. Nakamura K, Bossy-Wetzel E, Burns K et al. Changes in endoplasmic reticulum luminal environment affect cell sensitivity to apoptosis. J Cell Biol 2000; 150:731-740.
37. Liu H, Bowes RC III, Van de Water B et al. Endoplasmic reticulum chaperones GRP78 and calreticulin prevent oxidative stress, Ca^{2+} disturbances, and cell death in renal epithelial cells. J Biol Chem 1997; 272:21751-21759.

38. Hirota J, Furuichi T, Mikoshiba K. Inositol 1,4,5-trisphosphate receptor type 1 is a substrate for caspase-3 and is cleaved during apoptosis in a caspase-3-dependent manner. J Biol Chem 1999; 274:34433-34437.
39. Camacho P, Lechleiter JD. Calreticulin inhibits repetitive intracellular Ca^{2+} waves. Cell 1995; 82:765-771.
40. Nakamura K, Zuppini A, Arnaudeau S et al. Functional specialization of calreticulin domains. J Cell Biol 2001; 154:961-972.
41. Opas M, Dziak E, Fliegel L et al. Regulation of expression and intracellular distribution of calreticulin, a major calcium binding protein of nonmuscle cells. J Cell Physiol 1991; 149:160-171.
42. John LM, Lechleiter JD, Camacho P. Differential modulation of SERCA2 isoforms by calreticulin. J Cell Biol 1998; 142:963-973.
43. Enyedi P, Szabadkai G, Krause K-H et al. Inositol 1,4,5-trisphosphate binding sites copurify with the putative Ca-storage protein calreticulin in rat liver. Cell Calcium 1993; 14:485-492.
44. Simpson PB, Mehotra S, Lange GD et al. High density distribution of endoplasmic reticulum proteins and mitochondria at specialized Ca2+ release sites in oligodendrocyte processes. J Biol Chem 1997; 272:22654-22661.
45. Simpson PB, Mehotra S, Langley D et al. Specialized distributions of mitochondria and endoplasmic reticulum proteins define Ca2+ wave amplification sites in cultured astrocytes. J Neurosci Res 1998; 52:672-683.
46. Golovina VA, Blaustein MP. Spatially and functionally distinct Ca^{2+} stores in sarcoplasmic and endoplasmic reticulum. Science 1997; 275:1643-1648.
47. Blaustein MP, Golovina VA. Structural complexity and functional diversity of endoplasmic reticulum Ca(2+) stores. Trends Neurosci 2001; 24:602-608.
48. Johnson JD, Chang JP. Function- and agonist-specific Ca2+ signalling: the requirement for and mechanism of spatial and temporal complexity in Ca2+ signals. Biochem Cell Biol 2000; 78:217-240.
49. Ostwald TJ, MacLennan DH. Isolation of a high affinity calcium binding protein from sarcoplasmic reticulum. J Biol Chem 1974; 249:974-979.
50. Mesaeli N, Nakamura K, Opas M et al. Endoplasmic reticulum in the heart, a forgotten organelle? Mol Cell Biochem 2001; 225:1-6.
51. Mesaeli N, Nakamura K, Zvaritch E et al. Calreticulin is essential for cardiac development. J Cell Biol 1999; 144:857-868.
52. Opas M. The intracellular distribution and expression of calreticulin. In: Michalak M, ed. Calreticulin. Georgetown: Landes Bioscience, 1996:31-41.
53. Conrad ME, Umbreit JN, Moore EG. Rat duodenal iron-binding protein mobilferrin is a homologue of calreticulin. Gastroenterology 1993; 104:1700-1704.
54. Nunez MT, Osorio A, Tapia V et al. Iron-induced oxidative stress up-regulates calreticulin levels in intestinal epithelial (Caco-2) cells. J Cell Biochem 2001; 82:660-665.
55. Conway EM, Liu L, Nowakowski B et al. Heat shock-sensitive expression of calreticulin. In vitro and in vivo up-regulation. J Biol Chem 1995; 270:17011-17016.
56. Heal R, McGivan J. Induction of calreticulin expression in response to amino acid deprivation in Chinese hamster ovary cells. Biochem J 1998; 329:389-394.
57. Nguyen TQ, Capra JD, Sontheimer RD. Calreticulin is transcriptionally upregulated by heat shock, calcium and heavy metals. Mol Immunol 1996; 33:379-386.
58. Coppolino MG, Woodside MJ, Demaurex N et al. Calreticulin is essential for integrin-mediated calcium signalling and cell adhesion. Nature 1997; 386:843-847.
59. Coppolino MG, Dedhar S. Ligand-specific, transient interaction between integrins and calreticulin during cell adhesion to extracellular matrix proteins is dependent upon phosphorylation dephosphorylation events. Biochem J 1999; 340:41-50.
60. Johnson S, Michalak M, Opas M et al. The ins and outs of calreticulin: from the ER lumen to the extracellular space. Trends Cell Biol 2001; 11:122-129.
61. Xiao GQ, Chung TF, Pyun HY et al. KDEL proteins are found on the surface of NG108-15 cells. Mol Brain Res 1999; 72:121-128.
62. Treves S, Zorzato F, Pozzan T. Identification of calreticulin isoforms in the central nervous system. Biochem J 1992; 287:579-581.
63. Jethmalani SM, Henle KJ, Gazitt Y et al. Intracellular distribution of heat-induced stress glycoproteins. J Cell Biochem 1997; 66:98-111.
64. Henson PM, Bratton DL, Fadok VA. The phosphatidylserine receptor: a crucial molecular switch? Nat Rev Mol Cell Biol 2001; 2:627-633.
65. Ogden CA, deCathelineau A, Hoffmann PR et al. C1q and mannose binding lectin engagement of cell surface calreticulin and cd91 initiates macropinocytosis and uptake of apoptotic cells. J Exp Med 2001; 194:781-796.

66. Booth C, Koch GLE. Perturbation of cellular calcium induces secretion of luminal ER proteins. Cell 1989; 59:729-737.
67. Llewellyn DH, Roderick HL, Rose S. KDEL receptor expression is not coordinatedly up-regulated with ER stress-induced reticuloplasmin expression in HeLa cells. Biochem Biophys Res Comm 1997; 240:36-40.
68. Peterson KL, Zhang W, Lu PD et al. The C1q-binding cell membrane proteins cC1q-R and gC1q-R are released from activated cells: subcellular distribution and immunochemical characterization. Clin Immunol Immunopathol 1997; 84:17-26.
69. Casciola-Rosen LA, Anhalt G, Rosen A. Autoantigens targeted in systemic lupus erythematosus are clustered in two populations of surface structures on apoptotic keratinocytes. J Exp Med 1994; 179:1317-1330.
70. Eggleton P, Llewellyn DH. Pathophysiological roles of calreticulin in autoimmune disease. Scand J Immunol 1999; 49:466-473.
71. Kishore U, Sontheimer RD, Sastry KN et al. The systemic lupus erythematosus (SLE) disease autoantigen—Calreticulin inhibit C1q association with immune complexes. Clin Exp Immunol 1997; 108:181-190.
72. Van den Berg RH, Siegert CEH, Faber-Krol MC et al. Anti-C1q receptor/calreticulin autoantibodies in patients with systemic lupus erythematosus (SLE). Clin Exp Immunol 1998; 111:359-364.
73. Eggleton P, Ward FJ, Johnson S et al. Fine specificity of autoantibodies to calreticulin: epitope mapping and characterization. Clin Exp Immunol 2000; 120:384-391.
74. Yao L, Pike SE, Tosato G. Laminin binding to the calreticulin fragment vasostatin regulates endothelial cell function. J Leukoc Biol 2002; 71:47-53.
75. Yao L, Pike SE, Setsuda J et al. Effective targeting of tumor vasculature by the angiogenesis inhibitors vasostatin and interleukin-12. Blood 2000; 96:1900-1905.

Chapter 6

Calnexin and Calreticulin, Molecular Chaperones of the Endoplasmic Reticulum

Michael R. Leach and David B. Williams

Abstract

In this chapter we present the evidence that calnexin (CNX) and calreticulin (CRT) function as molecular chaperones to assist in the folding and subunit assembly of the majority of Asn-linked glycoproteins that pass through the endoplasmic reticulum. Mechanistic insights into how this function is accomplished have been provided through diverse approaches which include interfering with the recognition of glycoproteins through CNX/CRT's lectin site, expression of CNX/CRT and model substrates in heterologous systems, gene disruption, and reconstitution of function with purified components in vitro. Furthermore, the domain organization and locations of functional sites have been revealed through mutagenesis and the recent determination of the structure of the ER luminal domain of CNX and a portion of CRT. The controversial issue of whether CNX/CRT function solely as lectins or also as "classical" chaperones that recognize the unfolded polypeptide portion of glycoproteins is presented and the evidence supporting current models is discussed in detail.

Introduction

In 1991, CNX was discovered virtually simultaneously by three groups as a protein that interacts with partially assembled class I histocompatibility molecules,[1] with partial complexes of T cell receptors and membrane immunoglobulins,[2] and also as a microsomal membrane protein that can be phosphorylated in vitro.[3] Since then, CNX has been shown to interact transiently with a wide array of newly synthesized membrane or soluble proteins that pass though the ER.[4-6] Given the substantial sequence identity between CNX and CRT it was not long before CRT was demonstrated to share with CNX the ability to bind transiently to diverse nascent proteins.[7,8] In many cases, CNX and CRT were demonstrated to associate with folding or assembly intermediates but not with native conformers. For example, CNX binds to incompletely disulfide-bonded forms of influenza hemagglutinin (HA)[9] and transferrin[6,10] but dissociates at about the time these proteins become fully oxidized. In other cases it binds to individual subunits of proteins such as major histocompatibility complex (MHC) class I[11] or class II[12] molecules, the insulin receptor[13] or integrins[14] and dissociates at the time of oligomeric assembly. CRT behaves in similar fashion, binding primarily to partially oxidized HA[7] or to myeloperoxidase prior to heme assembly.[8] These early studies suggested that CNX and CRT are molecular chaperones, i.e., proteins that bind to non-native protein conformers by recognition of exposed hydrophobic segments and, through cycles of binding and release, prevent aggregation thereby allowing productive folding/assembly to occur more efficiently.

Another important finding was that CNX and CRT exhibit prolonged interaction with misfolded or incompletely assembled proteins and that this interaction correlates with extended residence of the non-native proteins within the ER.[4,7,11,14] These prolonged interactions

Calreticulin, 2nd ed., edited by Paul Eggleton and Marek Michalak.

suggested that CNX and CRT might be components of the ER quality control system that prevents non-native proteins from being exported from the ER. Indeed both molecules have subsequently been shown directly to participate in quality control.[15-17] Since the topic of quality control is discussed elsewhere in this volume (see Helenius and Ellgaard entry) we will focus on the roles of CNX and CRT as molecular chaperones by examining their structures, ligand binding properties, protein binding specificities, the evidence that they assist protein folding and assembly, and the possible mechanisms whereby they effect this latter function.

Structure and Ligand Binding Properties of CNX and CRT

Mammalian CNX is a ~570 residue type I membrane protein of the ER[3,4] whereas CRT is a ~400 amino acid soluble protein[18,19] that resides primarily within the ER lumen (Fig. 6.1). They share ~39% overall sequence identity with highest identity occurring in a central segment consisting of two tandemly repeated sequence motifs. Motif 1 [I-DP(D/E)A-KPEDWD(D/E)] is repeated four times in CNX followed by four copies of motif 2 [G-W--P-IN-P-Y]. In CRT, there are three copies of each motif. Both proteins bind Ca^{2+} with high affinity at a site within the tandem repeats and also have multiple sites for low affinity Ca^{2+} binding within the highly acidic N- and C-terminal regions of CNX[20] and the C-terminal region of CRT.[21] CRT also possesses two Zn^{2+} binding sites in its N-terminal region.[22] Both CNX and CRT bind ATP although no ATPase activity has been detected as yet.[23-26] Furthermore, as demonstrated by chemical cross-linking[27] and by direct binding experiments,[28,29] CNX and CRT interact with ERp57, a thiol oxidoreductase of the ER. CRT also binds to protein disulfide isomerase under conditions of low Ca^{2+} concentration.[30]

Perhaps the most distinctive property of CNX and CRT is that they are both lectins with specificity for a monoglucosylated oligosaccharide present on Asn-linked glycoproteins. A lectin function was initially suggested for CNX based on the observation that treatment of human hepatoma cells with the Asn-linked glycosylation inhibitor tunicamycin prevented the association of CNX with most newly synthesized proteins.[6] Subsequent experiments demonstrated that inhibitors of glucosidases I and II, ER enzymes that sequentially remove the three glucose residues from the initially attached $Glc_3Man_9GlcNAc_2$ oligosaccharide (see Fig. 6.2), also inhibited the binding of CNX[31] and CRT[7] to most glycoproteins. This finding, coupled with the demonstration that oligosaccharides with a single terminal glucose residue are present on glycoproteins bound to CNX or CRT, led to the suggestion that the $Glc_1Man_9GlcNAc_2$ oligosaccharide is the specific oligosaccharide recognized by these lectins.[31] This was subsequently confirmed by direct binding experiments in vitro using purified, immobilized CNX or CRT and various radiolabeled oligosaccharides containing 0-3 glucose residues ($Glc_{0\text{-}3}Man_9GlcNAc_2$). Only the monoglucosylated species bound to the immobilized proteins.[32-34] Additional binding specificity studies involving progressive removal of mannose residues revealed that the $Glc_1Man_{5\text{-}9}GlcNAc_2$ species were capable of binding but binding of the $Glc_1Man_4GlcNAc_2$ species was undetectable, indicating that both the innermost α1-6 branched mannose and the terminal glucose were important for recognition by CNX and CRT. Furthermore, binding competition experiments using monglucosylated di-, tri-, and tetrasaccharides demonstrated that the lectin sites of CNX and CRT recognize the entire glucosylated arm of the oligosaccharide, i.e., Glcα1-3Manα1-2Manα1-2Man (bold residues in Fig. 6.2).[34] The presence of Ca^{2+} was found to be essential for the lectin functions of both CNX and CRT.[34]

Recently, the structure of the ER luminal domain of canine CNX (residues 41-438, numbered as in Fig. 6.1) was solved at 2.9 Å resolution by X-ray crystallography.[35] The structure consists of two distinct domains: a globular β sandwich domain (residues 41-242 and 395-438) containing two antiparallel β sheets and an elongated arm domain (residues 250-394) that extends 140 Å away from the globular domain (Fig. 6.3). The globular domain resembles both legume lectins and galectins, and, consistent with this similarity, soaking the crystal in 50 mM α-D-glucose revealed monosaccharide binding to the globular domain near the base of the arm. A bound Ca^{2+} ion was also present within the globular domain (Fig. 6.3) which represents

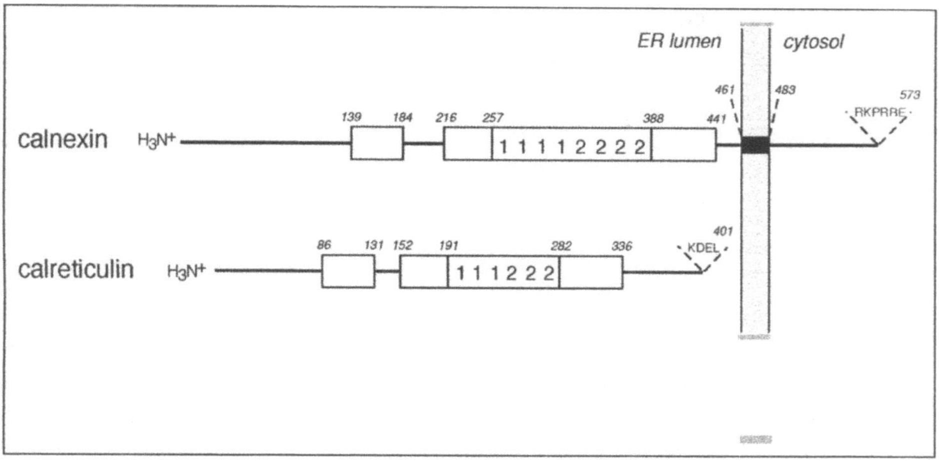

Figure 6.1. Features of the primary structures of canine calnexin and rabbit calreticulin. Regions of the two proteins that share substantial sequence identity are indicated by the white rectangles. The numbers 1 and 2 represent the two tandemly repeated sequence motifs. ER localization sequences are depicted at the C-termini of CNX and CRT using single letter amino acid symbols.

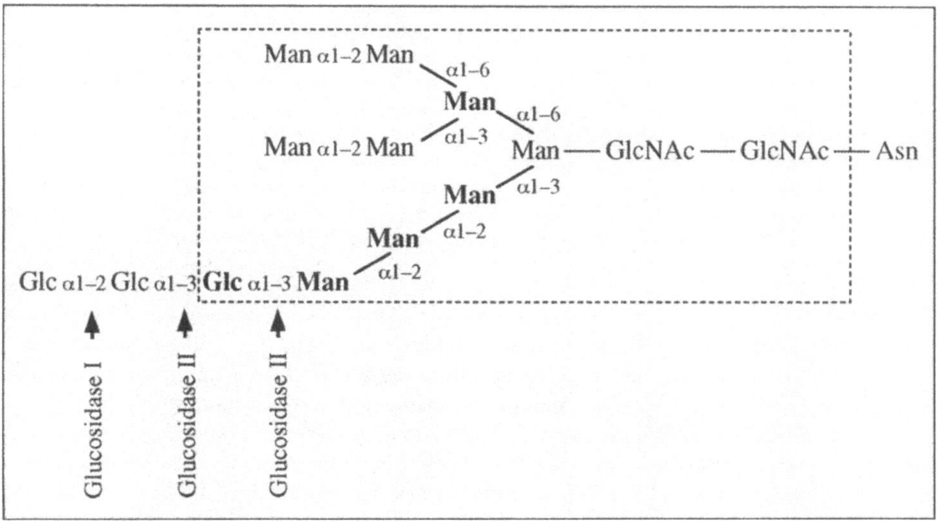

Figure 6.2. Oligosaccharide binding specificity of CNX and CRT. Shown is the $Glc_3Man_9GlcNAc_2$ oligosaccharide that is initially transferred to Asn residues of nascent polypeptide chains. This is subsequently processed by the sequential action of ER glucosidases I and II to produce the $Glc_1Man_9GlcNAc_2$ oligosaccharide (depicted by dashed rectangle) that is recognized by CNX and CRT. Binding specificity studies have revealed that the sugar residues depicted in bold type are important for CNX/CRT binding.

a distinct site compared to previous mapping studies that localized high affinity Ca^{2+} binding to the repeat motifs.[34] The arm consists of the repeat motifs in an extended hairpin loop with the four copies of motif 1 forming one strand of the loop and the four copies of motif 2 folding back on the motif 1 repeats to complete the hairpin. Each motif 1 interacts with a corresponding motif 2 in a head-to-tail orientation to form four distinct modules. The structure of the repeat segment, or P domain, of rat CRT was also recently solved by NMR (residues 188-288,

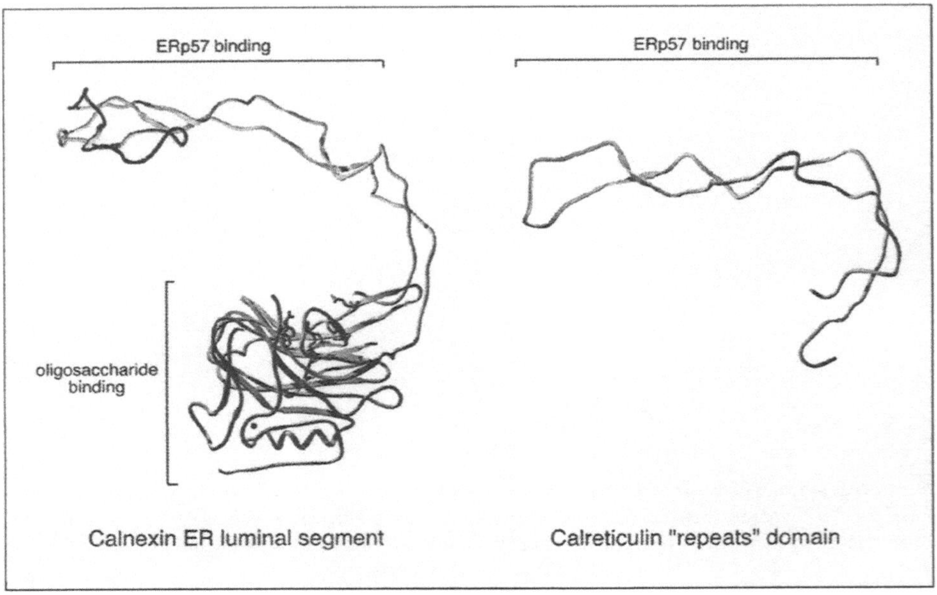

Figure 6.3. Structures of the ER luminal segment of calnexin and of the "repeats" or arm domain of calreticulin. The structures shown correspond to residues 41-438 of CNX and residues 188-288 of CRT (numbered as in Fig. 6.1). The CNX structure consists of a globular domain and an extended loop domain, the latter corresponding to the tandemly repeated sequence motifs. Residues involved in contacts with bound glucose are depicted in stick form and are located in a depression at the top of the globular domain. A single bound Ca^{2+} ion is represented as a black sphere in the globular domain. The CRT structure consists only of the tandemly repeated sequence motifs. Ligand mapping studies have revealed that the binding site for the $Glc_1Man_9GlcNAc_2$ oligosaccharide is located in the globular domain of CNX. In contrast, the ERp57 binding sites are located in distal regions of the arm domains of CNX and CRT.

numbered as in Fig. 6.1).[36] It also exists as a hairpin with the three copies of motif 1 interacting with the three copies of motif 2 to form three modular units (Fig. 6.3). Indeed, the last copy of motif 1 and first copy of motif 2 that together form the most distal module near the loop of the hairpins in CNX and CRT are nearly superimposable in the two structures.[35]

The extended arm domain is an obvious candidate for a protein interaction site. In a series of mapping experiments, we prepared deletion constructs consisting of the individual globular and arm domains fused to GST. When tested for binding to radiolabeled ERp57, the globular domain failed to interact whereas the most distal three repeat modules of the CNX arm bound ERp57 as did the two most distal modules of CRT (Fig. 6.3).[37] ERp57 binding to the tip of the CRT arm domain has also been reported by Ellgaard and co-workers (see Chapter 3). Since glucose is not the physiological ligand for binding to the lectin sites of CNX and CRT, we tested the single domain constructs for binding to radiolabeled $Glc_1Man_9GlcNAc_2$ oligosaccharide. In this case, the globular domains of CNX and CRT retained the bulk (~70%) of the oligosaccharide binding capability of the full length proteins whereas the arm domains consistently exhibited about 10-15% binding (Fig. 6.3).[37] This clearly confirms that the globular domain contains the lectin site for the physiologically relevant oligosaccharide. Furthermore, the persistent low level binding by the arm domain, while unexplained, helps to clarify previous reports that erroneously mapped the lectin site to this segment of CNX and CRT.[34,38]

Differences in Binding Specificity of CNX and CRT for Newly Synthesized Glycoproteins

Two-dimensional isoelectric focusing/SDS-PAGE analysis of glycoproteins that co-immunoisolate with CNX or CRT reveals that both molecules interact with roughly 50-100 newly synthesized proteins.[4,5,39] Indeed it is likely that most if not all glycoproteins bind to CNX, CRT or both at some stage in their biogenesis within the ER. However, even by one-dimensional SDS-PAGE analysis it is obvious that overlapping but distinctly different sets of glycoproteins interact with CNX versus CRT.[7,17] Many individual glycoproteins have been examined for their interactions with CNX or CRT and it is clear that no specific topological category of glycoprotein is preferentially bound by either chaperone, i.e., soluble, type I or type II membrane spanning, or polytopic glycoproteins can be found associated with either CNX or CRT (reviewed in ref. 40). Some glycoproteins such as the vesicular stomatitis virus G glycoprotein[7] and nicotinic acetylcholine receptor[41] bind to CNX but not CRT whereas others such as influenza HA[7] or the α and β subunits of the T cell receptor (TCR) interact with both.[39] In some instances, simultaneous interactions of CNX and CRT with an individual glycoprotein molecule have been reported.[42,43] There are also examples of temporal differences in chaperone interaction, as exemplified by the human MHC class I molecule. The free class I heavy chain (H chain) initially binds exclusively to CNX but, upon H chain assembly with the β_2-microglobulin subunit, CNX dissociates and is rapidly replaced by CRT. CRT then remains bound during assembly of a muti-component complex that facilitates loading of peptide ligands onto the class I molecule for subsequent display at the cell surface to cytotoxic T cells.[44,45]

Some studies suggest that it is the distinct topological relationship between CNX, CRT and the oligosaccharide chains of the various glycoproteins they bind that influences substrate selection. When CRT was expressed as a membrane-anchored protein in human hepatoma cells, the pattern of interacting glycoproteins resembled that of CNX.[46] Similar results were obtained in a separate study in which CRT was expressed as a membrane-anchored protein in mouse L cells and a CNX-like pattern of interacting proteins was obtained. Conversely, when CNX was expressed as a soluble protein in L cells, its substrate specificity switched to resemble that of CRT.[17] In a comprehensive study examining the effect of altering oligosaccharide location on a substrate glycoprotein, influenza HA, it was observed that CRT interacted preferentially with the rapidly folding top/hinge domain of HA which is presumably more accessible to the ER lumen. However, CNX was less discriminating in its interactions, binding to both the top/hinge domain and the membrane-proximal stem domain.[43] Collectively, these findings are consistent with the view that the distinct membrane versus soluble topologies of the lectin sites of CNX and CRT play a role in substrate selection. Interestingly, the substrate preferences of CNX and CRT can be overcome under some circumstances. For example, although free MHC class I H chains normally bind exclusively to CNX in mouse cells, when co-expressed in *Drosophila* cells with mammalian CRT but not CNX, CRT can substitute for the chaperone and quality control functions of CNX.[17]

Molecular Chaperone Functions of CNX and CRT

Several approaches have been used to study the involvement of CNX and CRT in glycoprotein folding and subunit assembly. The most common is to use inhibitors of ER glucosidases I and II, such as castanospermine (CAS) or deoxynojirimycin (DNJ), to prevent the formation of monoglucosylated oligosaccharides. This approach does not permit an examination of the individual functions of CNX or CRT. It is also limited in that the oligosaccharides of all cellular glycoproteins are affected and hence the possibility exists that any observed alteration in glycoprotein folding may not be a direct consequence of impaired CNX or CRT binding. Nevertheless, the accumulated data are consistent with a role for these molecules in enhancing correct folding of many glycoproteins. For example, treatment of dog pancreas microsomes with CAS doubled the rate of disulfide oxidation and oligomerization of influenza

HA, but decreased overall folding efficiency by increasing aggregate formation and enhancing degradation.[47] In the case of MHC class I molecules, CST treatment increased aggregate formation and reduced assembly efficiency in murine cells[48] and slowed disulfide formation in human cells.[49] CST or DNJ treatment also abolished expression of tyrosinase activity in Cos 7[50] or B16 melanoma cells,[51] caused premature dimerization and misfolding of the insulin receptor,[13] inhibited folding of the VSV G glycoprotein[52] and HIV gp120 glycoprotein,[53] and decreased folding, assembly and surface expression of the nicotinic acetylcholine receptor.[54]

An independent approach that does not utilize glucosidase inhibitors involves heterologous expression of mouse MHC class I subunits in *Drosophila melanogaster* cells in the absence or presence of mammalian CNX or CRT. It was found that co-expression of CNX increased folding efficiency of the H chain subunit, stabilized it against rapid degradation, and enhanced its assembly with the β_2-microglobulin subunit by as much as five-fold.[48] In a subsequent study, using the same approach, CRT was shown to exert similar effects as those observed for CNX.[17]

Finally, the functions of CNX and CRT have been examined by disrupting expression of the corresponding genes in a variety of cell lines and organisms. Surprisingly, in a CNX-deficient human leukemia cell line, there was no observable phenotype in MHC class I assembly, intracellular transport, or antigen presentation function.[55,56] However, this might be explained by compensatory action of CRT since, as shown in the *Drosophila* experiments above, the functions of CNX and CRT were largely interchangeable at least during early stages of class I folding and assembly. In contrast, in fibroblasts derived from CRT-deficient mice, newly synthesized class I molecules were prematurely released from the ER and were profoundly deficient in assembling with their peptide ligands[57] (also see entry by T. Elliott in this volume). This suggests that CNX may be less flexible than CRT in assuming a solo role in enhancing class I assembly and participating in quality control.

CRT-deficient mice have been produced which exhibit an embryonic lethal phenotype. Severe defects in heart development were observed which may be more related to CRT's role in Ca^{2+} homeostasis than to a molecular chaperone function.[58] Recently, both the CNX and CRT genes were disrupted in the amoeba *Dictyostelium discoideum*.[59] The double mutants were viable, exhibiting a moderately reduced growth rate, and were capable of chemotactic responses to cAMP. The most notable defect was a severe impairment in phagocytosis. However, since phagocytosis is strongly dependent on cytosolic Ca^{2+} concentration it is unclear if the defect is due to a lack of CNX/CRT's chaperone functions or a loss of their Ca^{2+} storage capacity.[59] In yeast cells, only the CNX gene is present and gene disruption experiments have demonstrated that CNX is essential for viability in *Schizosaccharomyces pombe*[60,61] whereas growth is normal in CNX-deficient *Saccharomyces cerevisiae* cells.[62] The basis for the lethal phenotype in *S. pombe* is unclear. It appears not to be due to a lack of lectin-mediated interactions of CNX with monoglucosylated glycoproteins since various mutations that prevent the formation of monoglucosylated oligosaccharides in this organism do not show a discernable phenotype under normal growth conditions.[63,64] There is some evidence that CNX deficiency in *S. cerevisiae* affects chaperone/quality control function since the cell-surface expression of the normally ER-retained ste2-3p allele of the α-pheromone receptor is increased as is the secretion of heterologously expressed mammalian α_1-antitrypsin.[62]

Mechanisms of Chaperone Action—The "Lectin Only" versus "Dual Binding" Controversy

There is a debate concerning how CNX and CRT interact with folding glycoproteins which centers on whether the association is solely lectin-oligosaccharide based or if there is an additional protein-protein interaction. The two models are depicted in Figure 6.4. In the "lectin-only" model originally proposed by Helenius and co-workers,[31,65] cycles of CNX/CRT binding and release are controlled by the availability of the terminal glucose residue on monoglucosylated Asn-linked oligosaccharides. Initial binding occurs following the trimming of the precursor

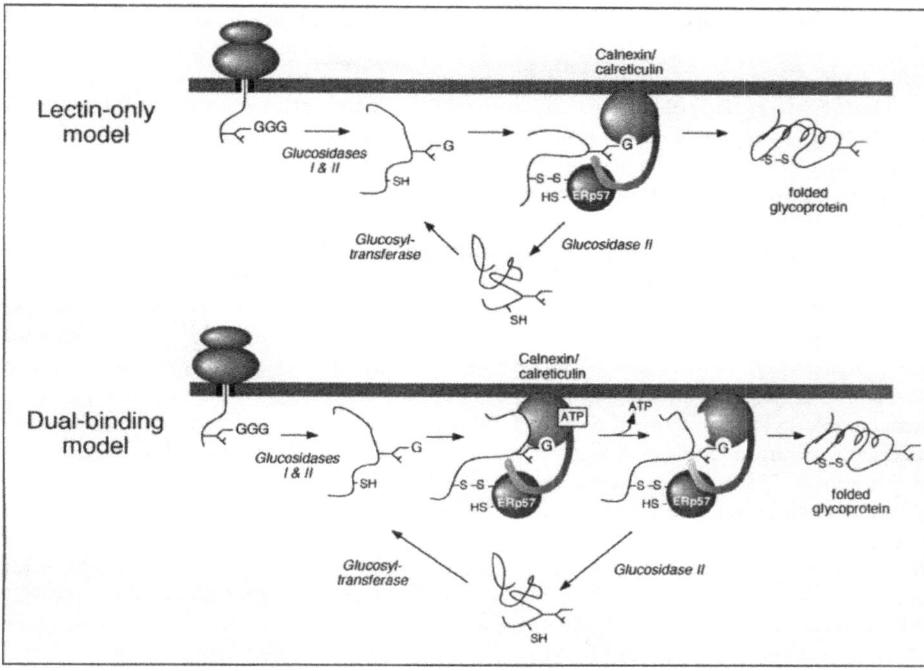

Lectin-only model
GGG
Glucosidases I & II
G
SH
Calnexin/ calreticulin
G
S–S
HS
ERp57
S–S
folded glycoprotein
Glucosyl-transferase
Glucosidase II
SH
Dual-binding model
GGG
Glucosidases I & II
G
SH
Calnexin/ calreticulin
ATP
G
S–S
HS
ERp57
ATP
G
S–S
HS
ERp57
S–S
folded glycoprotein
Glucosyl-transferase
Glucosidase II
SH

central difference between the models is that CNX and CRT function as classical molecular chaperones that suppress aggregation in addition to being capable of recruiting folding factors such as ERp57.

There is a large body of evidence that argues both for and against the two models. Support for the lectin-only model comes from the finding that cells lacking glucosidases I and II or treatment of cells with glycosylation or glucosidase inhibitors usually results in a dramatic reduction in the amounts of glycoproteins co-immunoisolating with CNX or CRT.[6-8,31,42,48,50,69,70] In addition, treatment of cells with glucosidase inhibitors after complexes are formed impairs complex dissociation supporting the view that glucosidase II is important for complex dissociation.[10,65,71] Furthermore, cycles of deglucosylation and reglucosylation have been clearly demonstrated in microsomal and cellular systems and have been demonstrated to be important for efficient glycoprotein folding.[10,65,71,72] However, what is frequently overlooked in reviewing these studies is that there is quite a spectrum of effects observed. For example, under conditions where glycosylation or glucosidase activity are inhibited, complexes are not detected between CNX or CRT and the α and β subunits of the T cell receptor,[5] influenza HA,[31] VSV G glycoprotein,[52] RNase B,[73] myeloperoxidase,[8] cruzipain,[74] and tyrosinase.[50] However, complexes can readily be detected at normal or reduced levels with the ε and δ subunits of the T cell receptor,[16,75] P glycoprotein,[76] erythrocyte AE1,[77] acid phosphatase,[78] MHC class II α and β chains,[79] MHC class II invariant chain,[80] MHC class I H chain,[81] and HIV gp160.[42] Interestingly, CAS treatment almost completely prevented the formation of complexes between CNX and coagulation factors V and VIII but only partially inhibited the formation of complexes with CRT.[82] Furthermore, CAS prevented the formation of complexes between CNX and the α subunit of the acetylcholine receptor in one study but had little apparent effect on complex formation in another study in which a different detergent was employed for cell lysis and recovery of CNX-α-subunit complexes.[41,54] When the entire spectrum of CNX or CRT associated proteins were examined variable results have also been apparent. For example, Kearse et al. observed strong association of many proteins with CNX following CAS treatment or in the glucosidase II-deficient Pha^{R}2.7 cell line, even though associations with TCRα and TCRβ were virtually eliminated.[5] In contrast, Helenius and co-workers observed an almost complete elimination of CNX- or CRT-associated proteins in Pha^{R}2.7 cells or in CAS-treated cells.[7,70] Therefore, it appears that there are significant differences in the extent to which individual glycoproteins may bind to CNX or CRT via lectin-oligosaccharide independent interactions and that varying results can arise depending on the specific cell lysis and immune isolation conditions employed.

There have been two reports in which the interactions of CRT and/or CNX were studied with different conformational forms of monoglucosylated RNase B.[73,83] These studies, conducted either with purified components in vitro or in a microsomal system with in vitro translated RNase B, demonstrated that binding to CNX and CRT was absolutely dependent on the presence of monoglucosylated oligosaccharide whereas the conformational status of the polypeptide chain did not affect the interaction. These studies have been highly cited in support of the lectin-only model but they suffer from one major drawback. A hydropathy plot of RNase reveals that this protein lacks hydrophobic segments considered essential for the binding of molecular chaperones that recognize substrates via protein-protein interactions.[84] Consistent with this lack of hydrophobic character, RNase fails to aggregate even upon heating to 100°C.

There are a number of lines of evidence to support the concept that CNX and CRT are capable of recognizing glycoproteins via protein-protein interactions, i.e., the dual-binding model. First, pre-formed complexes between CNX and either membrane-bound (MHC class I and II molecules) or soluble glycoproteins (α_1-antitrypsin) could not be dissociated by enzymatic removal of oligosaccharides.[32,79,85] However, it has been speculated that the observed lack of dissociation may be due to the trapping of the two species within the same detergent micelle.[73,83,86] Such an argument cannot be applied to the interaction with α_1-antitrypsin but rather it has been suggested that this substrate, being non-native, might become insoluble upon dissociation and thereby associate with CNX non-specifically.[73,83,86] Second, there are many examples of CNX or CRT interacting at normal or reduced levels with proteins that either completely lack Asn-linked oligosaccharides or, as described above, with glycoproteins lacking monoglucosylated oligosaccharides through glucosidase deficiency or inhibition.[16,41,75,76,78,80,82,87,88] These studies, particularly those with non-glycosylated proteins, have been criticized on the basis that the substrate may aggregate and trap CNX or CRT non-specifically.[73,83,86] Indeed CNX has been detected in association with aggregates of non-glycosylated VSV G protein.[89] Third, both CNX and CRT have been shown to bind specifically to non-glycosylated peptides both in vitro and in vivo.[90-93] In one study, the binding of 39 different peptides to CRT was examined and a marked preference for hydrophobic peptides lacking acidic residues was noted. There also appeared to be a minimum length requirement of ~ 10 residues.[91] Fourth, and perhaps most compelling, is that the purified ER luminal domain of CNX (S-CNX) and CRT were capable of functioning as molecular chaperones in vitro to suppress thermally-induced aggregation not only of glycoproteins bearing monoglucosylated oligosaccharides but also of non-glycosylated proteins such as citrate synthase (CS) and malate dehydrogenase (MDH).[25,26] As expected for molecular chaperones, S-CNX and CRT discriminated between native and non-native conformers of CS and MDH, forming stable complexes with unfolded forms but not the enzymatically active species. Aggregation suppression of both glycosylated and non-glycosylated proteins was enhanced in the presence of ATP but not ADP, consistent with a role for ATP in the dual binding model (Fig. 6.4). S-CNX and CRT were also shown to participate in the refolding of denatured CS by maintaining the non-native protein in a refolding-competent conformation. These experiments demonstrated that S-CNX and CRT do indeed utilize a polypeptide-based mode of substrate interaction to function as bona fide molecular chaperones in vitro.

Subsequent studies compared the relative potencies of the ER Hsp70 chaperone, BiP, and S-CNX to suppress aggregation and promote folding of monoglucosylated glycoproteins and non-glycoproteins.[94] S-CNX was just as potent as BiP at suppressing the aggregation of non-glycosylated CS but was much more effective than BiP when presented with monoglucosylated jack bean α-mannosidase or chicken IgY. Upon deglycosylation of the substrates, S-CNX lost its advantage but still could suppress aggregation, consistent with a dual mode of interaction with the monoglucosylated glycoproteins. This latter study indicates that S-CNX (and presumably CRT) are more potent molecular chaperones for monoglucosylated glycoproteins than is an Hsp70 chaperone that is restricted solely to polypeptide-based interactions.[94] Presumably a dual mode of substrate binding increases overall binding avidity relative to other ER chaperones such as BiP or Grp94.

Proponents of the lectin-only model have questioned the in vitro chaperone experiments in terms of their relevance to the in vivo situation.[86,95] To address this issue, Danilczyk et al. developed an extremely mild immunoisolation procedure in an effort to detect polypeptide-based CNX-substrate interactions in lysates of radiolabeled cells. It was reasoned that if a dual mode of CNX-substrate interaction exists in living cells and one interferes with the lectin-oligosaccharide component (e.g., by CAS treatment), then the remaining protein-protein interaction might be relatively weak and lost using more typical isolation conditions.[81] It was demonstrated that in glucosidase I or II-deficient cells or in CAS-treated wild type cells the interaction of CNX with many newly synthesized proteins was preserved whereas binding to other proteins was either reduced or eliminated. Analysis of complexes with specific glycoproteins revealed that CAS-treatment did not eliminate CNX binding to a human MHC class I molecule or to the MHC class II invariant chain. Furthermore, removal of all glycosylation sites from a mouse MHC class I molecule failed to ablate CNX binding. In each of these cases, sedimentation studies revealed that the specific substrate was neither insoluble nor present in aggregated form.[81] Consequently, there appears to be sufficient evidence to support a dual mode of CNX and CRT binding to at least certain glycoproteins both in vitro and in living cells.

Concluding Remarks

A decade has passed since the discovery of CNX and intensive study on the functions of this protein and those of CRT have clearly established their roles as molecular chaperones that assist glycoprotein folding and participate in ER quality control. The extent to which the cell relies on the functions of CNX and CRT relative to other ER chaperones has been difficult to assess. Certainly the lectin-oligosaccharide component of the interaction is dispensable for viability since glucosidase I and glucosidase II deficient mammalian and yeast cells grow normally. Mixed results have been obtained when the CRT and CNX genes have been disrupted either singly or in combination, with phenotypes ranging from subtle to essential. Much of the complexity can be attributed to the redundant nature of ER chaperones wherein the synthesis of BiP or GRp94 is upregulated as a compensatory response to impairments in the CNX/CRT system.[69,96] Also, the role of CNX and CRT in ER Ca^{2+} homeostasis in addition to their chaperone functions complicates interpretation of results. However, the most contentious issue is still the relative roles of lectin-oligosaccharide versus protein-protein modes of substrate interaction in vivo. With portions of the lectin site well-defined and ongoing progress in delineating ERp57, peptide and ATP binding sites, there will be much interest in examining the in vitro and in vivo functions of CNX and CRT mutants that are selectively deficient in the binding of each ligand.

References

1. Degen E, Williams DB. Participation of a novel 88-kD protein in the biogenesis of murine class I histocompatibility molecules. J Cell Biol 1991; 112(6):1099-1115.
2. Hochstenbach F, David V, Watkins S et al. Endoplasmic reticulum resident protein of 90 kilodaltons associates with the T- and B-cell antigen receptors and major histocompatibility complex antigens during their assembly. Proc Natl Acad Sci USA 1992; 89(10):4734-4738.
3. Wada I, Rindress D, Cameron PH et al. SSR alpha and associated calnexin are major calcium binding proteins of the endoplasmic reticulum membrane. J Biol Chem 1991; 266(29):19599-19610.
4. David V, Hochstenbach F, Rajagopalan S et al. Interaction with newly synthesized and retained proteins in the endoplasmic reticulum suggests a chaperone function for human integral membrane protein IP90 (calnexin). J Biol Chem 1993; 268(13):9585-9592.
5. Kearse KP, Williams DB, Singer A. Persistence of glucose residues on core oligosaccharides prevents association of TCR alpha and TCR beta proteins with calnexin and results specifically in accelerated degradation of nascent TCR alpha proteins within the endoplasmic reticulum. EMBO J 1994; 13(16):3678-3686.
6. Ou WJ, Cameron PH, Thomas DY et al. Association of folding intermediates of glycoproteins with calnexin during protein maturation. Nature 1993; 364(6440):771-776.
7. Peterson JR, Ora A, Van PN et al. Transient, lectin-like association of calreticulin with folding intermediates of cellular and viral glycoproteins. Mol Biol Cell 1995; 6(9):1173-1184.
8. Nauseef WM, McCormick SJ, Clark RA. Calreticulin functions as a molecular chaperone in the biosynthesis of myeloperoxidase. J Biol Chem 1995; 270(9):4741-4747.
9. Tatu U, Hammond C, Helenius A. Folding and oligomerization of influenza hemagglutinin in the ER and the intermediate compartment. EMBO J 1995; 14(7):1340-1348.
10. Wada I, Kai M, Imai S et al. Promotion of transferrin folding by cyclic interactions with calnexin and calreticulin. EMBO J 1997; 16(17):5420-5432.
11. Degen E, Cohen-Doyle MF, Williams DB. Efficient dissociation of the p88 chaperone from major histocompatibility complex class I molecules requires both beta 2- microglobulin and peptide. J Exp Med 1992; 175(6):1653-1661.
12. Anderson KS, Cresswell P. A role for calnexin (IP90) in the assembly of class II MHC molecules. EMBO J 1994; 13(3):675-682.
13. Bass J, Chiu G, Argon Y et al. Folding of insulin receptor monomers is facilitated by the molecular chaperones calnexin and calreticulin and impaired by rapid dimerization. J Cell Biol 1998; 141(3):637-646.
14. Lenter M, Vestweber D. The integrin chains beta 1 and alpha 6 associate with the chaperone calnexin prior to integrin assembly. J Biol Chem 1994; 269(16):12263-12268.
15. Jackson MR, Cohen-Doyle MF, Peterson PA et al. Regulation of MHC class I transport by the molecular chaperone, calnexin (p88, IP90). Science 1994; 263(5145):384-387.

16. Rajagopalan S, Xu Y, Brenner MB. Retention of unassembled components of integral membrane proteins by calnexin. Science 1994; 263(5145):387-390.
17. Danilczyk UG, Cohen-Doyle MF, Williams DB. Functional relationship between calreticulin, calnexin, and the endoplasmic reticulum luminal domain of calnexin. J Biol Chem 2000; 275(17):13089-13097.
18. Smith MJ, Koch GL. Multiple zones in the sequence of calreticulin (CRP55, calregulin, HACBP), a major calcium binding ER/SR protein. EMBO J 1989; 8(12):3581-3586.
19. Fliegel L, Burns K, MacLennan DH et al. Molecular cloning of the high affinity calcium-binding protein (calreticulin) of skeletal muscle sarcoplasmic reticulum. J Biol Chem 1989; 264(36):21522-21528.
20. Tjoelker LW, Seyfried CE, Eddy RL, Jr. et al. Human, mouse, and rat calnexin cDNA cloning: identification of potential calcium binding motifs and gene localization to human chromosome 5. Biochemistry 1994; 33(11):3229-3236.
21. Baksh S, Michalak M. Expression of calreticulin in Escherichia coli and identification of its Ca2+ binding domains. J Biol Chem 1991; 266(32):21458-21465.
22. Baksh S, Spamer C, Heilmann C et al. Identification of the Zn2+ binding region in calreticulin. FEBS Lett 1995; 376(1-2):53-57.
23. Ou WJ, Bergeron JJ, Li Y et al. Conformational changes induced in the endoplasmic reticulum luminal domain of calnexin by Mg-ATP and Ca2+. J Biol Chem 1995; 270(30):18051-18059.
24. Corbett EF, Michalak KM, Oikawa K et al. The conformation of calreticulin is influenced by the endoplasmic reticulum luminal environment. J Biol Chem 2000; 275(35):27177-27185.
25. Ihara Y, Cohen-Doyle MF, Saito Y et al. Calnexin discriminates between protein conformational states and functions as a molecular chaperone in vitro. Mol Cell 1999; 4(3):331-341.
26. Saito Y, Ihara Y, Leach MR et al. Calreticulin functions in vitro as a molecular chaperone for both glycosylated and non-glycosylated proteins. EMBO J 1999; 18(23):6718-6729.
27. Oliver JD, Roderick HL, Llewellyn DH et al. ERp57 functions as a subunit of specific complexes formed with the ER lectins calreticulin and calnexin. Mol Biol Cell 1999; 10(8):2573-2582.
28. Zapun A, Darby NJ, Tessier DC et al. Enhanced catalysis of ribonuclease B folding by the interaction of calnexin or calreticulin with ERp57. J Biol Chem 1998; 273(11):6009-6012.
29. Corbett EF, Oikawa K, Francois P et al. Ca2+ regulation of interactions between endoplasmic reticulum chaperones. J Biol Chem 1999; 274(10):6203-6211.
30. Baksh S, Burns K, Andrin C et al. Interaction of calreticulin with protein disulfide isomerase. J Biol Chem 1995; 270(52):31338-31344.
31. Hammond C, Braakman I, Helenius A. Role of N-linked oligosaccharide recognition, glucose trimming, and calnexin in glycoprotein folding and quality control. Proc Natl Acad Sci USA 1994; 91(3):913-917.
32. Ware FE, Vassilakos A, Peterson PA et al. The molecular chaperone calnexin binds Glc1Man9GlcNAc2 oligosaccharide as an initial step in recognizing unfolded glycoproteins. J Biol Chem 1995; 270(9):4697-4704.
33. Spiro RG, Zhu Q, Bhoyroo V et al. Definition of the lectin-like properties of the molecular chaperone, calreticulin, and demonstration of its copurification with endomannosidase from rat liver Golgi. J Biol Chem 1996; 271(19):11588-11594.
34. Vassilakos A, Michalak M, Lehrman MA et al. Oligosaccharide binding characteristics of the molecular chaperones calnexin and calreticulin. Biochemistry 1998; 37(10):3480-3490.
35. Schrag JD, Bergeron JJ, Li Y et al. The structure of calnexin, an ER chaperone involved in quality control of protein folding. Mol Cell 2001; 8(3):633-644.
36. Ellgaard L, Riek R, Herrmann T et al. NMR structure of the calreticulin P-domain. Proc Natl Acad Sci USA 2001; 98(6):3133-3138.
37. Leach MR, Cohen-Doyle MF, Thomas DY et al. Localization of the lectin, ERp57 binding sites of calnexin and calreticulin. J Biol Chem 2002; 277(33):29686-29697.
38. Peterson JR, Helenius A. In vitro reconstitution of calreticulin-substrate interactions. J Cell Sci 1999; 112(Pt 16):2775-2784.
39. Van Leeuwen JEM, Kearse KP. The related molecular chaperones calnexin and calreticulin differentially associate with nascent T cell antigen receptor proteins within the endoplasmic reticulum. J Biol Chem 1996; 271(41):25345-25349.
40. Helenius A, Trombetta ES, Hebert DN et al. Calnexin, calreticulin and the folding of glycoproteins. Trends Cell Biol 1997; 7:193-200.
41. Keller SH, Lindstrom J, Taylor P. Inhibition of glucose trimming with castanospermine reduces calnexin association and promotes proteasome degradation of the alpha-subunit of the nicotinic acetylcholine receptor. J Biol Chem 1998; 273(27):17064-17072.

42. Otteken A, Moss B. Calreticulin interacts with newly synthesized human immunodeficiency virus type 1 envelope glycoprotein, suggesting a chaperone function similar to that of calnexin. J Biol Chem 1996; 271(1):97-103.
43. Hebert DN, Zhang JX, Chen W et al. The number and location of glycans on influenza hemagglutinin determine folding and association with calnexin and calreticulin. J Cell Biol 1997; 139(3):613-623.
44. Sadasivan B, Lehner PJ, Ortmann B et al. Roles for calreticulin and a novel glycoprotein, tapasin, in the interaction of MHC class I molecules with TAP. Immunity 1996; 5(2):103-114.
45. Solheim JC, Harris MR, Kindle CS et al. Prominence of beta 2-microglobulin, class I heavy chain conformation, and tapasin in the interactions of class I heavy chain with calreticulin and the transporter associated with antigen processing. J Immunol 1997; 158(5):2236-2241.
46. Wada I, Imai S, Kai M et al. Chaperone function of calreticulin when expressed in the endoplasmic reticulum as the membrane-anchored and soluble forms. J Biol Chem 1995; 270(35):20298-20304.
47. Hebert DN, Foellmer B, Helenius A. Calnexin and calreticulin promote folding, delay oligomerization and suppress degradation of influenza hemagglutinin in microsomes. EMBO J 1996; 15(12):2961-2968.
48. Vassilakos A, Cohen-Doyle MF, Peterson PA et al. The molecular chaperone calnexin facilitates folding and assembly of class I histocompatibility molecules. EMBO J 1996; 15(7):1495-1506.
49. Tector M, Salter RD. Calnexin influences folding of human class I histocompatibility proteins but not their assembly with beta 2-microglobulin. J Biol Chem 1995; 270(33):19638-19642.
50. Toyofuku K, Wada I, Hirosaki K et al. Promotion of tyrosinase folding in COS 7 cells by calnexin. J Biochem (Tokyo) 1999; 125(1):82-89.
51. Petrescu SM, Petrescu AJ, Titu HN et al. Inhibition of N-glycan processing in B16 melanoma cells results in inactivation of tyrosinase but does not prevent its transport to the melanosome. J Biol Chem 1997; 272(25):15796-15803.
52. Hammond C, Helenius A. Folding of VSV G protein: sequential interaction with BiP and calnexin. Science 1994; 266(5184):456-458.
53. Fischer PB, Karlsson GB, Butters TD et al. N-butyldeoxynojirimycin-mediated inhibition of human immunodeficiency virus entry correlates with changes in antibody recognition of the V1/V2 region of gp120. J Virol 1996; 70(10):7143-7152.
54. Chang W, Gelman MS, Prives JM. Calnexin-dependent enhancement of nicotinic acetylcholine receptor assembly and surface expression. J Biol Chem 1997; 272(46):28925-28932.
55. Scott JE, Dawson JR. MHC class I expression and transport in a calnexin-deficient cell line. J Immunol 1995; 155(1):143-148.
56. Sadasivan BK, Cariappa A, Waneck GL et al. Assembly, peptide loading, and transport of MHC class I molecules in a calnexin-negative cell line. Cold Spring Harb Symp Quant Biol 1995; 60:267-275.
57. Gao B, Adhikari R, Nakamura K et al. Calreticulin plays a specific role in the intracellular loading of class I MHC molecules with antigenic peptides. 2nd Int Wrkshp Antigen Processing & Presentation 1999; Bar Harbor, ME.
58. Mesaeli N, Nakamura K, Zvaritch E et al. Calreticulin is essential for cardiac development. J Cell Biol 1999; 144(5):857-868.
59. Muller-Taubenberger A, Lupas AN, Li H et al. Calreticulin and calnexin in the endoplasmic reticulum are important for phagocytosis. EMBO J 2001; 20(23):6772-6782.
60. Jannatipour M, Rokeach LA. The Schizosaccharomyces pombe homologue of the chaperone calnexin is essential for viability. J Biol Chem 1995; 270(9):4845-4853.
61. Parlati F, Dignard D, Bergeron JJ et al. The calnexin homologue cnx1+ in Schizosaccharomyces pombe, is an essential gene which can be complemented by its soluble ER domain. EMBO J 1995; 14(13):3064-3072.
62. Parlati F, Dominguez M, Bergeron JJ et al. Saccharomyces cerevisiae CNE1 encodes an endoplasmic reticulum (ER) membrane protein with sequence similarity to calnexin and calreticulin and functions as a constituent of the ER quality control apparatus. J Biol Chem 1995; 270(1):244-253.
63. Fanchiotti S, Fernandez F, D'Alessio C et al. The UDP-Glc:Glycoprotein glucosyltransferase is essential for Schizosaccharomyces pombe viability under conditions of extreme endoplasmic reticulum stress. J Cell Biol 1998; 143(3):625-635.
64. D'Alessio C, Fernandez F, Trombetta ES et al. Genetic evidence for the heterodimeric structure of glucosidase II. The effect of disrupting the subunit-encoding genes on glycoprotein folding. J Biol Chem 1999; 274(36):25899-25905.
65. Hebert DN, Foellmer B, Helenius A. Glucose trimming and reglucosylation determine glycoprotein association with calnexin in the endoplasmic reticulum. Cell 1995; 81(3):425-433.

66. Patil AR, Thomas CJ, Surolia A. Kinetics and the mechanism of interaction of the endoplasmic reticulum chaperone, calreticulin, with monoglucosylated (Glc1Man9GlcNAc2) substrate. J Biol Chem 2000; 275(32):24348-24356.
67. Sousa M, Parodi AJ. The molecular basis for the recognition of misfolded glycoproteins by the UDP-Glc:glycoprotein glucosyltransferase. EMBO J 1995; 14(17):4196-4203.
68. Trombetta ES, Helenius A. Conformational requirements for glycoprotein reglucosylation in the endoplasmic reticulum. J Cell Biol 2000; 148(6):1123-1129.
69. Balow JP, Weissman JD, Kearse KP. Unique expression of major histocompatibility complex class I proteins in the absence of glucose trimming and calnexin association. J Biol Chem 1995; 270(48):29025-29029.
70. Ora A, Helenius A. Calnexin fails to associate with substrate proteins in glucosidase- deficient cell lines. J Biol Chem 1995; 270(44):26060-26062.
71. Cannon KS, Helenius A. Trimming and readdition of glucose to N-linked oligosaccharides determines calnexin association of a substrate glycoprotein in living cells. J Biol Chem 1999; 274(11):7537-7544.
72. Van Leeuwen JE, Kearse KP. Reglucosylation of N-linked glycans is critical for calnexin assembly with T cell receptor (TCR) alpha proteins but not TCRbeta proteins. J Biol Chem 1997; 272(7):4179-4186.
73. Rodan AR, Simons JF, Trombetta ES et al. N-linked oligosaccharides are necessary and sufficient for association of glycosylated forms of bovine RNase with calnexin and calreticulin. EMBO J 1996; 15(24):6921-6930.
74. Labriola C, Cazzulo JJ, Parodi AJ. Trypanosoma cruzi calreticulin is a lectin that binds monoglucosylated oligosaccharides but not protein moieties of glycoproteins. Mol Biol Cell 1999; 10(5):1381-1394.
75. van Leeuwen JE, Kearse KP. Calnexin associates exclusively with individual CD3 delta and T cell antigen receptor (TCR) alpha proteins containing incompletely trimmed glycans that are not assembled into multisubunit TCR complexes. J Biol Chem 1996; 271(16):9660-9665.
76. Loo TW, Clarke DM. P-glycoprotein. Associations between domains and between domains and molecular chaperones. J Biol Chem 1995; 270(37):21839-21844.
77. Popov M, Reithmeier RA. Calnexin interaction with N-glycosylation mutants of a polytopic membrane glycoprotein, the human erythrocyte anion exchanger 1 (band 3). J Biol Chem 1999; 274(25):17635-17642.
78. Jannatipour M, Callejo M, Parodi AJ et al. Calnexin and BiP interact with acid phosphatase independently of glucose trimming and reglucosylation in Schizosaccharomyces pombe. Biochemistry 1998; 37(49):17253-17261.
79. Arunachalam B, Cresswell P. Molecular requirements for the interaction of class II major histocompatibility complex molecules and invariant chain with calnexin. J Biol Chem 1995; 270(6):2784-2790.
80. Zhang Q, Salter RD. Distinct patterns of folding and interactions with calnexin and calreticulin in human class I MHC proteins with altered N-glycosylation. J Immunol 1998; 160(2):831-837.
81. Danilczyk UG, Williams DB. The lectin chaperone calnexin utilizes polypeptide-based interactions to associate with many of its substrates in vivo. J Biol Chem 2001; 276(27):25532-25540.
82. Pipe SW, Morris JA, Shah J et al. Differential interaction of coagulation factor VIII and factor V with protein chaperones calnexin and calreticulin. J Biol Chem 1998; 273(14):8537-8544.
83. Zapun A, Petrescu SM, Rudd PM et al. Conformation-independent binding of monoglucosylated ribonuclease B to calnexin. Cell 1997; 88(1):29-38.
84. Parodi AJ. Protein glucosylation and its role in protein folding. Annu Rev Biochem 2000; 69:69-93.
85. Zhang Q, Tector M, Salter RD. Calnexin recognizes carbohydrate and protein determinants of class I major histocompatibility complex molecules. J Biol Chem 1995; 270(8):3944-3948.
86. Parodi AJ. Role of N-oligosaccharide endoplasmic reticulum processing reactions in glycoprotein folding and degradation. Biochem J 2000; 348 Pt 1:1-13.
87. Carreno BM, Schreiber KL, McKean DJ et al. Aglycosylated and phosphatidylinositol-anchored MHC class I molecules are associated with calnexin. Evidence implicating the class I- connecting peptide segment in calnexin association. J Immunol 1995; 154(10):5173-5180.
88. Kim PS, Arvan P. Calnexin and BiP act as sequential molecular chaperones during thyroglobulin folding in the endoplasmic reticulum. J Cell Biol 1995; 128(1-2):29-38.
89. Cannon KS, Hebert DN, Helenius A. Glycan-dependent and -independent association of vesicular stomatitis virus G protein with calnexin. J Biol Chem 1996; 271(24):14280-14284.
90. Basu S, Srivastava PK. Calreticulin, a peptide-binding chaperone of the endoplasmic reticulum, elicits tumor- and peptide-specific immunity. J Exp Med 1999; 189(5):797-802.

91. Jorgensen CS, Heegaard NH, Holm A et al. Polypeptide binding properties of the chaperone calreticulin. Eur J Biochem 2000; 267(10):2945-2954.
92. Nair S, Wearsch PA, Mitchell DA et al. Calreticulin displays in vivo peptide-binding activity and can elicit CTL responses against bound peptides. J Immunol 1999; 162(11):6426-6432.
93. Spee P, Subjeck J, Neefjes J. Identification of novel peptide binding proteins in the endoplasmic reticulum: ERp72, calnexin, and grp170. Biochemistry 1999; 38(32):10559-10566.
94. Stronge VS, Saito Y, Ihara Y et al. Relationship between calnexin and BiP in suppressing aggregation and promoting refolding of protein and glycoprotein substrates. J Biol Chem 2001; 276(43):39779-39787.
95. Ellgaard L, Molinari M, Helenius A. Setting the standards: quality control in the secretory pathway. Science 1999; 286(5446):1882-1888.
96. Pahl HL, Baeuerle PA. A novel signal transduction pathway from the endoplasmic reticulum to the nucleus is mediated by transcription factor NF-kappa B. EMBO J 1995; 14(11):2580-2588.
97. Molinari M, Helenius A. Glycoproteins form mixed disulphides with oxidoreductases during folding in living cells. Nature 1999; 402(6757):90-93.

Chapter 7

Roles of Calreticulin and Calnexin in Myeloperoxidase Synthesis

William M. Nauseef

Abstract

Polymorphonuclear leukocytes (PMNs) represent the essential cellular component of acute inflammation. As such, PMNs mediate a wide array of functions critical for effective antimicrobial activity and integral for noninfectious proinflammatory events. PMNs contribute to normal host defense using cellular responses that include reactive oxygen species, an array of granule enzymes, and many directly cytotoxic antimicrobial proteins. The potency of the oxygen-dependent system is amplified by the action of myeloperoxidase (MPO), a glycosylated hemeprotein located in the PMN azurophilic granule. Under normal circumstances, MPO synthesis is restricted to the promyelocyte stage of myeloid development in the bone marrow. The molecular chaperones calreticulin, calnexin, and ERp57 each interact with normal MPO precursors during their biosynthesis in the ER. The mechanisms for these associations and the basis for their selectivity are not known. Not only do these chaperones participate in normal MPO biosynthesis, but they also contribute to "quality control", demonstrated by their prolonged association with mutant species of MPO. However, not all MPO mutants are handled in an identical fashion, indicating that the chaperones have the capacity to be selective in their interactions. Understanding the structural basis for these interactions, both with normal and aberrant MPO species, and the functional implications of these apparently selective associations should provide important insights into the role of molecular chaperones in normal protein folding and quality control in the ER.

Introduction

Lectin chaperones in the endoplasmic reticulum, including calreticulin, calnexin, and ERp57, participate in the biosynthesis of a wide variety of glycoproteins.[1-5] As lucidly described in several of the preceding chapters, these molecular chaperones serve at least two complementary and integrated functions in the synthesis of proteins in the secretory pathway. First, they serve as molecular chaperones, interacting transiently and reversibly with nascent glycoproteins as they emerge from the translocon and into the ER. As outlined in earlier chapters, these associations block nonproductive intramolecular interactions in the nascent protein that would result in misfolding and thus compromise normal structure and function. In addition, these molecular chaperones contribute to monitoring quality control in the ER, discussed in detail by Helenius and Ellgaard earlier (Chapter 3). Nascent proteins irreversibly misfolded, often a manifestation of mutations in primary sequence, remain associated with specific molecular chaperones and subsequently degraded, in some cases by the cytosolic proteasome. The precise mechanisms by which the status of protein folding in the ER lumen is relayed to the proteasome remain to be elucidated, although the cytosolic extension of the transmembrane calnexin may provide a signal to transmit such information.

Calreticulin, 2nd ed., edited by Paul Eggleton and Marek Michalak.

The focus of this chapter is the contribution of these three ER chaperones to the biosynthesis of myeloperoxidase (MPO), a protein that contributes significantly to host defense by phagocytes and to a wide variety of noninfectious, inflammatory conditions.[6] A member of the animal family of peroxidases, MPO exhibits unique structural and functional features and has limited tissue distribution under normal conditions.[7] Nonetheless, the synthesis of normal MPO and the ER-associated degradation of mutant MPO precursors illustrate many of the features outlined for glycoprotein synthesis in general but provide exceptions as well.

Myeloperoxidase

Polymorphonuclear neutrophils (PMNs) represent the major circulating cellular component of the innate immune system[8] and, as such, serve an essential function as an early sentinel of infection. When PMNs encounter a particulate stimulus such as an invading microorganism or tumor cell, a cascade of complex cellular responses follows that culminates in internalization of the target and its compartmentalization in a phagosome. Under optimal conditions, the microorganism is killed and degraded, thus limiting the clinical consequences from invasion by the potential pathogen. An outcome of the human PMN-microbe interaction that is beneficial to the host requires integration of the products of multiple antimicrobial systems, including reactive oxygen species, an array of granule enzymes, and many directly cytotoxic antimicrobial proteins. The most efficient oxygen-dependent antimicrobial system reflects the synergy between hydrogen peroxide and other reactive oxygen species generated by the NADPH-dependent oxidase and MPO released from the azurophilic granule during its fusion with the phagosome.[9,10] Within the phagosome 5-10 nmoles/ml of superoxide anion are generated and ambient concentrations of MPO are 1-2 mM, resulting in as much as 50 nmoles of HOCl produced per million stimulated PMNs in 30 minutes.[11] Although the toxicity of HOCl for microbes is widely recognized, the precise prokaryotic target(s) for HOCl or its products has not been defined and is actively under study.[12]

Mature, native MPO is a heme-containing homodimeric protein composed of 785 amino acids and possessing four potential sites for N-linked glycosylation [reviewed in refs. 10,13]. Each half of the holoprotein has the same specific activity as does the native protein[14] and recombinant, unprocessed monomeric proMPO likewise has identical spectral and kinetic properties as does fully processed MPO,[15] suggesting that the functional domains of each half act independently. The identical halves are each composed of a heavy (466 amino acids) and light (108 amino acids) subunit and linked together by a single disulfide bond between a pair of cysteines in the two heavy subunits (vide infra). The crystal structure of human myeloperoxidase, recently determined at 1.8 Å, elucidated the nature of the heme-containing reactive site (see below) and confirmed its symmetrical organization along a dyad axis, although only three sites for N-linked glycosylation were identified.[16,17]

Studies of the biosynthesis of MPO by normal bone marrow precursors and cultured myeloid cell lines [reviewed in refs. 13,18,19] provide an overview of its structural maturation from the precursor form in the ER of promyelocytes to the mature species in the azurophilic granule of PMNs (Fig. 7.1). The 80-kDa primary translation product undergoes cotranslational N-linked glycosylation to yield a short-lived 93-kDa form of apoproMPO that has cleavage of terminal glucoses to generate the very stable 90-kDa apoproMPO. Glycosylated at five asparagines, apoproMPO resides in the ER for several hours before acquiring heme and becoming proMPO. However proMPO soon exits the ER and enters the secretory pathway where it has at least two distinct fates. The majority of proMPO undergoes cleavage of the pro region to yield a 74-78-kDa intermediate that contains only the light and heavy subunits as a single polypeptide. Although experimental evidence indicates that the processing and transport of MPO precursors and intermediates occur in an acidic subcellular compartment,[20] in vitro studies suggest that the 74-78 kDa intermediate species is relatively short-lived with very little accumulating in the cell and generated in a neutral pH compartment.[21] Subsequently there is removal of a small peptide between light and heavy subunits and C-terminal proteolysis of a

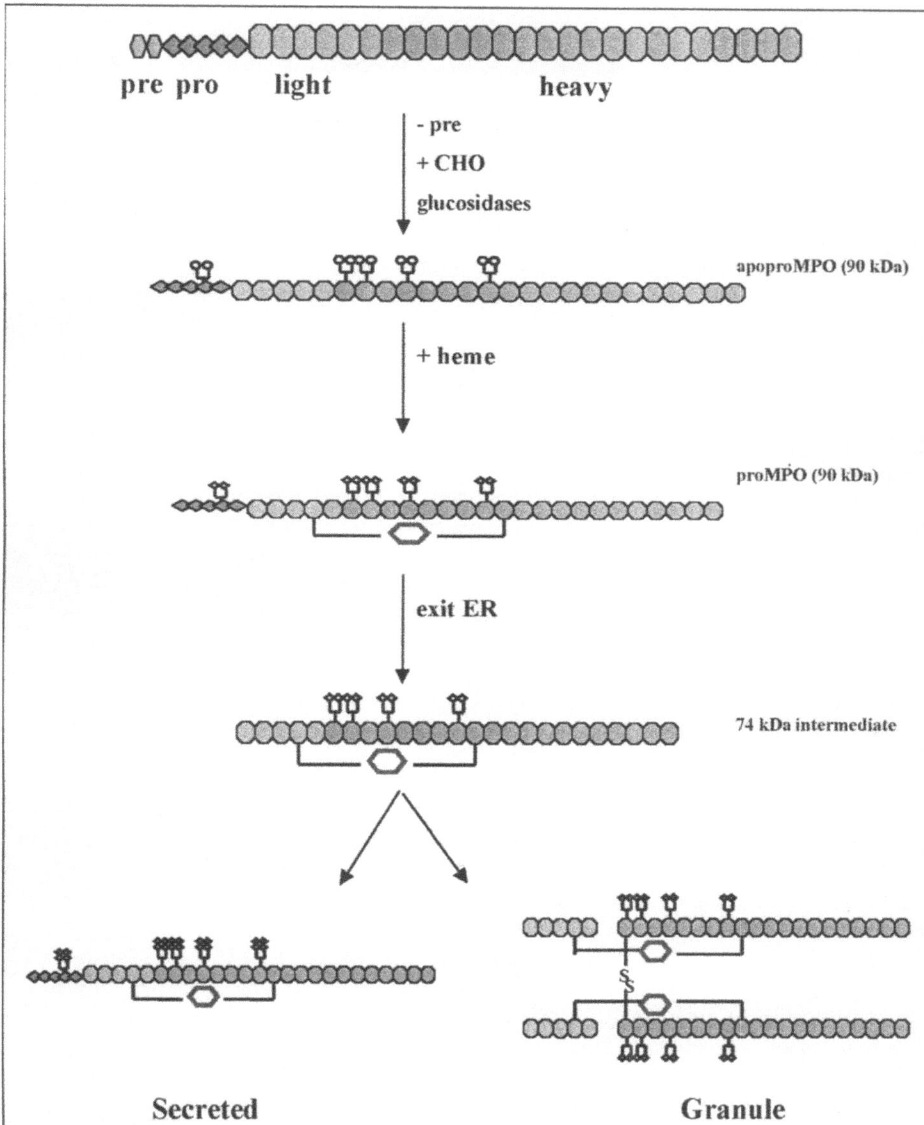

Figure 7.1. Proteolytic processing during myeloperoxidase biosynthesis. The MPO gene encodes a single, 80-kDa primary translation product that cotranslationally undergoes cleavage of its signal peptide and N-linked glycosylation at 5 sites to yield a relatively short-lived 93-kDa glycoprotein. ER glucosidases eliminate terminal glucoses on the oligosaccharide sidechains, producing a 90-kDa, enzymatically inactive apoproMPO. The half-life of apoproMPO in the ER is very long, allowing associations with CRT, ERp57, and CNX and the resultant conformational maturation needed to accommodate insertion of heme. The heme-containing proMPO is enzymatically active and rapidly exits the ER and enters the secretory pathway. The majority of proMPO undergoes a series of proteolytic processing steps, generating an intermediate 74-kDa species and then the heavy and light subunits of the mature enzyme. Within the azurophilic granule, pairs of heavy-light protomers dimerize to produce the native enzyme. Approximately 10% of the proMPO entering the secretory pathway is secreted into the medium after modification by oligotransferases in the Golgi.

single serine at the carboxy terminus of the heavy subunit.[22] Dimerization of the heavy-light protomers to form native MPO occurs in dense granules,[23] although the precise determinants of the disulfide bond formation in such a subcellular compartment are unknown. The N-linked carbohydrates on proMPO and the heavy subunit of mature MPO remain susceptible to endoglycosidase H,[24] indicating the presence of high-mannose groups despite exposure to oligosaccharide-modifying enzymes in the Golgi en route to the azurophil granule. In contrast, the carbohydrate sidechains on the ~ 10% of the proMPO produced by the cell that is released as monomer into the culture medium[25,26] are converted to complex oligosaccharides.[27,28] The structural and functional implications of this differential carbohydrate modification of intracellular and extracellular proMPO are not fully understood.

MPO is a member of the protein family of animal peroxidases,[7,29,30] which includes eosinophil peroxidase, thyroid peroxidase, and lactoperoxidase. Whereas all members function as peroxidases and oxidize certain halides,[31] only MPO has the capacity to catalyze the 2-electron oxidation of Cl^- to produce HOCl.[31-34] Paralleling this unique functional property, the ligation of heme in MPO is also peculiar among members of the animal peroxidase protein family. A derivative of protoporphyrin IX, the heme in MPO has three coordination sites in addition to the two histidine residues. There are ester linkages between methyl groups on the heme with carboxylic groups on residues D260 and E403 and a sulfonium ion bond to the sulfur atom of M409.* Whereas the ester linkages are conserved among other members of the peroxidase protein family, the sulfonium ion linkage is unique to MPO and is responsible for its characteristic spectral properties.[35-37]

Overall, the biosynthesis of mature MPO is relatively slow, with complete maturation into the dimeric form in cultured HL-60 cells having a half-life of approximately 36 hours.[23] Although the bulk of this delayed maturation reflects slow posttranslational modifications occurring after proMPO exits the ER, we reasoned molecular chaperones might participate in early events in MPO biosynthesis and may have a specialized role in the formation or acquisition of its very unusual heme group.

The Lectin Chaperones in the Biosynthesis of Normal MPO

Using pulse-radiolabeled PLB-985 cells, a cultured human promyelocytic cell line that actively synthesizes structurally and functionally normal MPO,[38,39] we recovered proteins coprecipitating with apoproMPO and proMPO. All three ER lectin chaperones, calreticulin, calnexin, and ERp57, associate transiently with biosynthetic intermediates of MPO, although each with different kinetics.[39,40] In order to assess the temporal sequence of the interactions of molecular chaperones with MPO precursors, we used puromycin, an agent that inhibits protein synthesis by mimicking tRNA and causing premature release of polypeptides from the ribosome, to synchronize protein synthesis when translation is reinitiated.[41]

After being cultured in the presence of 10 μM puromycin, PLB-985 cells were pulse-labeled with [^{35}S]-methionine for 10 minutes and were treated with cycloheximide to terminate protein synthesis. Lysed cells underwent sequential immunoprecipitation,[39,42,43] first under nondenaturing conditions with antibody to CRT or CNX and, after the immunoprecipitated complex was dissociated, with antibody to MPO. The MPO precipitation provides assessment of the total amount of MPO newly synthesized during the labeling period, whereas the immunoprecipitation with CRT or CNX recovers only the subset of MPO precursor associated with each chaperone. During this very brief labeling period, CRT-associated MPO precursor represented the bulk of newly synthesized MPO-related protein (Fig. 7.2). ERp57, the third ER protein in addition to CRT and CNX implicated as a molecular chaperone for glycoproteins,[44]

*Because of our interest in the precursor forms of MPO, our system for enumerating the amino acids in its primary sequence begins with the first methionine in the propeptide, resulting in the inclusion of 166 amino acids proximal to the first methionine in the mature enzyme. Consequently the nomenclature used to indicate critical residues identified from solution of the crystal structure differs from ours by 166. For example, the distal histidine identified by Fenna as H95 is H261 in our numbering scheme.

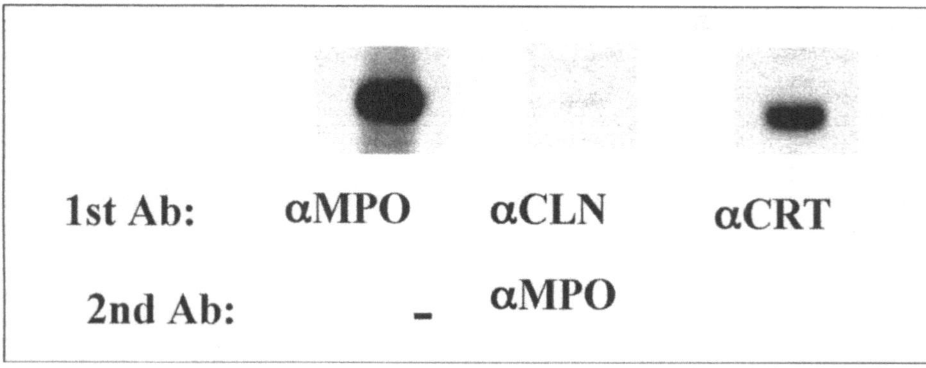

Figure 7.2. Chaperone-association of MPO precursors during synchronized biosynthesis. Protein synthesis in PLB-985 cells was synchronized with puromycin and cells labeled for 10 minutes. Lysates were sequentially immunoprecipitated under non denaturing conditions antibodies against MPO, CNX, or CRT (1st antibody). The MPO immunoprecipitate was solubilized for analysis by SDS-PAGE, whereas the other two samples were denatured by heating in SDS. After dilution (to reduce the SDS concentration), the samples were immunoprecipitated with antibody against MPO (2nd antibody) to recover CNX- and CRT-associated MPO, respectively [Reprinted with permission of the publisher, W.S. Maney & Son, Ltd.[40]].

also interacts with apoproMPO. Less than 5% of the apoproMPO made appears in a ternary complex with ERp57 and CRT and quickly dissociates within less than 60 minutes (data not shown). However very little MPO precursor associates with CNX during this early time in MPO synthesis, consistent with our hypothesis that CRT is the proximal ER chaperone interacting with apoproMPO. This sequence of interactions could reflect the orientation of the molecular chaperones in the ER,[2,5] as both CRT and ERp57 are soluble proteins in the ER lumen whereas CNX is a transmembrane protein, and/or have specific functional implications. Unique peptide regions of the nascent MPO precursor may be selectively recognized by CRT/CNX, or specific functions, such as heme insertion or delivery of misfolded variants to proteasomes (see below) may be restricted to particular ER chaperones and/or particular regions of the ER.

In contrast to the interactions of CRT and ERp57 with apoproMPO, CNX associates with apoproMPO much later in biosynthesis and ~ 6% is recovered in a ternary complex with CRT and apoproMPO. CNX also coprecipitates with the heme-containing proMPO, although only 13% of the proMPO was recovered associated with CNX.[39] It is possible that the CNX-proMPO complex is relatively unstable, thus resulting in the low recovery of proMPO coprecipitating with CNX, or that this subpopulation represents a functionally important intermediate species. For example, the interaction of CNX with apoproMPO may influence its conformation in such a way as to facilitate heme acquisition, with subsequent dissociation of the complex and transport of the proMPO from the ER. Alternatively the CNX may interact only with proMPO in the ER and mediate its transfer to a transport protein that advances it into the secretory pathway. At this time, we do not understand the functional significance of the associations of CNX with apoproMPO and proMPO for normal MPO biosynthesis.

Despite significant structural similarity, CRT and CNX are both specific and selective in their interactions with glycoproteins in the ER,[45-47] as illustrated with apoproMPO and proMPO. The mechanism by which CNX or CRT interacts with nascent glycoproteins has been the object of considerable study and the basis for two prominent models (see Chapters 3 and 6). The first model proposes that the lectin domains of CNX or CRT associate with the monoglucosylated intermediates of target glycoproteins.[45] The first step in N-linked glycosylation is the en bloc transfer of $Glc_3Man_9GlcNAc_2$ to target asparagines in the peptide backbone.[48] Soon after, ER glucosidases I and II remove the terminal two glucose residues to generate the monoglucosylated form ($Glc_1Man_9GlcNAc_2$) recognized by CNX or CRT.[45] According to the "lectin only" hypothesis,[2] CRT and CNX interact with their targets exclusively through their lectin domain and the terminal Glc_1 of glycoprotein intermediates. Removal of the remaining glucose residues results in dissociation of the CRT/CNX-glycoprotein complex but reassociation occurs if a single glucose is replaced by UDP-glucose: glycoprotein glucosyltransferase (UGGT). UGGT glucosylates only unfolded proteins[49] and thus serves to monitor the folding state of nascent glycoproteins. Thus unfolded proteins first become reglucosylated and subsequently reassociate with CNX or CRT. In contrast, correctly folded proteins are not reglucosylated, thereby remaining free of CNX or CRT and able to escape from the ER. Substantial data from experimental systems in which formation of monoglucosylated glycoprotein is blocked, either by glucosidase inhibitors or in mutant cell lines lacking endogenous glucosidase activity in their ER [reviewed in ref. 2], support the "lectin only" model.

However CRT and CNX interact with many glycoprotein intermediates in the ER in the presence of glucosidase inhibitors, bind to nonglycosylated proteins, and discriminate between native and denatured conformations of nonglycosylated proteins in vitro.[50-59] Such studies thus indicate that CNX and CRT each have the capacity to bind polypeptide sites independent of their lectin domains.[60,61] A recent study[62] directly implicates the N-domain of CRT as contributing to folding of the bradykinin receptor. Based on such data, Williams has proposed, as an alternative to the "lectin only" model, that CRT and CNX interact with target unfolded glycoproteins via binding to oligosaccharide and peptide sites.[1] In this "dual binding" model, dissociation of CRT or CNX with the target glycoprotein requires the action of glucosidase II as well as a conformational change. Interactions would be expected to cease when the glycoprotein achieves a conformation, presumably the native state, in which neither UGGT nor the peptide-binding region of the chaperone recognizes it.

To determine which of these two models better described interactions between MPO precursors and CRT and CNX, we examined the impact of glucosidase inhibition on coprecipitation of CRT/CNX with apoproMPO. We reasoned that inhibition of ER glucosidase activity would preserve the $Glc_3Man_9GlcNAc_2$ structures on apoproMPO and thus block the association of ER chaperones with nascent apoproMPO. Inclusion of deoxynojirimycin, an inhibitor of glucosidases I and II,[63] in pulse labeling of PLB-985 resulted in biosynthesis of a larger form of apoproMPO (93-kDa), consistent with the presence of additional glucose residues on the oligosaccharide sidechains secondary to inhibited glucosidase activity (Fig. 7.3). However approximately 30% of the total apoproMPO synthesized during the labeling interval, whether wild-type or the more glucosylated 93-kDa form, coprecipitated with CRT. Similar results were obtained when coprecipitates with CNX were assessed or in the presence of castanospermine[40] or bromoconduritol, two other inhibitors of ER glucosidase activity. These data suggest that the interactions of apoproMPO with CRT or CNX are not exclusively lectin-mediated but rather are substantially a product of protein-protein interactions. Supporting this interpretation are studies wherein the CRT-apoproMPO complex remains intact after removal of the high-mannose oligosaccharides by digestion with endoglycosidase H.[40] Thus our data describing interactions of CRT and CNX with MPO precursors best fit the "dual binding model", although the precise peptide motifs in MPO precursors recognized specifically by CRT or CNX have not been identified.

Quality Control in MPO Biosynthesis

As detailed earlier, molecular chaperones not only participate in normal glycoprotein biosynthesis in the ER but also contribute to quality control, diverting misfolded proteins from the secretory pathway and, in some cases, to the proteasome for degradation.[64-66] The same principles apply when specific mutants of MPO are expressed in vitro. For these studies we transfected K562 cells, a human hematopoietic cell line that lacks endogenous MPO, with cDNA encoding wild type or mutant MPO. Several genotypes have been reported to result in MPO deficiency,[40,42,67-69] the most commonly described being a missense mutation resulting in replacement of an arginine at codon 569 with a tryptophan (R569W).[67,70,71] In contrast to

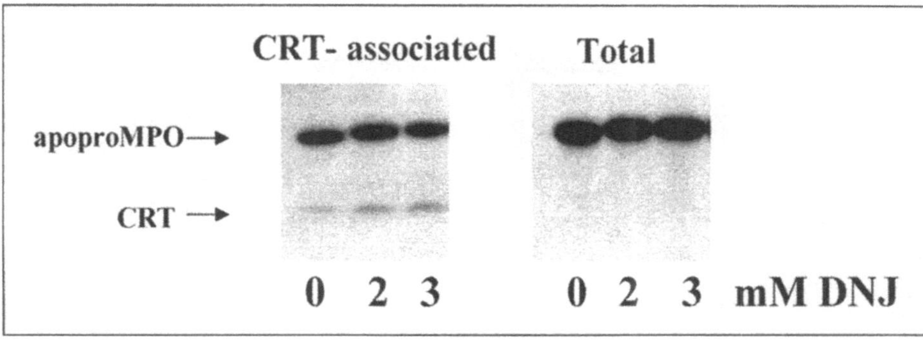

Figure 7.3. Association of CRT with apoproMPO in the presence of deoxynojirimycin. PLB-985 cells were cultured for 2 hours in the presence of 0, 2, or 3 mM deoxynojirimycin (DNJ) before labeling for 30 minutes with ^{35}S-methionine. Cells were lysed and immunoprecipitated under nondenaturing conditions with antibody against CRT, recovering CRT-associated MPO, or MPO, providing an index of the total MPO synthesized during the interval. In the presence of DNJ, the apoproMPO migrated as a 93-kDa protein, reflecting the inhibition of the action of glucosidase I in the ER. Approximately 30% of the apoproMPO synthesized during the pulse period coprecipitated with CRT under control conditions and in the presence of DNJ. Identical results have been obtained when castanospermine[40] or bromoconduritol were used to inhibit glucosidase activity.

transfectants expressing normal MPO, R569W cells synthesize a fully glycosylated precursor that fails to incorporate heme, exit the ER, or undergo proteolytic processing to mature MPO subunits.[70] The association of R569W precursors with CRT or CNX is greatly prolonged in comparison to complexes with wild type MPO, consistent with the heme-free R569W precursor being retained in the ER.[39]

We identified more recently a novel missense mutation that results in replacement of tyrosine at codon 173 with cysteine (Y173C) and causes MPO deficiency.[42] Y173C alters disulfide bond formation and, like R569W, results in an arrest in MPO maturation in the biosynthetic pathway, such that neither proteolytic processing nor lysosomal targeting occur. Unlike R569W, Y173C incorporates heme with the same relative efficiency as does wild type MPO. However both PMNs isolated from patients with Y173C and K562 cells transfected with Y173C cDNA lack peroxidase activity. We believe this apparent discrepancy (i.e., heme incorporation but no peroxidase activity) suggests that the proMPO-Y173C is extremely short-lived, rapidly undergoing degradation in the ER-associated proteasome[42] and thus never achieving a concentration sufficient for detection by the enzymatic assays.

It is important to note that not all mutations in MPO result in proteasome-mediated degradation[40] (Fig. 7.4). Y173C degradation is proteasome-dependent whereas that of R569W proteasome-independent, and the mechanisms by which CNX selects and delivers misfolded MPO precursors such as Y173C to the cytosolic proteasome for degradation have not been determined. The distinct differences in the proteasome-mediated degradation of different mutant proteins may reflect surveillance by different ER retrieval systems, perhaps related to monitoring disulfide bond formation or some other functionally important posttranslational modification.[72]

Mature MPO present in the azurophilic granule is a dimer (Fig. 7.1), linked together by an intramolecular disulfide bond between the C319 residue in each heavy subunit.[16] To examine the importance of dimerization, we replaced cysteine at codon 319 with alanine (C319A), thereby eliminating the residues mediating dimerization.[16] K562 cells stably expressing C319A lacked peroxidase activity, as judged by a spectrophotometric assay or in an activity gel.[73] Furthermore, MPO precursors of C319A failed to undergo proteolytic processing into mature MPO subunits or to be transported to the granule compartment (data not shown). Like Y173C, C319A was retained in the ER by prolonged association with CNX. In sequential immunoprecipitations nearly one fifth of the total radiolabeled C319A MPO precursor was still associated

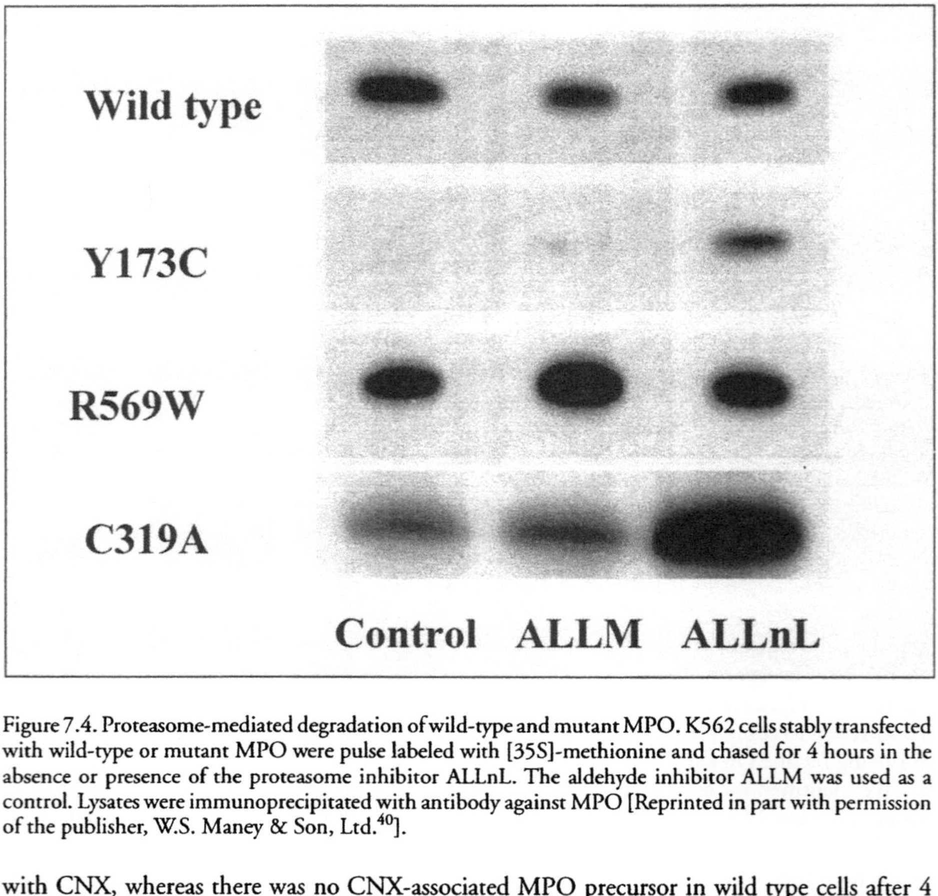

Figure 7.4. Proteasome-mediated degradation of wild-type and mutant MPO. K562 cells stably transfected with wild-type or mutant MPO were pulse labeled with [35S]-methionine and chased for 4 hours in the absence or presence of the proteasome inhibitor ALLnL. The aldehyde inhibitor ALLM was used as a control. Lysates were immunoprecipitated with antibody against MPO [Reprinted in part with permission of the publisher, W.S. Maney & Son, Ltd.[40]].

with CNX, whereas there was no CNX-associated MPO precursor in wild type cells after 4 hours of chase. To assess the role of the cytosolic proteasome in degradation of the C319A precursor, pulse-chase experiments in the absence or presence of the proteasome inhibitor acetyl-leu-leu-norleucinal (ALLnL)[74] were performed. As a control, parallel samples were incubated with acetyl-leu-leu-methional (ALLM), a related aldehyde with cysteine protease inhibitory properties similar to those of ALLnL but only ~ 1/10 the activity against proteasome activity.[75] The proteasome inhibitor ALLnL, but not the control aldehyde inhibitor ALLM, blocked the degradation of C319A precursor during the chase period[40] and these findings were confirmed with lactacystin, a more specific inhibitor of proteasomal activity.[76]

Degradation in the cytosolic proteasome of aberrantly folded MPO precursors in the ER poses a topological problem, since MPO and its precursors are soluble proteins in the ER lumen. CNX is a type I transmembrane protein with an 89 amino acid tail extending into the cytoplasm.[1] Because this cytoplasmic extension includes three potential sites for phosphorylation, it has been suggested that conformational and/or posttranslational changes in the cytosolic domain of CNX might provide a signal to recruit the proteasome for delivery of misfolded proteins.[77] When stable transfectants expressing wild type or Y173C were biosynthetically labeled with ^{32}P and lysates immunoprecipitated sequentially for CNX (non-denaturing conditions) and subsequently CNX or MPO, the amount of ^{32}P-CNX precipitated was 2.2-fold greater in Y173C cells relative to cells expressing wild type MPO. Not only was the total amount of ^{32}P-CNX increased but also there was 3-fold more Y173C precursor associated with CNX in comparison to the wild-type complex with CNX.[42] As proMPO is phosphorylated in the Golgi,[78] there was no ^{32}P-MPO recovered from the CNX-MPO precursor complexes, presumably restricted to the ER. Taken together these data suggest that the increased phosphorylation of CNX correlated with the presence of misfolded Y173C in the ER and with its prolonged association in a complex with CNX. Studies are underway to test more directly the hypothesis that the phosphorylation state of CNX provides the signal that a misfolded protein in the ER lumen requires transport to and subsequent degradation in the cytosolic proteasome.

Summary

Taken together, these data indicate that the ER molecular chaperones interact in a coordinated and sequential fashion with glycosylated MPO precursors. The relationship of these interactions to monitoring the fidelity of protein synthesis in the ER and the structural basis for the selectivity of CRT and CNX for specific forms of MPO precursors are not precisely understood. For example, differential interactions of CRT with apoproMPO and CNX with both apoproMPO and proMPO suggest that CRT and CNX may have the capacity to discriminate between the two forms based on structural determinants. This discriminatory ability may have important functional implications, if the CNX-mediated association directly influences heme acquisition by the apoenzyme. Identification of the structural basis for these interactions should provide insights into the mechanisms of chaperone associations in the ER with normal and mutated glycoproteins.

Acknowledgement

Work supported in part by grants AI 34879 and HL 53592 from the National Institutes of Health and a Merit Review Award from the Department of Veterans Affairs.

References

1. Williams DB. Calnexin: a molecular chaperone with a taste for carbohydrate. Biochem Cell Biol 1995; 73:123-132.
2. Helenius A, Trombetta ES, Hebert DN et al. Calnexin, calreticulin and the folding of glycoproteins. Trends Cell Biol 1997; 7:193-200.
3. Michalak M, Milner RE, Burns K et al. Calreticulin. Biochem J 1992; 285:681-692.
4. Zapun A, Darby NJ, Tessier DC et al. Enhanced catalysis of ribonuclease B folding by the interaction of calnexin or calreticulin with ERp57. J Biol Chem 1998; 273:6009-6012.
5. Molinari M, Helenius A. Chaperone selection during glycoprotein translocation into the endoplasmic reticulum. Science 2000; 288:331-333.
6. Nauseef WM. Contributions of myeloperoxidase to proinflammatory events: more than an antimicrobial system. Intl J Hematol 2001; 74:125-133.
7. The Peroxidase Multigene Family of Enzymes: biochemical basis and clinical applications. First ed. Berlin: Springer-Verlag, 2000.
8. Nauseef WM, Clark RA. Granulocytic phagocytes. In: Mandell GL, Bennett JE, Dolin R, eds. Principles and Practice of Infectious Diseases. Philadelphia: Churchill-Livingstone, 2000:89-111.
9. Klebanoff SJ. Myeloperoxidase. Proc Assoc Am Physicians 1999; 111(5):383-389.
10. Winterbourn CC, Vissers M, Kettle AJ. Myeloperoxidase. Curr Opin Hematol 2000; 7:53-58.
11. Hampton MB, Kettle AJ, Winterbourn CC. Inside the neutrophil phagosome: oxidants, myeloperoxidase, and bacterial killing. Blood 1998; 92:3007-3017.
12. Rosen H, Michel BR, VanDevanter DR et al. Differential effects of myeloperoxidase-derived oxidants on Escherichia coli DNA replication. Infect Immun 1998; 66:2655-2659.
13. Dinauer MC, Nauseef WM, Newburger PE. Inherited disorders of phagocyte killing. In: Scriver CR, Beaudet AL, Valle D et al, eds. The Metabolic and Molecular Bases of Inherited Diseases. New York: McGraw-Hill Companies, 2001:4857-4887.
14. Andrews PC, Parnes C, Krinsky NI. Comparison of myeloperoxidase and hemi-myeloperoxidase with respect to catalysis, regulation, and bactericidal activity. Arch Biochem Biophys 1984; 228:439-442.
15. Furtmuller PG, Jantschko W, Regelsberger G et al. A transient kinetic study on the reactivity of recombinant unprocessed monomeric myeloperoxidase. FEBS Lett 2001; 503:147-150.
16. Zeng J, Fenna RE. X-ray crystal structure of canine myeloperoxidase at 3 Å resolution. J Mol Biol 1992; 226:185-207.

17. Blair-Johnson M, Fiedler T, Fenna R. Human myeloperoxidase: Structure of a cyanide complex and its interaction with bromide and thiocyanate substrates at 1.9 Å resolution. Biochemistry 2001; 40(46):13990-13997.
18. Gullberg U, Bengtsson N, Bülow E et al. Processing and targeting of granule proteins in human neutrophils. J Immunol Meth 1999; 232:201-210.
19. Gullberg U, Andersson E, Garwicz D et al. Biosynthesis, processing and sorting of neutrophil proteins: insight into neutrophil granule development. Eur J Haematol 1997; 58:137-153.
20. Akin DT, Kinkade JM Jr. Evidence for the involvement of an acidic compartment in the processing of myeloperoxidase in human promyelocytic leukemia HL-60 cells. Arch Biochem Biophys 1987; 255:428-436.
21. Akin DT, Kinkade JM Jr. Processing of a newly identified intermediate of human myeloperoxidase in isolated granules occurs at neutral pH. J Biol Chem 1986; 261:8370-8375.
22. Hashinaka K, Nishio C, Hur SJ et al. Multiple species of myeloperoxidase messenger RNAs produced by alternative splicing and differential polyadenylation. Biochemistry 1988; 27:5906-5914.
23. Taylor KL, Guzman GS, Burgess CA et al. Assembly of dimeric myeloperoxidase during posttranslational maturation in human leukemic HL-60 cells. Biochemistry 1990; 29:1533-1539.
24. Nauseef WM. Posttranslational processing of a human myeloid lysosomal protein, myeloperoxidase. Blood 1987; 70:1143-1150.
25. Yamada M, Hur S-J, Toda H. Isolation and characterization of extracellular myeloperoxidase precursor in HL-60 cell cultures. Biochem Biophys Res Commun 1990; 166:852-859.
26. Hur SJ, Toda H, Yamada M. Isolation and characterization of an unprocessed extracellular myeloperoxidase in HL-60 cell cultures. J Biol Chem 1989; 264:8542-8548.
27. Andersson E, Hellman L, Gullberg U et al. The role of the propeptide for processing and sorting of human myeloperoxidase. J Biol Chem 1998; 273(8):4747-4753.
28. Bülow E, Nauseef WM, Goedken M et al. Sorting for storage in myeloid cells of non-myeloid proteins and chimeras with the propeptide of myeloperoxidase precursor. J Leukoc Biol 2002; in press.
29. Daiyasu H, Toh H. Molecular evolution of the myeloperoxidase family. J Mol Evol 2000; 51:433-445.
30. Kimura S, Ikeda-Saito M. Human myeloperoxidase and thyroid peroxidase, two enzymes with separate and distinct physiological functions, are evolutionarily related members of the same gene family. Proteins 1988; 3:113-120.
31. Dunford HB. Heme Peroxidases. First ed. New York: Wiley-VCH, 1999.
32. Kettle AJ, Winterbourn CC. Myeloperoxidase: a key regulator of neutrophil oxidant production. Redox Report 1997; 3:3-15.
33. Harrison JE, Schultz J. Studies on the chlorinating activity of myeloperoxidase. J Biol Chem 1976; 251:1371-1374.
34. Weiss SJ, Test ST, Eckmann CM et al. Brominating oxidants generated by human eosinophils. Science 1986; 234:200-202.
35. Kooter IM, Moguilevsky N, Bollen A et al. Characterization of the Asp94 and Glu242 mutants in myeloperoxidase, the residues linking the heme group via ester bonds. Eur J Biochem 1999; 264(1):211-217.
36. Kooter IM, Koehler BP, Moguilevsky N et al. The Met243 sulfonium ion linkage is responsible for the anomalous magnetic circular dichroism and optical spectral properties of myeloperoxidase. JBIC 1999; 4:688-691.
37. Kooter IM, Moguilevsky N, Bollen A et al. The sulfonium ion linkage in myeloperoxidase. J Biol Chem 1999; 274:26794-26802.
38. Nauseef WM, McCormick SJ, Clark RA. Calreticulin functions as a molecular chaperone in the biosynthesis of myeloperoxidase. J Biol Chem 1995; 270:4741-4747.
39. Nauseef WM, McCormick SJ, Goedken M. Coordinated participation of calreticulin and calnexin in the biosynthesis of myeloperoxidase. J Biol Chem 1998; 273:7107-7111.
40. Nauseef WM, McCormick S, Goedken M. Impact of missense mutations on biosynthesis of myeloperoxidase. Redox Report 2000; 5:197-206.
41. Benoist F, Grand-Perret T. Co-translational degradation of apolipoprotein B100 by the proteasome is prevented by microsomal triglyceride transfer protein. J Biol Chem 1997; 272(33):20435-20442.
42. DeLeo FR, Goedken M, McCormick SJ et al. A novel form of hereditary myeloperoxidase deficiency linked to endoplasmic reticulum/proteasome degradation. J Clin Invest 1998; 101:2900-2909.
43. DeLeo FR, Burritt JB, Yu L et al. Processing and maturation of flavocytochrome b558 includes incorporation of heme as a prerequisite for heterodimer assembly. J Biol Chem 2000; 275:13986-13993.

44. High S, Lecomte FJL, Russell SJ et al. Glycoprotein folding in the endoplasmic reticulum: a tale of three chaperones? FEBS Lett 2000; 476(1-2):38-41.
45. Peterson JR, Ora A, Van PN et al. Transient, lectin-like association of calreticulin with folding intermediates of cellular and viral glycoproteins. Mol Biol Cell 1995; 6:1173-1184.
46. Wada I, Imai S, Kai M et al. Chaperone function of calreticulin when expressed in the endoplasmic reticulum as the membrane-anchored and soluble forms. J Biol Chem 1995; 270(35):20298-20304.
47. Van Leeuwen JEM, Kearse KP. Calnexin associates exclusively with individual CD3d and T cell antigen receptor (TCR) a proteins containing incompletely trimmed glycans that are not assembled into multisubunit TCR complexes. J Biol Chem 1996; 271:9660-9665.
48. Kornfeld S, Mellman I. The biogenesis of lysosomes. Annu Rev Cell Biol 1989; 5:483-525.
49. Sousa M, Parodi AJ. The molecular basis for the recognition of misfolded glycoproteins by the UDP-Glc:glycoprotein glucosyltransferase. EMBO J 1995; 14:4196-4203.
50. Svaerke C, Houen G. Chaperone properties of calreticulin. Acta Chem Scand 1998; 52:942-949.
51. Ware FE, Vassilakos A, Peterson PA et al. The molecular chaperone calnexin binds $Glc_1Man_9GlcNAc_2$ oligosaccharide as an initial step in recognizing unfolded glycoproteins. J Biol Chem 1995; 270(9):4697-4704.
52. Zhang Q, Tector M, Salter RD. Calnexin recognizes carbohydrate and protein determinants of class I major histocompatibility complex molecules. J Biol Chem 1995; 270(8):3944-3948.
53. Arunachalam B, Cresswell P. Molecular requirements for the interaction of class II major histocompatibility complex molecules and invariant chain with calnexin. J Biol Chem 1995; 270(6):2784-2790.
54. Rajagopalan S, Xu Y, Brenner MB. Retention of unassembled components of integral membrane proteins by calnexin. Science 1994; 263:387-390.
55. Carreno BM, Schreiber KL, McKean DJ et al. Aglycosylated and phosphatidylinositol-anchored MHC class I molecules are associated with calnexin. J Immunol 1995; 154:5173-5180.
56. Kim PS, Arvan P. Calnexin and BiP act as sequential molecular chaperones during thyroglobulin folding in the endoplasmic reticulum. J Cell Biol 1995; 128:29-38.
57. Basu S, Srivastava PK. Calreticulin, a peptide-binding chaperone of the endoplasmic reticulum, elicits tumor- and peptide-specific immunity. J Exp Med 1999; 189(5):797-802.
58. Ihara Y, Cohen-Doyle MF, Saito Y et al. Calnexin discriminates between protein conformational states and functions as a molecular chaperone in vitro. Mol Cell 1999; 4:331-341.
59. Saito Y, Ihara Y, Leach MR et al. Calreticulin functions in vitro as a molecular chaperone for both glycosylated and non-glycosylated proteins. EMBO J 1999; 18:6718-6729.
60. Danilczyk UG, Williams DB. The lectin chaperone calnexin utilizes polypeptide-based interactions to associate with many of its substrates in vivo. J Biol Chem 2001; 276(27):25532-25540.
61. Seo HG, Fujii J, Soejima H et al. Heme requirement for production of active endothelial nitric oxide synthase in baculovirus-infected insect cells. Biochem Biophys Res Commun 1995; 208(1):10-18.
62. Nakamura K, Zuppini A, Arnaudeau S et al. Functional specialization of calreticulin domains. J Cell Biol 2001; 154(5):961-972.
63. Takeuchi KH. Biochemical and immunological identification of human neutrophil elastase on nitrocellulose membranes. Stain Technol 1991; 66:324-329.
64. Hayes SA, Dice JF. Roles of molecular chaperones in protein degradation. J Cell Biol 1996; 132:255-258.
65. Kuznetsov G, Nigam SK. Folding of secretory and membrane proteins. New Engl J Med 1998; 339:1688-1695.
66. Brodsky JL, McCracken AA. ER-associated and proteasome-mediated protein degradation: how two topologically restricted events came together. Trends Cell Biol 1997; 7:151-156.
67. Nauseef WM, Brigham S, Cogley M. Hereditary myeloperoxidase deficiency due to a missense mutation of arginine 569 to tryptophan. J Biol Chem 1994; 269:1212-1216.
68. Romano M, Dri P, Dadalt L et al. Biochemical and molecular characterization of hereditary myeloperoxidase deficiency. Blood 1997; 90:4126-4134.
69. Nauseef WM, Petrides PE. Peroxidases and human disease: a meeting of minds. Molecular Medicine Today 1999; 5:58-60.
70. Nauseef WM, Cogley M, McCormick S. Effect of the R569W missense mutation on the biosynthesis of myeloperoxidase. J Biol Chem 1996; 271(16):9546-9549.
71. Nauseef WM, Cogley M, Bock S et al. Pattern of inheritance in hereditary myeloperoxidase deficiency associated with the R569W missense mutation. J Leukoc Biol 1998; 63:264-269.
72. Cabral CM, Liu Y, Sifers RN. Dissecting glycoprotein quality control in the secretory pathway. Trends Biochem Sci 2001; 26:619-624.

73. Van Dalen CJ, Whitehouse MW, Winterbourn CC et al. Thiocyanate and chloride as competing substrates for myeloperoxidase. Biochem J 1997; 327:487-492.
74. Lee DH, Goldberg AL. Proteasome inhibitors: valuable new tools for cell biologists. Trends Cell Biol 1998; 8:397-403.
75. Jensen TJ, Loo MA, Pind S et al. Multiple proteolytic systems, including the proteasome, contribute to CFTR processing. Cell 1995; 83:129-135.
76. Fenteany G, Schreiber SL. Lactacystin, proteasome function, and cell fate. J Biol Chem 1998; 273:8545-8548.
77. Chevet E, Wong HN, Gerber D et al. Phosphorylation by CK2 and MAPK enhances calnexin association with ribosomes. EMBO J 1999; 18:3655-3666.
78. Nauseef WM, McCormick S, Yi H. Roles of heme insertion and the mannose-6-phosphate receptor in processing of the human myeloid lysosomal enzyme, myeloperoxidase. Blood 1992; 80:2622-2633.

Chapter 8

Calreticulin-Mediated Nuclear Protein Export

Ben E. Black and Bryce M. Paschal

Abstract

The role of calreticulin (CRT) as a molecular chaperone that functions in the endoplasmic reticulum (ER) is well established. This involves transient binding of CRT to hydrophobic residues and carbohydrate chains in polypeptides undergoing folding reactions in the lumen of the ER. The issue of CRT distribution and function outside of the ER, though controversial for several years, has now been addressed by rigorous biochemical fractionation and cell biological analysis. Cytosolic CRT, which refers to the non-ER form of the protein that shuttles between the cytoplasm and nucleus, can function as a receptor that mediates nuclear export of the glucocorticoid receptor (GR). The signal recognized by CRT is contained within the DNA binding domain (DBD) of GR. In this chapter, we introduce the topic of nuclear export and summarize the characterization of cytosolic CRT as an export receptor. We also review the evidence that the DBD functions as a signal for export of GR. The DBD is likely to function as the export signal for other members of the nuclear receptor (NR) superfamily, which is the largest family of transcription factors in higher eukaryotes. Our working model is that the non-ER form of CRT contributes to the regulation of multiple cellular pathways through a nuclear export-based mechanism.

Nucleocytoplasmic Transport Pathways

Nuclear import and export pathways generally use cis-acting signals to direct cargoes to the nucleus and cytoplasm.[1,2] These signals are recognized and bound by specific receptors that facilitate translocation through large channels in the nuclear envelope, termed nuclear pore complexes (NPCs).[3-5] The leucine-rich or hydrophobic nuclear export signal (NES), which is the most common signal for export, was first identified in the HIV-1 Rev protein and protein kinase inhibitor (PKI).[6,7] The leucine-rich NES in Rev (LPPLERLTL) and PKI (LALKLAGLDIN) is recognized by the export receptor Crm1, a member of the importin β (also called karyopherin β) family of nuclear transport receptors.[8,9] Crm1 binds the NES and mediates export to the cytoplasm.[10,11] This export pathway is regulated by the GTPase Ran, which, in its GTP-bound form, assembles into a stoichiometric complex with Crm1 and the NES cargo.[12] Following translocation, the export complex is disassembled in the cytoplasm through the action of several factors including the Ran GTPase activating protein.

The functions of a number of proteins in the cell require both nuclear import and nuclear export, a process referred to as nucleocytoplasmic shuttling. Nuclear transport of some shuttling proteins relies on separate signals for import and export, and bidirectional transport depends on interactions with both import and export receptors. Examples of this type of shuttling protein include the proteins p53 and NFAT.[13-16] Other shuttling proteins, such as the hnRNA A1 protein that assembles into RNP complexes, contain a single transport signal that mediates both import and export.[2] Nucleocytoplasmic shuttling is also a property of many, if not all, steroid hormone receptors. This endows the cell with the ability to regulate transcription by controlling the distribution of steroid hormone receptors. The pathway for nuclear

Calreticulin, 2nd ed., edited by Paul Eggleton and Marek Michalak.

import of steroid hormone receptors, which has been studied extensively with GR, is initiated in the cytoplasm by ligand binding. This induces a conformational change that releases chaperones and exposes the NLS, which is recognized by the nuclear import receptor importin-β. The pathway for nuclear export of steroid hormone receptors has, until recently, remained obscure. Steroid hormone receptors do not contain a leucine-rich NES, and there is clear evidence that nuclear export of these proteins is not mediated by the export receptor Crm1. This includes the finding that steroid hormone receptor export is insensitive to Leptomycin B, a compound that specifically inhibits Crm1 function and blocks leucine-rich NES-dependent export.[17] As discussed below, our laboratory found that nuclear export of GR is mediated by the Ca^{2+} -binding protein CRT. This export pathway appears to be used by other members of the nuclear receptor (NR) superfamily.

Purification of CRT Using an Export Assay

Because multiple receptors and pathways are used for nuclear import,[9] we reasoned that cells should contain export receptors in addition to Crm1. To test this hypothesis, our laboratory developed an assay that reconstitutes the nuclear export of the NES-containing protein PKI.[18] The PKI export assay is carried out in digitonin-permeabilized cells, the most widely used model system for analysis of nuclear transport in vitro.[14,18,19] PKI is loaded into the nuclei of permeabilized cells during the import phase, and cytosol is added to stimulate nuclear export during the export phase. The cytosol-dependence of the export phase allowed us to make several observations that were consistent with the presence of multiple export factors.[18] The key observation was that quantitative depletion of Crm1 from cytosol by treatment with Phenyl-Sepharose resulted in only partial depletion of total export activity, as measured in the PKI export assay.

We devised a purification scheme that involved ammonium sulfate precipitation, ion exchange, and gel filtration chromatography steps, using PKI export as the assay. This resulted in purification of a single protein, with an apparent molecular weight of ~60 kDa on silver-stained gels, that was sufficient to stimulate PKI export in permeabilized cells. Mass spectrometry identified the protein as CRT. Definitive evidence that the export activity could be attributed to CRT, and not to a minor contaminant, was obtained by showing that recombinant CRT was sufficient to promote export in PKI assay. The mechanism of CRT-dependent PKI export, like that of Crm1, requires Ran-GTP.[20] Mutations in the NES that inhibit recognition by Crm1 also inhibit recognition by CRT. Although the CRT and Crm1 export pathways for NES proteins display clear functional similarities, the proteins are unrelated at the sequence level.

Subcellular Distribution of CRT

CRT contains an amino terminal signal sequence and a carboxyl terminal KDEL retention signal, hallmarks of an ER protein. Immunofluorescence microscopy clearly shows that CRT is localized to the ER, the organelle from which the protein was first isolated.[21] Nevertheless, multiple laboratories have reported finding CRT in locations outside of the ER, including the nucleus. The apparent localization of CRT to the nucleus seemed consistent with previous data indicating that CRT could suppress transcriptional activity of steroid hormone receptors. Unfortunately, the localization of endogenous CRT by immunofluorescence is technically difficult because the high concentration of CRT in the ER obscures detection of the non-ER pool of the protein.[22-24]

We chose instead to analyze the distribution of CRT in HeLa cells by classical sub-cellular fractionation and immunoblotting, using well-established marker proteins to define the compartments. We found that CRT is present in the microsomal compartment, coincident with ER marker proteins, and in the soluble compartment, coincident with soluble marker proteins.[20] Proteinase K digestion was used to show that the microsomal pool of CRT is contained within vesicles and susceptible to digestion only in the presence of detergent, and that the soluble pool of CRT is degraded by proteinase K even in the absence of detergent. This provides clear evidence that CRT is found in both ER and non-ER compartments. These data also explain how our purification scheme, which started with a detergent-free, high-speed extract devoid of organelles, resulted in the isolation of CRT. The biosynthetic pathway that generates the non-ER, cytosolic form of CRT is under investigation in our laboratory.

CRT Is the Export Receptor for GR

Several years prior to our isolation of CRT as an export factor, CRT was shown to interact with GR and other members of the NR superfamily.[25-27] Transfection of CRT inhibited the transactivation mediated by NRs, and recombinant CRT inhibited NR binding to DNA response elements in gel shift assays. The latter result provided a potential molecular explanation for the inhibitory effect of CRT on transcription. Nonetheless, our results showing that CRT mediates nuclear export of NES-containing proteins led us to consider whether CRT might also function as an export receptor for NRs. We viewed this as an attractive hypothesis because nuclear export would inhibit the transactivation by relocating NRs to the cytoplasm, where NRs reassemble into multi-subunit complexes containing heat shock proteins including hsp90.[28] It should be noted that even in the presence of their respective ligand, NRs including GR are actively shuttling between the nucleus and cytoplasm, indicating that cells have robust mechanism for nuclear export for these proteins. As mentioned above, the absence of a leucine-rich NES in NRs and the insensitivity of NR export to Leptomycin B appeared to rule out the Crm1 pathway.

The system we initially chose for examining whether CRT mediates GR export was the digitonin-permeabilized cell assay. By using a green fluorescent protein fusion (GFP) of GR, we were able to show that addition of recombinant CRT is sufficient to stimulate GR export from the nucleus.[20] This result was corroborated in vivo using CRT-deficient cells isolated from embryos of CRT knockout mice.[29] The in vivo assay involves ligand addition and subsequent withdrawal to allow for GR-GFP export (Fig. 8.1A). The $crt^{-/-}$ cells were found to be deficient for GR export, and the transport defect was corrected by back-transfection of CRT (Fig. 8.1B). We also found that recombinant CRT could potently stimulate GR export when microinjected into the hamster cell line BHK.[20] These experiments established, for the first time, that CRT mediates the nuclear to cytoplasmic localization of GR in a pathway that is independent of Crm1.

Identification of the Export Signal in GR

We set out to characterize the signal within GR that is recognized by CRT. For this analysis, we constructed a GFP reporter that would reveal the export activity of sequences from GR in a fluorescence microscopy assay.[30] The GFP reporter contained the ligand binding domain of GR that facilitates dexamethasone (Dex)-inducible nuclear import. Because previous work suggested the DBD was a strong candidate as an export signal, we transplanted the 69 residues that include the GR DBD to the GFP reporter and tested it in the assay. The GFP reporter is mostly nuclear up to six hours after removal of ligand (Fig. 8.2, No DBD). This contrasts with the distribution of the GFP reporter that contains the GR DBD, which undergoes nuclear export and is clearly cytoplasmic by four hours (Fig. 8.2, GR-DBD (418-486)).

The DBD of GR, like that of all other NR superfamily members, contains two zinc-binding loops and makes sequence-specific contacts with its corresponding DNA response element.[31,32] We predicted that these zinc-binding loops of GR might be important for CRT-recognition, since they are critical for DBD structure. Surprisingly, cysteine-to-alanine point mutations that are known to disrupt the structure of either the first or second zinc-binding loop caused only a modest reduction in export activity.[30] We analyzed the effects of alanine mutations within the 15 amino acid region between the two zinc-binding loops, a region identified in peptide-binding experiments as a CRT-binding site.[25,26] While several of these mutations reduced the export activity of the DBD, the most striking defect in export was caused by mutating two adjacent

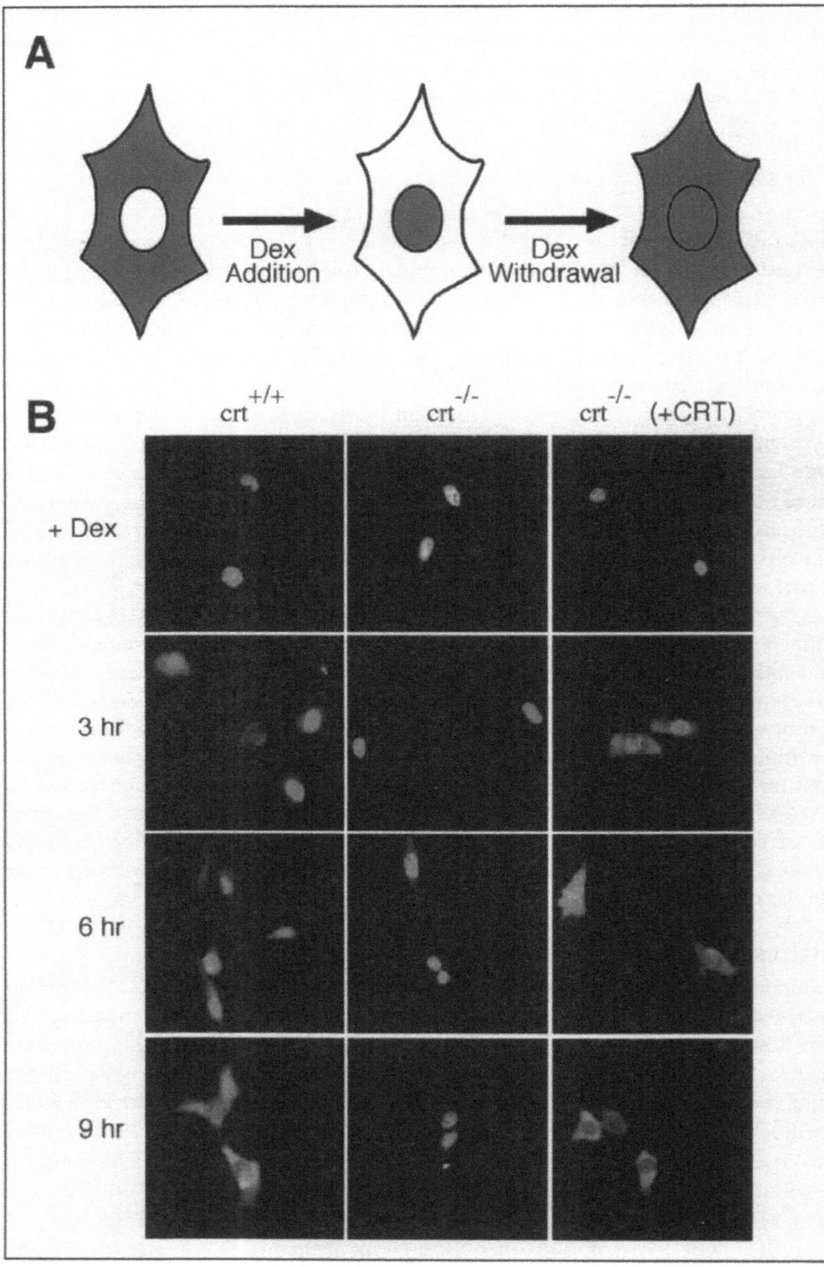

Figure 8.1. CRT mediates the nuclear export of GR in vivo. (A) In vivo assay for GR export. Import is induced by the addition of Dex. Following Dex withdrawal, cells are incubated and imaged at the indicated time-points. (B) Nuclear export of GR is impaired in the absence of CRT, and is restored by CRT expression. A plasmid encoding GR-GFP was transfected into the indicated cell lines, and nuclear accumulation of the reporter was induced with Dex. After agonist removal, the cells were examined at 3 hours intervals to monitor nuclear export. Nuclear export of the GR-GFP reporter was observed in WT (crt $^{+/+}$) and CRT-transfected (crt $^{-/-}$ [CRT]) cells, but not in CRT-deficient (crt$^{-/-}$) cells. Reprinted with permission from the J Cell Biol[20] and Curr Biol.[30]

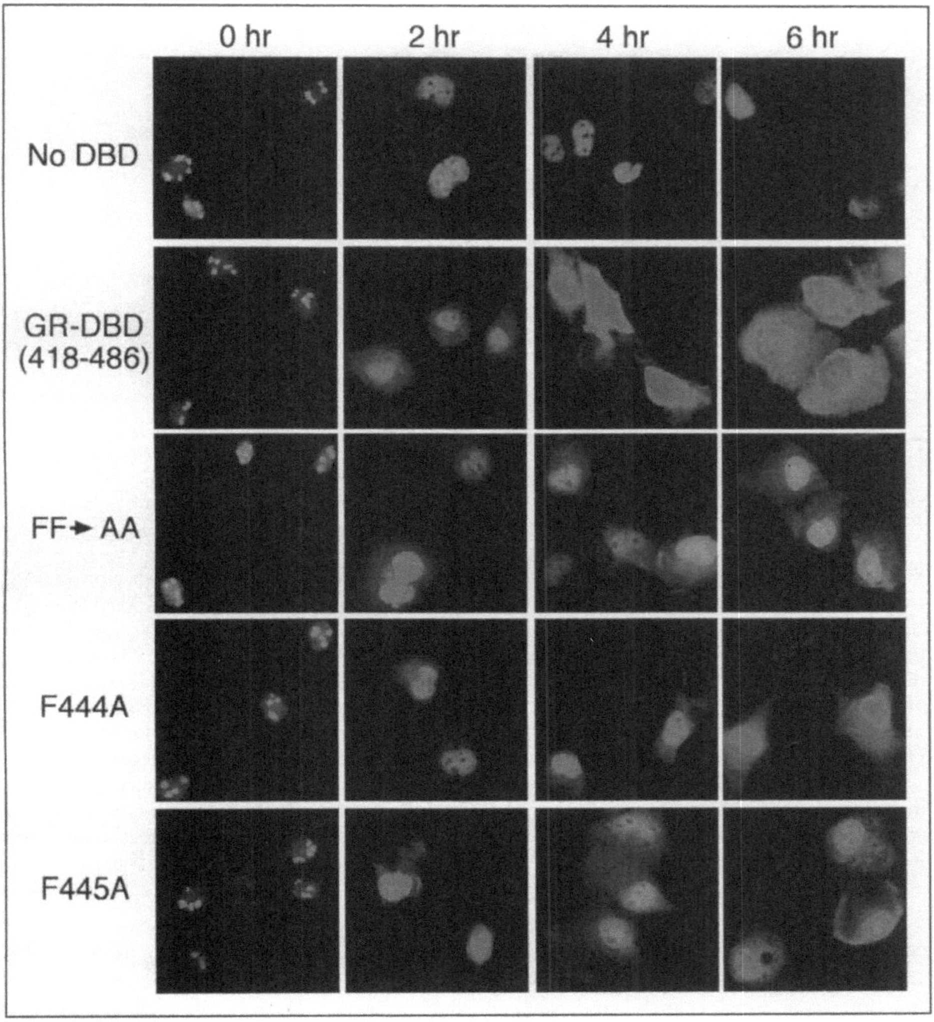

Figure 8.2. Two adjacent phenylalanines in the GR DBD are both required for a functional NES. The GFP reporter alone (no DBD) remains nuclear during the course of the experiment, while including the GR DBD (WT DBD) in the GFP reporter confers export. The FF→AA mutation has been shown to abolish the activity of the DBD NES.[20,30] Mutation of either phenylalanine (F444A, F445A) also blocks DBD-mediated export, indicating that each is critical for the activity of the NES.

phenylalanines.[20,30] We mutated each phenylalanine individually to determine if both are required for NES function. Both single F→A point mutations (F444A and F445A) led to a major reduction in export activity that appeared similar to the double mutant (FF→AA), suggesting that both phenylalanines are important for nuclear export (Fig. 8.2). These particular phenylalanines are invariant residues in the DBD of all NR superfamily members, and are near the middle of the DNA recognition helix.

The DBD Is Necessary for Export

To determine if the DBD is necessary for GR export, we used a heterokaryon shuttling assay that scores export from a donor nucleus and import into an acceptor nucleus (Fig. 8.3).[30]

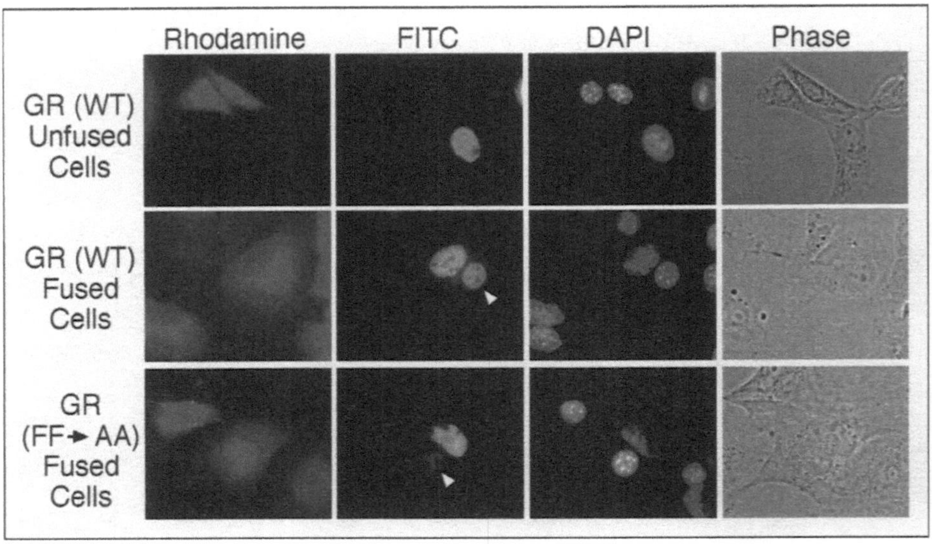

Figure 8.3. The DBD export signal is necessary for GR shuttling in vivo. Interspecies heterokaryon shuttling assays were performed with Cos cells transfected with full-length GR fused to GFP (FITC) and NIH 3T3 cells labeled with the dye CellTracker CMTMR (Rhodamine). When co-seeded on coverslips and fused by brief (30 seconds) incubation in polyethylene glycol (Roche; 50% vol:vol) the Cos and 3T3 cells fuse and fluoresce red. Nucleocytoplasmic shuttling (export and import) of the GFP reporter results in equilibration of green fluorescence between the donor Cos cell nuclei and the acceptor 3T3 cell nuclei within the heterokaryon. Acceptor cell nuclei are also distinguished by centromeric foci that stained brightly with DAPI. The DBD export signal is necessary for nucleocytoplasmic shuttling in the context of full-length hormone receptors. Full-length WT or mutant (FF→AA) GR was tested for nucleocytoplasmic shuttling in the presence of 1 μM Dex. The WT receptor equilibrates between the nuclei of a heterokaryon (acceptor cell nuclei are denoted by white arrowheads), however, the FF→AA mutation inhibits GR shuttling. Reprinted with permission from Curr Biol.[30]

This type of assay has been used to demonstrate that a variety of proteins, which appear to be constitutively nuclear, actually undergo nucleocytoplasmic shuttling.[33,34] For example, GR appears to be constitutively nuclear in the presence of its ligand, however, the heterokaryon shuttling assay reveals that GR undergoes constant movement between the nucleus and cytoplasm.[33] In the assay, cells expressing GR-GFP are treated with ligand to induce nuclear import, and then fused with cells labeled with a red fluorescent dye. The appearance of GR-GFP in the nuclei of multi-nucleate cells that contain the red dye demonstrates that the reporter has undergone nuclear export from the donor nuclei and nuclear import into acceptor nuclei (Fig. 8.3). The FF→AA mutation that abolishes the export activity in the context of the isolated DBD, also abolishes the export of full-length GR, expressed as a fusion to GFP (Fig. 8.3). Thus, the DBD of GR, which is sufficient for nuclear export, is also required for nuclear export.

Our data indicated that both an intact DBD and the protein CRT are necessary for nuclear export of GR. This export pathway requires a physical interaction between these proteins, based on the following observations. First, CRT binds directly to the DBD, and the FF→AA mutation that inhibits nuclear export also inhibits the binding of CRT.[30] Second, CRT introduced by microinjection can promote nuclear export of GR that contains a functional DBD, but not of GR that contains mutations in the DBD.[30] Third, the presence of excess DBD is sufficient to competitively-inhibit CRT-dependent export, but the competition occurs only if the competing DBD can be recognized by CRT.[30] Our finding that DBD-dependent export is saturable in vivo provides evidence that NRs compete for a limiting component for nuclear export. CRT is one of the rate limiting components for this export pathway.

Regulating GR Export

While some aspects of CRT-mediated recognition of the export signal in GR have now been established, the regulation of this export pathway is largely unexplored. By analogy with nuclear import pathways, the CRT-dependent export pathway for GR and other NRs could be regulated at several different levels. First, the accessibility of the DBD as the export signal is a potential point of regulation. Since the DBD will not be exposed when the NR is directly bound to DNA, accessibility of the DBD to CRT will be determined by the rate of NR dissociation from DNA. Second, the assembly of NRs into large, macromolecular transcription complexes, or partitioning into the insoluble nuclear matrix, or both, might result in nuclear retention. Since nuclear retention could be dominant over nuclear export, controlling the release from nuclear retention would be another potential point of regulation. Third, it is possible that CRT itself might be subject to positive or negative regulation. Positive regulation could be achieved by raising the concentration of the cytosolic pool of CRT, or by covalent modification of CRT by phosphorylation,[35] or by increasing the activity of CRT through Ca^{2+}-binding. Fourth, CRT could be regulated by the Ran GTPase, which we have shown is a necessary component of export complexes that contain CRT and proteins with a leucine-rich NES.[20]

We have performed experiments that address two of the potential regulatory mechanisms described above. These are regulation by the Ran GTPase, and regulation by Ca^{2+} binding. In contrast to the critical role that Ran plays in leucine-rich NES export for both the Crm1 and CRT export pathways, Ran does not appear to be an essential factor for CRT-dependent GR export. This was tested by examining GR export in permeabilized cells under conditions where CRT was rate-limiting and Ran was present in excess. In these assays, wild-type Ran does not stimulate CRT-dependent export of GR, and mutant forms of Ran that interfere with its GTPase cycle neither stimulate nor inhibit CRT-dependent export of GR.[36]

We examined whether Ca^{2+} is important for CRT-dependent export of GR by two experimental approaches. In the first approach, Ca^{2+} was stripped from CRT using the chelator EGTA, and the Ca^{2+}-free CRT was tested for GR export in a permeabilized cell assay. Significantly, this revealed that Ca^{2+} is necessary for CRT-dependent GR export. Binding assays performed in parallel confirmed that Ca^{2+} removal from CRT inhibits binding to the DBD. In the second approach, C-terminal deletion mutants of CRT that lack the high capacity, low affinity Ca^{2+}-binding sites were tested in the GR export and DBD binding assays. The results from these experiments indicated that these Ca^{2+}-binding sites are not essential for CRT export activity. Rather, the low affinity, high capacity Ca^{2+}-binding sites appear to regulate the activity of CRT, since Ca^{2+}-binding to these sites is necessary in the context of full-length CRT. Ca^{2+}-binding induces a change in the structure of CRT from an extended conformation to a more compact conformation, the latter of which is active for DBD binding and nuclear export.[36] It has been shown previously that Ca^{2+} binding to CRT is important for its chaperone functions as well.[37]

Common Pathways for NR Transport

A general mechanism is thought to account for the nuclear import of virtually all NRs. This mechanism involves the assembly of the NR into a cytoplasmic complex with several factors including hsp90, which maintain the NR in a conformation that is competent for ligand-binding.[28] Ligand binding initiates a series of events including dissociation of hsp90, exposure of the NLS, recognition by the nuclear import machinery, and import into the nucleus. These molecular events are best understood for GR, in part because this NR shows efficient relocalization to the cytoplasm when ligand is removed from the system. That NRs such as the estrogen receptor (ER) are predominantly nuclear in the absence of ligand may be a consequence of nuclear retention or ligand-independent regulation of nuclear import.

Our studies suggest that a general mechanism may also account for nuclear export of most NRs. After finding that GR export is mediated by its DBD, we tested the DBDs of nine additional NRs for nuclear export activity. The motivation for these experiments derived from

```
                                                                                         % Identity
                                                                                          vs. hGR
      hGR PKLCLVCSDEASGCHYGVLTCGSCKVFFKRAVEGQH--NYL-CAGRNDCIIDKIRRKNCPACRYRKCLQAGM      100
      rAR QKTCLICGDEASGCHYGALTCGSCKVFFKRAAEGKQ--KYL-CASRNDCTIDKFRRKNCPSCRLRKCYEAGM       77
     rERα TRYCAVCNDYASGYHYGVWSCEGCKAFFKRSIQGHN--DYM-CPATNQCTIDKNRRKSCQACRLRKCYEVGM       55
    hLXRα NELCSVCGDKASGFHYNVLSCEGCKGFFRRSVIKGA--HYI-CHSGGHCPMDTYMRRKCQECRLRKCRQAGM       49
      hPR QKICLICGDEASGCHYGVLTCGSCKVFFKRAMEGQH--NYL-CAGRNDCIVDKIRRKNCPACRLRKCCQAGM       88
    hRARα YKPCFVCQDKSSGYHYGVSACEGCKGFFRRSIQKNM--VYT-CHRDKNCIINKVTRNRCQYCRLQKCFEVGM       45
hRevErbα VLLCKVCGDVASGFHYGVLACEGCKGFFRRSIQQNI--QYKRCLKNENCSIVRINRNRCQQCRFKKCLSVGM       48
    hRXRα KHICAICGDRSSGKHYGVYSCEGCKGFFKRTVRKDL--TYT-CRDNKDCLIDKRQRNRCQYCRYQKCLAMGM       49
     hTRβ DELCVVCGDKATGYHYRCITCEGCKGFFRRTIQKNLHPSYS-CKYEGKCVIDKVTRNQCQECRFKKCIYVGM       43
     hVDR PRICGVCGDRATGFHFNAMTCEGCKGFFRRSMKRKA--LFT-CPFNGDCRITKDNRRHCQACRLKRCVDIGM       41
```

Figure 8.4. Alignment of DBDs from NRs used in this study and the percent identity to the DBD of human GR. Highly conserved residues (bold) including the cysteines that coordinate zinc binding (green) and the pair of phenylalanines that are present in the DNA recognition helix of all NRs (red) are indicated. Reprinted with permission from Curr Biol.[30]

the fact that the sequence and structure of the DBD is highly conserved among NRs (Fig. 8.4). The additional DBDs tested were three other steroid receptors (AR, ER, and progesterone receptor [PR]); four non-steroid receptors (RAR, RXR, thyroid hormone receptor [TR], and vitamin D receptor [VDR]), and two orphan receptors (liver X receptor [LXR] and RevErb). Each of the ten DBDs is sufficient for nuclear export activity, indicating that NR DBDs define a new type of export signal.[30] The DBDs are structurally similar and appear to use a common export pathway since the DBDs from two different NRs, GR and VDR, compete for nuclear export in vitro and in vivo.

Why Do Nuclear Receptors Undergo Export?

Nucleocytoplasmic shuttling of NRs should be taken into account when considering the function, regulation, and activity of these transcription factors. Nuclear export can be viewed as an absolute mechanism for turning off transcription since it removes the NR from its primary site of action. Nuclear export has, in fact, been found to be an evolutionarily conserved mechanism for regulating activity of multiple transcription factors.[38] Some well-studied examples include the Ca^{2+} -regulated trafficking of NF-AT in mammals and the phosphate-regulated trafficking of PHO4 in yeast. In addition, since NRs can also regulate the activity of a variety of co-activators and repressors, nuclear export of NRs is an important pathway that impacts on the activity of proteins outside of the NR superfamily. GR is known to negatively regulate the transcription factor NF-κB by transcriptional interference.[39,40] Transcriptional interference involves the sequestration of shared co-factors by steroid receptors, and this usually occurs within the nucleus. The GR-mediated regulation of the NF-κB pathway is critical for the anti-inflammatory effects of glucocorticoids.

Nucleocytoplasmic shuttling may also be linked to the turnover of NRs. Artificially accelerating nuclear export increases ubiquitin-dependent degradation of GR.[41] A similar observation regarding turnover has been made with the tumor suppressor p53. Blocking Crm1-dependent nuclear export with Leptomycin B was found to inhibit p53 turnover, resulting in the accumulation of p53 in the nucleus.[42] Likewise, it has been shown that blocking nuclear export of IκB-α prevents its turnover, since its degradation occurs in the cytoplasm.[43]

Finally, nucleocytoplasmic shuttling of NR superfamily members is important for their non-genomic activities, which involve ligand-induced signaling events in the cytoplasm.[44,45] Non-genomic effects of ligands are manifest through NRs within minutes of addition, and include direct interactions with cytoplasmic components of the Src/Map Kinase and phosphatidylinositol-3-OH (PI-3) kinase signaling pathways.[46-48] NRs identified in this type of signaling include ER and TR, receptors that appear mostly nuclear in the presence or absence of ligand. Nucleocytoplasmic shuttling ensures that a sufficient supply of these NRs is available in the cytoplasm to participate in signaling pathways.

Concluding Remarks

The identification of CRT as a receptor for nuclear export of leucine-rich NES-containing proteins and previous links between CRT and NR activity have provided an unanticipated entrée into analyzing NR function. Recent progress in this area has included identification of the DBD as an export signal, and demonstration that the DBD is necessary and sufficient for nuclear export of both GR and AR.[20,30,49] These observations provide a framework for future experiments that should address how CRT physically contacts the DBD, how these proteins translocate through the NPC, and how these interactions may be regulated by conditions that influence growth and development.

References

1. Gorlich D, Kutay U. Transport between the cell nucleus and the cytoplasm. Annu Rev Cell Dev Biol 1999; 15:607-660.
2. Nakielny S, Dreyfuss G. Transport of proteins and RNAs in and out of the nucleus. Cell 1999; 99:677-690.
3. Stoffler D, Fahrenkrog B, Aebi U. The nuclear pore complex: from molecular architecture to functional dynamics. Curr Opin Cell Biol 1999; 11:391-401.
4. Wente SR. Gatekeepers of the nucleus. Science 2000; 288:1374-1377.
5. Vasu SK, Forbes DJ. Nuclear pores and nuclear assembly. Curr Opin Cell Biol 2001; 13:363-375.
6. Fischer U, Huber J, Boelens WC et al. The HIV-1 Rev activation domain is a nuclear export signal that accesses an export pathway used by specific cellular RNAs. Cell 1995; 82:475-83.
7. Wen W, Meinkoth JL, Tsien RY et al. Identification of a signal for rapid export of proteins from the nucleus. Cell 1995; 82:463-473.
8. Fornerod M, van Deursen J, van Baal S et al. The human homologue of yeast CRM1 is in a dynamic subcomplex with CAN/Nup214 and a novel nuclear pore component Nup88. EMBO J 1997; 16:807-816.
9. Pemberton LF, Blobel G, Rosenblum JS. Transport routes through the nuclear pore complex. Curr Opin Cell Biol 1998; 10:392-399.
10. Fornerod M, Ohno M, Yoshida M et al. CRM1 is an export receptor for leucine-rich nuclear export signals. Cell 1997; 90:1051-60.
11. Stade K, Ford CS, Guthrie C et al. Exportin 1 (Crm1p) is an essential nuclear export factor. Cell 1997; 90:1041-1050.
12. Steggerda SM, Paschal BP. Regulation of nuclear import and export by the GTPase Ran. Int Rev Cytol 2002; 217:41-91.
13. Klemm JD, Beals CR, Crabtree GR. Rapid targeting of nuclear proteins to the cytoplasm. Curr Biol 1997; 7:638-644.
14. Kehlenbach RH, Dickmanns A, Gerace L. Nucleocytoplasmic shuttling factors including Ran and Crm1 mediate nuclear export of NFAT in vitro. J Cell Biol 1998; 141:863-874.
15. Roth J, Dobbelstein M, Freedman DA et al. Nucleo-cytoplasmic shuttling of the hdm2 oncoprotein regulates the levels of the p53 protein via a pathway used by the human immunodeficiency virus rev protein. EMBO J 1998; 17:554-564.
16. Stommel JM, Marchenko ND, Jimenez GS et al. A leucine-rich nuclear export signal in the p53 tetramerization domain: regulation of subcellular localization and p53 activity by NES masking. EMBO J 1999; 18:1660-1672.
17. Wolff B, Sanglier JJ, Wang Y. Leptomycin B is an inhibitor of nuclear export: inhibition of nucleo-cytoplasmic translocation of the human immunodeficiency virus type 1 (HIV-1) Rev protein and Rev-dependent mRNA. Chem Biol 1997; 4:139-147.
18. Holaska JM, Paschal BM. A cytosolic activity distinct from Crm1 mediates nuclear export of protein kinase inhibitor in permeabilized cells. Proc Natl Acad Sci USA 1998; 95:14739-14744.
19. Adam SA, Sterne-Marr RE, Gerace L. Nuclear protein import in permeabilized mammalian cells requires soluble cytoplasmic factors. J Cell Biol 1990; 111:807-816.
20. Holaska JM, Black BE, Love DC et al. Calreticulin is a receptor for nuclear export. J Cell Biol 2001; 152:127-140.
21. Ostwald TJ, MacLennan DH. Isolation of a high affinity calcium-binding protein from sarcoplasmic reticulum. J Biol Chem 1974; 249:974-979.
22. Michalak M, Burns K, Andrin C et al. Endoplasmic reticulum form of calreticulin modulates glucocorticoid-sensitive gene expression. J Biol Chem 1996; 271:29436-29445.
23. Jethmalani SM, Henle KJ, Gazitt Y et al. Intracellular distribution of heat-induced stress glycoproteins. J Cell Biochem 1997; 66:98-111.

24. Roderick HL, Campbell AK, Llewellyn DH. Nuclear localisation of calreticulin in vivo is enhanced by its interaction with glucocorticoid receptors. FEBS Lett 1997; 405:181-185.
25. Burns K, Duggan B, Atkinson EA et al. Modulation of gene expression by calreticulin binding to the glucocorticoid receptor. Nature 1994; 367:476-480.
26. Dedhar S, Rennie PS, Shago M et al. Inhibition of nuclear hormone receptor activity by calreticulin. Nature 1994; 367:480-483.
27. Wheeler DG, Horsford J, Michalak M et al. Calreticulin inhibits vitamin D3 signal transduction. Nucleic Acids Res 1995; 23:3268-3274.
28. Buchner J. Hsp90 & Co.-a holding for folding. Trends Biochem Sci 1999; 24:136-141.
29. Mesaeli N, Nakamura K, Zvaritch E et al. Calreticulin is essential for cardiac development. J Cell Biol 1999; 144:857-868.
30. Black BE, Holaska JM, Rastinejad F et al. DNA binding domains in diverse nuclear receptors function as nuclear export signals. Curr Biol 2001; 11:1749-1758.
31. Evans RM. The steroid and thyroid hormone receptor superfamily. Science 1988; 240:889-895.
32. Luisi BF, Xu WX, Otwinowski Z et al. Crystallographic analysis of the interaction of the glucocorticoid receptor with DNA. Nature 1991; 352:497-505.
33. Madan AP, DeFranco DB. Bidirectional transport of glucocorticoid receptors across the nuclear envelope. Proc Natl Acad Sci USA 1993; 90:3588-3592.
34. Michael WM, Choi M, Dreyfuss G. A nuclear export signal in hnRNP A1: a signal-mediated, temperature-dependent nuclear protein export pathway. Cell 1995; 83:415-422.
35. Singh NK, Atreya CD, Nakhasi HL. Identification of calreticulin as a rubella virus RNA binding protein. Proc Natl Acad Sci USA 1994; 91:12770-12774.
36. Holaska JM, Black BE, Rastinejad FR et al. Ca^{2+}-dependent nuclear export mediated by calreticulin. Mol Cell Biol 2002; 22:6286-6297.
37. Vassilakos A, Michalak M, Lehrman MA et al. Oligosaccharide binding characteristics of the molecular chaperones calnexin and calreticulin. Biochemistry 1998; 37:3480-3490.
38. Komeili A, O'Shea EK. Nuclear transport and transcription. Curr Opin Cell Biol 2000; 12:355-360.
39. Gottlicher M, Heck S, Herrlich P. Transcriptional cross-talk, the second mode of steroid hormone action. J Mol Med 1998; 76:480-489.
40. Karin M, Chang L. AP-1/glucocorticoid receptor crosstalk taken to a higher level. J Endocrinol 2001; 169:447-451.
41. Liu J, DeFranco DB. Protracted nuclear export of glucocorticoid receptor limits its turnover and does not require the exportin 1/CRM1-directed nuclear export pathway. Mol Endocrinol 2000; 14:40-51.
42. Freedman DA, Levine AJ. Nuclear export is required for degradation of endogenous p53 by MDM2 and human papillomavirus E6. Mol Cell Biol 1998; 18:7288-7293.
43. Rodriguez MS, Thompson J, Hay RT et al. Nuclear retention of IkB-a protects it from signal-induced degradation and inhibits nuclear factor kB transcriptional activation. J Biol Chem 1999; 274:9108-9115.
44. Falkenstein E, Tillmann HC, Christ M et al. Multiple actions of steroid hormones-a focus on rapid, nongenomic effects. Pharmacol Rev 2000; 52:513-555.
45. Manolagas SC, Kousteni S. Perspective: nonreproductive sites of action of reproductive hormones. Endo 2001; 142:2200-2204.
46. Migliaccio A, Castoria G, Di Domenico M et al. Steroid-induced androgen receptor-oestradiol receptor b-Src complex triggers prostate cancer cell progression. EMBO J 2000; 19:5406-5417.
47. Simoncini T, Hafezi-Moghadam A, Brazil DP et al. Interaction of oestrogen receptor with the regulatory subunit of phosphatidylinositol-3-OH kinase. Nature 2000; 407:538-541.
48. Kousteni S, Bellido T, Plotkin LI et al. Nongenotropic, sex-nonspecific signaling through the estrogen or androgen receptors: dissociation from transcriptional activity. Cell 2001; 104:719-730.
49. DeFranco DB. DNA-binding domains find a surprising partner. Curr Biol 2001; 11:R1036-R1037.

CHAPTER 9

The Role of Calnexin and Calreticulin in MHC Class I Assembly

Raju Adhikari and Tim Elliott

Abstract

Assembly of Major Histocompatibility Complex (MHC) class I heavy chain (HC) with β2-microglobulin (β2m) and subsequent acquisition of optimal peptides is necessary for class I antigen presentation to cytotoxic T cells (CTLs). Calnexin and calreticulin are two major chaperones involved in the assembly of class I. Recent findings suggest calnexin is important in early stages of class I assembly where it recruits Erp57 to facilitate disulphide bond formation in class I HC and protects class I HC from degradation prior to their assembly with β2m. In addition, assembly of class I HC with β2m is reduced in the absence of calnexin suggesting a direct involvement of calnexin in the assembly of HC with β2m. Calreticulin, on the other hand, is involved in the later stages of class I assembly. Studies of a calreticulin-deficient cell line has demonstrated its critical role in the peptide loading of class I molecules.

Introduction to Class I Assembly

MHC class I molecules are expressed on the surface of virtually all nucleated cells where they serve to present antigenic peptides to cytotoxic T cells. The assembly of MHC class I molecules with antigenic peptides of 8-10 amino acids in length is a co-ordinated and regulated process involving a host of chaperones and cofactors resident in the ER.[1]

Assembly of MHC class I molecules is a two stage process which begins in the ER soon after the synthesis of class I heavy chain (HC). In the first stage, newly synthesized HC bind to calnexin, during which time folding begins and intramolecular disulphide bonds form.[2,3] Moreover, several studies have demonstrated transient association of calnexin with newly synthesized class I HC.[2] It has been suggested that calnexin recruits Erp57, a protein disulphide isomerase (PDI) ortholog,[4] which is predicted to promote class I heavy chain disulphide bond formation. Calnexin also protects class I heavy chain from degradation until β2-microglobulin (β2m) has associated.[5,6]

Upon HC binding to β2-m, calnexin dissociates from the class I HC probably as a result of class I HC undergoing a conformational change.[7-10] Thus, whereas very little HC:calnexin complex can be seen in normal cells, cell lines[11] or mice[12] that lack β2-m fail to express MHC class I molecules on the cell surface and a prolonged association between calnexin with class I HC is observed.[13] Assembly of class I molecules with β2-m is not sufficient for cell surface expression: this requires a supply of peptides. Thus, class I molecules are retained in the early secretory pathway in cells that are unable to supply antigenic peptides to the ER (i.e., cells lacking a functional Transporter associated with Antigen Processing (TAP)).[14]

The assembly of an "empty" (i.e., peptide receptive) class I molecule therefore marks the end of the first stage of class I biogenesis. These molecules are unstable and dissociate with a half-life of less than an hour at 37°C: a direct result of not having bound to a high affinity

Calreticulin, 2nd ed., edited by Paul Eggleton and Marek Michalak.

(stablizing) peptide ligands. Even though at this stage the class I molecule appears fully folded according to most criteria (such as its recognition by conformation sensitive antibodies and failure to be recognized by antibodies raised to denatured class I Heavy chains), its release from the ER is not permitted.

The second stage in class I assembly involves the binding of peptides and their subsequent optimization. If peptides—even those with a low average affinity—are allowed to bind to class I, progression beyond the checkpoint marking the end of stage 1 is permitted. This was illustrated by a class I point mutant at position 134 (Threonine to lysine or T134K) which, like wild-type molecules was only allowed to enter the secretory pathway from the ER in TAP-competent cells. However, the mutant was transported to the cell surface as an unstable, peptide-receptive molecule implying that it had become loaded with peptide cargo but that this cargo had not been optimized. T134K also failed to interact with the cofactors TAP, tapasin, calreticulin and ERp57 but retained its interaction with calnexin. Work with this mutant therefore led to the notion that peptide loading occurred in two steps: peptide binding followed by peptide optimization. This optimization process involves the replacement (or conversion) of non-stabilizing peptides for stabilizing ones and has been shown to be dependent on the action of tapasin, calreticulin and Erp57; although their precise roles are still unknown. In the latter section of this chapter, we will consider the role of calreticulin in this process. For a full review of the process of class I assembly, see Williams et al.[15]

The position and number of glycosylations in the class I HC seems to be critically important in maintaining the division of labour between calnexin and calreticulin. Calnexin is not detected in association with human class I HC once bound to β2m.[16-18] In contrast, calnexin remains associated with mouse class I HC even after HC associates with β2m.[17] One significant difference between human and mouse class I is that human class I HC is glycosylated only at asparagine 86, but mouse HC is glycosylated at asparagines both 86 and 176 (and for some alleles additionally at position 256 e.g., H-2K^d, D^b and L^d, see Fig. 9.1). It is therefore possible that newly synthesized class I HC may first associate with calnexin by virtue of its higher affinity than calreticulin for the glycosylation at position 86 of the free class I HC. As disulphide bonds are formed and class I HC folds, the glycan at N86 may become sterically less accessible to calnexin thus favouring an association with calreticulin. This situation could be achieved by virtue of the fact that calnexin is membrane bound, but calreticulin is luminal. In support of this, Zhang and Salter have shown that addition of a second asparagine-linked glycan to the human class I molecule A*0201 at position 176, a site present in mouse, increased binding to calnexin and reduced interaction with calreticulin and TAP relative to wild-type A*0201 bearing a single asparagine-glycan at position 86.[9] Similarly, removal of the H-2L^d α1 domain glycosylation site by site-directed mutagenesis resulted in poor association of calreticulin although calnexin still bound strongly.[10]

A model for class I assembly, demonstrating the pivotal role of calnexin and calreticulin is shown in Figure 9.2.

Functions of Calnexin in Class I Assembly

Calnexin Performs Several Distinct Functions with Respect to Class I Assembly:

Erp57 Recruitment

Calnexin is associated with class I HC soon after its synthesis and before the formation of intra-chain disulfide bonds within the α2 and α3 domains. Several studies have used a cross-linking approach to identify ERp57 as a co-chaperone of calnexin that interacts with glycoproteins including class I forming a complex with calnexin.[2,19-21] The interaction of ERp57 with a nondisulfide-bonded population at earlier time points suggests a role for ERp57, in association with calnexin, in the folding and disulfide bond formation of heavy chain.[3] Moreover, it has been demonstrated that only the fully disulfide-bonded form of HC is assembled

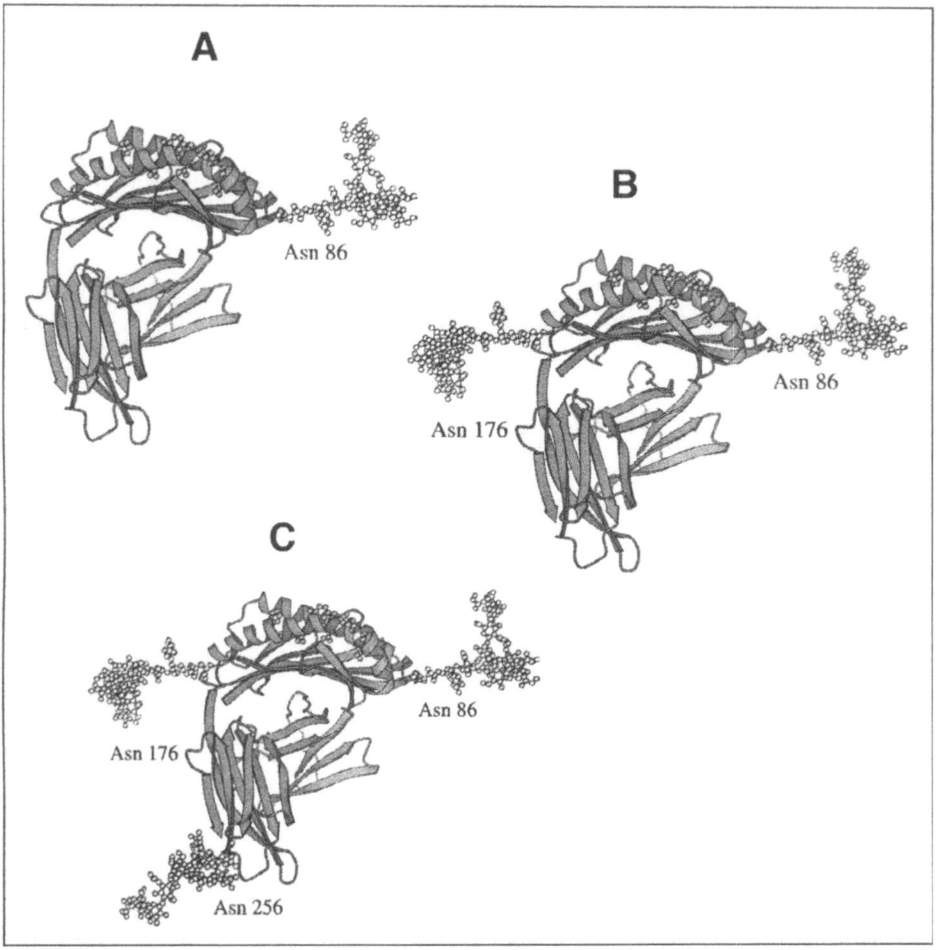

Figure 9.1. Differential glycosylation of different MHC class I alleles. All class I molecules are glycosylated in the alpha 1 domain on asparagine 86. Mouse, but not human class I is also glycosylated in the alpha 2 domain on asparagine 176. In addition, some mouse alleles are glycosylated in the alpha 3 domain on asparagine 276.

with β2m, emphasizing that the formation of disulfide bonds is a requirement for correct folding and assembly.[22]

Protection from Degradation

Calnexin may have a pivotal role in stabilizing class I HC and preventing their premature degradation.[23] When heavy chain is expressed in *Drosophila* cells in the presence of β2m, assembly occurs; however, co-expression of calnexin prevents heavy chain degradation and thereby increases the yield of MHC Class I molecules.[24] Co-expression of calnexin with K^b and D^b in *Drosophila* cells has been shown to extend the half-life of class I molecules by four- to fivefold, again suggesting that calnexin protects the heavy chains from rapid intracellular degradation.[5] Direct evidence for the role of calnexin in prevention of degradation of class I HC comes from a mammalian semipermeabilized (SP) cell system that faithfully reconstitutes the proteasome-mediated degradation of class I heavy chain.[6] In this study, heavy chain was translated in vitro in the presence of SP calnexin-competent (CEM) or calnexin-deficient (CEM-NK^R)

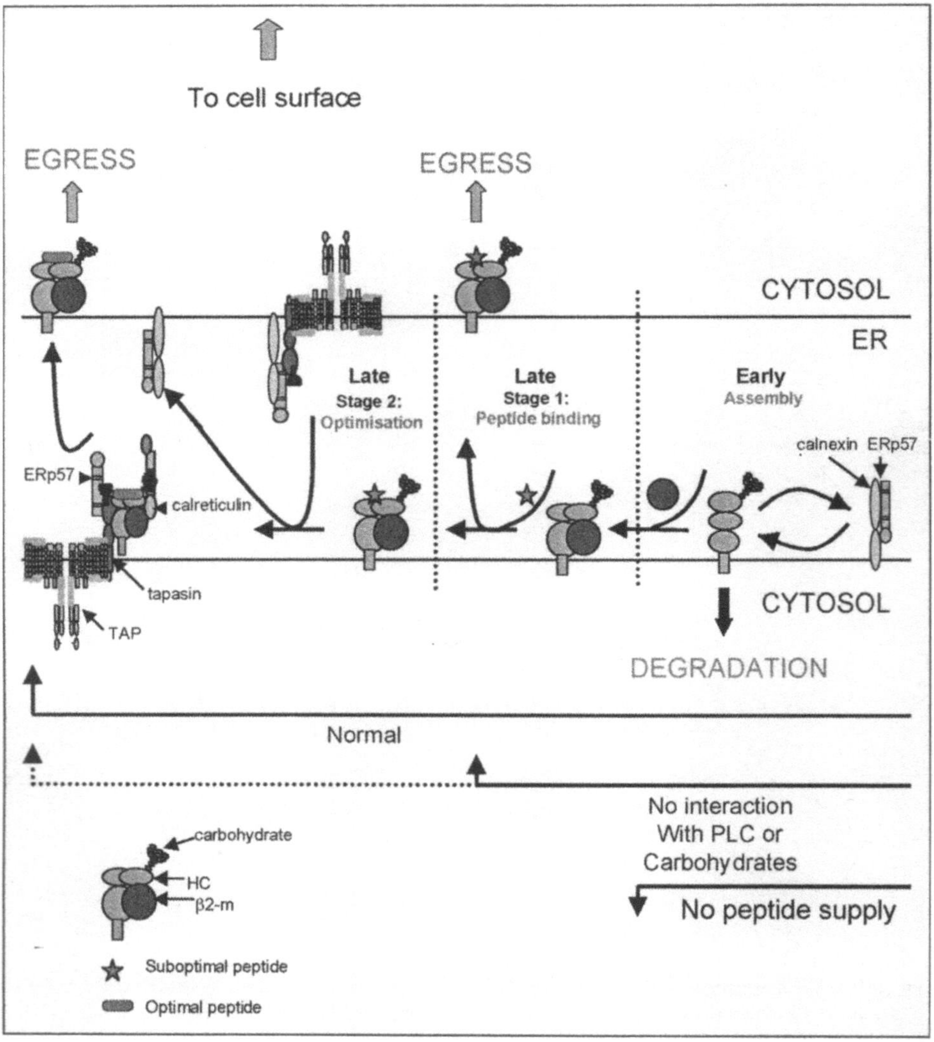

Figure 9.2. Assembly of a functional MHC class I molecule is a multistep process. In the early stage of class I assembly, calnexin binds to newly synthesized class I HC and protects from degradation allowing HC folding. Calnexin also recruits Erp57, which promotes disulfide bond formation within the HC. Calnexin remains associated with HC until β2m binds. Upon β2m binding, at least in human class I alleles, which have just one glycosylation site at asparagine 86, class I undergoes a conformational change[7] and calnexin dissociates from class I HC. This signals the end of the early stage and the beginning of the late stage, which is concerned primarily with peptide binding and optimization. Initially, class I molecules bind to whatever peptides are available in the ER—most of which are suboptimal with respect to their ability to stabilize the HC:b2-m complex. Upon peptide binding, class I are competent to egress from the ER. This process probably occurs independenly of any cofactor association. Class I molecules and calreticulin then bind to the "preloading" complex comprizing of TAP, tapasin, ERp57 and calnexin which results in the displacement of calnexin (in human cells) forming the "loading" complex.[41] It is here that the cargo of peptides bound to class I molecules is optimized. Optimally loaded class I molecules then egress to the cell surface for presentation to the T cells, and those which are sub-optimally loaded are either degraded or returned to the ER for further optimization (see text).

cells. Rapid degradation of heavy chain was observed in SP CEM-NK^R, which could be blocked by lactacystin, a specific inhibitor of proteasome. Rapid degradation was also observed for heavy chain products translated in the presence of castanospermine (a drug that prevents generation of monoglucosylated glycoproteins to which calnexin and calreticulin bind).[6]

HC Assembly with β2m

Despite the accelerated degradation of HC in SP CEM.NK^R, Class I surface expression is normal,[25] suggesting perhaps that other ER chaperones could substitute for calnexin in CEM.NK^R.[25] Sadasivan et al showed normal cell surface expression of class I and similar amount of recognition of class I by conformation-dependent antibodies in CEM.NK^R and CEM.NK^R transfected with calnexin.[26] Moreover, T cell recognition was also normal for certain human and mouse class I alleles in CEM.NK^R

We have investigated the assembly of class I in CEM.NK^R in more detail. When CEM and CEM.NK^R were infected with recombinant vaccinia virus encoding HLA-A2.1, the same amount of class I HC was synthesized in both cells and we did not observe differences in their rate of degradation. However, the total amount of class I that assembled with β2-m during the pulse-label was significantly higher in CEM. Less than half of A2.1 assembled in CEM.NK^R (25%) compared to 57% in CEM. Calreticulin is upregulated in CEM.NK^R by 50% (our unpublished observations), raising the possibility that it might be able to substitute for calnexin by binding to free HC in the absence of any competition from calnexin—perhaps overcoming low affinity with increased expression. However this cannot be the case, because the same relative difference in assembly was seen for the class I point mutant T134K, which does not associate with calreticulin.

A recent experiment by Paulsson et al has explored the relationship between calnexin association, β2-m association and HC degradation.[27] A pulse-labelled cohort of newly synthesized HC in a β2-m-negative cell line dissociated from calnexin with a half-life of 10 minutes (whereupon it was rapidly degraded), whereas in cells that expressed β2-m but lacked TAP, HC dissociation from calnexin had a half-time of 30 min. (whereupon it assembled with β2-m). None of the HC bound to calnexin was also bound to β2-m. This observation suggests therefore that binding of HC to β2-m protects it from intracellular degradation—allowing it to rebind to calnexin upon dissociation of β2-m thereby apparently prolonging the interaction between HC and calnexin. As we have already mentioned, binding of HC to calnexin also protects it from degradation. It is unclear whether HC binds to β2-m while it is associated with calnexin leading to its rapid dissociation, or whether β2-m binds only to free HC, but the overall effect seems to be to sequester free HC from proteolysis.

Peptide Optimization

It is possible that calnexin may play an indirect role in the optimization of class I molecules. Sadasivan et al have eluted peptides from MHC class I expressed on CEM and CEM.NK^R cells.[26] Their data showed differences in the HPLC profiles of peptides eluted from MHC class I molecules expressed on CEM and CEM.NK^R cells suggesting differential loading of peptides in calnexin-positive and calnexin-negative cells. We have also found an increase in cell-surface class I expression in cells that over express calnexin under an inducible promoter (our unpublished observations). How calnexin may mediate peptide optimization is unclear but as a calcium-binding protein, it may modulate the activity of calcium-dependent trimmases in the ER. ER resident gp96 has recently been identified as a calcium-dependent aminopeptidase capable of timming long precursor peptides in vitro.[28]

Role of Calreticulin in Class I Assembly

Calreticulin is an important chaperone involved in the second stage of class I assembly and binds only to class I HC that are assembled with β2-m. Calreticulin is a component of the loading complex, which includes TAP, tapasin and Erp57.

The availability of calreticulin-deficient mouse fibroblast cell line[29,30] has made it possible to dissect the precise role of calreticulin in the regulated assembly and loading of class I.

We[31] observed a reduction (50-70%) in the cell surface expression of class I in the absence of calreticulin despite the same amount of HC synthesis.[31] Class I assembly with β2-m was normal in the mutant cells, but they failed to become loaded with optimal peptides and were rapidly exported to the cell surface in a peptide-receptive state. These experiments indicate that calreticulin function is critical in the latter stages of class I assembly—during the optimization of peptide cargo. As a direct consequence of defective peptide loading, we observed impaired presentation of endogenously derived T cell epitopes to CTL in mutant cells, attesting to the physiological significance of this biochemical phenomenon.

The Exact Role That Calreticulin Plays in This Process Is Not Clear, Although There Are Several Possibilities:

Stabilizing the Loading Complex

It is possible that loading complexes that include calreticulin are more stable and therefore more effective in optimizing the peptide cargo of bound class I molecules. We have found that the absence of calreticulin does not prevent newly assembled class I molecules from associating with TAP:tapasin:ERp57,[31] but we cannot rule out a more subtle effect of calreticulin on stabilizing the complex. In support of this, castanospermine, which prevents generation of the monoglucosyl-glycan recognized by calreticulin inhibits the interaction between TAP and class I molecules.[8,32] Moreover, HLA-A2.1 molecules lacking the monoglucosyl-glycan recognized by calreticulin interact with TAP complex very poorly (Adhikari, manuscript in preparation). Furthermore, on close inspection of co-immunoprecipitates made from calreticulin negative and wild-type cells, we observed slightly lower incorporation of class I molecules in the TAP complex in mutant cells.

Assembly of Class I

Calreticulin may have a role in promoting the assembly of class I HC with β2-m. We have observed slightly better assembly of wild-type A2.1 (57%) compared to A2.1T134K (45%), which fails to interact with calreticulin, in CEM cells (our unpublished observation). Moreover, a slightly lower amount of heavy chain is recovered in calreticulin-deficient K42 cells[31] despite identical levels of mRNA. Recovery of heavy-chain levels could be achieved by culturing K42 cells with exogenously added, stabilizing peptides.

Peptide Loading

Calreticulin may be directly involved in the loading of peptides delivered through TAP and into the peptide-binding groove of class I, by acting as a peptide chaperone. There is evidence from peptide cross-linking experiments that calreticulin can bind to peptides delivered to the ER via TAP.[33] Moreover, two independent studies suggest the possibility that CD91 mediated uptake of immunogenic peptides bound to calreticulin, might be taken up by antigen presenting cell and loaded onto class I for presentation to T cells.[34,35] In addition, peptides eluted from calreticulin after its purification from tumor cells, also elicit T cell response.[35] It is possible, therefore, that calreticulin could chaperone peptides between TAP and class I, although there is no direct evidence for this and it is hard to reconcile this function with the fact that, in calreticulin-negative cells, class I molecules are transported to the cell surface faster than in wild type cells and not slower as might be predicted from this putative function.

Modulation of ER Enzyme Activities

Free ER calcium is reduced in calreticulin-deficient cells.[30] Moreover, free ER calcium has been shown to regulate activities of ER resident chaperones and other proteins.[36] Therefore, it is possible that through calcium regulation, calreticulin could modulate the activity of ER-resident proteases which are involved in modification or trimming of longer peptides within the ER. Interestingly, the ER resident chaperone grp96 has been identified as a calcium-dependent aminopeptidase capable of trimming a precursor T cell epitope in vitro.[28]

Optimization of Peptide Cargo

It is possible that calreticulin may act in concert with tapasin to effect peptide cargo optimization.[37,38] Peptide optimization could be achieved by the ability of calreticulin to retain class I molecules in the ER or retrieve them from the early secretory pathway (Fig. 9.3). Recycling is an important quality control step to ensure proper folding of several glycoproteins and is mediated by the recognition of either a cytoplasmic KKXX motif or a carboxy-terminal KDEL motif.[39] Class I molecules, however, lack these recycling recognition motifs. It is possible that class I that has lost its peptide ligand en route to the cell surface as a result of binding a suboptimal sequence could be retrieved to the ER by virtue of an interaction with calreticulin and the KDEL receptor. If this were the case, calreticulin probably has the ability to distinguish between a class I molecule that is "empty" and one that is sufficiently stable to exit to the cell surface. In light of this, it is to be noted that Li et al found that calreticulin predominantly associated with empty class I molecules while tapasin associated with both empty and peptide loaded class I molecules.[40]

Concluding Remarks

The role of lectin chaperones in MHC class I assembly illustrates the diverse functions of these proteins in glycoproteins biogenesis, and highlights the extent to which a division of labour can be achieved between calnexin and calreticulin. In terms of presenting foreign antigens to T cells, the different roles observed for calnexin and calreticulin can be assigned clear

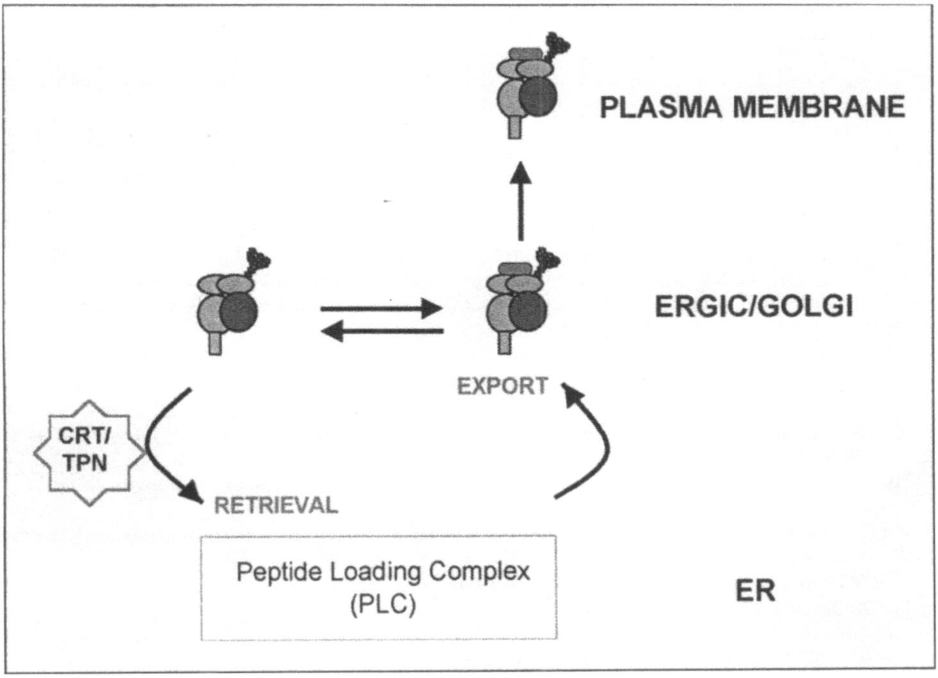

Figure 9.3. A model for calreticulin function in the optmization of peptide cargo bound to MHC class I molecules. It is possible that calreticulin functions to retrieve "empty" class I molecules to the ER if they lose their peptide-cargo en route to the cell surface (see text).

physiological purpose. Calnexin is involved primarily in the early folding stages and disulphide bond formation of class I HC. Calreticulin will not act as a substitute for this function even in cells that lack calnexin. Conversely, calreticulin is involved in the more specific aspects of loading HC:b2-m with high affinity peptide cargo. It is interesting to note that calreticulin associates with folded class I HC-β2m heterodimer and therefore recognizes conformational-sensitive determinants and not unfolded determinants. In this regard, it is different to calreticulin association during folding of vesicular stomatitis virus G (VSVG) protein and hemaglutinin (HA). This function appears to be specific for calreticulin since "soluble" forms of calnexin (i.e., calnexin in which the transmembrane domain is replaced by KDEL) cannot act as a substitute for this function even in cells that lack calreticulin.

The application of knowledge gained from studying the assembly of class I molecules will undoubtedly inform investigations into the roles of calnexin and calreticulin in the biogenesis of other proteins.

Acknowledgement

We would like to thank The Wellcome Trust for supporting our work.

References

1. Pamer E, Cresswell P. Mechanisms of MHC class I—restricted antigen processing. Annu Rev Immunol 1998; 16:323-358.
2. Tector M, Salter RD. Calnexin influences folding of human class I histocompatibility proteins but not their assembly with beta 2-microglobulin. J Biol Chem 1995; 270(33):19638-19642.
3. Farmery MR, Allen S, Allen AJ et al. The role of ERp57 in disulfide bond formation during the assembly of major histocompatibility complex class I in a synchronized semipermeabilized cell translation system. J Biol Chem 2000; 275(20):14933-14938.
4. Zapun A, Darby NJ, Tessier DC et al. Enhanced catalysis of ribonuclease B folding by the interaction of calnexin or calreticulin with ERp57. J Biol Chem 1998; 273(11):6009-6012.
5. Jackson MR, Cohen-Doyle MF, Peterson PA et al. Regulation of MHC class I transport by the molecular chaperone, calnexin (p88, IP90). Science 1994; 263(5145):384-387.
6. Wilson CM, Farmery MR, Bulleid NJ. Pivotal role of calnexin and mannose trimming in regulating the endoplasmic reticulum-associated degradation of major histocompatibility complex class I heavy chain. J Biol Chem 2000; 275(28):21224-21232.
7. Elliott T, Cerundolo V, Elvin J et al. Peptide-induced conformational change of the class I heavy chain. Nature 1991; 351(6325):402-406.
8. Sadasivan B, Lehner PJ, Ortmann B et al. Roles for calreticulin and a novel glycoprotein, tapasin, in the interaction of MHC class I molecules with TAP. Immunity 1996; 5(2):103-114.
9. Zhang Q, Salter RD. Distinct patterns of folding and interactions with calnexin and calreticulin in human class I MHC proteins with altered N-glycosylation. J Immunol 1998; 160(2):831-837.
10. Harris MR, Yu YY, Kindle CS et al. Calreticulin and calnexin interact with different protein and glycan determinants during the assembly of MHC class I. J Immunol 1998; 160(11):5404-5409.
11. Seong RH, Clayberger CA, Krensky AM et al. Rescue of Daudi cell HLA expression by transfection of the mouse beta 2- microglobulin gene. J Exp Med 1988; 167(2):288-299.
12. Claesson MH, Endel B, Ulrik J et al. Antibodies directed against monomorphic and evolutionary conserved self epitopes may be generated in 'knock-out' mice. Development of monoclonal antibodies directed against monomorphic MHC class I determinants. Scand J Immunol 1994; 40(2):257-264.
13. Perarnau B, Siegrist CA, Gillet A et al. Beta 2-microglobulin restriction of antigen presentation. Nature 1990; 346(6286):751-754.
14. Townsend A, Ohlen C, Bastin J et al. Association of class I major histocompatibility heavy and light chains induced by viral peptides. Nature 1989; 340(6233):443-448.
15. Williams A, Peh CA, Elliott T. The cell biology of MHC class I antigen presentation. Tissue Antigens 2002; 59(1):3-17.
16. Ortmann B, Androlewicz MJ, Cresswell P. MHC class I/beta 2-microglobulin complexes associate with TAP transporters before peptide binding. Nature 1994; 368(6474):864-867.
17. Nossner E, Parham P. Species-specific differences in chaperone interaction of human and mouse major histocompatibility complex class I molecules. J Exp Med 1995; 181(1):327-337.

18. Sugita M, Brenner MB. An unstable beta 2-microglobulin: major histocompatibility complex class I heavy chain intermediate dissociates from calnexin and then is stabilized by binding peptide. J Exp Med 1994; 180(6):2163-2171.
19. Oliver JD, van der Wal FJ, Bulleid NJ et al. Interaction of the thiol-dependent reductase ERp57 with nascent glycoproteins. Science 1997; 275(5296):86-88.
20. Rupp K, Birnbach U, Lundstrom J et al. Effects of CaBP2, the rat analog of ERp72, and of CaBP1 on the refolding of denatured reduced proteins. Comparison with protein disulfide isomerase. J Biol Chem 1994; 269(4):2501-2507.
21. Oliver JD, Roderick HL, Llewellyn DH et al. ERp57 functions as a subunit of specific complexes formed with the ER lectins calreticulin and calnexin. Mol Biol Cell 1999; 10(8):2573-2582.
22. Wang H, Capps GG, Robinson BE et al. Ab initio association with beta 2-microglobulin during biosynthesis of the H-2Ld class I major histocompatibility complex heavy chain promotes proper disulfide bond formation and stable peptide binding. J Biol Chem 1994; 269(35):22276-22281.
23. Williams DB, Watts TH. Molecular chaperones in antigen presentation. Curr Opin Immunol 1995; 7(1):77-84.
24. Vassilakos A, Cohen-Doyle MF, Peterson PA et al. The molecular chaperone calnexin facilitates folding and assembly of class I histocompatibility molecules. Embo J 1996; 15(7):1495-1506.
25. Scott JE, Dawson JR. MHC class I expression and transport in a calnexin-deficient cell line. J Immunol 1995; 155(1):143-148.
26. Sadasivan BK, Cariappa A, Waneck GL et al. Assembly, peptide loading, and transport of MHC class I molecules in a calnexin-negative cell line. Cold Spring Harb Symp Quant Biol 1995; 60:267-275.
27. Paulsson KM, Wang P, Anderson PO et al. Distinct differences in association of MHC class I with endoplasmic reticulum proteins in wild-type, and beta 2-microglobulin- and TAP- deficient cell lines. Int Immunol 2001; 13(8):1063-1073.
28. Menoret A, Li Z, Niswonger ML et al. An endoplasmic reticulum protein implicated in chaperoning peptides to major histocompatibility of class I is an aminopeptidase. J Biol Chem 2001; 276(36):33313-33318.
29. Mesaeli N, Nakamura K, Zvaritch E et al. Calreticulin is essential for cardiac development. J Cell Biol 1999; 144(5):857-868.
30. Nakamura K, Zuppini A, Arnaudeau S et al. Functional specialization of calreticulin domains. J Cell Biol 2001; 154(5):961-972.
31. Gao B, Adhikari R, Howarth M et al. Assembly and Antigen-Presenting Function of MHC Class I Molecules in Cells Lacking the ER Chaperone Calreticulin. Immunity 2002; 16(1):99-109.
32. Lewis JW, Sewell A, Price D et al. HLA-A*0201 presents TAP-dependent peptide epitopes to cytotoxic T lymphocytes in the absence of tapasin. Eur J Immunol 1998; 28(10):3214-3220.
33. Spee P, Neefjes J. TAP-translocated peptides specifically bind proteins in the endoplasmic reticulum, including gp96, protein disulfide isomerase and calreticulin. Eur J Immunol 1997; 27(9):2441-2449.
34. Basu S, Srivastava PK. Calreticulin, a peptide-binding chaperone of the endoplasmic reticulum, elicits tumor- and peptide-specific immunity. J Exp Med 1999; 189(5):797-802.
35. Nair S, Wearsch PA, Mitchell DA et al. Calreticulin displays in vivo peptide-binding activity and can elicit CTL responses against bound peptides. J Immunol 1999; 162(11):6426-6432.
36. Corbett EF, Michalak M. Calcium, a signaling molecule in the endoplasmic reticulum? Trends Biochem Sci 2000; 25(7):307-311.
37. Lehner PJ, Trowsdale J. Antigen presentation: coming out gracefully. Curr Biol 1998; 8(17):R605-608.
38. Peh CA, Burrows SR, Barnden M et al. HLA-B27-restricted antigen presentation in the absence of tapasin reveals polymorphism in mechanisms of HLA class I peptide loading. Immunity 1998; 8(5):531-542.
39. Yamamoto K, Fujii R, Toyofuku Y et al. The KDEL receptor mediates a retrieval mechanism that contributes to quality control at the endoplasmic reticulum. Embo J 2001; 20(12):3082-3091.
40. Li S, Paulsson KM, Sjogren HO et al. Peptide-bound major histocompatibility complex class I molecules associate with tapasin before dissociation from transporter associated with antigen processing. J Biol Chem 1999; 274(13):8649-8654.
41. Diedrich G, Bangia N, Pan M et al. A role for calnexin in the assembly of the MHC class I loading complex in the endoplasmic reticulum. J Immunol 2001; 166(3):1703-1709.

Chapter 10

Calreticulin and the Endoplasmic Reticulum in Plant Cell Biology

Paola Mariani, Lorella Navazio and Anna Zuppini

Abstract

Calreticulin is ubiquitously expressed in plants. The plant homologue shares with its animal counterpart a similar structural organization and basic functioning. A wide range of developmental and environmental stimuli differentially affect the expression of calreticulin in plant cells, highlighting its importance in cell physiology. Nevertheless, current knowledge on calreticulin's relevance in plant physiology is rather limited compared with animal systems. The contribution of the endoplasmic reticulum to Ca^{2+} homeostasis and signalling, and the multifunctional role of calreticulin in plant cellular events are rapidly emerging areas of study in plant biology.

Introduction

Plant calreticulin appeared late in the cell biology field. Early information on this protein involved biochemical characterization and DNA sequencing, mainly in different species of higher plants. It was rapidly established that calreticulin is ubiquitous in plant cells. All green organisms in the evolutionary tree, from algae to higher plants, express calreticulin. Within the complex body of higher plants, all cell types examined to date, both meristematic and mature, constitutively express calreticulin. Its relative abundance may be related to the greater extension of the endoplasmic reticulum in some specialized cells. Calreticulin is one of the most abundant proteins resident in the endoplasmic reticulum[1] and it is highly stable with a relatively long lifetime (half-time about 26 h).[2]

In comparison with the rapid growth of information coming from the animal world, which stresses an increasingly complex role for calreticulin in cell physiology, the acquisition of knowledge regarding plant calreticulin is proceeding more slowly. It is conceivable that calreticulin in plant cells has the same functions as in all eukaryotic cells. It is less easy to assign to calreticulin any role exclusive to plants, and to correlate calreticulin with specific plant metabolism and behavior. This results, in part, from the lack of conclusive knockout experiments and from the lack of any strong evidence supporting a specific function for calreticulin in plants. Nevertheless, some insights are now emerging. In particular, the largely common involvement of Ca^{2+} as a second messenger in regulating the interactions of plants with their environment highlights the crucial participation of the endoplasmic reticulum in Ca^{2+} homeostasis and signalling, either together with, or alternatively to, the vacuole.

Characteristics of Plant Calreticulin

Plant calreticulin does not strictly follow the rule "one protein, one gene" assessed for its animal counterpart[3] but is encoded by a small copy-number gene family.[4-8] Its amino acid sequence and molecular structure are highly conserved among plants (about 80% similarity),

Calreticulin, 2nd ed., edited by Paul Eggleton and Marek Michalak.

whereas the similarity between animal and plant calreticulins is somewhat lower, at about 50%. The three domain organization and the overall biochemical characteristics, in particular the Ca^{2+}-binding and lectin-binding properties, remain unchanged (castor bean calreticulin binds 15 mol Ca^{2+}/mol protein).[6] The major difference between animal and plant calreticulins is that potential N-glycosylation site(s) are actually occupied by glycan chain(s) in many plant species, and several potential phosphorylation consensus sequences for protein kinase CK2 ('casein kinase-2') are phosphorylated efficiently in plant calreticulin, in vitro.

N-Glycosylation

A comparison of available calreticulin sequences indicates that several, but not all, plant calreticulins have consensus site(s) for N-glycosylation. The most conserved N-glycosylation site is located at position 32 in the N-domain. Additional N-glycosylation sites in the same region are present in *Prunus armeniaca, Ricinus communis, Beta vulgaris, Nicotiana tabacum, Brassica napus* and *Arabidopsis thaliana*. Moreover, calreticulin from *Arabidopsis* shows a third consensus site for N-glycosylation in the C-domain. Distinctively, in calreticulin from the algae *Euglena* and *Chlamydomonas* the N-glycosylation consensus sites are lacking. Evidence from N-glycan structural analyses,[9,10] endoglycosidase H sensitivity,[2,10] and Concanavalin A binding[5,11,12] suggest that the N-glycans have a high mannose structure, compatible with localization of the protein in the endoplasmic reticulum. It is not known why some (but not all) plant calreticulins are N-glycosylated and what functional role can be assigned to the glycosylation. It may represent an additional property that favors calreticulin's folding during biosynthesis, and increases its stability.[13] Clearly, the N-linked glycan chain(s) should not hinder calreticulin from acquiring its correct three-dimensional structure or from binding specific substrates at its lectin domain.[14]

The detectable presence, in *Liriodendron tulipifera* L. ovary, of calreticulin glycoforms bearing complex carbohydrate chains suggests that in this species calreticulin can travel up to the medial and trans-Golgi where the protein acquires specific sugar residues. Evidence for this traffic comes from immunodetection assays with anti-(1,2)xylose antibodies (Fig. 10.1; Faye and Fitchette-Lainè, personal communication). The monosaccharide composition of the N-linked glycan chains of *L. tulipifera* calreticulin has been recently investigated (Navazio et al. 2002, note added in proof). Both tobacco[2] and maize[10] calreticulins have been shown to acquire competence for N-glycan maturation inside the Golgi compartment when treatment with brefeldin A induces redistribution of Golgi enzymes into the endoplasmic reticulum. These results indicate that calreticulin N-glycans are accessible to glycan-processing enzymes resident in the Golgi. The limited amount of data so far available does not allow an evaluation of the extent of the actual in vivo occurrence of complex glycan chain(s) on plant calreticulin.

Phosphorylation

It has been demonstrated that calreticulin from spinach leaves and *L. tulipifera* ovary is phosphorylated, in vitro, by both exogenous and endogenous protein kinase CK2. The optimal consensus sites for phosphorylation by CK2 are located mainly in the C-terminus.[15] Unpublished observations from our group indicate a similar behavior by calreticulin from tobacco and carrot suspension cultured cells and tobacco pollen tubes. Under the same experimental conditions, calreticulin from *Euglena*[16] and *Chlamydomonas*[17] is not a substrate for CK2; in both these algal calreticulins the potential phosphorylation sites are hindered by either basic or proline residues in proximity to the phospho-acceptor residue. Results to date suggest that phosphorylation by CK2 is limited to calreticulin in higher plants. Currently, there is no evidence for in vivo phosphorylation of calreticulin by CK2, and CK2 has not been localized within the plant endoplasmic reticulum. However, Cala[18] has reported that in insect cells the reticuloplasmin GRP94, which is a substrate for CK2 in vitro, is also phosphorylated by this kinase in vivo. His data support the possibility that a CK2 isoform or a CK2-like protein is localized in the endoplasmic reticulum, strengthening the suggestion that phosphorylation of reticuloplasmins by CK2 may have physiological significance.

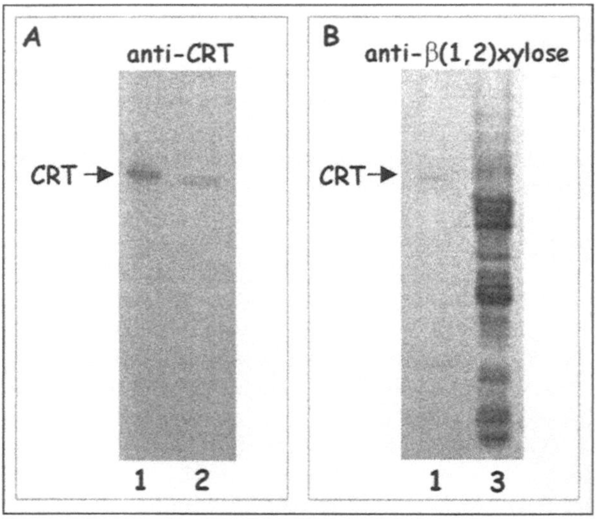

Figure 10.1. Immunodetection with anti-spinach calreticulin (A) and anti-β(1,2)xylose (B) antibodies on protein extracts from *Liriodendron tulipifera* ovary (lane 1, 20 μg), spinach leaves (lane 2, 0.5 μg) and bean seeds (lane 3, 50 μg). CRT, calreticulin.

The possible relevance of calreticulin phosphorylation by CK2 (if confirmed in vivo) is at the moment purely speculative. Phosphorylation events could control and modulate the biological activity of the protein and/or be involved in the complex cellular signalling network. Droillard et al[19] have demonstrated that the phosphorylation of tobacco calreticulin is modulated both in vitro and in vivo during signalling induced by elicitors such as cell wall pectic fragments. However, they did not identify which protein kinase(s) is responsible for the phosphorylation of calreticulin. In vitro experimental evidence indicates that phosphorylation of calreticulin by CK2 is significantly reduced at Ca^{2+} concentrations which nearly fully saturate the binding capacity of calreticulin without affecting normal CK2 activity (Baldan et al unpublished results). These results suggest that the Ca^{2+}-binding activity of calreticulin could be negatively regulated through possible conformational control of its C-terminal tail, where both the phosphorylation sites and the low affinity, high capacity Ca^{2+}-binding sites are located.

Intracellular Localization of Calreticulin

The retention of calreticulin in the endoplasmic reticulum largely depends on the C-terminal specific retention/retrieval signal. All plant calreticulins so far cloned contain the HDEL sequence, with the exception of *Euglena* calreticulin which has KDEL.[16] In plant cells, as in yeast and mammalian cells, proteins that are resident in the endoplasmic reticulum can exit and recycle back via the K/HDEL-dependent retrieval mechanism, which is mediated by the membrane-bound ERD2 receptor.[20] Recently, calreticulin has been detected in COPI-coated vesicles,[21] confirming that an efficient mechanism for the retrieval of endoplasmic reticulum proteins from the Golgi compartment functions in plant cells. Apparently calreticulin can become competent for export from the endoplasmic reticulum, since a form of calreticulin, minus the HDEL motif, is transported in a COPII-dependent anterograde pathway.[22] The ability of reticuloplasmins to interact and complex with other proteins resident in the endoplasmic reticulum, forming a large network, is considered to be partially responsible for their retention in this compartment. Calreticulin has been found in association with BiP in tobacco cells. These proteins form stable complexes with different molecular weights, in a Ca^{2+}-independent way, probably with the participation of other reticuloplasmins.[23]

There are some indications that in plant cells endoplasmic reticulum domains can be specifically enriched in calreticulin. Recently, using immunocytochemistry in maize roots, calreticulin has been localized to the plasmodesmata, gateable cell-to-cell cytoplasmic channels which are unique to the plant body.[24,25] These structures, frequently grouped into pit fields, span the cell wall, are lined by the plasma membrane and contain a cytoplasmic sleeve, coaxial to a central endoplasmic reticulum strand (desmotubule). A cytoplasmic continuum between adjacent cells is established through the plasmodesmata, and cell-to-cell communications are made possible. Endoplasmic reticulum is also continuous between cells, but along the entire length of the plasmodesmata the endoplasmic reticulum membranes are appressed and the lumen eliminated: the desmotubule is essentially a solid strand of lipids. Transport through the plasmodesmata is supposed to be regulated by an actin-myosin-based mechanism. The enrichment, in calreticulin, of the endoplasmic reticulum elements that are associated with the plasmodesmata, and its co-localization with myosin VIII, is suggestive that calreticulin participates in regulating plasmodesmal gating through modulation of local Ca^{2+} levels.[25] Further support for this notion comes from the demonstration that Ca^{2+} is involved in the regulation of plasmodesmata permeability: the elevation of cytosolic Ca^{2+} concentration ($[Ca^{2+}]_{cyt}$) that results from cold shock[26] or mastoparan[27] induces a rapid closure of plasmodesmata. In tip-growing cells, such as pollen tubes and root hairs, a tip-high gradient of $[Ca^{2+}]_{cyt}$ is generated and maintained during polarized cell growth. The gradient is regulated by influx of Ca^{2+} through channels located on the apical plasma membrane and by Ca^{2+} sequestration in endoplasmic reticulum elements that act as internal Ca^{2+} buffering stores. Calreticulin has been found to accumulate in the apical zone of maize growing root hairs[28] and of *Petunia* pollen tubes,[29] where the endoplasmic reticulum is very abundant and densely arranged.

Localization of calreticulin outside the endoplasmic reticulum has been reported in plant cells, specifically in the Golgi compartment (with a high abundance), in several small patches on the plasma membrane of *Nicotiana plumbaginifolia* protoplasts,[30] and in protein bodies/ protein storage vacuoles of rice endosperm cells.[31] Calreticulin is totally absent from the vacuole, the major Ca^{2+} store in plant cells.[32]

Inducible Expression of Calreticulin

In plants, a number of different stimuli have been found to increase endogenous basal levels of expression of both calreticulin mRNA and protein. Moreover, calreticulin is regulated at the transcriptional level during different developmental stages of the life cycle.

The first suggestion that calreticulin may be important during fertilization events came from Chen et al[4] who observed increased expression of the calreticulin gene in barley ovaries one day after pollination and during the early stages of embryogenesis. In subsequent developmental steps, the level of calreticulin mRNA returns to that in unpollinated ovaries. Similar data have been obtained in tobacco,[1] maize,[33,34] *Arabidopsis thaliana,*[7] *Ricinus communis*[6] and *N. plumbaginifolia.*[30] The high levels of calreticulin observed in maize cells after fertilization, and in the immature embryos and floral tissues of tobacco, *Arabidopsis* and *Ricinus*, highlights modulation of calreticulin expression during plant reproduction. Northern blot assays have shown elevated expression of calreticulin mRNA during the early developmental stages of somatic embryos and ovules after fertilization of *N. plumbaginifolia.*[30] Moreover, the unicellular chlorophyte *Chlamydomonas reinhardtii* shows enhanced levels of both calreticulin transcripts and protein in gametes compared with vegetative cells.[17] It is unknown how calreticulin participates in plant reproductive processes, but cytosolic Ca^{2+} fluxes recognized to occur during animal reproduction enable the suggestion that calreticulin may act as a Ca^{2+}-buffer in regulating $[Ca^{2+}]_{cyt}$. Indeed, during gamete differentiation *Chlamydomonas* cells accumulate Ca^{2+} in intracellular stores[35] and maize sperm cells exposed to Ca^{2+} rapidly internalize the ion.[34] Thus, Ca^{2+} seems to be necessary for gamete activation. Interestingly, during the differentiation of *Chlamydomonas* gametes it is possible to distinguish a pre-gamete phase when the cells are not able to mate (induced by a withdrawal of the nitrogen source), in which over-expression of calreticulin occurs with a simultaneous increase in BiP mRNA and protein. In the mature gametes, despite a further increase in expression of calreticulin, BiP expression remains constant.[17] The finding that a transient cytosolic Ca^{2+} increase triggers plant post-fertilization phases[36] and that this corresponds with an up-regulation of the calreticulin gene, suggests important role(s) for calreticulin in both pre-fertilization and post-fertilization events in plant sexual reproduction.

The up-regulation of calreticulin expression is not limited to reproductive processes. An induction of calreticulin gene expression has been observed in proliferating and secreting tissues.[33,6,7] Significant accumulation of calreticulin transcripts has been found in meristematic regions such as root tips, nodes and leaf base,[1,33,6] suggesting a possible role for calreticulin in plant cell division. Moreover, the high abundance of calreticulin mRNA in cells that are active in secretion has led to the hypothesis that calreticulin probably acts as a molecular chaperone in assisting the assembly of newly synthesized enzymes and/or secreted (glyco)proteins.[1,6,7,30]

The expression of the calreticulin gene in plant cells can be affected by different stresses. Denecke et al[1] and Borisjuk et al[30] demonstrated modulation of calreticulin expression by treatment with exogenous phytohormones. Barley aleurone cells treated with gibberellic acid have enhanced levels of the calreticulin transcript.[1] *N. plumbaginifolia* cells show auxin-dependent changes in the amount of calreticulin,[30] with increased protein expression in the presence of α-naphthaleneacetic acid and decreased transcript levels in the presence of 2,4-dichlorophenoxyacetic acid. Furthermore, studies on the regeneration of rice cultured suspension cells have shown increased transcription of the calreticulin gene related to the growth factors naphthaleneacetic acid and 6-benzyladenine.[8] Interestingly, tunicamycin treatment of tobacco cells does not affect calreticulin expression, whereas BiP and PDI transcriptional levels are enhanced.[1]

Calreticulin has been implicated in signalling pathways specific to plants, such as the differential growth linked to the perception of gravity.[37] Gravistimulation in maize plants has been found to induce several-fold increase in calreticulin and calmodulin transcripts, which preferentially accumulate in the stem pulvinus cells induced to respond to the gravity stimulus. An increased recruitment of calreticulin and calmodulin transcripts onto polyribosomes has also been observed, implying increased synthesis of these proteins and suggesting a role for them during the early stages of the gravity response.[37] Pathogen attack causes plant defence responses which aim to combat the invader and prevent further invasion. This process occurs with the production of a range of defence-related proteins, most synthesized in the rough endoplasmic reticulum. It has been shown that expression of some lumenal proteins of the endoplasmic reticulum, including calreticulin, is induced during plant-pathogen interaction, probably as an early response necessary to enable the synthesis of pathogenesis-related proteins.[1,38] Since both gravitropism and pathogen-activated signalling are known to be mediated by calcium ions,[39,40] it is conceivable that the role played by calreticulin can be attributed to its potential to affect cellular Ca^{2+} homeostasis.

Altered growth conditions can interfere with endoplasmic reticulum functions leading to the up-regulation of genes encoding endoplasmic reticulum proteins. In maize roots, mannitol-induced osmotic stress and aluminum treatment both cause the deposition of callose at plasmodesmata pit fields with an increased expression of calreticulin at the sites of callose deposition.[24,41] Calreticulin Ca^{2+} buffering and signalling might then be essential for structural and functional properties of plant cell plasmodesmata.

Overall, the spatial and temporal analyses of calreticulin's expression pattern highlight its importance as both a chaperone and a Ca^{2+}-buffering and signalling protein.

Endoplasmic Reticulum in Plant Cell Physiology

Quality Control

As in all eukaryotic cells, plant endoplasmic reticulum provides a specialized environment promoting the folding, oxidation and oligomeric assembly of proteins. Plant endoplasmic reticulum is equipped with several folding enzymes, molecular chaperones, and folding sensors largely similar to those operating in all eukaryotes. The endoplasmic reticulum enables newly synthesized and properly folded proteins to access subsequent steps of the secretory pathway.

A very efficient quality control system that inhibits export of incompletely folded or misfolded proteins from the endoplasmic reticulum is active in plants. For example, endoplasmic reticulum quality control is involved in the proper maturation of phaseolin, the vacuolar storage glycoprotein of the common bean. When correctly assembled in a trimeric form, phaseolin is targeted to the vacuole. However, a defectively assembled form of phaseolin remains confined to the endoplasmic reticulum, extensively associated with BiP, and is eventually degraded.[42] Similar evidence comes from experiments with a mutated form of the pea storage protein vicilin,[43] and with zein polypeptides expressed in transgenic plants.[44]

The unfolded protein response results in the transcriptional up-regulation of a set of endoplasmic reticulum chaperones, and some other target genes. This response may also be triggered in plants: a variety of stresses leading to the accumulation of misfolded proteins in the endoplasmic reticulum increase the transcription of the BiP gene. Furthermore, over-expression of BiP in tobacco cells mitigates the endoplasmic reticulum stress.[45]

Among endoplasmic reticulum chaperones that have been identified in plants, BiP is one of the best characterized:[46] its function(s) under both normal growth conditions and endoplasmic reticulum stress are well documented. A possible role as a molecular chaperone has also been attributed, in several circumstances, to calreticulin (see above in this chapter). In fact, the activity of the non classical chaperones[47] (calreticulin and its related partner calnexin) is linked to their lectin binding properties. Evidence for the calnexin/calreticulin cycle in glycoprotein folding, which is so well characterized in mammals,[14] is only circumstantial in plant cells: assembly of phaseolin is affected by its degree of glycosylation, as shown by the faster assembly rate observed when glucose-trimming by endoplasmic reticulum glucosidases is inhibited.[48]

Plant Endoplasmic Reticulum As a Ca^{2+} Store

The plant cell has several potential sites for Ca^{2+} accumulation (Fig. 10.2), although not all of them can be considered to be rapidly exchangeable Ca^{2+} pools.

In the cell wall, Ca^{2+}-binding sites are mainly located on the pectic polymers and the $[Ca^{2+}]$ is estimated to be in the millimolar range.[49] The $[Ca^{2+}]$ in chloroplasts and in the nucleus is controlled independently of the Ca^{2+} level in the cytosol: both chloroplasts and the nucleus generate their own Ca^{2+} signals, which are expected to regulate Ca^{2+}-dependent processes within the two compartments.[50,51,52] Information about the contribution of plant mitochondria to the Ca^{2+} network is still very scarce. Nevertheless, the recent imaging in animal cells of close contacts between mitochondria and the endoplasmic reticulum[53] opens up the possibility that even in plant cells the spatial distribution of these organelles may allow microdomains of Ca^{2+} sensing.

The vacuole has, so far, been considered as the main intracellular Ca^{2+} store because of its large volume and key role in ion homeostasis in the plant cell. Evidence for several Ca^{2+} transporters and Ca^{2+} release channels in the vacuolar membrane (for a review see ref. 54) has further reinforced appreciation of the vacuole as a major stimulus-releasable reservoir of Ca^{2+}. In the vacuole, owing to the low pH (pH 3-6) of the vacuolar sap, the Ca^{2+} buffering role may be carried out by organic or inorganic ions, and/or by Ca^{2+}-binding proteins with properties different from Ca^{2+}-binding reticuloplasmins. A low affinity, high capacity Ca^{2+}-binding protein has been recently characterized in radish vacuole; its deduced amino acid sequence does not show any significant similarity with either calreticulin or other Ca^{2+}-binding proteins. In

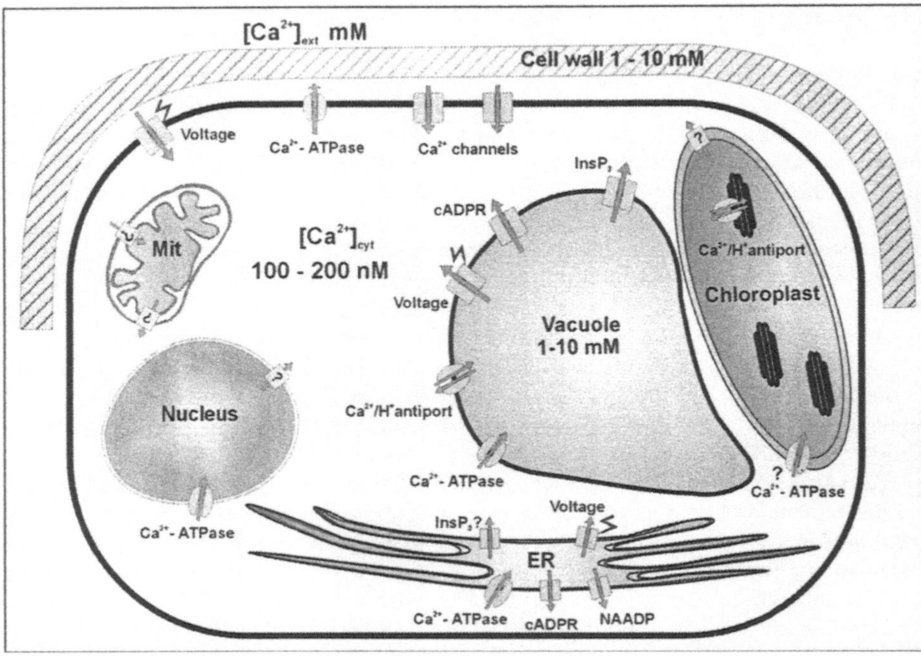

Figure 10.2. Calcium stores in plant cells.

view of its properties, this protein can be considered as a good candidate for Ca^{2+} buffering in the vacuole.[55]

Alongside the vacuole, the endoplasmic reticulum is increasingly being seen as an intracellular Ca^{2+} store that plays a potentially important role in Ca^{2+} signalling in plants. Ca^{2+} ATPases, inhibited by cyclopiazonic acid but not by thapsigargin, and differentially regulated by calmodulin[56,57] are located in plant endoplasmic reticulum membranes. Different classes of Ca^{2+} permeable channels have been reported in plant endoplasmic reticulum, i.e., voltage-gated[58,59] and ligand-gated, activated by the pyridine nucleotide derivatives nicotinic acid adenine dinucleotide phosphate (NAADP)[60] and cyclic ADP-ribose (cADPR)[61] and, possibly, by inositol 1,4,5-trisphosphate ($InsP_3$).[62] The occurrence of multiple Ca^{2+} release pathways suggests that the endoplasmic reticulum is not just a Ca^{2+} repository for the plant cell, but can be implicated in cell signalling as a mobilizable Ca^{2+} store. In keeping with this, plant endoplasmic reticulum contains calreticulin, an effective Ca^{2+} buffer that may allow the transient storage of the ion and its prompt mobilization when Ca^{2+} release is triggered.

Endoplasmic reticulum membranes form a dynamic, three-dimensional network, the distribution of which within the cell may fulfil localized requests for Ca^{2+}. In plant cells the cortical endoplasmic reticulum, i.e., the endoplasmic reticulum underlying the plasma membrane, is highly developed and may function as a semi-immobile polygonal network along which movement of the Golgi stacks are driven by actin cables.[63] Reuzeau et al[64] have proposed that plant cortical endoplasmic reticulum physically attaches to the plasma membrane at adhesion sites through cytoskeletal proteins and transmembrane integrin-like proteins. Ion channels and signal receptors may also be clustered around these adhesion sites. Indeed, the close proximity and association between the cortical endoplasmic reticulum and the plasma membrane would allow ready access to signals emanating from the plasma membrane.

The source and/or location of Ca^{2+} signals helps to determine their specificity. In aequorin-transformed tobacco seedlings, signalling induced by cold shock triggers Ca^{2+} fluxes primarily at the plasma membrane, whereas mechanical stimulation involves elevations in $[Ca^{2+}]$

which derive from intracellular Ca^{2+} stores.[65,66] The conclusive assessment of the direct participation of the endoplasmic reticulum in specific signal transduction pathways awaits accurate and reliable measurements of the Ca^{2+} concentration in the lumen of the endoplasmic reticulum ($[Ca^{2+}]_{ER}$) and of its variations during signalling.

Plant endoplasmic reticulum may be involved in the generation of Ca^{2+} oscillations in some specialized cell types, such as stomata guard cells and pollen tubes, in response to a wide range of stimuli.[67] Repetitive Ca^{2+} release and Ca^{2+} re-uptake by the endoplasmic reticulum have been proposed to generate repetitive Ca^{2+} spikes in a unicellular green alga.[68,69] In contrast to animal cells, there are only a few examples of Ca^{2+} waves in plants.[54] In all cases, Ca^{2+} waves have been observed to propagate through regions containing endoplasmic reticulum but no large vacuoles, suggesting that the presence of a huge central vacuole in many plant cells may hamper propagation and detection of waves of elevated $[Ca^{2+}]$.[70]

Calreticulin and Ca^{2+} Signalling

The interrelationships between calreticulin and Ca^{2+} in the endoplasmic reticulum have been deeply investigated in animal cells. The emerging picture is that of a complex sensing-signalling network involving more than one role for calreticulin, including a lectin-like chaperone activity, interactions with other endoplasmic reticulum chaperones, regulation of $[Ca^{2+}]_{ER}$ and participation in the endoplasmic reticulum signalling network.[71] Nothing, or very little, is known about these issues in plant cells. Although the Ca^{2+} binding properties of plant calreticulin suggest a potential role in intracellular Ca^{2+} homeostasis, as in animal cells, conclusive evidence that calreticulin affects the Ca^{2+} status of the plant endoplasmic reticulum has only recently been obtained. Persson et al[72] demonstrated that the over-expression of calreticulin in tobacco suspension cells affects the endoplasmic reticulum Ca^{2+} pool. Elevation of the calreticulin level, in microsomes enriched with endoplasmic reticulum membranes, resulted in increased ATP-dependent Ca^{2+} accumulation, and in increased Ca^{2+} release and Ca^{2+} retention after ionophore treatment. At present it is not known whether this effect is exerted via modulation of the activity of the endoplasmic reticulum Ca^{2+} ATPases and/or agonist-triggered Ca^{2+} channels, as shown in animal cells.[73,74]

Over-expression of calreticulin *in planta* enhances the survival of transgenic plants grown in a limiting, low Ca^{2+} medium.[72] Furthermore, expression of the C-domain of calreticulin (targeted to the endoplasmic reticulum) in *Arabidopsis* enhances survival of seedlings on Ca^{2+}-depleted medium[75], supporting the hypothesis that the key factor helping cells to maintain their Ca^{2+} homeostasis under altered growth conditions is an increased Ca^{2+} buffering ability stemming from over-production of calreticulin.

Although calreticulin is highly conserved, constitutively present and ubiquitously distributed, the roles of the protein in plant cells have not been fully elucidated. However, despite the fundamental differences between plants and animals in their cellular organization, body plan and life style, a convergence of the physiological behavior of calreticulin as a multifunctional player in the eukaryotic kingdom is increasingly becoming apparent.

Note Added in Proof

The results concerning the characterization of glycan chains of *L. tulipifera* calreticulin, reported on page 96, have been recently published: Navazio L, Miuzzo M, Royle L et al. Monitoring endoplasmic reticulum-to-Golgi traffic of a plant calreticulin by protein glycosylation analysis. Biochemistry 2002; 41:14141-14149.

Acknowledgements

We are grateful to L. Faye (Mt. St. Aignan, France) for making available results on our collaborative work and to F. Meggio (Padova, Italy) for helpful discussion on calreticulin phosphorylation. Research in the authors' laboratory is supported by grants from Ministero dell' Universitá e della Ricerca Scientifica e Tecnologica.

References

1. Denecke J, Carlsson LE, Vidal S et al. The tobacco homolog of mammalian calreticulin is present in protein complexes in vivo. Plant cell 1995; 7:391-406.
2. Crofts AJ, Leborgne-Castel N, Hillmer S et al. Saturation of the endoplasmic reticulum retention machinery reveals anterograde bulk flow. Plant cell 1999; 11:2233-2247.
3. Michalak M, Corbett EF, Mesaeli N et al. Calreticulin: one protein, one gene, many functions. Biochem J 1999; 344:281-92.
4. Chen F, Hayes PM, Mulrooney DM et al. Identification and characterization of cDNA clones encoding plant calreticulin in barley. Plant cell 1994; 6:835-843.
5. Napier RM, Trueman S, Henderson J et al. Purification, sequencing and functions of calreticulin from maize. J Exp Bot 1995; 46:1603-1613.
6. Coughlan SJ, Hastings C, Winfrey R. Cloning and characterization of the calreticulin gene from Ricinus communis L. Plant Mol Biol 1997; 34:897-911.
7. Nelson DE, Glaunsinger B, Bohnert HJ. Abundant accumulation of the calcium-binding molecular chaperone calreticulin in specific floral tissues of Arabidopsis thaliana. Plant Physiol 1997; 114:29-37.
8. Li Z, Komatsu S. Molecular cloning and characterization of calreticulin, a calcium-binding protein involved in the regeneration of rice cultured suspension cells. Eur J Biochem 2000; 267:737-745.
9. Navazio L, Baldan B, Mariani P et al. Primary structure of N-linked carbohydrate chains of calreticulin from spinach leaves. Glycoconjugate J 1996; 13:977-983.
10. Pagny S, Cabanes-Macheteau M, Gillikin JW et al. Protein recycling from the Golgi apparatus to the endoplasmic reticulum in plants and its minor contribution to calreticulin retention. Plant cell 2000; 12:739-755.
11. Navazio L, Baldan B, Dainese P et al. Evidence that spinach leaves express calreticulin but not calsequestrin. Plant Physiol 1995; 109:983-990.
12. Navazio L, Sponga L, Dainese P et al. The calcium binding protein calreticulin in pollen of Liriodendron tulipifera L. Plant Sci 1998; 131:35-42.
13. Helenius A, Aebi M. Intracellular functions of N-linked glycans. Science 2001; 291:2364-2369.
14. Ellgaard L, Helenius A. ER quality control: towards an understanding at the molecular level. Curr Opin Cell Biol 2001; 13:431-437.
15. Baldan B, Navazio L, Friso A et al. Plant calreticulin is specifically and efficiently phosphorylated by protein kinase CK2. Biochem Biophys Res Commun 1996; 221:498-502.
16. Navazio L, Nardi MC, Pancaldi S et al. Functional conservation of calreticulin in Euglena gracilis. J Euk Microbiol 1998; 45:307-313.
17. Zuppini A, Barbato R, Bergantino E et al. Ca^{2+} binding protein calreticulin in Chlamydomonas reinhardtii (Chlorophyta): biochemical characterization, differential expression during sexual reproduction, and phylogenetic analysis. J Phycol 1999; 35:1224-1232.
18. Cala SE. GRP94 hyperglycosylation and phosphorylation in Sf21 cells. Biochem Biophys Acta 2000; 1496:296-310.
19. Droillard MJ, Güclü J, Le Caer J-P et al. Identification of calreticulin-like protein as one of the phosphoproteins modulated in response to oligogalacturonides in tobacco cells. Planta 1997; 202:341-348.
20. Sanderfoot AA, Raikhel NV. The specificity of vesicle traffiking: coat proteins and SNAREs. Plant cell 1999; 11:629-641.
21. Pimpl P, Movafeghi A, Coughlan S et al. In situ localization and in vitro induction of plant COPI-coated vesicles. Plant cell 2000; 12:2219-2235.
22. Phillipson BA, Pimpl P, Pinto daSilva LL et al. Secretory bulk flow of soluble proteins is efficient and COPII dependent. Plant cell 2001; 13:2005-2020.
23. Crofts AJ, Leborgne-Castel N, Pesca M et al. BiP and calreticulin form an abundant complex that is independent of endoplasmic reticulum stress. Plant cell 1998; 10:813-823.
24. Baluska F, Samaj J, Napier R et al. Maize calreticulin localizes preferentially to plasmodesmata in root apex. Plant J 1999; 19:481-488.
25. Baluska F, Cvrcková F, Kendrick-Jones J et al. Sink plasmodesmata as gateways for phloem unloading. Myosin VIII and calreticulin as molecular determinants of sink strength? Plant Physiol 2001; 126:39-46.
26. Holdaway-Clarke TL, Walker NA, Hepler PK et al. Physiological elevations in cytoplasmic free calcium by cold or iron injection result in transient closure of higher plant plasmodesmata. Planta 2000; 210:329-335.
27. Tucker EB, Boss WF. Mastoparan-induced intracellular Ca^{2+} fluxes may regulate cell-to-cell communication in plants. Plant Physiol 1996; 111:459-467.

28. Baluska F, Salaj J, Mathur J et al. Root hair formation: F-actin-dependent tip growth is initiated by local assembly of profilin-supported F-actin meshworks accumulated within expansin-enriched bulges. Dev Biol 2000; 227:618-632.
29. Lenartowska M, Karas K, Marshall J et al. Immunocytochemical evidence of calreticulin-like protein in pollen tubes and styles of Petunia hybrida Hort. Protoplasma 2002; 219:23-30.
30. Borisjuk N, Sitailo L, Adler K et al. Calreticulin expression in plant cells: developmental regulation, tissue specificity and intracellular distribution. Planta 1998; 206:504-514.
31. Torres E, Gonzales-Melendi P, Stöger E et al. Native and artificial reticuloplasmins co-accumulate in distinct domains of the endoplasmic reticulum and in post-endoplasmic reticulum compartments. Plant Physiol 2001; 127:1212-1223.
32. Opas M, Tharin S, Milner RE et al. Identification and localization of calreticulin in plant cells. Protoplasma 1996; 191:164-171.
33. Dresselhaus T, Hagel C, Lörz H et al. Isolation of a full-length cDNA encoding calreticulin from a PCR library of in vitro zygotes of maize. Plant Mol Biol 1996; 31:23-34.
34. Williams CM, Zhang G, Michalak M et al. Calcium-induced protein phosphorylation and changes in levels of calmodulin and calreticulin in maize sperm cells. Sex Plant Reprod 1997; 10:83-88.
35. Harris HH. The Chlamydomonas sourcebook: a comprehensive guide to biology and laboratory use. Harcourt Brace Jovanovich, eds.Academic Press Inc. S. Diego, 1989.
36. Faure J-E. Double fertilization in flowering plants: discovery, study methods and mechanisms. Life Sci 2001; 324:551-558.
37. Heilmann I, Shin J, Huang J et al. Transient dissociation of polyribosomes and concurrent recruitment of calreticulin and calmodulin transcripts in gravistimulated maize pulvini. Plant Physiol 2001; 127: 1193-1203.
38. Jelitto-Van Dooren EPWM, Viadl S, Denecke J. Anticipating endoplasmic reticulum stress: a novel early response before pathogenesis-related gene induction. Plant cell 1999; 11:1935-1943.
39. Sinclair W, Trewavas AJ. Calcium in gravitropism: a re-examination. Planta 1997; 203:S85-S90.
40. John M, Röhring H, Shmidt J et al. Cell signalling by oligosaccharides. Trends Plant Sci 1997; 2:111-115.
41. Sivaguru M, Fujiwara T, Samaj J et al. Aluminum-induced 1→3-β-D-glucan inhibits cell-to-cell traffiking of molecules through plasmodesmata. A new mechanism of aluminum toxicity in plants. Plant Physiol 2000; 124:991-1005.
42. Pedrazzini E, Giovinazzo G, Bielli A et al. Protein quality control along the route to the plant vacuole. Plant cell 1997; 9:1869-1880.
43. Kermode AR, Fisher SA, Polishchuk E et al. Accumulation and proteolytic processing of vicilin deletion-mutant proteins in the leaf and seed of transgenic tobacco. Planta 1995; 197:501-513.
44. Coleman CE, Herman EM, Takasaki K et al. The maize γ-zein sequesters α-zein and stabilizes its accumulation in protein bodies of transgenic tobacco endosperm. Plant cell 1996; 8:2335-2345.
45. Leborgne-Castel N, Jelitto-Van Dooren EPWM, Crofts AJ et al. Overexpression of BiP in tobacco alleviates endoplasmic reticulum stress. Plant cell 1999; 11:459-469.
46. Pedrazzini E, Vitale A. The binding protein (BiP) and the synthesis of secretory proteins. Plant Physiol Biochem 1996; 34:207-216.
47. Chevet E, Cameron PH, Pelletier MF et al. The endoplasmic reticulum: integration of protein folding, quality control, signaling and degradation. Curr Opin Struct Biol 2001; 11:120-124.
48. Lupattelli F, Pedrazzini E, Bollini R et al. The rate of phaseolin assembly is controlled by the glucosylation state of its N-linked oligosaccharide chains. Plant cell 1997; 9:597-609.
49. Trewavas AJ, Malhó R. Ca^{2+} signalling in plant cells: the big network! Curr Opin Plant Biol 1998; 1: 428-433.
50. Johnson CH, Knight MR, Kondo T et al. Circadian oscillations of cytosolic and chloroplastic free calcium in plants. Science 1995; 269:1863-1865.
51. van der Luit AH, Olivari C, Haley A et al. Distinct calcium signaling pathways regulate calmodulin gene expression in tobacco. Plant Phys 1999; 121:705-714.
52. Pauly N, Knight MR, Thuleau P et al. Control of free calcium in plant cell nuclei. Nature 2000; 405:754-755.
53. Rizzuto R, Pinton P, Carrington W et al. Close contacts with the endoplasmic reticulum as determinants of mitochondrial Ca^{2+} responses. Science 1998; 280:1763-1766.
54. Sanders D, Brownlee C, Harper JF. Communicating with calcium. Plant cell 1999; 11:691-706.
55. Yuasa K, Maeshima M. Purification, properties, and molecular cloning of a novel Ca^{2+}-binding protein in radish vacuoles. Plant Physiol 2000; 124:1069-1078.
56. Liang F, Sze H. A high affinity Ca^{2+} pump, ECA1, from the endoplasmic reticulum is inhibited by cyclopiazonic acid but not by thapsigargin. Plant Physiol 1998;817-825.

57. Hong BA, Ichida S, Wang Y et al. Identification of a calmodulin-regulated Ca^{2+} ATPase in the ER. Plant Physiol 1999; 119; 1165-1176.
58. Klüsener B, Boheim G, Liss H et al. Gadolinium-sensitive, voltage-dependent calcium release channels in the endoplasmic reticulum of a higher plant mechanoreceptor organ. EMBO J 1995; 14:2708-2714.
59. Klüsener B, Weiler EW. A calcium-selective channel from root-tip endomembranes of garden cress. Plant Phys 1999 119:1399-1405.
60. Navazio L, Bewell MA, Siddiqua A et al. Calcium release from the endoplasmic reticulum of higher plants elicited by the NADP metabolite nicotinic acid adenine dinucleotide phosphate. Proc Natl Acad Sci USA 2000; 8693-8698.
61. Navazio L, Mariani P, Sanders D. Mobilization of Ca^{2+} by cyclic ADP-ribose from the endoplasmic reticulum of cauliflower florets. Plant Physiol 2001; 125:2129-2138.
62. Muir SR, Sanders D. Inositol 1,4,5-trisphosphate-sensitive Ca^{2+} release across nonvacuolar membranes in cauliflower. Plant Physiol 1997; 11:1511-1521.
63. Boevink P, Oparka K, Santa Cruz S et al. Stacks on tracks: the plant Golgi apparatus traffics on an actin/ER network. Plant J 1998; 15:441-447.
64. Reuzeau C, McNally JG, Pickard B. The endomembrane sheath: a key structure for understanding the plant cell? Protoplasma 1997; 200:1-9.
65. Knight H, Trewavas AJ, Knight MR. Cold calcium signaling in Arabidopsis involves two cellular pools and a change in calcium signature after acclimation. Plant cell 1996; 8:489-503
66. Knight MR, Smith SM, Trewavas AJ. Wind-induced plant motion immediately increases cytosolic calcium. Proc Natl Acad Sci 1992; 89:4967-4971.
67. Evans NH, McAinsh MR, Hetherington AM. Calcium oscillations in higher plants. Curr Opin Plant Biol 2001; 4:415-420.
68. Bauer CS, Plieth C, Hansen U-P et al. Repetitive Ca^{2+} spikes in a unicellular green alga. FEBS Lett 1997; 405:390-393.
69. Bauer CS, Plieth C, Bethmann B et al. Strontium-induced repetitive calcium spikes in a unicellular green alga. Plant Physiol 1998; 117:545-557.
70. Malhó R, Moutinho A, van der Luit A et al. Spatial characteristics of calcium signalling: the calcium wave as a basic unit in plant cell calcium signalling. Phil Trans R Soc Lond 1998; 353:1463-1473.
71. Corbett EF, Michalak M. Calcium, a signalling molecule in the endoplasmic reticulum? Trends Biol Sci 2000; 25:307-311.
72. Persson S, Wyatt SE, Love J et al. The Ca^{2+} status of the endoplasmic reticulum is altered by induction of calreticulin expression in transgenic plants. Plant Physiol 2001; 126:1092-1104.
73. Camacho P, Lechleiter JD. Calreticulin inhibits repetitive intracellular Ca^{2+} waves. Cell 1995; 82:765-771.
74. John LM, Lechleiter JD, Camacho P. Differential modulation of SERCA2 isoforms by calreticulin. J Cell Biol 1998; 142:963-973.
75. Wyatt SE, Tsou PL, Robertson D. Expression of the high capacity calcium-binding domain of calreticulin increases bioavailable calcium stores in plants. Transgenic Res 2002; 11:1-10.

Chapter 11

Modulation of Calcium Homeostasis by the Endoplasmic Reticulum in Health and Disease

György Szabadkai, Mounia Chami, Paolo Pinton and Rosario Rizzuto

Abstract

The endoplasmic reticulum (ER) is the main intracellular agonist-sensitive Ca^{2+} store, and is involved in the regulation of a wide range of cellular functions depending on cytosolic Ca^{2+}. In addition, it has recently been recognized that Ca^{2+} regulates also processes occurring in the ER lumen, such as protein synthesis and trafficking, and cellular responses to stress. Accordingly, perturbation of ER Ca^{2+} homeostasis appears to be a key component in the development of several pathological situations. In this chapter, after providing an overview of the Ca^{2+} signaling components of the ER, we briefly summarize their role in basic pathophysiological processes and specific diseases.

Regulation of Endoplasmic Reticulum [Ca^{2+}]

Virtually in all eukaryotic cells the dynamic regulation of cytosolic calcium concentration ($[Ca^{2+}]_c$) is fundamental for cell life, controlling extremely diverse functions as muscle contraction, hormone secretion, neuronal circuits, immune responses and gene expression.[1,2] Together with the plasma membrane the intracellular calcium stores play an essential role in completing this regulation[3] The most important intracellular calcium store is represented by the endoplasmic reticulum (ER) in non-muscle cells, and by its specialized counterpart, the sarcoplasmic reticulum (SR), in muscle cells.[4] Our chapter is mainly limited to the discussion of ER Ca^{2+} homeostasis, but since the ER and SR share the basic characteristics of intracellular Ca^{2+} stores capable of rapid Ca^{2+} exchange, and extensive amount of information has been collected from muscle cells concerning the pathophysiological role of the Ca^{2+} stores, we will refer also to some details of SR Ca^{2+} signaling.

In the resting cell the $[Ca^{2+}]_c$ is low, with values around 100 nM, whereas in the ER the $[Ca^{2+}]$ ($[Ca^{2+}]_{er}$) is more than three orders of magnitude higher (100-800 μM, for review see ref. 5). During the stimulation of cells by Ca^{2+} mobilizing agonists (acting through G-protein mediated activation of phospholipase C and consequent generation of IP3) there is a transient increase in $[Ca^{2+}]_c$ (to low μM range) with a parallel transient reduction in the $[Ca^{2+}]_{er}$, showing that released Ca^{2+} plays a fundamental role in the initiation of the Ca^{2+} signal. If the stimulation is sustained, Ca^{2+} influx from the extracellular space maintains a steady state $[Ca^{2+}]_{er}$ level, which in turn serves for continuous or oscillating release of Ca^{2+} until the agonist is removed. Similarly, in muscle cells, electrical activation of the plasma membrane is followed by direct or Ca^{2+} mediated activation of Ca^{2+} release from the SR, followed by its refilling.

The potential of the ER/SR to function as a rapidly exchanging Ca^{2+} store is due to the presence of three main components: i) ATP dependent pumps for Ca^{2+} uptake (called SERCAs:

Calreticulin, 2nd ed., edited by Paul Eggleton and Marek Michalak.

Sarco/Endoplasmic reticulum Ca^{2+} ATPases), ii) channels for Ca^{2+} release such as the ubiquitous inositol 1,4,5-trisphosphate receptor (IP3R) and the ryanodine receptor (RyR), and iii) Ca^{2+} binding proteins for Ca^{2+} storage, the best characterized being calreticulin and calsequestrin.

Ca^{2+}-Uptake

The Ca^{2+}-ATPases present in the ER/SR membrane are able to maintain the resting Ca^{2+} gradient (counteracting the significant passive Ca^{2+} leak) and to refill the store after its emptying by consuming the energy of ATP hydrolysis. Tissue specific expression of different Ca^{2+}-pump isoforms has been demonstrated: SERCA1 or fast type Ca^{2+}-ATPase (with two alternatively spliced variants: 1a and 1b) present exclusively in fast-twitch skeletal muscle; SERCA2 or slow type Ca^{2+}-ATPase (existing in two variants 2a and 2b) expressed in slow-twitch skeletal and cardiac muscle (2a) and in non-muscle cells (2b); SERCA3 (expressed in three different isoforms 3a, 3b and 3c) which is also a slow type Ca^{2+}-ATPase that is found only in non-muscle cells, generally coexpressed with SERCA2b, representing the most divergent isoform (for reviews see 4,6). The regulation of SERCA activity is complex including both the ER luminal and cytosolic $[Ca^{2+}]$, and subtype-specific regulatory proteins, e.g., phospholamban for the cardiac SERCA2b (see below), and has an important role in the generation of the complex cytosolic Ca^{2+} signaling pattern.[7] Different Ca^{2+}-uptake systems are present in other intracellular stores such as the Golgi apparatus or secretory vesicles, but since they appear to be unrelated to ER Ca^{2+} homeostasis itself, they will not be discussed here.[8,9]

Ca^{2+}-Release

The two well-characterized families of Ca^{2+} channels responsible for the rapid release of Ca^{2+} from the ER in response to extracellular stimuli are the families of IP3Rs and RyRs, coexpressed often in many cell types.[10] Coordinated activation of these receptors is responsible for the generation of elementary and global (oscillating or persistent) Ca^{2+} signals observed in the cytosol.[11,12,13] Interestingly, their basic structural and functional properties are similar, i.e., they are both composed of a Ca^{2+} selective pore including transmembrane segments and a large cytoplasmic regulatory subunit, they both form tetramers, and they are both controlled by $[Ca^{2+}]_c$ and $[Ca^{2+}]_{er}$. Despite these similarities they appear to have significantly different regulation by second messengers and also by associated proteins (not discussed here, see refs. 14,15). In the case of the IP3Rs the molecule responsible for the opening of the channel is IP3, generated from $PtdInsP_2$ by PLC after its activation by diverse G-protein coupled receptors (GPCRs).[16] Three subtypes of IP_3Rs are known, all showing a wide tissue distribution, although the type 1 receptor appears to be predominant in the central nervous system. These receptors, in contrast to RyRs, can also form heterotetramers.[17,18] As for the RyR family, cADPr is proposed to be a physiological second messenger opening these channels[19], but whether it directly acts on the receptor or it exerts its activating effect through RyR regulating proteins such as the FK506 immunophilin binding protein FKBP12.6 is still a matter of debate.[20,21] Nevertheless, the most documented roles of RyRs are (i) its direct activation by the voltage sensor dihydropyridine receptor (DHPR, skeletal muscle L-type voltage dependent Ca^{2+} channel) during the excitation/contraction coupling of the skeletal muscle, (ii) generation of Ca^{2+} induced Ca^{2+} release (CICR) as a result of its high sensitivity to $[Ca^{2+}]_c$ elevation, a role proved in both muscle and non-muscle cells. As to tissue distribution, the RyR1 is representative of skeletal muscle, RyR2 is present in cardiac muscle, RyR3 has a wide tissue distribution, but recently the expression of all three subtypes was shown in different brain areas.[22]

Recently, the existence of two additional Ca^{2+} release channels has been proposed in the ER. First, the NAADP sensitive channel, the molecular nature of which is still unresolved, has been proposed to play a role in initiating Ca^{2+} oscillations in some cell type from a separate IP_3 and ryanodine insensitive Ca^{2+} pool.[23,24] The second is polycystin-2, a Ca^{2+} activated ER membrane channel, recognized by its mutation in the inheritant polycystic kidney disease, which seems to have a role in amplifying IP_3 induced Ca^{2+} release in the cytosolic domain below the plasma membrane.[25,26]

Calcium Binding Proteins

Ca^{2+} binding proteins resident in the ER buffers most of the Ca^{2+} content of the ER. The most relevant feature of the Ca^{2+} binding proteins of ER lumen, which distinguish them from the cytosolic ones, is their low affinity and high capacity Ca^{2+} binding. The consequences of these properties are that comparing to the cytosol (i) they buffer large amounts of readily releasable Ca^{2+} (ii) they allow higher rate of Ca^{2+} diffusion and fast equilibration of $[Ca^{2+}]$ gradient throughout the interconnected network. The two main representatives of the ER/SR Ca^{2+} binding proteins are calreticulin and calsequestrin, respectively, the latter being almost entirely restricted to muscle cells. Since the detailed properties and complex functions of these proteins are extensively discussed in the other chapters of this book, to avoid repetitions, we do not discuss them in detail in this section.

The Regulatory Role of ER Luminal $[Ca^{2+}]$

In addition to the role of ER in distributing the Ca^{2+} signal to the entire cell interior, oscillations of $[Ca^{2+}]_{er}$ itself participate in the regulation of normal cell function at multiple levels[2]. First, the ER resident components of the machinery of Ca^{2+} homeostasis are regulated by the luminal $[Ca^{2+}]$. Regulation of the IP_3Rs and RyRs and SERCA2b by $[Ca^{2+}]_{er}$ has been demonstrated.[27,28] Moreover, the conformational changes in IP_3R structure has been proposed to have a role also in the control of store operated Ca^{2+} influx (SOC) through the plasmamembrane, allowing the SOC to be activated by $[Ca^{2+}]_{er}$ decrease. Several signaling hypothesis has been proposed, including diffusible messengers and direct coupling of the ER to plasma membrane (for review see refs. 29,30). Second, apart from controlling Ca^{2+} fluxes, $[Ca^{2+}]_{er}$ has regulatory functions in other cellular processes. Newly synthetized membrane and secretory proteins, as well as denaturated proteins are folded and processed further in the ER, and $[Ca^{2+}]_{er}$ dependent interactions of calreticulin with other ER luminal or membrane proteins (calnexin, erp57, BiP) appear to be fundamental in this function.[31-33] As a consequence, these synthetic and folding processes oscillate depending on the filling state of the ER lumen. Moreover, in pathological conditions, sustained depletion of the ER Ca^{2+} stores leads to accumulation of unfolded proteins in the ER lumen, leading to the activation of two highly conserved stress responses, the ER overload response (ERO)[34] and the unfolded protein response (UPR).[35,36] ERO involves the activation of the transcription factor NF-κB, which in turn, triggers the expression of different target genes. UPR triggers two downstream processes: (i) activates the expression of genes coding for ER-resident stress proteins, and in parallel (ii) causes the suppression of the initiation of protein synthesis (for discussion of ER role in apoptosis see section on "ER Calcium homeostasis". Following these lines, recent work demonstrates a complex gene transcription system, working also in physiological conditions, regulated by signals originating from $[Ca^{2+}]_{er}$ changes.[37-39]

The ER As Central Component of Compartmentalized Ca^{2+} Signaling

The final shape of the cytosolic Ca^{2+} signal is determined by the tuning of ER channel opening, as well as by the diffusion of Ca^{2+} from the microdomains at mouth of the channel to the rest of the cytosol. The microenvironment of ER Ca^{2+} channels includes other intracellular organelles, such as mitochondria, exocytotic sites, synaptic apparatus, the cell nucleus, and the plasma membrane, that act both as targets of local signals and as modulators of channel activity. These local signals control events of paramount importance (e.g., secretion and mitochondrial metabolism) and thus it can be inferred that the property of distributing and releasing Ca^{2+} at the site of interest is a primary function of the ER Ca^{2+} store. This role is particularly important during sustained Ca^{2+} signals, when the only source for maintaining the Ca^{2+} activating processes is the Ca^{2+} content of the extracellular space at a long distance from particular locations of the cell interior. To discuss this role of the ER first we first briefly review the heterogeneity of the ER Ca^{2+} store, then we look at its individual contact partners.

Heterogeneity and Continuity of the ER Ca^{2+} Store

The non-random distribution of the proteins involved in uptake, release and luminal binding of Ca^{2+} is responsible for the heterogeneity of Ca^{2+} storage and exchange in the ER of non-muscle cells (reviewed in ref. 40). Different subtypes of SERCA can be localized in different ER locations,[41] and clustering of Ca^{2+} release channels has been noticed in several cell types.[42,43] Moreover, evidence from functional studies revealed separate Ca^{2+} releasable pools, which are differentially sensitive to Ca^{2+}-ATPase blockers and Ca^{2+} release channel agonists. Interestingly, the distribution of Ca^{2+} binding proteins has been shown to be more widespread in the ER network, in sharp contrast to the uneven distribution of total (bound + free) Ca^{2+} content of ER, showing that still unidentified components may contribute to this issue. Even so, there is a general agreement on the significance of clustered assembly of ER Ca^{2+} signaling apparatus in generating spatially directed signals both in excitable and non-excitable cells. A further important, but still unresolved issue in this respect is whether the ER consists in one continuous vesicle system or is fragmented in isolated compartments. Evidences showing different Ca^{2+} concentrations and independent loading of separate ER regions argue for the existence individual compartments, such as a specialized subplasmalemmal ER domain in neurons (for review in favour of this hypothesis, see ref. 44). On the other hand, wide-ranging continuity has been shown in several cell types allowing diffusion of proteins, dyes and even Ca^{2+} itself, and according to these results, the ER could function as a readily equilibrating tunnel system, serving for Ca^{2+} signal transduction cable inside cells, types ranging from the extensively studied pancreatic acinar cells to neurons (for review see ref. 45).

The Mitochondrial Network: Spatial and Temporal Restriction of the Ca^{2+} Transient

The first evidences showing the intimate connection between the ER and mitochondria came out studying the mitochondrial Ca^{2+} uptake in intact cells. The direct measurement of the Ca^{2+} concentration in the mitochondrial matrix ($[Ca^{2+}]_m$) of living cells[46-49] demonstrated that mitochondrial calcium spikes parallel cytosolic calcium transients elicited by IP3-generating agonists. Subsequently, other studies extended this finding to a variety of cell types and alternative calcium release pathways.[50-52] The amplitude of the $[Ca^{2+}]_m$ rise varies within different cell types, but is generally well above the bulk cytosolic $[Ca^{2+}]$ increase, reaching nearly millimolar values in chromaffin cells.[51] These new observations implied that presumably $[Ca^{2+}]$ increased much more in the vicinity of the uniporter than in the bulk cytosol. The requisite for this possibility is that close contacts occur between the mitochondria and the ER. In fact, regions where ER and mitochondrial membranes come in close apposition have been documented by electron microscopy[53,54] and even point contacts between the outer mitochondrial membrane and ER have been observed using electron tomography.[55] Wide-field fluorescence microscopy combined with digital deconvolution has revealed that mitochondria form a largely interconnected dynamic network in living cells.[56] By expressing spectral variants of GFP targeted to the mitochondria and endoplasmic reticulum, 5-20% of the mitochondrial surface was estimated to be in close apposition to the ER in HeLa cells. In the same cell type, specific expression of aequorin at the outer surface of the mitochondrial inner membrane directly demonstrated that mitochondria are exposed to much larger $[Ca^{2+}]$ increases than the bulk cytosol. Another strong argument for the existence of microdomains of high $[Ca^{2+}]$ generated by IP3R or ryanodine receptors (RyR) and sensed by neighbouring mitochondria is that in the presence of 100-200 μM EGTA, agonist-induced $[Ca^{2+}]_m$ transients are still observed while $[Ca^{2+}]_c$ increases are abolished.[52,57]

The above findings led to a new concept of ER/mitochondrial interaction depicted as 'synaptic-like connection' between these organelles,[57] and first pointing to the role of ER as tunnelling Ca^{2+} to place of need. On the other hand, these results also gave rise to the reciprocal question of the participation of the organelle in global calcium signaling. Although commonly believed to be marginal until the last few years, the work of numerous groups now demonstrated that mitochondrial Ca^{2+} handling influences amplitude, duration, localization and propagation of cytosolic calcium transients,[58-61] originating either from release from internal stores or from Ca^{2+} influx from the extracellular space. Most of these works identified mitochondria as buffers, which takes up released Ca^{2+} during the peak phase of Ca^{2+} signal, and consequently recycle it during the sustained phase. Moreover, it has been recently discovered that mitochondria play also in the refilling process of the ER/SR, driven by ATP-dependent SERCA) present in their membranes. Mitochondrial inhibitors dramatically slow the refilling of the ER after bradykinin challenge in BHK-21 cells.[62] Similarly, it was shown in HeLa cells,[63] that blocking of Ca^{2+} cycling in the mitochondrial inner membrane leads to more substantial emptying of the ER. These effects may partly be due to a local ATP depletion impairing SERCA activity, and to an active participation of mitochondria in the recycling of Ca^{2+} following its release. These results point to the existence of microdomains not only of Ca^{2+} but presumably also of ATP between these organelles.

As to the molecular nature of these local signaling units involving the ER, very little is known so far.[64] A reasonable prediction is that ER Ca^{2+} release channels participate in the presumed cluster, building up the close contact. Two lines of evidence support this possibility. The first is the demonstration that mitochondrial Ca^{2+} uptake modulates the feed-back effect of Ca^{2+} on the IP_3R, implying that a great deal of channels are indeed located in the vicinity of mitochondria (see above). The second is based on morphological studies revealing high density of IP_3Rs/RyRs in SR/ER domains facing the mitochondria (for review see ref. 65). Elevation of intracellular Ca^{2+} has a trigger role in this clustering.[66,67] Investigating the nature of these domains, Wang et al[68] using electron and confocal microscopy, described a specific smooth ER domain contacting the mitochondria, which can be distinguished from other ER compartments by the presence of specific marker proteins, such as the autocrine motility factor receptor. The existence of this specialized domain was not unexpected, as other processes, such as phospholipid transport, are known to occur between defined ER domains and mitochondria.[69] Interestingly, the co-localization of this compartment with mitochondria is Ca^{2+}-dependent, i.e., low (<100nM) $[Ca^{2+}]$ allows unidentified cytosolic factors to disrupt the connection.[66] As for the mitochondrial side, we recently reviewed[64] the participation of proteins maintaining and regulating mitochondrial morphology in the formation of ER/mitochondrial interaction. Moreover, we showed that the voltage dependent anion channel (VDAC) of the outer mitochondrial membrane plays an important role in Ca^{2+} diffusion toward the mitochondrial uptake system (E. Rapizzi et al manuscript submitted). Still the contribution of yet unidentified components is presumable, giving an interesting task for further investigation.

The ER-Plasma Membrane Connection

There is accumulating evidence showing that the connection of ER to the plasma membrane plays an important role in the ER and global cellular Ca^{2+} homeostasis. Models of Ca^{2+} signaling in several cell types comprise these connections in respect to localized Ca^{2+} release and localized refilling of the ER store. In pancreatic acinar cells O.H. Petersen's group showed that at the initial phase of the physiological activation of the cells focal Ca^{2+} release occurs exclusively at the secretory pole, serving as trigger for exocytosis, and leading to local emptying of the Ca^{2+} store.[43] Given that the ER in these cells proved to form a continuous network, the resulting luminal $[Ca^{2+}]$ gradient causes rapid diffusion of Ca^{2+} from the basolateral part of the cells, where store operated Ca^{2+} entry (SOC) is exclusively localized.[43] By analogy, in neurons it was proposed (but experimentally not yet confirmed) that subplasmalemmal ER cisternae of the cell body may be responsible for Ca^{2+} refilling from the extracellular space, and a continuous ER network would transport Ca^{2+} to the site of release in dendritic spines.[70] In addition, also in other cell types, similarly to the skeletal muscle and cardiac excitation-contraction coupling system, the co-localization of release and SOC events in adjacent ER and plasma membrane regions (PL-ER junctions) was shown.[44]

Similarly to the case of mitochondrial connection little is known about the structural factors keeping the release and refilling mechanisms in a well-localized compartment. Theoretically, to resolve the problem of localized release, we can imagine either a localized signal arriving to the ER release sites and/or a barrier keeping the released Ca^{2+} in the defined compartment. Indeed, taking examples from the above-described model of pancreatic acinar cells or from the similarly specialized system of the neuronal synaptic system, both solutions seem to exist. In the postsynaptic density, looking for determinants of synaptic plasticity Worley's group described a family of proteins serving as anchor for the components of glutamate induced signaling cascade.[71] These proteins, called Homers, by forming dimers using their coiled-coil domains, seem to be able to cross link the metabotropic GluR (type 1 and 5: involved in Ca^{2+} signaling) to the IP3R (or possibly also to RyRs), restricting the route of IP3 generated by receptor activation.[72] Even if the exact role of these proteins in the Ca^{2+} signaling pathway is not yet clear, and they may play a role in localizing the receptor operated Ca^{2+} entry pathway (see below), they opened the way for the paradigm of localized activation of Ca^{2+} release sites. On the other hand, turning back to acinar cells, after its initiation, the Ca^{2+} wave is confined to the secretory pole by the mitochondrial barrier, localized as a ring over the secretory vesicles toward the basolateral pole of the cells.[73] Indeed, inhibiting mitochondrial Ca^{2+} uptake, the localized signal in these cells becomes global and reaches the whole cytosol. A similar spatial barrier role of mitochondria has been also proposed in dendritic Ca^{2+} signal transmission in neurons.[74]

As to the question of localized refilling of the Ca^{2+} stores, both the secretion like and the conformational coupling model of the activation of SOCs involves direct interaction between the ER and plasma membrane components. Moreover, in both model there are evidences for coupling to preferential sites of the plasma membrane. The secretion like coupling model, originally proposed by C. Fasolato,[75] was recently strengthened by molecular data, showing that exocytotic anchoring proteins, such as SNAP25, are required for store operated Ca^{2+} influx.[76] Recently, even the most favoured conformational coupling model , originally only implying an interaction between the IP3 receptors and the SOC channels, has been widened to imply a more general interaction between the components of Ca^{2+} signaling pathway, particularly the phospholipase C (PLC).[77] In this model, analogously to the *Drosophila* light signal transduction system by the INAD protein[78] (binding *trp* channels to PLC),[79,80] the existence of a bridging protein has been proposed, keeping in close contact the above mentioned protein components, and consequently restricting a cytosolic microdomain containing saturating concentrations of IP3, produced by thē basal PLC activity, since bindig of IP3 seems to be also required for the interactions between the ER and plasmamembrane channels.

Nuclear Ca^{2+} Signaling

Several data suggest that Ca^{2+} is responsible for the regulation of many nuclear events, thus regulation of nucleoplasmic $[Ca^{2+}]$ ($[Ca^{2+}]_n$) is of high importance. The nuclear envelope (NE), the lumen of which appears to be continuous with the ER lumen, may serve as a specialized compartment and may have the capacity for independent Ca^{2+} and inositide cycling. Indeed, data from several cell types suggested the existence of an independent Ca^{2+} release and uptake apparatus in the nuclear envelope as well as a standing Ca^{2+} gradient between the cytosol and the nuclear matrix (for review see refs. 81,82). However, studies based on the direct measurement of $[Ca^{2+}]_n$ with targeted probes and fluorescent dyes[83,84] showed no difference between cytosolic and nuclear resting and stimulated $[Ca^{2+}]$ in agreement with the notion of the high Ca^{2+} permeability of the nuclear pore complex. The possible significance of elementary release events in nuclear Ca^{2+} signal and their relevance to previous findings has been reviewed recently.[85]

ER Calcium Homeostasis, Regulation of Cellular Proliferation and Apoptosis

Apoptosis

Calcium signals are enormously versatile and are implicated in the regulation of both cell proliferation and apoptosis.[86] Participation of calcium signaling in the control of apoptosis implies a multitude of complex mechanisms (for reviews see refs. 36,87), which can be initiated from different intracellular organelles and lead to final executing mechanisms involving mitochondrial membrane permeabilization (MMP) and/or caspase activation. Cytosolic Ca^{2+} has been implicated as a pro-apoptotic second messenger involved in both triggering apoptosis and in regulating death-specific enzymes.[88] In parallel, mitochondrial calcium overload has been proposed to be the trigger of MMP, including opening of PTP and release of activators of caspases.[89]

Recently it has been recognized, that modification of the ER Ca^{2+} pool may itself be involved in the initiation and regulation of the apoptotic process. As to Ca^{2+} release, it has been shown that owing to the efficient Ca^{2+} signal transmission between ER and mitochondria, ER Ca^{2+} release can sensitize to ceramide-induced apoptosis.[90] The involvement of IP3Rs in sensitization to apoptotic stimuli has been shown also by transgenic and anti-sense techniques. Thus, $InsP_3R1$-deficient lymphocytes are resistant to a large panel of apoptosis inducers[91] and anti-sense oligonucleotide-mediated downregulation of $InsP_3R3$ decreases the cell death of glucocorticoid-treated T cells[92] and NGF-depleted neurons.[93]

Moreover, now it is clear, that changes in the steady-state ER Ca^{2+} level itself have a significant influence in the apoptotic pathways. However, this issue appears to be extremely complicated, owing to the complexity of processes in the ER connected to its Ca^{2+} homeostasis. UPR has several implications in apoptotic processes, and $[Ca^{2+}]_{er}$ has been linked to this complex process at several points, such as up regulation of Bip/Grp78, calreticulin and the CHOP/GADD153 transcription factor leading to decreased Bcl-2 expression (reviewed in ref. 36). Another signal for ER apoptosis is the ER-associated caspase-12, which has been described first in mice. Caspase-12 is strictly localized on the cytoplasmic side of the ER and is specifically activated upon ER stress, e.g., after treatment with brefeldin-A, tunicamycin or thapsigargin[94] in a Ca^{2+} dependent way. Elevation of $[Ca^{2+}]_c$ and sequential calpain activation also leads to proteolytic activation of caspase-12.[95]

Modulation and deregulation of ER handling proteins is considered as a powerful tool in the understanding of $[Ca^{2+}]_{er}$ participation in apoptosis, but still there is no consensus on the basic question, what is the relation between $[Ca^{2+}]_{er}$ and cellular sensitivity to apoptogenic factors? Here we shortly discuss the data concerning this issue. On one hand, it is well established that large ER Ca^{2+} depletion, leading to ER stress response and fragmentation[96], induce apoptosis. Indeed, the SERCA inhibitor thapsigargin is a potent apoptotic stimulus.[97,98] Similarly, reduction of ER Ca^{2+} content was shown to potentiate the apoptotic effect of various stimuli, as H_2O_2, ocadaic acid, and growth factor starvation.[99-101] On the other hand, numerous lines of evidence, including our work on the anti-apoptotic Bcl-2 protein, indicate that Ca^{2+} may act as a mediator of apoptosis, and a partial reduction of ER Ca^{2+} levels (such as that caused by Bcl-2) reduces the efficacy of apoptotic stimuli.[102,103] In agreement with theses observation, the expression of a cytosolic calcium buffer, such as calbindin, or a diminution of ER calcium release is associated to a delay in apoptosis induced by glucocorticoide in lymphatic cells.[104] Moreover, overexpression of SERCA protein (and the consequent ER Ca^{2+} overload) has been shown to induce apoptotic cell death.[105]

Bcl-2 Protein and ER Ca^{2+} Handling

Three experimental observations strongly suggested that the oncoprotein Bcl-2 could affect ER Ca^{2+} handling. The first is the demonstration that the purified protein, inserted into lipid bilayers, can act as an ion channel.[106-108] The second is the distribution of Bcl-2 to ER and mitochondrial membranes.[109,110] Its ER localization has been proved to be important, since a Bcl-2 mutant exclusively localized in the ER is able to inhibit apoptosis induced by Myc in Rat-1 fibroblast cell line.[111] Third, results obtained in lymphoma cells overexpressing Bcl-2 showed an attenuation in the size of the Tg releasable pool and in the amplitude of agonist dependent Ca^{2+} signals.[112]

The capacity of Bcl-2 to alter ER Ca^{2+} homeostasis was recently demonstrated by directly measuring $[Ca^{2+}]_{er}$ in Bcl-2 overexpressing cells. These measurements were carried out by us, using targeted chimaeras of the Ca^{2+} sensitive photoprotein aequorin[103] and by the group of K.H. Krause using GFP based (cameleon) probes targeted to the ER,[113] obtaining very similar results. In brief, it was shown that Bcl-2 reduced the state of filling of intracellular Ca^{2+} stores (ER and Golgi apparatus) by about 30%, increasing the passive Ca^{2+} leak from the organelle (in agreement with the possibility the Bcl-2 acts as an ion channel). Consequently, stimulus-dependent $[Ca^{2+}]$ increases were reduced both in the cytoplasm and in the mitochondria, a potential target for an apoptogenic effect of Ca^{2+}. The reduction in steady state $[Ca^{2+}]_{er}$ caused by Bcl-2 overexpression might cause a substantial activation of capacitative Ca^{2+} influx[114] with consequent increase in $[Ca^{2+}]_c$. Interestingly, we have observed that Bcl-2 downregulates capacitative Ca^{2+} influx probably as an adaptive consequence to the prolonged reduction in steady state $[Ca^{2+}]_{er}$[103] avoiding the deleterious effects of a prolonged increase of the $[Ca^{2+}]_c$.

These data are only apparently in contrast with previous reports showing that the SERCA inhibition by different compounds and mechanisms (thapsigargin, tBuBHQ, and cyclopiazonic acid) is followed by apoptosis.[115] With these treatments, the Ca^{2+} depletion is complete and rapid, while in Bcl-2 transfected cells the drop in $[Ca^{2+}]_{er}$ is modest and develops slowly. As discussed before, a drastic reduction in the level of $[Ca^{2+}]_{er}$ might interfere with the basic activity of ER such as for example the regulation of ER protein folding and chaperone interactions. Conversely, a partial depletion in $[Ca^{2+}]_{er}$, as measured after Bcl-2 over-expression should decrease the impact of a pro-apoptotic signal. Indeed, in an elegant series of experiments G. Szalai and coworkers demonstrated that in the presence of apoptotic agents, a physiological Ca^{2+} signal causes the opening of the mitochondrial transition pore with ensuing organelle swelling and release of caspase cofactors.[116]

However, the possibility that Ca^{2+} depletion induced by Bcl-2 has a role in its anti-apoptotic activity awaited direct experimental support. To address this issue, we carried out experiments in order to verify whether the alteration of ER Ca^{2+} handling caused by Bcl-2 could per se affect the sensitivity to apoptotic agents (namely ceramide, that is known to induce apoptosis through a pathway that is inhibited by Bcl-2). For this purpose, we mimicked the Bcl-2 effect on $[Ca^{2+}]_{er}$ by different pharmacological and molecular approaches. All conditions that lowered $[Ca^{2+}]_{er}$ (at values similar to those of Bcl-2 over-expressing cells) protected the cells from the effect of ceramide, while treatments that increased $[Ca^{2+}]_{er}$ had the opposite effect.[102] Overexpression of calreticulin drastically reduced cell survival upon ceramide treatment, indicating that rather than the decrease in the luminal $[Ca^{2+}]$, the key parameter is the amount of Ca^{2+} that is released, and thus the signal conveyed to cytosolic nuclear or mitochondrial effectors. Indeed in calreticulin overexpressing cells the $[Ca^{2+}]_{er}$ was shown to be similar[117,118] or lower[119] than in controls, while the amount of releasable Ca^{2+} is significantly increased.[120,121] These results well agree with previous data indicating that calreticulin deficient cell lines are more resistant to apoptosis.[122]

Confirming the hypothesis that the release of Ca^{2+} from Ca^{2+} stores is important to trigger apoptosis, we have observed that ceramide caused a progressive release of Ca^{2+} from intracellular stores, causing a $[Ca^{2+}]_c$ rise and maintained a long-lasting $[Ca^{2+}]_c$ plateau by activation of capacitative Ca^{2+} influx[102] that is down regulated by Bcl-2 (see above). As a result, Bcl-2, by reducing both the amount of Ca^{2+} release (by lowering $[Ca^{2+}]_{er}$) and the influx of Ca^{2+} (by down regulation of SOC), could prevent the cell death induced by increase of $[Ca^{2+}]_c$. Interestingly, we have observed that ceramide treatment induced a slowly developing and prolonged increase also in $[Ca^{2+}]_m$, maintained for much longer than a typical physiological challenge and this increase was followed by dramatic alterations of mitochondrial morphology. Moreover, the experimental maneuvers that reduced $[Ca^{2+}]_{er}$ prevented also the changes in mitochondrial morphology.

On the light of these observations it is possible to conclude that ER Ca^{2+} depletion caused by Bcl-2 over-expression is important for the anti-apoptotic effect of this oncoprotein and that the ER-mitochondrial cross-talk (that proved to be important for upregulating mitochondrial metabolism in stimulated cells) plays a role also in decoding Ca^{2+} mediated apoptotic signals.

Regulation of Proliferation and Differentiation, a Potential Role of SERCA Proteins

Wide experimental evidence shows a correlation between the level of expression and/or isoform selection of SERCA proteins and the control of cellular proliferation and differentiation.

A role of SERCA proteins in the control of the cell cycle was first demonstrated in various system using SERCA inhibitors.[98,123] It was also demonstrated that downregulation of SERCA gene expression was associated to a delay in G1 to S progression and was accompanied by increased mRNA levels of p21, a universal cell cycle inhibitor, up-regulation of p53 mRNA and underphosphorylation of the retinoblastoma protein.[124] Finally, reduction of the expression of SERCA2 protein was demonstrated in INF γ (interferon) and TNF α (tumor necrosis factor) growth arrested human salivary gland cells.[125]

Although the significance of the existence of different SERCA isoforms is not completely understood, molecular, biochemical, tissue and cell distribution data suggest a modulation of the expression of individual calcium pumps as a function of fine modulation of calcium events in different cellular regions and at different times during the cell differentiation program. Thus, the expression of specific SERCA isoforms was associated to the control of the differentiation-proliferation balance. The activation of several T-lymphocyte cell lines is associated to the down regulation of SERCA PL/IM (SERCA3b) and the up-regulation of SERCA2b[126] and similarly, in vitro differentiation of a human myeloid/promyelocytic cell line along the neutrophil/granulocytic lineage was associated with a decrease in SERCA2b and a parallel increase in SERCA PL/IM.[127] In the same line of experiments, an increase in the SERCA2a pump expression was observed in PDGF-treated, aortic smooth muscle cells undergoing cell proliferation,[128] and it was subsequently shown that this SERCA2a up-regulation was related to G1/S transition in the cell cycle.[129] These results are consistent with a potential role of SERCA2 a/b isoforms in cellular proliferation and of SERCA PL/IM in cellular differentiation. However, the situation is not so clear, because the myogenic differentiation of mouse tumor cells BC3H1 was associated with a switch from the predominant isoform SERCA2b to SERCA2a[130] suggesting in this cell type a role for SERCA2a in cellular differentiation. In fact, more data are necessary to relate properly SERCA isoform switching to a specialized cellular function.

A direct implication of SERCA1 protein in the control of cell proliferation and viability was demonstrated in a case of clonal mutation of SERCA1 gene selectively occurring in a human hepatocellular carcinoma. This mutation is due to the integration of the Hepatitis B virus (HBV) DNA into SERCA1 gene in the liver cell genome. This integration causes cis-activation of chimeric HBV-X/SERCA1 transcripts, with splicing of SERCA1 exon 11, encoding C-terminally truncated SERCA1 proteins unable to pump calcium.[131] Expression of non-chimeric SERCA1 truncated proteins was also demonstrated in normal liver and in different transformed cell lines. Overexpression of these SERCA1 truncated proteins, in vitro, induces a large depletion of ER calcium content and is associated to apoptotic cell death.[132] Different oncogenic proteins have been shown to interfere with apoptosis depending on the level of expression and the system used.[133] It is plausible that SERCA1 truncated proteins, by acting on ER calcium homeostasis, play a role in the regulation of both apoptosis and proliferation. All together, these data converge to show that ER calcium content is a major regulator of cell growth and apoptosis. In this context, we can imagine that a dysfunction of this mechanism might play a role in cell transformation.

Diseases Associated with Ca^{2+} Signaling Components of the ER

As we discussed above, Ca^{2+} signal plays a profound role in basic cellular processes, thus elimination of the function of major signaling components leading to massive changes in the pattern of Ca^{2+} signal has generally fatal consequences in the life of cells, organs and individuals. Nevertheless, also given the redundancy of the signaling machinery (each element occurs in multiple isoforms), minor changes can be compensated, leading to a disordered but viable state. In this section, we will discuss some of the data obtained by studying the consequences of artificial deletions of Ca^{2+} signaling components, and the involvement of ER Ca^{2+} signaling in different naturally occurring diseases. Since ER/SR Ca^{2+} signaling plays a prominent role in the regulation of nervous and muscle function, most of the diseases affect the skeletal and cardiac muscle and the nervous system, although in some cases other tissues (e.g., the epithelial and endocrine systems) are involved (Table 11.1).

Skeletal Muscle Disorders

Mutations of RyR1 lead to dysfunction of excitation/contraction coupling of the skeletal muscle. In the case of malignant hyperthermia, which is a genetic abnormality with autosomal dominant inheritance,[134,135] the dysfunction is apparent only during application of anaestethics and depolarizing muscle relaxants. The abnormal RyR channels, having mutations in the regulatory part of the channel, are more sensitive to activators, they release Ca^{2+} at higher rates and do not close readily.[136] These changes lead to hyperactivation and spasm of the muscles, accompanied by excessive heat production and general disturbances in bulk homeostasis of the body. In some cases mutations also in the dihydropiridine receptor channel can lead to similar phenotype,[137] and similar mutations of the pig RyR1 leads to a similar syndrome, called porcine stress syndrome.[138] In contrast to malignant hyperthermia patients, mutations in the central cytoplasmic domain or in the transmembrane part of the RyR1[139,140] lead to spontaneous increase in Ca^{2+} release in central core disease, a skeletal muscle myopathy.[141,142] In these cases the myoplasm is continuously exposed to higher $[Ca^{2+}]$, resulting in cramps, then loss of mitochondria, decreased ATP production, thus finally in weakness and atrophy of the muscle.

Loss of SERCA1a, the fast twitch skeletal muscle sarcoplasmic reticulum isoform causes exercise-induced impairment of the relaxation of skeletal muscle, the major diagnostic feature in Brody disease.[143] Ca^{2+} uptake and Ca^{2+} ATPase activities are reduced in the sarcoplasmic reticulum, thus Ca^{2+} is cleared less efficiently from the myoplasm and muscles relaxes more slowly. This symptom really suggest that Brody disease results from defects in the SERCA1a gene, but in fact was proven only in its form with autosomal recessive inheritance.[144]

Cardiac Pathology

Changes in the cellular Ca^{2+} handling are associated with cardiac hypertrophy and associated cardiomyopathies and heart failure. In cardiac muscle the cooperation of RyR2, SERCA2a and its regulator phospholamban (the inhibitory function of which is switched off by protein kinase A or Ca^{2+}-calmodulin dependent kinase II driven phosphorylation) is responsible for the regulation of the continuous contraction/relaxation cycle. Thus dysfunction of either component can contribute to the pathogenesis of cardiac diseases. Indeed numerous studies show reduced Ca^{2+} uptake into the SR in cardiac dysfunction, which can be a consequence of decreased SERCA2a expression, as shown in animal models. Moreover, adenovirus-mediated gene transfer of SERCA2a is able to restore normal function in these models (for reviews see refs. 145,146). On the other hand, studies with knockout and transgenic mice show that gain or inhibitory function of phospholamban alters cardiac contractility and could be a causal feature in some cardiomyopathies.[147,148] Finally, regarding ER Ca^{2+} binding proteins (that will be discussed in more detail in other parts of the book) overexpression of cardiac calsequestrin leads to cardiomyopathy,[149] while ablation of calreticulin alters cardiac development.[150]

These alterations, which lead to reduced SR Ca^{2+} uptake could explain changes in the relaxation phase of the heart cycle, but can not account for alterations in Ca^{2+} release. In this

Table 11.1. Diseases associated with Ca^{2+} signalling components of the ER

Pathology	Disease Characteristics	Pathogenic Mutations	Associated Dysfunction	Calcium Homeostasis	Reference
Skeletal muscle disorders	Malignant hypertherimia (MH) (human autosomal dominant inheritance)	RYR1	Excitation/contraction coupling	Higher ER Ca^{2+} release (inhibited closing of the channel)	134,135
	Central core disease (CCD) (human autosomal dominant inheritance)	RYR1	Hypotonia and proximal muscle weakness	Spontaneous increase in ER Ca^{2+} release	141,142
	Brody disease (human autosomal recessive or dominant inheritance)	SERCA1a	Relaxation of squeletal muscle	Reduced Ca^{2+} uptake and diminished SERCA activity	143
Cardiac disorders	Cardiac hypertrophy (experimental animal models)	calsequestrin	Decreased contractility	Reduced SR Ca^{2+} content	149
	Cardiac hypertrophy and cardiomyopathy (experimental animal models)	phospholamban SERCA2a RYR2	Depression of contractile parameters	Alteration in SR Ca^{2+} uptake and release	146-148 151-158
	Cardiac development (experimental animal models)	calreticulin	Heart failure	Reduced SR Ca^{2+} content	150
Epithelial disorders	Darier's disease (human autosomal dominant inheritance)	SERCA2b epithelial cells	Loss of adhesion between	Inhibition of SERCA2 activity	186, 187
	Polysyctic kidney disease	Polycystin-2 Ca^{2+} release channel	Fluid-filled cysts in the kidney and liver Renal failure	Reduction of Ca^{2+} induced Ca^{2+} release	25

Continued on next page

Table 11.1. Diseases associated with Ca^{2+} signalling components of the ER (continued)

Pathology	Disease Characteristics	Pathogenic Mutations	Associated Dysfunction	Calcium Homeostasis	Reference
Endocrine pathology	Non-insulin-dependant diabets mellitus (NIDDM) (experimental animal model)	SERCA3	Abnormal insulin secretion	Defective glucose dependant ER Ca^{2+} sequestration	188
	Human Type II diabetes	SERCA3	Abnormal insulin secretion	-	189,190
Central nervous system disorders	Opisthonos (naturally occurring IP3R1 mouse mutants)	IP_3R1	Brain ataxia and epileptic seizure	-	159, 162, 163
	Alzheimer disease	Presenilin type 1 and 2	Neuronal cell death	Overfilling of Ca^{2+} stores and enhancement of ER Ca^{2+} release	161, 164-185
	Experimental animal model	RYR3	Learning and synaptic plasticity	Small and brief local ER Ca^{2+} release	161,164

respect, it was shown that in hypertrophy and heart failure reduced activity of the Ca^{2+} pump is accompanied by a down-regulation of RyR2,[151,152] or defect in coupling between plasma membrane Ca^{2+} channels and the RyR. This loss of function can be partly compensated by IP_3Rs,[153] but even in these cases the reduction of Ca^{2+} release is apparent. Another interesting feature of RyRs from a pathological point of view is their inhibition by bound FKBP12.6.[154] This interaction can be disrupted by the application of immunosuppressive drugs like rapamycin or FK506 in humans,[155] or by knocking out the FKBP12 gene in mice.[156] The disruption leads to increased Ca^{2+} release from the SR, which consequently results in hypertrophic cardiomyopathy and heart failure. Similarly, in the failing human heart, RyR2 hyperphosphorylation leads to increased sensitivity of the receptor Ca^{2+} induced Ca^{2+} release.[157] Ca^{2+} transporters outside of ER were also assumed to take part of cardiac pathology (see in 145). As for the role of SERCA3 in the cardiovascular pathology some results point to the involvement of this Ca^{2+} pump in the pathogenesis of some cases of hypertonia.[158]

Disorders in the Central Nervous System

Since the recognition of ER and ER-driven Ca^{2+} signaling as a core component of structural and functional regulation of neuronal circuits, recently several studies investigated its role either from the viewpoint of neuronal development and synaptic plasticity or in the pathological conditions of neurodegenerative diseases. The ER is present virtually in all parts of the neuron bearing either IP_3R or RyR mediated Ca^{2+} release, and there is accumulating evidence that it regulates neurite outgrowth in the developing brain.[159,160] Moreover, in the adult brain long term synaptic regulation appears to be dependent on Ca^{2+} release from the ER, initiating LTD (long term depression), regulating expression of genes (e.g., AP1, NF-κB) involved in the maintenance of LTP (long term potentiation) or causing structural changes in postsynaptic dendritic spines (for review see ref. 161).

In accordance with the IP_3R1 knockout studies, opisthotonos mice, which misses two exons from the IP_3R1 gene, lacking almost all of the normal IP_3R1 population, develop ataxia and epilepsy and die immediately after birth, showing atrophy in cerebellar neurons.[162] Decreased IP_3R immunoreactivity was also shown in patients suffering from ataxia, indicating that these findings may also be relevant in humans.[163] However, the role of ER Ca^{2+} deregulation in neuronal pathologies is not limited to genetic alterations, but rather may play a prominent role in common brain disorders.

As discussed above, alterations of ER mediated Ca^{2+} homeostasis is sufficient to induce apoptosis in a variety of cell types, and recently it has been also demonstrated in neurons.[164] In parallel, the role of Ca^{2+} overload has been extensively investigated during excitotoxicity, induced by exposing neurons to excessive glutamate stimulation. In addition to glutamate mediated Ca^{2+} influx through ionotropic glutamate receptors, Ca^{2+} release from the ER has been shown to contribute to Ca^{2+} overload, and agents inhibiting Ca^{2+} release have a protective effect against glutamate toxicity, and consequent apoptosis.[161,165,166] Even larger evidence point to the involvement of ER Ca^{2+} in a series of neurodegenerative diseases. The most studied case is Alzheimer disease, characterized primarily by accumulation of misprocessed forms of amyloid β-peptide and ensuing neuronal cell death in brain areas involved in learning and memory processes. Inherited mutations of genes coding for presenilin proteins (type 1 and 2: PS1, PS2) account for almost half of the early onset familial Alzheimer's disease (FAD).[167] Interestingly, these proteins are localized in the ER, or in a specialized subregion between the ER and Golgi system.[168] Besides having a role in β-amyloid processing,[169] and the notch signaling pathway,[170] their mutations seem to modify also the ER and consequently the entire cellular Ca^{2+} homeostasis (for a comprehensible summary of presenilin mutations see ref. 171). So far the results concerning this issue unequivocally show an increased Ca^{2+} release induced by IP_3 generating agonists or IP_3 itself in cells and cell lines of neuronal origin,[172,173] in fibroblast cells of genetically modified animals[174] and in Xenopus oocyte models[175] having expressed mutant forms of either PS1 or PS2.[176] Nevertheless, the release channels and the mechanism underlying this augmentation of release are still unclear. Presenilins were shown to colocalize and interact with RyRs, thus causing either increase of RyR density in the ER membrane[173] or a potential regulatory interaction through a protein called sorcin,[177] a known regulator or RyR channel activity. Accordingly, release of Ca^{2+} after glutamate stimulation in hyppocampal neurons can be reversed by dantrolene, an inhibitor of RyRs.[178,179] Interestingly, dantrolene treatment also attenuates cell death conferred by PS mutations. So far, no interaction of presenilins with IP_3Rs has been demonstrated, thus the issue of whether the effect of mutated presenilins can be attributed only to RyR-mediated Ca^{2+} release remains open.

An even less clarified issue is the effect of mutant PSs on the state of ER filling and of store operated Ca^{2+} entry. Expression of mutant PSs leads to inhibition/down regulation of SOC, which in itself can lead to abnormal amyloid-β peptide (Aβ-42) production, and together with Ca^{2+} pool depletion (also caused by increased release) may be responsible for developing the pathological phenotype observed in Alzheimer's disease.[180] Overfilling of the ER Ca^{2+} store was observed in fibroblasts expressing mutant PSs,[181] leading to more shallow store emptying, which in turn may account for impaired activation of SOC. In summary, regardless of the points not clarified so far, modulation of ER Ca^{2+} homeostasis seems to have a profound role in the pathogenesis of FAD caused by mutations of PSs, either by contributing to the production of neurotoxic forms of β-amyloid peptide, or by modifying the filling state of the ER, thus rendering cells more sensitive to apoptosis. This latter aspect would correlate with the observations about the mechanism by which Bcl-2 reduces the efficacy of the apoptotic process. In this regard it is interesting to mention that PSs were shown to interact with Bcl-2.[182] Ultimately, two further observations about inhibitors of mutant PS induced pathology support the role of Ca^{2+} in the pathogenesis of this disease. (i) Expression of the Ca^{2+} binding protein calbindin D28k, by reducing the potentiated Ca^{2+} signaling, protects cells harboring PS mutants from death[183] (see also the effect of dantrolene above); (ii) an interactor of PSs, the Ca^{2+} binding protein/nuclear transcription factor calsenilin/DREAM,[184] is able to counteract the enhancement of Ca^{2+} release and the proapoptotic effect of the mutant PSs, presumably by directly binding to their C-terminal part.

Such a direct involvement of ER Ca^{2+} in other neurodegenerative disorders has not yet been described, although alterations of protein processing enzyme activity and increased Ca^{2+} release was reported also in Huntington's disease.[185] In the case of Parkinson's disease the main patophysiological phenomena include deficiency in mitochondrial complex I function and mutation of the protein α-synuclein, located in synaptic terminals. Therefore alterations in mitochondrial Ca^{2+} handling occur, leading to altered cellular homeostasis, NMDA receptor function, and finally excytotoxicity and cell death. Whether ER mediated Ca^{2+} signals contribute to these processes, has not yet been clarified.

Other Pathologies Associated to the ER Ca^{2+} Homeostasis

Darier's disease is an autosomal dominant skin disorder, characterized by a loss of adhesion between epithelial cells that is associated with breakdown of desmosomal filament complexes and premature keratinization. This disease is associated to mutations in the ATPA2 gene, which encodes the sarco-endoplasmic reticulum Ca^{2+}-ATPase type 2 isoform (SERCA2) and is highly expressed in keratinocytes. The pathophysiology of the disease suggests a role of SERCA2 gene in Ca^{2+}-signaling pathway regulating cell-to-cell adhesion and differentiation of the epidermis.[186] Since SERCA2 is also highly expressed in cardiac and skeletal muscle, it seems likely that in patients with Darier's disease the loss of SERCA2 function in these tissues is compensated by other Ca^{2+} pumps.[187]

Non-insulin-dependent diabetes mellitus (NIDDM) is a metabolic disease associated with abnormal insulin secretion. An animal mouse model of NIDDM revealed a defective glucose dependent ER Ca^{2+} sequestration in pancreatic β-cells.[188] Mutations of SERCA3 gene have been identified in patients with type II diabets indicating that the SERCA3 locus possibly contributes to the genetic susceptibility to this disease.[189] Consistently with these observations, the impaired function and expression of SERCA3 has been associated to B-cell apoptosis in diabetic animal models,[190] also in case of B-cell degeneration caused by chronic hyperglycemia.[191]

In a very recent work, Koulen at al. showed that polycystin-2, expressed abundantly in epithelial cells lead to an IP_3 dependent Ca^{2+} release.[25] Mutations in the genes coding either for polycystin-2 or its apparent plasma membrane partner polycystin-1 leads to polycystic kidney disease, characterized by numerous fluid-filled cysts in the kidney and liver resulting in renal failure in 50% of the cases. Therefore, although by a so far unidentified mechanism, the disturbance of ER driven Ca^{2+} signaling may lead also to this disease.[26]

References

1. Clapham DE. Calcium signaling. Cell 1995; 80:259-268.
2. Berridge MJ, Lipp P, Bootman MD. The versatility and universality of calcium signaling. Nat Rev Mol Cell Biol 2000; 1:11-21.
3. Carafoli E, Santella L, Branca D et al. Generation, control, and processing of cellular calcium signals. Crit Rev Biochem Mol Biol 2001; 36:107-260.
4. Pozzan T, Rizzuto R, Volpe P et al. Molecular and cellular physiology of intracellular calcium stores. Physiol Rev 1994; 74:595-636.
5. Elliott AC. Recent developments in non-excitable cell calcium entry. Cell Calcium 2001; 30:73-93.
6. East JM. Sarco(endo)plasmic reticulum calcium pumps: recent advances in our understanding of structure/function and biology (review). Mol Membr Biol 2000; 17:189-200.
7. Lechleiter JD, John LM, Camacho P. Ca^{2+} wave dispersion and spiral wave entrainment in Xenopus laevis oocytes overexpressing Ca^{2+} ATPases. Biophys Chem 1998; 72:123-129.
8. Missiaen L, Van Acker K, Parys JB et al. Baseline cytosolic Ca2+ oscillations derived from a non-endoplasmic reticulum Ca^{2+} store. J Biol Chem 2001; 276:39161-39170.
9. Mitchell KJ, Pinton P, Varadi A et al. Dense core secretory vesicles revealed as a dynamic Ca^{2+} store in neuroendocrine cells with a vesicle-associated membrane protein aequorin chimaera. J Cell Biol 2001; 155:41-51.
10. da Silva CP, Guse AH. Intracellular Ca^{2+} release mechanisms: multiple pathways having multiple functions within the same cell type? Biochim Biophys Acta 2000; 1498:122-133.
11. Berridge MJ. Elementary and global aspects of calcium signaling. J Exp Biol 1997; 200(Pt 2):315-319.
12. Johnson JD, Chang JP. Function- and agonist-specific Ca^{2+} signaling: the requirement for and mechanism of spatial and temporal complexity in Ca^{2+} signals. Biochem Cell Biol 2000; 78:217-240.
13. Bootman MD, Lipp P, Berridge MJ. The organization and functions of local Ca^{2+} signals. J Cell Sci 2001; 114:2213-2222.
14. Koulen P, Thrower EC. Pharmacological modulation of intracellular Ca^{2+} channels at the single-channel level. Mol Neurobiol 2001; 24:65-86.
15. Shoshan-Barmatz V, Ashley RH. The structure, function, and cellular regulation of ryanodine-sensitive Ca^{2+} release channels. Int Rev Cytol 1998; 183:185-270.
16. Berridge MJ, Irvine RF. Inositol phosphates and cell signaling. Nature 1989; 341:197-205.
17. Monkawa T, Miyawaki A, Sugiyama T et al. Heterotetrameric complex formation of inositol 1,4,5-trisphosphate receptor subunits. J Biol Chem 1995; 270:14700-14704.
18. Patel S, Joseph SK, Thomas AP. Molecular properties of inositol 1,4,5-trisphosphate receptors. Cell Calcium 1999; 25:247-264.
19. Guse AH. Cyclic ADP-ribose. J Mol Med 2000; 78:26-35.
20. Noguchi N, Takasawa S, Nata K et al. Cyclic ADP-ribose binds to FK506-binding protein 12.6 to release Ca^{2+} from islet microsomes. J Biol Chem 1997; 272:3133-3136.
21. Thomas JM, Masgrau R, Churchill GC et al. Pharmacological characterization of the putative cADP-ribose receptor. Biochem J 2001; 359:451-457.
22. Meldolesi J. Rapidly exchanging Ca^{2+} stores in neurons: molecular, structural and functional properties. Prog Neurobiol 2001; 65:309-338.
23. Patel S, Churchill GC, Galione A. Coordination of Ca^{2+} signaling by NAADP. Trends Biochem Sci 2001; 26:482-489.
24. Patel S, Churchill GC, Sharp T et al. Widespread distribution of binding sites for the novel Ca^{2+}-mobilizing messenger, nicotinic acid adenine dinucleotide phosphate, in the brain. J Biol Chem 2000; 275:36495-36497.
25. Koulen P, Cai Y, Geng L et al. Polycystin-2 is an intracellular calcium release channel. Nat Cell Biol 2002; 4:191-197.
26. Gonzalez-Perret S, Kim K, Ibarra C et al. Polycystin-2, the protein mutated in autosomal dominant polycystic kidney disease (ADPKD), is a Ca^{2+}-permeable nonselective cation channel. Proc Natl Acad Sci USA 2001; 98:1182-1187.

27. John LM, Lechleiter JD, Camacho P. Differential modulation of SERCA2 isoforms by calreticulin. J Cell Biol 1998; 142:963-973.
28. Meldolesi J, Pozzan T. The endoplasmic reticulum Ca^{2+} store: a view from the lumen. Trends Biochem Sci 1998; 23:10-14.
29. Putney JW Jr, Broad LM, Braun FJ et al. Mechanisms of capacitative calcium entry. J Cell Sci 2001; 114:2223-2229.
30. Putney JW Jr, Ribeiro CM. Signaling pathways between the plasma membrane and endoplasmic reticulum calcium stores. Cell Mol Life Sci 2000; 57:1272-1286.
31. Ellgaard L, Helenius A. ER quality control: towards an understanding at the molecular level. Curr Opin Cell Biol 2001; 13:431-437.
32. Corbett EF, Michalak M. Calcium, a signaling molecule in the endoplasmic reticulum? Trends Biochem Sci 2000; 25:307-311.
33. Michalak M, Corbett EF, Mesaeli N et al. Calreticulin: one protein, one gene, many functions. Biochem J 1999; 344 Pt 2:281-292.
34. Pahl HL, Baeuerle PA. The ER-overload response: activation of NF-kappa B. Trends Biochem Sci 1997; 22:63-67.
35. Soboloff J, Berger SA. Sustained ER Ca^{2+} Depletion Suppresses Protein Synthesis and Induces Activation-enhanced Cell Death in Mast Cells. J Biol Chem 2002; 277:13812-13820.
36. Ferri KF, Kroemer G. Organelle-specific initiation of cell death pathways. Nat Cell Biol 2001; 3:E255-E263.
37. Michalak M, Burns K, Andrin C et al. Endoplasmic reticulum form of calreticulin modulates glucocorticoid- sensitive gene expression. J Biol Chem 1996; 271:29436-29445.
38. Opas M, Szewczenko-Pawlikowski M, Jass GK et al. Calreticulin modulates cell adhesiveness via regulation of vinculin expression. J Cell Biol 1996; 135:1913-1923.
39. Kaufman RJ. Molecular chaperones and the heat shock response. Biochim Biophys Acta 1999; 1423:R13-R27.
40. Meldolesi J, Pozzan T. The heterogeneity of ER Ca^{2+} stores has a key rôle in nonmuscle cell signaling and function. J Cell Biol 1998; 142:1395-1398.
41. Lee MG, Xu X, Zeng W et al. Polarized expression of Ca^{2+} pumps in pancreatic and salivary gland cells. Role in initiation and propagation of $[Ca^{2+}]i$ waves. J Biol Chem 1997; 272:15771-15776.
42. Kasai H, Petersen OH. Spatial dynamics of second messengers: IP3 and cAMP as long-range and associative messengers. Trends Neurosci 1994; 17:95-101.
43. Petersen OH, Burdakov D, Tepikin AV. Polarity in intracellular calcium signaling. Bioessays 1999; 21:851-860.
44. Blaustein MP, Golovina VA. Structural complexity and functional diversity of endoplasmic reticulum Ca^{2+} stores. Trends Neurosci 2001; 24:602-608.
45. Petersen OH, Tepikin A, Park MK. The endoplasmic reticulum: one continuous or several separate Ca^{2+} stores? Trends Neurosci 2001; 24:271-276.
46. Rizzuto R, Simpson AW, Brini M et al. Rapid changes of mitochondrial Ca^{2+} revealed by specifically targeted recombinant aequorin. Nature 1992; 358:325-327.
47. Rizzuto R, Brini M, Murgia M et al. Microdomains with high Ca^{2+} close to IP3-sensitive channels that are sensed by neighboring mitochondria. Science 1993; 262:744-747.
48. Rizzuto R, Bastianutto C, Brini M et al. Mitochondrial Ca^{2+} homeostasis in intact cells. J Cell Biol 1994; 126:1183-1194.
49. Hajnoczky G, Robb-Gaspers LD, Seitz MB et al. Decoding of cytosolic calcium oscillations in the mitochondria. Cell 1995; 82:415-424.
50. Brini M, De Giorgi F, Murgia M et al. Subcellular analysis of Ca^{2+} homeostasis in primary cultures of skeletal muscle myotubes. Mol Biol Cell 1997; 8:129-143.
51. Montero M, Alonso MT, Carnicero E et al. Chromaffin-cell stimulation triggers fast millimolar mitochondrial Ca^{2+} transients that modulate secretion. Nat Cell Biol 2000; 2:57-61.
52. Szalai G, Csordas G, Hantash BM et al. Calcium signal transmission between ryanodine receptors and mitochondria. J Biol Chem 2000; 275:15305-15313.
53. Satoh T, Ross CA, Villa A et al. The inositol 1,4,5,-trisphosphate receptor in cerebellar Purkinje cells: quantitative immunogold labeling reveals concentration in an ER subcompartment. J Cell Biol 1990; 111:615-624.
54. Takei K, Stukenbrok H, Metcalf A et al. Ca^{2+} stores in Purkinje neurons: endoplasmic reticulum subcompartments demonstrated by the heterogeneous distribution of the InsP3 receptor, Ca^{2+}-ATPase, and calsequestrin. J Neurosci 1992; 12:489-505.
55. Mannella CA, Buttle K, Rath BK et al. Electron microscopic tomography of rat-liver mitochondria and their interaction with the endoplasmic reticulum. Biofactors 1998; 8:225-228.

56. Rizzuto R, Pinton P, Carrington W et al. Close contacts with the endoplasmic reticulum as determinants of mitochondrial Ca^{2+} responses. Science 1998; 280:1763-1766.
57. Csordas G, Thomas AP, Hajnoczky G. Quasi-synaptic calcium signal transmission between endoplasmic reticulum and mitochondria. EMBO J 1999; 18:96-108.
58. Friel DD, Tsien RW. An FCCP-sensitive Ca^{2+} store in bullfrog sympathetic neurons and its participation in stimulus-evoked changes in $[Ca^{2+}]i$. J Neurosci 1994; 14:4007-4024.
59. Jouaville LS, Ichas F, Mazat JP. Modulation of cell calcium signals by mitochondria. Mol Cell Biochem 1998; 184:371-376.
60. Hajnoczky G, Hager R, Thomas AP. Mitochondria suppress local feedback activation of inositol 1,4, 5- trisphosphate receptors by Ca^{2+}. J Biol Chem 1999; 274:14157-14162.
61. Duchen MR. Contributions of mitochondria to animal physiology: from homeostatic sensor to calcium signaling and cell death. J Physiol 1999; 516(Pt 1):1-17.
62. Landolfi B, Curci S, Debellis L et al. Ca^{2+} homeostasis in the agonist-sensitive internal store: functional interactions between mitochondria and the ER measured In situ in intact cells. J Cell Biol 1998; 142:1235-1243.
63. Arnaudeau S, Kelley WL, Walsh JV Jr et al. Mitochondria recycle Ca^{2+} to the endoplasmic reticulum and prevent the depletion of neighboring endoplasmic reticulum regions. J Biol Chem 2001; 276:29430-29439.
64. Vandecasteele G, Szabadkai G, Rizzuto R. Mitochondrial calcium homeostasis: mechanisms and molecules. IUBMB Life 2001; 52:213-219.
65. Hajnoczky G, Csordas G, Madesh M et al. The machinery of local Ca^{2+} signaling between sarco-endoplasmic reticulum and mitochondria. J Physiol 2000; 529 Pt 1:69-81.
66. Wilson BS, Pfeiffer JR, Smith AJ et al. Calcium-dependent clustering of inositol 1,4,5-trisphosphate receptors. Mol Biol Cell 1998; 9:1465-1478.
67. Katayama E, Funahashi H, Michikawa T et al. Native structure and arrangement of inositol-1,4,5-trisphosphate receptor molecules in bovine cerebellar Purkinje cells as studied by quick-freeze deep-etch electron microscopy. EMBO J 1996; 15:4844-4851.
68. Wang HJ, Guay G, Pogan L et al. Calcium regulates the association between mitochondria and a smooth subdomain of the endoplasmic reticulum. J Cell Biol 2000; 150:1489-1498.
69. Shiao YJ, Balcerzak B, Vance JE. A mitochondrial membrane protein is required for translocation of phosphatidylserine from mitochondria-associated membranes to mitochondria. Biochem J 1998; 331 (Pt 1):217-223.
70. Park MK, Petersen OH, Tepikin AV. The endoplasmic reticulum as one continuous Ca^{2+} pool: visualization of rapid Ca^{2+} movements and equilibration. EMBO J 2000; 19:5729-5739.
71. Xiao B, Tu JC, Worley PF. Homer: a link between neural activity and glutamate receptor function. Curr Opin Neurobiol 2000; 10:370-374.
72. Xiao B, Tu JC, Petralia RS et al. Homer regulates the association of group 1 metabotropic glutamate receptors with multivalent complexes of homer-related, synaptic proteins. Neuron 1998; 21:707-716.
73. Park MK, Ashby MC, Erdemli G et al. Perinuclear, perigranular and sub-plasmalemmal mitochondria have distinct functions in the regulation of cellular calcium transport. EMBO J 2001; 20:1863-1874.
74. Sabatini BL, Maravall M, Svoboda K. Ca^{2+} signaling in dendritic spines. Curr Opin Neurobiol 2001; 11:349-356.
75. Fasolato C, Hoth M, Penner R. A GTP-dependent step in the activation mechanism of capacitative calcium influx. J Biol Chem 1993; 268:20737-20740.
76. Yao Y, Ferrer-Montiel AV, Montal M et al. Activation of store-operated Ca^{2+} current in Xenopus oocytes requires SNAP-25 but not a diffusible messenger. Cell 1999; 98:475-485.
77. Broad LM, Braun FJ, Lievremont JP et al. Role of the phospholipase C-inositol 1,4,5-trisphosphate pathway in calcium release-activated calcium current and capacitative calcium entry. J Biol Chem 2001; 276:15945-15952.
78. Tsunoda S, Zuker CS. The organization of INAD-signaling complexes by a multivalent PDZ domain protein in Drosophila photoreceptor cells ensures sensitivity and speed of signaling. Cell Calcium 1999; 26:165-171.
79. Montell C. Physiology, phylogeny, and functions of the TRP superfamily of cation channels. Review. Sci STKE 2001; 90:RE1.
80. Li HS, Montell C. TRP and the PDZ protein, INAD, form the core complex required for retention of the signalplex in Drosophila photoreceptor cells. J Cell Biol 2000; 150:1411-1422.
81. Petersen OH, Gerasimenko OV, Gerasimenko JV et al. The calcium store in the nuclear envelope. Cell Calcium 1998; 23:87-90.
82. Badminton MN, Kendall JM, Rembold CM et al. Current evidence suggests independent regulation of nuclear calcium. Cell Calcium 1998; 23:79-86.

83. Brini M, Murgia M, Pasti L et al. Nuclear Ca^{2+} concentration measured with specifically targeted recombinant aequorin. EMBO J 1993; 12:4813-4819.
84. Connor JA. Intracellular calcium mobilization by inositol 1,4,5-trisphosphate: intracellular movements and compartmentalization. Cell Calcium 1993; 14:185-200.
85. Bootman MD, Thomas D, Tovey SC et al. Nuclear calcium signaling. Cell Mol Life Sci 2000; 57:371-378.
86. Berridge MJ. Calcium signaling and cell proliferation. Bioessays 1995; 17:491-500.
87. Duchen MR. Mitochondria and calcium: from cell signaling to cell death. J Physiol 2000; 529(Pt 1):57-68.
88. Nicotera P, Orrenius S. The role of calcium in apoptosis. Cell Calcium 1998; 23:173-180.
89. Bernardi P, Scorrano L, Colonna R et al. Mitochondria and cell death. Mechanistic aspects and methodological issues. Eur J Biochem 1999; 264:687-701.
90. Hajnoczky G, Csordas G, Madesh M et al. Control of apoptosis by IP(3) and ryanodine receptor driven calcium signals. Cell Calcium 2000; 28:349-363.
91. Jayaraman T, Marks AR. T cells deficient in inositol 1,4,5-trisphosphate receptor are resistant to apoptosis. Mol Cell Biol 1997; 17:3005-3012.
92. Khan AA, Soloski MJ, Sharp AH et al. Lymphocyte apoptosis: mediation by increased type 3 inositol 1,4,5- trisphosphate receptor. Science 1996; 273:503-507.
93. Blackshaw S, Sawa A, Sharp AH et al. Type 3 inositol 1,4,5-trisphosphate receptor modulates cell death. FASEB J 2000; 14:1375-1379.
94. Nakagawa T, Zhu H, Morishima N et al. Caspase-12 mediates endoplasmic-reticulum-specific apoptosis and cytotoxicity by amyloid-beta. Nature 2000; 403:98-103.
95. Nakagawa T, Yuan J. Cross-talk between two cysteine protease families. Activation of caspase-12 by calpain in apoptosis. J Cell Biol 2000; 150:887-894.
96. Subramanian K, Meyer T. Calcium-induced restructuring of nuclear envelope and endoplasmic reticulum calcium stores. Cell 1997; 89:963-971.
97. Graber MN, Alfonso A, Gill DL. Ca^{2+} pools and cell growth: arachidonic acid induces recovery of cells growth-arrested by Ca^{2+} pool depletion. J Biol Chem 1996; 271:883-888.
98. Short AD, Bian J, Ghosh TK et al. Intracellular Ca^{2+} pool content is linked to control of cell growth. Proc Natl Acad Sci USA 1993; 90:4986-4990.
99. Baffy G, Miyashita T, Williamson JR et al. Apoptosis induced by withdrawal of interleukin-3 (IL-3) from an IL-3- dependent hematopoietic cell line is associated with repartitioning of intracellular calcium and is blocked by enforced Bcl-2 oncoprotein production. J Biol Chem 1993; 268:6511-6519.
100. Distelhorst CW, Lam M, McCormick TS. Bcl-2 inhibits hydrogen peroxide-induced ER Ca^{2+} pool depletion. Oncogene 1996; 12:2051-2055.
101. Preston GA, Barrett JC, Biermann JA et al. Effects of alterations in calcium homeostasis on apoptosis during neoplastic progression. Cancer Res 1997; 57:537-542.
102. Pinton P, Ferrari D, Rapizzi E et al. The Ca^{2+} concentration of the endoplasmic reticulum is a key determinant of ceramide-induced apoptosis: significance for the molecular mechanism of Bcl-2 action. EMBO J 2001; 20:2690-2701.
103. Pinton P, Ferrari D, Magalhaes P et al. Reduced loading of intracellular Ca^{2+} stores and downregulation of capacitative Ca^{2+} influx in Bcl-2-overexpressing cells. J Cell Biol 2000; 148:857-862.
104. Dowd DR, MacDonald PN, Komm BS et al. Stable expression of the calbindin-D28K complementary DNA interferes with the apoptotic pathway in lymphocytes. Mol Endocrinol 1992; 6:1843-1848.
105. Ma TS, Mann DL, Lee JH et al. SR compartment calcium and cell apoptosis in SERCA overexpression. Cell Calcium 1999; 26:25-36.
106. Schendel SL, Xie Z, Montal MO et al. Channel formation by antiapoptotic protein Bcl-2. Proc Natl Acad Sci USA 1997; 94:5113-5118.
107. Minn AJ, Velez P, Schendel SL et al. Bcl-x(L) forms an ion channel in synthetic lipid membranes. Nature 1997; 385:353-357.
108. Schlesinger PH, Gross A, Yin XM et al. Comparison of the ion channel characteristics of proapoptotic BAX and antiapoptotic BCL-2. Proc Natl Acad Sci USA 1997; 94:11357-11362.
109. Rudner J, Lepple-Wienhues A, Budach W et al. Wild-type, mitochondrial and ER-restricted Bcl-2 inhibit DNA damage- induced apoptosis but do not affect death receptor-induced apoptosis. J Cell Sci 2001; 114:4161-4172.
110. Wang NS, Unkila MT, Reineks EZ et al. Transient expression of wild-type or mitochondrially targeted Bcl-2 induces apoptosis, whereas transient expression of endoplasmic reticulum-targeted Bcl-2 is protective against Bax-induced cell death. J Biol Chem 2001; 276:44117-44128.

111. Lee ST, Hoeflich KP, Wasfy GW et al. Bcl-2 targeted to the endoplasmic reticulum can inhibit apoptosis induced by Myc but not etoposide in Rat-1 fibroblasts. Oncogene 1999; 18:3520-3528.
112. He H, Lam M, McCormick TS et al. Maintenance of calcium homeostasis in the endoplasmic reticulum by Bcl- 2. J Cell Biol 1997; 138:1219-1228.
113. Foyouzi-Youssefi R, Arnaudeau S, Borner C et al. Bcl-2 decreases the free Ca^{2+} concentration within the endoplasmic reticulum. Proc Natl Acad Sci USA 2000; 97:5723-5728.
114. Hofer AM, Fasolato C, Pozzan T. Capacitative Ca^{2+} entry is closely linked to the filling state of internal Ca^{2+} stores: a study using simultaneous measurements of ICRAC and intraluminal $[Ca^{2+}]$. J Cell Biol 1998; 140:325-334.
115. Bian X, Hughes FM Jr, Huang Y et al. Roles of cytoplasmic Ca^{2+} and intracellular Ca^{2+} stores in induction and suppression of apoptosis in S49 cells. Am J Physiol 1997; 272:C1241-C1249.
116. Szalai G, Krishnamurthy R, Hajnoczky G. Apoptosis driven by IP(3)-linked mitochondrial calcium signals. EMBO J 1999; 18:6349-6361.
117. Fasolato C, Pizzo P, Pozzan T. Delayed activation of the store-operated calcium current induced by calreticulin overexpression in RBL-1 cells. Mol Biol Cell 1998; 9:1513-1522.
118. Xu W, Longo FJ, Wintermantel MR et al. Calreticulin modulates capacitative Ca^{2+} influx by controlling the extent of inositol 1,4,5-trisphosphate-induced Ca^{2+} store depletion. J Biol Chem 2000; 275:36676-36682.
119. Camacho P, Lechleiter JD. Calreticulin inhibits repetitive intracellular Ca^{2+} waves. Cell 1995; 82:765-771.
120. Bastianutto C, Clementi E, Codazzi F et al. Overexpression of calreticulin increases the Ca^{2+} capacity of rapidly exchanging Ca^{2+} stores and reveals aspects of their lumenal microenvironment and function. J Cell Biol 1995; 130:847-855.
121. Mery L, Mesaeli N, Michalak M et al. Overexpression of calreticulin increases intracellular Ca^{2+} storage and decreases store-operated Ca^{2+} influx. J Biol Chem 1996; 271:9332-9339.
122. Nakamura K, Bossy-Wetzel E, Burns K et al. Changes in endoplasmic reticulum luminal environment affect cell sensitivity to apoptosis. J Cell Biol 2000; 150:731-740.
123. Simon VR, Moran MF. SERCA activity is required for timely progression through G1/S. Cell Prolif 2001; 34:15-30.
124. Cheng G, Liu BF, Yu Y et al. The exit from G(0) into the cell cycle requires and is controlled by sarco(endo)plasmic reticulum Ca^{2+} pump. Arch Biochem Biophys 1996; 329:65-72.
125. Meehan S, Wu AJ, Kang EC et al. Interferon-gamma induces a decrease in the intracellular calcium pump in a human salivary gland cell line. Am J Physiol 1997; 273:C2030-C2036.
126. Launay S, Bobe R, Lacabaratz-Porret C et al. Modulation of endoplasmic reticulum calcium pump expression during T lymphocyte activation. J Biol Chem 1997; 272:10746-10750.
127. Launay S, Gianni M, Kovacs T et al. Lineage-specific modulation of calcium pump expression during myeloid differentiation. Blood 1999; 93:4395-4405.
128. Magnier C, Papp B, Corvazier E et al. Regulation of sarco-endoplasmic reticulum Ca^{2+}-ATPases during platelet-derived growth factor-induced smooth muscle cell proliferation. J Biol Chem 1992; 267:15808-15815.
129. Magnier-Gaubil C, Herbert JM, Quarck R et al. Smooth muscle cell cycle and proliferation. Relationship between calcium influx and sarco-endoplasmic reticulum Ca^{2+}ATPase regulation. J Biol Chem 1996; 271:27788-27794.
130. De Smedt H, Eggermont JA, Wuytack F et al. Isoform switching of the sarco(endo)plasmic reticulum Ca^{2+} pump during differentiation of BC3H1 myoblasts. J Biol Chem 1991; 266:7092-7095.
131. Chami M, Gozuacik D, Saigo K et al. Hepatitis B virus-related insertional mutagenesis implicates SERCA1 gene in the control of apoptosis. Oncogene 2000; 19:2877-2886.
132. Chami M, Gozuacik D, Lagorce D et al. SERCA1 truncated proteins unable to pump calcium reduce the endoplasmic reticulum calcium concentration and induce apoptosis. J Cell Biol 2001; 153:1301-1314.
133. Tsuneoka M, Mekada E. Ras/MEK signaling suppresses Myc-dependent apoptosis in cells transformed by c-myc and activated ras. Oncogene 2000; 19:115-123.
134. MacLennan DH, Duff C, Zorzato F et al. Ryanodine receptor gene is a candidate for predisposition to malignant hyperthermia. Nature 1990; 343:559-561.
135. MacLennan DH, Phillips MS. Malignant hyperthermia. Science 1992; 256:789-794.
136. Tong J, McCarthy TV, MacLennan DH. Measurement of resting cytosolic Ca^{2+} concentrations and Ca^{2+} store size in HEK-293 cells transfected with malignant hyperthermia or central core disease mutant Ca^{2+} release channels. J Biol Chem 1999; 274:693-702.
137. Monnier N, Procaccio V, Stieglitz P et al. Malignant-hyperthermia susceptibility is associated with a mutation of the alpha 1-subunit of the human dihydropyridine-sensitive L-type voltage-dependent calcium-channel receptor in skeletal muscle. Am J Hum Genet 1997; 60:1316-1325.

138. Fujii J, Otsu K, Zorzato F et al. Identification of a mutation in porcine ryanodine receptor associated with malignant hyperthermia. Science 1991; 253:448-451.
139. Brandt A, Schleithoff L, Jurkat-Rott K et al. Screening of the ryanodine receptor gene in 105 malignant hyperthermia families: novel mutations and concordance with the in vitro contracture test. Hum Mol Genet 1999; 8:2055-2062.
140. Lynch PJ, Tong J, Lehane M et al. A mutation in the transmembrane/luminal domain of the ryanodine receptor is associated with abnormal Ca^{2+} release channel function and severe central core disease. Proc Natl Acad Sci USA 1999; 96:4164-4169.
141. Zhang Y, Chen HS, Khanna VK et al. A mutation in the human ryanodine receptor gene associated with central core disease. Nat Genet 1993; 5:46-50.
142. Quane KA, Healy JM, Keating KE et al. Mutations in the ryanodine receptor gene in central core disease and malignant hyperthermia. Nat Genet 1993; 5:51-55.
143. Odermatt A, Taschner PE, Khanna VK et al. Mutations in the gene-encoding SERCA1, the fast-twitch skeletal muscle sarcoplasmic reticulum Ca^{2+} ATPase, are associated with Brody disease. Nat Genet 1996; 14:191-194.
144. Odermatt A, Taschner PE, Scherer SW et al. Characterization of the gene encoding human sarcolipin (SLN), a proteolipid associated with SERCA1: absence of structural mutations in five patients with Brody disease. Genomics 1997; 45:541-553.
145. Missiaen L, Robberecht W, Van den BL et al. Abnormal intracellular Ca^{2+} homeostasis and disease. Cell Calcium 2000; 28:1-21.
146. Periasamy M, Huke S. SERCA pump level is a critical determinant of Ca^{2+} homeostasis and cardiac contractility. J Mol Cell Cardiol 2001; 33:1053-1063.
147. Zvaritch E, Backx PH, Jirik F et al. The transgenic expression of highly inhibitory monomeric forms of phospholamban in mouse heart impairs cardiac contractility. J Biol Chem 2000; 275:14985-14991.
148. Zhai J, Schmidt AG, Hoit BD et al. Cardiac-specific overexpression of a superinhibitory pentameric phospholamban mutant enhances inhibition of cardiac function in vivo. J Biol Chem 2000; 275:10538-10544.
149. Jones LR, Suzuki YJ, Wang W et al. Regulation of Ca^{2+} signaling in transgenic mouse cardiac myocytes overexpressing calsequestrin. J Clin Invest 1998; 101:1385-1393.
150. Mesaeli N, Nakamura K, Zvaritch E et al. Calreticulin is essential for cardiac development. J Cell Biol 1999; 144:857-868.
151. Matsui H, MacLennan DH, Alpert NR et al. Sarcoplasmic reticulum gene expression in pressure overload-induced cardiac hypertrophy in rabbit. Am J Physiol 1995; 268:C252-C258.
152. Rannou F, Sainte-Beuve C, Oliviero P et al. The effects of compensated cardiac hypertrophy on dihydropyridine and ryanodine receptors in rat, ferret and guinea-pig hearts. J Mol Cell Cardiol 1995; 27:1225-1234.
153. Go LO, Moschella MC, Watras J et al. Differential regulation of two types of intracellular calcium release channels during end-stage heart failure. J Clin Invest 1995; 95:888-894.
154. Lam E, Martin MM, Timerman AP et al. A novel FK506 binding protein can mediate the immunosuppressive effects of FK506 and is associated with the cardiac ryanodine receptor. J Biol Chem 1995; 270:26511-26522.
155. Atkison P, Joubert G, Barron A et al. Hypertrophic cardiomyopathy associated with tacrolimus in paediatric transplant patients. Lancet 1995; 345:894-896.
156. Shou W, Aghdasi B, Armstrong DL et al. Cardiac defects and altered ryanodine receptor function in mice lacking FKBP12. Nature 1998; 391:489-492.
157. Marx SO, Reiken S, Hisamatsu Y et al. PKA phosphorylation dissociates FKBP12.6 from the calcium release channel (ryanodine receptor): defective regulation in failing hearts. Cell 2000; 101:365-376.
158. Oshima T, Young EW, McCarron DA. Abnormal platelet and lymphocyte calcium handling in prehypertensive rats. Hypertension 1991; 18:111-115.
159. Dent MA, Raisman G, Lai FA. Expression of type 1 inositol 1,4,5-trisphosphate receptor during axogenesis and synaptic contact in the central and peripheral nervous system of developing rat. Development 1996; 122:1029-1039.
160. Lankford KL, Rand MN, Waxman SG et al. Blocking Ca^{2+} mobilization with thapsigargin reduces neurite initiation in cultured adult rat DRG neurons. Brain Res Dev Brain Res 1995; 84:151-163.
161. Mattson MP, LaFerla FM, Chan SL et al. Calcium signaling in the ER: its role in neuronal plasticity and neurodegenerative disorders. Trends Neurosci 2000; 23:222-229.
162. Street VA, Bosma MM, Demas VP et al. The type 1 inositol 1,4,5-trisphosphate receptor gene is altered in the opisthotonos mouse. J Neurosci 1997; 17:635-645.

163. Zecevic N, Milosevic A, Ehrlich BE. Calcium signaling molecules in human cerebellum at midgestation and in ataxia. Early Hum Dev 1999; 54:103-116.
164. Toescu EC. Apoptosis and cell death in neuronal cells: where does Ca^{2+} fit in? Cell Calcium 1998; 24:387-403.
165. Khodorov BI. Mechanisms of destabilization of Ca^{2+}-homeostasis of brain neurons caused by toxic glutamate challenge. Membr Cell Biol 2000; 14:149-162.
166. Nicholls DG, Budd SL. Mitochondria and neuronal survival. Physiol Rev 2000; 80:315-360.
167. Tanzi RE, Bertram L. New frontiers in Alzheimer's disease genetics. Neuron 2001; 32:181-184.
168. Kim SH, Lah JJ, Thinakaran G et al. Subcellular localization of presenilins: association with a unique membrane pool in cultured cells. Neurobiol Dis 2000; 7:99-117.
169. Scheuner D, Eckman C, Jensen M et al. Secreted amyloid beta-protein similar to that in the senile plaques of Alzheimer's disease is increased in vivo by the presenilin 1 and 2 and APP mutations linked to familial Alzheimer's disease. Nat Med 1996; 2:864-870.
170. Fortini ME. Notch and presenilin: a proteolytic mechanism emerges. Curr Opin Cell Biol 2001; 13:627-634.
171. Fraser PE, Yang DS, Yu G et al. Presenilin structure, function and role in Alzheimer disease. Biochim Biophys Acta 2000; 1502:1-15.
172. Guo Q, Furukawa K, Sopher BL et al. Alzheimer's PS-1 mutation perturbs calcium homeostasis and sensitizes PC12 cells to death induced by amyloid beta-peptide. Neuroreport 1996; 8:379-383.
173. Chan SL, Mayne M, Holden CP et al. Presenilin-1 mutations increase levels of ryanodine receptors and calcium release in PC12 cells and cortical neurons. J Biol Chem 2000; 275:18195-18200.
174. Leissring MA, Parker I, LaFerla FM. Presenilin-2 mutations modulate amplitude and kinetics of inositol 1, 4,5-trisphosphate-mediated calcium signals. J Biol Chem 1999; 274:32535-32538.
175. Leissring MA, Paul BA, Parker I et al. Alzheimer's presenilin-1 mutation potentiates inositol 1,4,5-trisphosphate-mediated calcium signaling in Xenopus oocytes. J Neurochem 1999; 72:1061-1068.
176. Leissring MA, LaFerla FM, Callamaras N et al. Subcellular mechanisms of presenilin-mediated enhancement of calcium signaling. Neurobiol Dis 2001; 8:469-478.
177. Pack-Chung E, Meyers MB, Pettingell WP et al. Presenilin 2 interacts with sorcin, a modulator of the ryanodine receptor. J Biol Chem 2000; 275:14440-14445.
178. Parent A, Linden DJ, Sisodia SS et al. Synaptic transmission and hippocampal long-term potentiation in transgenic mice expressing FAD-linked presenilin 1. Neurobiol Dis 1999; 6:56-62.
179. Zaman SH, Parent A, Laskey A et al. Enhanced synaptic potentiation in transgenic mice expressing presenilin 1 familial Alzheimer's disease mutation is normalized with a benzodiazepine. Neurobiol Dis 2000; 7:54-63.
180. Yoo AS, Cheng I, Chung S et al. Presenilin-mediated modulation of capacitative calcium entry. Neuron 2000; 27:561-572.
181. Leissring MA, Akbari Y, Fanger CM et al. Capacitative calcium entry deficits and elevated luminal calcium content in mutant presenilin-1 knockin mice. J Cell Biol 2000; 149:793-798.
182. Araki W, Yuasa K, Takeda S et al. Pro-apoptotic effect of presenilin 2 (PS2) overexpression is associated with down-regulation of Bcl-2 in cultured neurons. J Neurochem 2001; 79:1161-1168.
183. Guo Q, Sopher BL, Furukawa K et al. Alzheimer's presenilin mutation sensitizes neural cells to apoptosis induced by trophic factor withdrawal and amyloid beta-peptide: involvement of calcium and oxyradicals. J Neurosci 1997; 17:4212-4222.
184. Leissring MA, Yamasaki TR, Wasco W et al. Calsenilin reverses presenilin-mediated enhancement of calcium signaling. Proc Natl Acad Sci USA 2000; 97:8590-8593.
185. Cross AJ, Crow TJ, Johnson JA et al. Loss of endoplasmic reticulum-associated enzymes in affected brain regions in Huntington's disease and Alzheimer-type dementia. J Neurol Sci 1985; 71:137-143.
186. Sakuntabhai A, Ruiz-Perez V, Carter S et al. Mutations in ATP2A2, encoding a Ca^{2+} pump, cause Darier disease. Nat Genet 1999; 21:271-277.
187. Peacocke M, Christiano AM. Bumps and pumps, SERCA 1999. Nat Genet 1999; 21:252-253.
188. Roe MW, Philipson LH, Frangakis CJ et al. Defective glucose-dependent endoplasmic reticulum Ca^{2+} sequestration in diabetic mouse islets of Langerhans. J Biol Chem 1994; 269:18279-18282.
189. Varadi A, Lebel L, Hashim Y et al. Sequence variants of the sarco(endo)plasmic reticulum Ca^{2+}-transport ATPase 3 gene (SERCA3) in Caucasian type II diabetic patients (UK Prospective Diabetes Study 48). Diabetologia 1999; 42:1240-1243.
190. Varadi A, Molnar E, Ostenson CG et al. Isoforms of endoplasmic reticulum Ca^{2+}-ATPase are differentially expressed in normal and diabetic islets of Langerhans. Biochem J 1996; 319(Pt 2):521-527.
191. Jonas JC, Sharma A, Hasenkamp W et al. Chronic hyperglycemia triggers loss of pancreatic beta cell differentiation in an animal model of diabetes. J Biol Chem 1999; 274:14112-14121.

Chapter 12

Calnexin and Calreticulin, ER Associated Modulators of Calcium Transport in the ER

Patricia Camacho, Linu John, Yun Li, R. Madelaine Paredes and H. Llewelyn Roderick

Abstract

Calreticulin (CRT) and calnexin (CNX) are members of a family of endoplasmic reticulum (ER) chaperones that fold newly synthesized polypeptides. Aside from their role as foldases in the ER, our laboratory has shown that all members of this family of proteins modulate Ca^{2+} oscillations. In *Xenopus* oocytes and other cells, stimulation by G-protein and tyrosine coupled receptors results in Ca^{2+} release from the Inositol 1,4,5 trisphosphate receptor (IP_3R) located in the ER. Following release, Ca^{2+} is re-sequestered into the ER by Ca^{2+} ATPases of the SERCA family. CRT and CNX overexpression inhibit Ca^{2+} oscillations when co-expressed with SERCA2b or when oocytes are treated with pyruvate malate to induce oscillations. By domain deletion mutagenesis of CRT we have determined that the N and P domains are necessary for the inhibition of Ca^{2+} oscillations. The mechanism of inhibition may involve a lectin-like interaction since mutagenesis of a lumenal asparagine to alanine in SERCA2b destroys the inhibitory effect. Coexpression of SERCA2a (which lacks the luminal asparagine) with either CRT or CNX does not inhibit Ca^{2+} oscillations, consistent with the notion that a lectin interaction may be involved. Unlike CRT, which is entirely lumenal, CNX has a cytosolic domain that is phosphorylated by multiple kinases. Mutagenesis of two PKC/PDK residues in CNX indicated that S562 supports phosphorylation. Expression of SERCA2b with a mutated CNX in S562 prevents the inhibition of Ca^{2+} oscillations suggesting that this residue serves as a phosphorylatable regulatory switch controlling the interaction of CNX with SERCA2b. Indeed, immunoprecipitations with a CNX specific antibody of oocytes treated with or without IP_3 and preloaded with [γ-^{32}P]-ATP demonstrated that S562 is phosphorylated at rest and dephosphorylated in response to IP_3. Phosphorylation-mediated control of the interaction of CNX with SERCA2b is of significance since it suggests a bi-directional mode of communication between the Ca^{2+} signaling system and the folding machinery in the ER to maintain Ca^{2+} homeostasis in the organelle. The maintenance of Ca^{2+} homeostasis in the ER is then essential for protein folding.

Introduction

In the past six years the goal of our laboratory has been to understand the role of ER chaperone proteins in regulating Ca^{2+} signaling. Our studies have primarily focused on Ca^{2+} re-uptake into the ER lumen by the family of sarco-endoplasmic reticulum Ca^{2+} ATPases (SERCAs). Prior to our work, SERCAs were largely perceived as housekeeping enzymes, with very little dynamic regulation. In addition, ER chaperone proteins were largely thought to be involved only in nascent protein folding. We discovered that at least two ER chaperone proteins, CRT and CNX, can modulate Ca^{2+} ATPase activity. This regulation occurs with ER

Calreticulin, 2nd ed., edited by Paul Eggleton and Marek Michalak.

resident mature proteins such as SERCA2b and is consistent with the need to maintain ER Ca^{2+} concentrations at levels that are optimal for protein folding. These findings underscore the necessity to expand our classical views of Ca^{2+} signaling and ER chaperones. In this chapter, we review the evidence that has lead to this new model of ER Ca^{2+} signaling.

Our initial studies arose from the observations that increments in the state of Ca^{2+} store refilling correlated with the sensitivity of the inositol 1,4,5 trisphosphate receptor (IP_3R) to release Ca^{2+}.[1] These reports suggested that the luminal Ca^{2+} concentration could regulate Ca^{2+} release. To investigate this problem, we chose to manipulate ER luminal Ca^{2+} concentrations by overexpressing calreticulin (CRT), a prototypical Ca^{2+} storage protein[2] that is expressed at millimolar concentrations in the ER.[3,4] Our hypothesis was simply that higher Ca^{2+} levels in the lumen of the ER would increase Ca^{2+} oscillations via the IP_3R (to increase release) or via SERCA (to increase uptake). As presented below, we successfully determined that IP_3 induced Ca^{2+} release could in fact be regulated by altering the expression levels of CRT. In addition, overexpression of CNX, a closely related, but membrane anchored family member of CRT also regulated Ca^{2+} release. However, our approach also led to several unexpected findings that changed our understanding of the underlying physiology. Namely, that high capacity Ca^{2+} storage in CRT was not responsible for the effects observed, as discussed later. Before we discuss these data, we briefly review our experimental system.

Xenopus Oocytes As an Expression System

Xenopus oocytes were used throughout these studies. A complete description of this system is available.[5] Briefly, we used IP_3-induced Ca^{2+} wave activity to assay both Ca^{2+} release and Ca^{2+} uptake. The rising phase of individual waves reflects the activity of the IP_3Rs and the decay phase reflects the uptake processes contributed mainly by Ca^{2+} ATPases of the SERCA family. Mitochondria can also contribute to the decay phase in the presence of energizing respiratory chain substrates.[6] Cytosolic Ca^{2+} is imaged with Ca^{2+} indicator dyes and confocal microscopy. The advantage of this approach is that all experimental measurements can be carried out in vivo. The disadvantage is that contributions to the Ca^{2+} signal from Ca^{2+} release vs Ca^{2+} uptake cannot be precisely distinguished. Consequently, the functional Ca^{2+} oscillations assay is complemented by molecular and biochemical techniques.

Calreticulin and Calnexin Have an Inhibitory Effect on Ca^{2+} Oscillations

Overexpression of CRT[7] or CNX[8] in oocytes reduces the number of oocytes displaying IP3-mediated Ca^{2+} oscillations. In the remaining oocytes that exhibit repetitive activity, the amplitude and frequency of Ca^{2+} oscillations was significantly lower.[7,9] These observations were consistent with an effect of CRT or CNX on the endogenous SERCA pump to inhibit uptake or on the IP3R to favor Ca^{2+} release. We decided to test the "SERCA" hypothesis first and this is what we have focused in this chapter. We had previously reported that overexpression of SERCA pumps leads to an increase in the frequency of Ca^{2+} oscillations.[10] To determine whether the inhibitory effect of CRT and CNX on Ca^{2+} oscillations was due to inhibition of pump activity, we co-expressed CRT[7,9] and CNX[8] with SERCA2b, the ubiquitous Ca^{2+} ATPase of the ER. In this high background of Ca^{2+} oscillations, CRT still had a strong inhibitory effect. There was a significant decrease in the number of oocytes displaying Ca^{2+} oscillations, and in those oocytes that had repetitive Ca^{2+} release, there was a decrease in period between waves and a corresponding decrease in the decay time of individual oscillations (measured as the 50% decrease in T1/2).[8,9] Energization of mitochondria by respiratory substrates (pyruvate-malate, PM) also modulates cytosolic Ca^{2+} by eliciting robust, low-frequency synchronized Ca^{2+} oscillations.[6] In yet another control experiement, we demonstrated that overexpression of either CRT[7] or CNX [8] in oocytes treated with PM also reduces the number of oocytes that have Ca^{2+} oscillations and lowered the amplitude and frequency of oscillations in those oocytes that still had repetitive activity. We concluded from these experiments that the most parsimonious explanation of the data was that Ca^{2+} oscillations were decreased by an inhibition of endogenous SERCA2b.[7,8] Furthermore, since both CRT and CNX have only the conserved P-domain in common, inhibition of Ca^{2+} oscillations was likely being mediated by a lectin or chaperone interaction with the SERCA substrate as discussed below.

The observed inhibitory effects of CRT and CNX may be due to a physical interaction with the pump. Both chaperones contribute to the folding and oligomerization of glycoproteins in the ER.[11-14] The lectin domain was thought to be in the central portion of the P-domain.[15] However more recently it is the N domain that is considered to contain the lectin binding domain.[16] In CRT, this domain has a low affinity Ca^{2+} binding site and is responsible for binding to the substrate during protein folding. Recent evidence suggests that the P-domain is flexible and is responsible for binding to the thioreductase ERp57, which together with CRT and CNX forms inter or intra disulfide bonds with the substrate.[16-19] The Ca^{2+} storage domain is in the C-domain of CRT.[2] We performed domain deletion mutagenesis of CRT to determine whether the inhibition of Ca^{2+} oscillations was due to Ca^{2+} storage or lectin interaction with SERCA2b, which has a consensus site for glycosylation on the 11th transmembrane segment. Mutant ΔC contains the N+P domains, but lacks the C domain. Similarly, the ΔP mutant has the N+C domain and lacks the P-domain, while the ΔPC mutant only carries the N-domain. All mutants were made with a KDEL ER retention signal and we confirmed their localization to the ER by confocal immunofluorescence.[7] We found that inhibition of Ca^{2+} oscillations in CRT overexpressing oocytes requires the N+P domain.[7] Our initial bias had been that overexpression of CRT would increase the amount of releasable Ca^{2+} from the ER by working as a Ca^{2+} storage protein. However, deletion of the C-domain did not remove the inhibitory action of CRT on Ca^{2+} oscillations, clearing demonstrating that the Ca^{2+} storage properties of CRT were not responsible for our observed effects. Rather, we thought that the inhibitory activity of CRT on Ca^{2+} oscillations may be mediated by a lectin-like interaction between CRT and the pump since the N+P domains are required. This model is also consistent with the inhibitory effect of CNX on Ca^{2+} oscillations as discussed further below.

Inhibition of Ca^{2+} Oscillations Is Mediated by the COOH Terminus of SERCA2b

The SERCA2 gene generates two alternative spliced products that are expressed in a tissue and developmental specific manner.[20-24] SERCA2a, the heart isoform[24] is shorter having only the prototypical 10 transmembrane (TM) segments. SERCA2b is ubiquitously expressed [24] and deviates structurally from other Ca^{2+} ATPases by having an 11th TM segment. The COOH terminus of SERCA terminates in the ER lumen with a glycosylation consensus signal at asparagine N1036 [25]. To test the hypothesis that glycosylation was required for pump inhibition, we first performed experiments in which SERCA2a or SERCA2b were co-expressed with CRT.[9] Our prediction was that CRT would interact with the COOH terminus of SERCA2b causing inhibition of Ca^{2+} oscillations whereas oscillations should not be affected by co-expression of the CRT with SERCA2a, which only has 10 TM segments. Indeed this was the case. Ca^{2+} oscillations were similar in all respects regardless of whether the oocytes overexpressed either SERCA2a alone or SERCA2a + CRT.[9] On the other hand Ca^{2+} oscillations were inhibited when SERCA2b was expressed with CRT.[7,9] This critical finding suggested that a direct interaction between CRT and the COOH terminus of SERCA2b was responsible for the luminal modulation of the pump. To test this hypothesis further, we created a site directed mutant of SERCA2b (SERCA2b-N1036A) in which the asparagine was mutated to alanine. Two groups of oocytes were overexpressed, those coexpressing SERCA2b-N1036A + ΔC mutant and oocytes expressing SERCA2b-N1036A alone.[9] We found that N1036 was absolutely required for the inhibitory effect of CRT. These results, together with the findings from the another group of investigators demonstrate that progressive deletion of the COOH terminus of SERCA2b converts SERCA2b to a SERCA2a[22,23] suggest that the COOH terminus of SERCA2b is critical in determining the differences between the two SERCA2 Ca^{2+} ATPases. Furthermore the findings suggest that an interaction of CRT and CNX at the COOH terminus of SERCA2b may require N1036 glycosylation.

Interaction of CNX with the COOH Terminus of SERCA2b

CNX behaves similarly to CRT in the control of Ca^{2+} oscillations.[8] Since it has a similar P-like domain in the lumen, but no Ca^{2+} binding domain there, the inhibition of Ca^{2+} oscillations is consistent with a glycosylation mediated effect. To test this hypothesis, we determined first whether SERCA2b was indeed glycosylated. We generated constructs for expression of the TM9 to TM11 of SERCA2b (SERCA2b/TM9-11) as well as the equivalent mutant with the 1036A mutation. In addition, we generated a SERCA2a/TM9-10 mutant. Correct polytopic insertion in the ER membrane of similar constructs was previously demonstrated.[25] We performed in vitro translations in rabbit reticulosite lysate in the presence of canine pancreatic microsomes. All constructs ran at the predicted molecular mass (SERCA2b and SERCA2b-N1036A/TM9-11, ~13.2 kD; and SERCA2a ~7.2 kD). When run on SDS-PAGE, we observed no migrational difference between SERCAb/TM9-11 and SERCA2b-N1036/TM9-11, despite the fact that a positive control, *S. cerevisiae* α factor displayed full glycosylation at three sites.[8] This indicated that N1036 was not glycosylated. To corroborate this finding, we treated *In vitro* translated products SERCAb/TM9-11 and SERCA2b-N1036/TM9-11 with endoglycosidase H (endo H). Addition of EndoH did not alter the mobility shift of SERCAb/TM9-11 and SERCA2b-N1036/TM9-11 despite the fact that the enzyme caused a complete downward shift of positive control *S. cerevisiae* α factor, which is glycosylated in three asparagines.[8] This finding may not yet rule out glycosylation at this N1036 and more sophisticated analysis is required to conclusively demonstrate glycosylation. Further more rigorous analysis needs to be completed to demonstrate whether SERCA2b is glycosylated at N1036. Irrespective of the state of glycosylation of N1036, it is still possible that CNX engages the SERCA2b substrate via a peptide-peptide interaction. Indeed, both CRT and CNX are known to bind to substrates via peptide-peptide interactions as traditional chaperones do.[26,27] To determine if an interaction existed between CNX and the COOH tail of SERCA2b, endogenous CNX from the microsomes were immunoprecipitated by a CNX-specific antibody from the in vitro translated products of SERCA2a/TM9-10 or SERCA2b/TM9-11. Co-immunoprecipitated proteins were subsequently detected by fluorography. We demonstrated an interaction of CNX with SERCA2b, a very reduced interaction with SERCA2b-N1036A and as expected, no interaction with SERCA2a.[8] Furthermore, the controls behaved as expected, i.e., *S. cerevisiae* α factor was shown to interact with CNX and the negative control (no mRNA supplemented in the in vitro translation reaction) did not show an interaction. Together, these results suggest that there is a specific interaction involving CNX and SERCA 2b but not with SERCA2a. Further, this interaction is localized at the COOH tail of SERCA2b. These observations combined with our imaging data suggest that the differences between SERCA2a and SERCA2b are due to an interaction of CNX with the COOH terminus of SERCA2b and strengthen our view that the inhibition of Ca^{2+} oscillations is dependent on this interaction.

A PKC Phosphorylation Site in CNX Regulates Inhibition of Ca^{2+} Oscillations

CNX is a Type I, single pass transmembrane segment protein with multiple phosphorylation sites in the cytosol. These sites fit the consensus sequences for protein kinase C (PKC/PDK) and casein kinase 2 (CK2).[28,29] The PKC sites provided the attractive possibility that activation of the IP_3R signaling system could directly regulate their phosphorylation by activating a Ca^{2+} sensitive PKC. The PKC/PDK sites would then serve as a regulatory switch in the control of Ca^{2+} oscillations or in the control of glycoprotein binding to substrates in the ER. Our strategy to test this hypothesis was to mutate the phosphorylation sites and overexpress the constructs in oocytes. Two PKC/PDK sites are present in CNX, S562 and S485. Mutations to alanine were made singly and at both sites. When co-expressed with SERCA2b, we found that

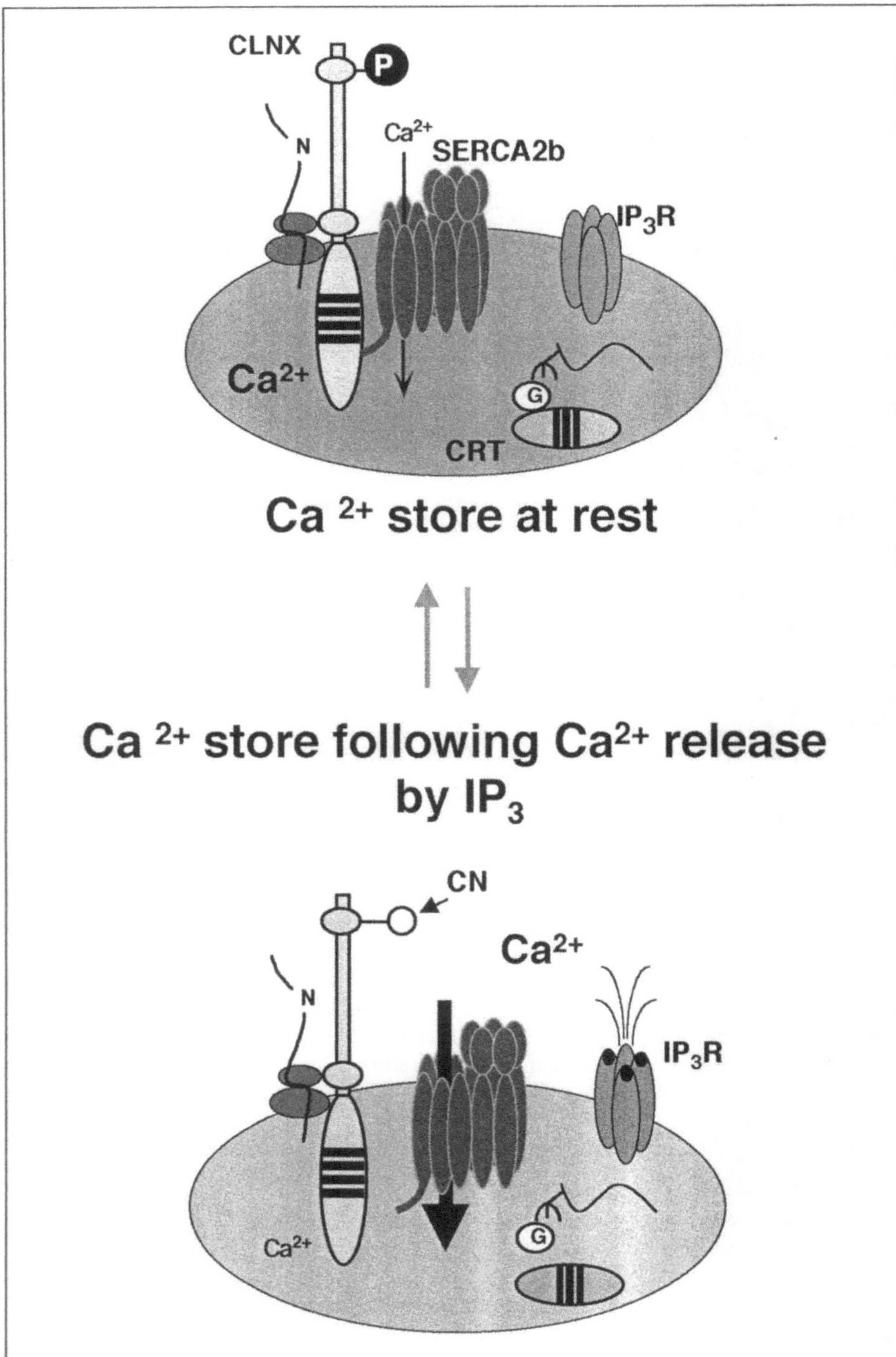

Figure12.1. Model of CRT/CNX regulation of ER Ca^{2+} signaling. Adapted from Roderick et al 2001.

the COOH terminal mutant (CNX-S562) did not inhibit Ca^{2+} oscillations. This result indicated that phosphorylation controlled the interaction of CNX with the pump and supported our model that phosphorylation of this site behaved as a regulatory switch controlling Ca^{2+} oscillations.[8] To corroborate this point, a cytosolic peptide spanning the S562 site was generated (CNXcyt) to compete with the endogenous kinase controlling the inhibition of Ca^{2+} oscillations. When this peptide was co-expressed with SERCA2b and CNX, it prevented the inhibition of Ca^{2+} oscillations, suggesting that it was efficiently acting as a pseudosubstrate for the endogenous kinase that otherwise was responsible for phosphorylating the PKC/PDK sites.[8] The identity of this kinase appears to be PDK.[29]

We determined the state of phosphorylation of CNX at rest and under conditions of IP_3-mediated mobilization of Ca^{2+}. CNX + SERCA2b or CNX-S562A mutant + SERCA2b, or CNX + SERCA2b + $CNX_{cyt.}$ were overexpressed in oocytes. We prelabelled oocytes with $[\gamma^{32}P]$ATP and injected them with either water (control) or IP_3 (300 nM final) to mobilize Ca^{2+} from intracellular stores. CNX was then immunoprecipitated with a specific antibody and found to be phosphorylated at rest and dephosphorylated by IP_3 Ca^{2+} release.[8] Importantly, immunoprecipitation of CNX from oocytes overexpressing CNX-S562A mutant + SERCA2b showed that CNX was minimally phosphorylated at rest and this level of phosphorylation was not changed by mutagenesis, suggesting that S562 supports phosphorylation and it is dephosphorylated by a Ca^{2+} dependent phosphatase.[8] Ongoing work in our laboratory has indicated that the identity of this phosphatase is calcineurin (CN), which was initially recognized as Ca^{2+} dependent by Klee and co-workers.[30] These data suggest that CN dephosphorylates CNX and that this dephosphorylation controls the dissociation of CNX with SERCA2b.

The studies described here have focused on the "SERCA" hypothesis. Based on our data, CRT (unpublished data) and CNX[8] are associated with SERCA2b and inhibit its activity when Ca^{2+} stores are full (i.e., under resting conditions). In this state, the pump is sufficiently active to maintain the ER lumen at full Ca^{2+} capacity. This environment is optimal for protein folding given the requirement of ER chaperone activity for Ca^{2+}.[31] When the IP_3 signaling pathway is activated several rapid changes occur. First, IP_3-mediated Ca^{2+} release depletes Ca^{2+} from the ER with a corresponding mirror image increase in the cytosol. These cytosolic increases cause activation of the Ca^{2+} dependent phosphatase CN, which de-phosphorylates CNX. This results in the dissociation of CNX from SERCA2b, removing pump inhibition. The return to maximum pumping activity rapidly refills the ER lumen and minimizes the potential risk of impaired protein folding during cytosolic Ca^{2+} signaling. In our view, the role of CRT/CNX regulation of SERCA2b activity is to minimize the duration of ER Ca^{2+} depletion.

References

1. Missiaen L, Taylor CW, Berridge MJ. Luminal Ca^{2+} promoting spontaneous Ca^{2+} release from inositol trisphosphate-sensitive stores in rat hepatocytes. J Physiol 1992; 455:623-40.
2. Baksh S, Michalak M. Expression of calreticulin in Escherichia coli and identification of its Ca2+ binding domains. J Biol Chem 1991; 266:21458-65.
3. Fliegel L, Burns K, MacLennan DH et al. Molecular cloning of the high affinity calcium-binding protein (calreticulin) of skeletal muscle sarcoplasmic reticulum. J Biol Chem 1989; 264(36):21522-8.
4. Milner RE, Baksh S, Shemanko C et al. Calreticulin, and not calsequestrin, is the major calcium binding protein of smooth muscle sarcoplasmic reticulum and liver endoplasmic reticulum. J Biol Chem 1991; 266(11):7155-65.
5. Camacho P, Lechleiter JD. Xenpus oocytes as a tool in calcium signaling research. In: Putney J, ed. Calcium Signaling. Boca Raton: CRC Press, 2000:157-81.
6. Jouaville LS, Ichas F, Holmuhamedov EL et al. Synchronization of calcium waves by mitochondrial substrates in Xenopus laevis oocytes. Nature 1995; 377(6548):438-41.
7. Camacho P, Lechleiter JD. Calreticulin inhibits repetitive intracellular Ca^{2+} waves. Cell 1995; 82(5):765-71.
8. Roderick HL, Lechleiter JD, Camacho P. Cytosolic phosphorylation of calnexin controls intracellular Ca(2+) oscillations via an interaction with SERCA2b. J Cell Biol 2000; 149(6):1235-48.
9. John LM, Lechleiter JD, Camacho P. Differential modulation of SERCA2 isoforms by calreticulin. J Cell Biol 1998; 142(4):963-73.

10. Camacho P, Lechleiter JD. Increased frequency of calcium waves in Xenopus laevis oocytes that express a calcium-ATPase. Science 1993; 260(5105):226-9.
11. Bergeron JJ, Brenner MB, Thomas DY et al. Calnexin: a membrane-bound chaperone of the endoplasmic reticulum. Trends Biochem Sci 1994; 19(3):124-8.
12. Ellgaard L, Helenius A. ER quality control: towards an understanding at the molecular level. Curr Opin Cell Biol 2001; 13(4):431-7.
13. Jakob CA, Chevet E, Thomas DY et al. Lectins of the ER quality control machinery. Results Probl Cell Differ 2001; 33:1-17.
14. Helenius A, Trombetta ES, Hebert DN et al. Calnexin, Calreticulin, and the folding of glycoproteins. Trends Cell Biol 1997; 7:193-200.
15. Michalak M, Milner RE, Burns K et al. Calreticulin. Biochem J 1992; 285(Pt 3):681-92.
16. Schrag JD, Bergeron JJ, Li Y et al. The Structure of calnexin, an ER chaperone involved in quality control of protein folding. Mol Cell 2001; 8(3):633-44.
17. Ellgaard L, Riek R, Herrmann T et al. NMR structure of the calreticulin P-domain. Proc Natl Acad Sci USA 2001; 98(6):3133-8.
18. Ellgaard L, Riek R, Braun D et al. Three-dimensional structure topology of the calreticulin P-domain based on NMR assignment. FEBS Letters 2001; 488(1-2):69-73.
19. Frickel EM, Riek R, Jelesarov I et al. TROSY-NMR reveals interaction between ERp57 and the tip of the calreticulin P-domain. Proc Natl Acad Sci USA 2002; 99(4):1954-9.
20. Gunteski-Hamblin AM, Greeb J, Shull GE. A novel Ca2+ pump expressed in brain, kidney, and stomach is encoded by an alternative transcript of the slow-twitch muscle sarcoplasmic reticulum Ca-ATPase gene. Identification of cDNAs encoding Ca^{2+} and other cation-transporting ATPases using an oligonucleotide probe derived from the ATP-binding site. J Biol Chem 1988; 263(29):15032-40.
21. Lytton J, Westlin M, Burk SE et al. Functional comparisons between isoforms of the sarcoplasmic or endoplasmic reticulum family of calcium pumps. J Biol Chem 1992; 267(20):14483-9.
22. Verboomen H, Wuytack F, De Smedt H et al. Functional difference between SERCA2a and SERCA2b Ca2+ pumps and their modulation by phospholamban. Biochem J 1992; 286(Pt 2):591-5.
23. Verboomen H, Wuytack F, Van den Bosch L et al. The functional importance of the extreme C-terminal tail in the gene 2 organellar Ca(2+)-transport ATPase (SERCA2a/b). Biochem J 1994; 303(Pt 3):979-84.
24. Wu KD, Lee WS, Wey J et al. Localization and quantification of endoplasmic reticulum Ca(2+)-ATPase isoform transcripts. Am J Physiol 1995; 269(3Pt 1):C775-84.
25. Bayle D, Weeks D, Sachs G. The membrane topology of the rat sarcoplasmic and endoplasmic reticulum calcium ATPases by in vitro translation scanning. J Bio Chem 1995; 270(43):25678-84.
26. Ihara Y, Cohen-Doyle MF, Saito Y et al. Calnexin discriminates between protein conformational states and functions as a molecular chaperone in vitro. Mol Cell 1999; 4(3):331-41.
27. Saito Y, Ihara Y, Leach MR et al. Calreticulin functions in vitro as a molecular chaperone for both glycosylated and non-glycosylated proteins. EMBO J 1999; 18(23):6718-29.
28. Tjoelker LW, Seyfried CE, Eddy RL Jr et al. Human, mouse, and rat calnexin cDNA cloning: identification of potential calcium binding motifs and gene localization to human chromosome 5. Biochemistry 1994; 33(11):3229-36.
29. Wong HN, Ward MA, Bell AW et al. Conserved in vivo phosphorylation of calnexin at casein kinase II sites as well as a protein kinase C/proline-directed kinase site. J Biol Chem 1998; 273(27):17227-35.
30. Klee CB, Ren H, Wang X. Regulation of the calmodulin-stimulated protein phosphatase, calcineurin. J Biol Chem 1998; 273(22):13367-70.
31. Corbett EF, Oikawa K, Francois P et al. Ca^{2+} regulation of interactions between endoplasmic reticulum chaperones. J Biol Chem 1999; 274(10):6203-11.

Chapter 13

ER Calcium and ER Chaperones: New Players in Apoptosis?

Nicolas Demaurex, Maud Frieden and Serge Arnaudeau

Abstract

By using calcium ions as an intracellular messenger, cells walk a tight rope between life and death. Because critical cellular functions depend on the precise delivery of Ca^{2+} at the right time and place, calcium ions must navigate at all times between intracellular calcium stores and target proteins located in the cytosol, the mitochondria, or the nucleus. Due to the toxicity of high Ca^{2+} concentrations, even slight disruption of the elaborate calcium signaling machinery can have devastating consequences on cell functions: too much or too little calcium at the wrong time and place might lead to rapid cell death by necrosis, or to the induction of the cell death program of apoptosis. ER chaperones, and most notably calreticulin, play a key role in the making and decoding of both normal and pathological calcium signals. Calreticulin is the main Ca^{2+}-binding protein residing in the ER, and as such contributes most of the ER Ca^{2+} buffering capacity. Calreticulin also acts as a chaperone for several ER Ca^{2+} transport proteins, and thus indirectly modulates Ca^{2+} fluxes across the ER membrane. Accordingly, over- or underexpression of calreticulin leads to rapid and severe alterations in ER Ca^{2+} homeostasis. Calreticulin expression levels are controlled by the ER Ca^{2+} levels, thus enabling cells to mount an appropriate response during long-term perturbations in ER Ca^{2+} storage. However, calreticulin levels are also increased by a variety of cellular stress conditions, and this upregulation might contribute to the Ca^{2+} signaling defects leading to apoptosis. In this chapter, we will review the role of calreticulin and of other ER chaperones in the control of Ca^{2+}-mediated apoptosis.

Introduction

Apoptosis, a process first described in 1972 by Kerr et al[1] has changed radically our perception of cell death. In this elaborated form of cellular suicide, cells sacrifice themselves for the well being of the whole organism by dying in a quiet manner, without undergoing cell lysis. In contrast, during necrosis cell membranes are disrupted and the release of the intracellular contents contributes to the generation of inflammation and tissue damage. Apoptosis is in fact a highly regulated process of cell deletion involved during development, during normal cell turnover and during cell elimination following injury. This programmed cell death is defined morphologically by a typical sequence of events: cytoplasmic shrinkage, loss of intercellular contacts, organelle compaction and chromatine condensation and, finally, cytoplasmic blebbing with generation of apoptotic bodies that are phagocytosed by neighboring cells. A family of cysteine proteases known as caspases are tightly involved in this process. These enzymes appear to be essential components in both the initial signaling events and the downstream proteolytic cleavage that results in the apoptotic phenotype. Ca^{2+} seems to modulate the role of some proteases like caspase 3[2] and is also implicated in the activation of other players of apoptosis

Calreticulin, 2nd ed., edited by Paul Eggleton and Marek Michalak.

like calpain and Ca^{2+} dependent endonucleases. The first evidence that Ca^{2+} was involved in triggering apoptosis came from the evidence that glucocorticoid-stimulated apoptosis was associated with enhanced Ca^{2+} influx.[3] Afterwards, Ca^{2+} ionophores were shown to mimick the cytolytic effects of glucocorticoids on lymphocytes[4,5] and overexpression of the Ca^{2+}-binding protein calbindin in thymoma cells was able to prevent ionophore induced apoptosis.[6] Since these initial observations, numerous studies have shown that conditions that preclude cytosolic Ca^{2+} elevations, such as removing Ca^{2+} from the external medium, buffering the intracellular $[Ca^{2+}]_{cyt}$, or inhibiting plasma membrane Ca^{2+} channels, protect cells from apoptosis.[7-10] Conversely, the SERCA inhibitor thapsigargin, which generates long-lasting Ca^{2+} elevations by depleting Ca^{2+} stores and activating capacitative Ca^{2+} entry, triggers all the morphological and biochemical events of apoptosis in numerous cell types.[11-13] All these data illustrate the pivotal role of Ca^{2+} in the apoptotic process. However the precise mechanism leading to the apoptotic response are not yet understood. While excessive $[Ca^{2+}]_{cyt}$ elevations are pro-apoptotic, moderate $[Ca^{2+}]_{cyt}$ elevations appear to be anti-apoptotic.[14] The beneficial effects of Ca^{2+} signals on cell survival might involve the activation of Ca^{2+}/calmodulin-dependent kinase kinase, protein kinase B and phosphorylation of Bad.[15]

Role of ER Calcium in Apoptosis

Ca^{2+} is a ubiquitous intracellular messenger involved in many cellular processes ranging from muscle contraction to hormone secretion, synaptic transmission, and gene transcription. The bewildering array of functions controlled by this simple ion stems from the complexity and versatility of intracellular calcium signals, which can be encoded in time, space, frequency, and amplitude. This plasticity allows cells to generate subtle and diverse patterns of Ca^{2+} signals, both on a local (sparks and puffs) or global scale (transients and waves). To generate such complex Ca^{2+} signals, cells rely on the rapid release of the Ca^{2+} stored within the endo/sarcoplasmic reticulum (ER/SR) as well as on the controlled influx of Ca^{2+} from the extracellular medium. Opening of Ca^{2+} release or Ca^{2+} influx channels transiently increases the averaged cytosolic Ca^{2+} concentration, from ~100 nM to ~1 μM, but much higher values can be achieved close to the mouth of the channels. Because of the toxicity of such high Ca^{2+} concentrations, high Ca^{2+} levels are reached only transiently and Ca^{2+} is rapidly removed from the cytosol by Ca^{2+} pumps and exchangers. Both the plasma membrane Ca^{2+}-ATPase (PMCA) and the Na^+/Ca^{2+} exchanger drive Ca^{2+} out to the external medium, whereas the sarco-endoplasmic reticulum Ca^{2+}-ATPase (SERCA) recycle Ca^{2+} to the ER/SR in order to replenish the stores. This ensures that a high Ca^{2+} concentration is maintained at all times within the lumen of the ER, a condition that is crucial not only for the generation of further Ca^{2+} signals but also for the proper function of ER-resident proteins. The release of Ca^{2+} from the ER stores occurs by the opening of Ca^{2+} release channels belonging to two families: the inositol 1, 4, 5 trisphosphate receptor (IP_3R) and the ryanodine receptor (RyR). Activation of these intracellular channels generate local Ca^{2+} signals (puffs and sparks, respectively) which, by a positive feed-back mechanism of Ca^{2+}-induced Ca^{2+}-release, further activate the Ca^{2+}-release channels to produce regenerative Ca^{2+} oscillations and waves (for review see ref. 16).

Recently, mitochondria have also emerged as bona-fide Ca^{2+} signaling organelles, able to encode and decode Ca^{2+} signals. Mitochondria are often located close to the ER, and therefore exposed to the Ca^{2+} released by the IP_3R[17] and the RyR.[18] The high Ca^{2+} levels achieved at these contact sites favors Ca^{2+} uptake into mitochondria, via a Ca^{2+} uniporter driven by the very negative mitochondrial membrane potential (-150 mV). Indeed, Ca^{2+}-clamp experiments in permeabilized hepatocytes have revealed that Ca^{2+} influx into mitochondria occurs at cytosolic Ca^{2+} concentrations exceeding 300 μM. Mitochondrial Ca^{2+} transients, nicknamed Ca^{2+} "marks", can be observed during elementary Ca^{2+} release events,[19] indicating that such high Ca^{2+} concentration are indeed achieved around individual mitochondria during physiological Ca^{2+} signals. The Ca^{2+} taken up by mitochondria is subsequently released to the cytosol where it can diffuse locally or return back to the ER, allowing mitochondria to shape cytosolic Ca^{2+}

signals and to prevent the depletion of the ER Ca^{2+} stores.[20,21] Thus, mitochondria located close to Ca^{2+} release or influx channels handle a large part of the Ca^{2+} used for signaling. Mitochondria also play a critical role in apoptosis, by providing the energy required for the ordered execution of cells and by delivering apoptogenic proteins. One crucial step in the apoptotic process is the irreversible opening of the mitochondrial permeability transition pore (in its high conductance state), and the collapse of the mitochondria membrane potential, $\Delta\Psi_m$. This phenomenon, controlled by members of the Bcl-2 family,[22] is associated with an increased permeability of the outer mitochondrial membrane and a swelling of the mitochondria inner membrane. As a result, soluble proteins are released from the intermembrane space such as the proapoptotic cytochrome c,[23,24] procaspase 2, 3 and 9,[25,26] the apoptosis inducing factor (AIF)[27] and Smac/Diablo,[28,29] initiating the caspase cascade leading to the cleavage of a large quantity of proteins and eventually to the ordered disassembly of the cell.

Because of their tight coupling to ER Ca^{2+} stores, mitochondria are highly susceptible to abnormalities in Ca^{2+} signaling. Recent evidences suggest that the amount of Ca^{2+} going through mitochondria is crucial in triggering Ca^{2+}-dependent apoptotic responses. The amount of Ca^{2+} sensed by mitochondria depends on several factors, notably:

1. the activity of Ca^{2+}-release channels in the ER membrane (IP_3R or RyR), which controls the flux across the ER membrane,
2. the ER Ca^{2+} load, which determines the total amount of Ca^{2+} that can be released from the ER,
3. the free ER Ca^{2+} concentration, $[Ca^{2+}]_{ER}$, which determine the driving force for Ca^{2+} release, and
4. the proximity between mitochondria and the ER, which determines the magnitude of the Ca^{2+} microdomain sensed by mitochondria.[17,30]

Increase in any of these parameters will increase the Ca^{2+} flowing through mitochondria, and induce a switch from the cell survival to the cell death program. Increased expression of the type 3 IP_3R has been reported in lymphocytes undergoing cell death,[31] and was also observed during developmental apoptosis in several post natal tissues.[32] The apoptosis associated with IP_3-dependent Ca^{2+} signals in lymphocytes appears to be mediated by calcineurin, a Ca^{2+}-regulated phosphatase that can dephosphorylate and activate the pro-apoptotic factor Bad.[33,34] Numerous procedures that reduce the ER Ca^{2+} load, such as lowering extracellular Ca^{2+}, depleting the stores with low doses of the SERCA inhibitor tBuBHQ, or overexpressing the plasma membrane Ca^{2+} pump, protect HeLa cells from ceramide-induced apoptosis.[35] Similarly, deletion of the calreticulin gene, by removing the major Ca^{2+}-bufferring protein from the ER lumen, reduces the total amount of Ca^{2+} stored in the ER and increases cell survival.[36] Importantly, the free ER Ca^{2+} concentration, $[Ca^{2+}]_{ER}$, was not altered in the calreticulin knock-out cells, but the ability of these cells to generate Ca^{2+} transients upon stimulation with agonists was markedly reduced. Conversely, overexpression of calreticulin or of SERCA ATPases increased both the total ER Ca^{2+} load as well as $[Ca^{2+}]_{ER}$, and enhanced the sensitivity of cells to ceramide-induced apoptosis.[35,37]

In the presence of ceramide, a classical apoptotic stimulus, even physiological IP_3-dependent Ca^{2+} signals are able to trigger the apoptotic process, probably by the opening of a sensitized state of the mPTP. The switch from the life to the death program might involve coincident detection of pro-apoptotic stimuli and calcium signals,[38] reviewed in.[39] As mentioned above, the opening of the mPTP promotes the release of apoptotic factors, most notably cytochrome c, which forms a complex with pro-caspase 9, Apaf-1 and dATP.[40,41] This results in the activation of caspase 9, which dissociates from the complex and activates other executioner caspases such as caspase 3. By integrating Ca^{2+} and apoptotic stimuli, mitochondria thus function as a central checkpoint in determining cell survival or cell death. The release of cytochrome c from mitochondria is prevented by the anti-apoptotic factor Bcl-2, one of the most widely studied proto-oncogene, whose mechanism of action is still debated. Several Bcl-2 family members have been identified: anti-apoptotic factors such as Bcl-2, $Bcl\text{-}X_L$ and Mcl-1 pro-apoptotic factors such as Bax, Bad and Bid. Bcl-2 can prevent the opening of the mPTP and protect cell from apoptosis, whereas Bax has the opposite effect. Most Bcl-2 family members are anchored by a hydrophobic stretch of amino-acids in the outer mitochondrial membrane,[42] but Bcl-2 has also been detected in the membrane of the ER, suggesting that this organelle is also implicated in the apoptosis program. Recently, the expression of recombinant Bcl-2 was shown to reduce the ER Ca^{2+} concentration by increasing the passive leak across the ER membrane.[43,44] This effect of Bcl-2 is consistent with the three-dimensional structure of Bcl-X_L proteins, which is reminiscent of pore forming bacterial toxins,[45,46] and with the observation that Bcl-2 can function as an ion channel in artificial lipid bilayers.[47,48] In this model, Bcl-2 insertion in the ER membrane increases the passive ER Ca^{2+} permeability, thus reducing both the total amount of stored Ca^{2+} and the free ER Ca^{2+} concentration, $[Ca^{2+}]_{ER}$. The ensuing Ca^{2+} depletion of the ER is an integral part of the mechanism of action of Bcl-2.

All these results are consistent with the hypothesis that a moderate reduction in ER Ca^{2+} protects cells from apoptotic stimuli, by decreasing the amount of toxic Ca^{2+} ions sensed by the cytosol and mitochondria. In contrast, an ER Ca^{2+} overload sensitizes cells to apoptosis by the opposite mechanism. This scheme is in apparent contradiction with the well-known pro-apoptotic effects of agents such as the SERCA inhibitor thapsigargin or the Ca^{2+} ionophore A23187, which induce a massive Ca^{2+} store depletion. These conflicting observations can be reconciled by postulating that a severe ER Ca^{2+} depletion, in itself, is sensed as a stress signal by the cell and causes apoptosis.[49] Alternatively, the massive and long-lasting increase in cytosolic Ca^{2+} caused by these agents might bypass the protective effect of the ER Ca^{2+} depletion and trigger apoptosis. In the former scenario, transduction of the ER -induced apoptosis signal might be mediated by caspase 12, an ER-associated caspase. This particular ER caspase is activated by a variety of ER stress conditions, including not only the disruption of ER Ca^{2+} homeostasis but also the accumulation of excess protein in the ER.[50]

Role of ER Chaperones in Apoptosis

Besides its role as the most prominent intracellular Ca^{2+} store, the ER compartment plays a crucial role in protein maturation, folding, transport, and storage. These two roles are closely intricated, and alterations in ER functions that perturb either the protein folding process or change the Ca^{2+} level in the ER lead to a situation called ER stress response. Depending on the severity of such ER stress, this process can path either to an adaptive response or to apoptosis. In general, the ER stress is due to an accumulation of misfolded proteins in the lumen of the ER that, in turn, yields to a phenomenon referred as "unfolded protein response" (UPR). The UPR is characterized by a general decrease of protein synthesis whereas the expression of specific sets of proteins, mainly the ER resident chaperones, is increased. Both events constitute responsive processes that will eventually result in the normalization of the folding process machinery, or, if unsuccessful, trigger apoptosis (for reviewed see ref. 51,52).

Chaperones are found in every cell compartment and aid to protein maturation in two ways: chaperones bind to unfolded proteins in order to prevent their further aggregation and degradation and, in addition, actively promote protein folding.[51] Many ER-resident proteins function as molecular chaperones that belong to the glucose-regulated proteins (GRP) family. The major ER chaperones are GRP 78/BiP GRP94, the protein disulfide isomerase (PDI) and its homologue ERp57, as well as the two lectin-like proteins calreticulin and calnexin that bind newly synthesized glycosylated proteins. Several of the ER chaperones, including calreticulin, calnexin, GRP78/BiP, and GRP94 are major ER Ca^{2+} binding proteins, able to bind large quantities of Ca^{2+} ions. The GRPs are constitutively expressed but their transcription can be enhanced by various stimuli that disrupt ER homeostasis, such as an exhaustive ER Ca^{2+} depletion by Ca^{2+} ionophores or SERCA inhibitors, inhibition of N-glycosylation by tunacamycin, and the prevention of ER-Golgi protein trafficking by brefeldin A. Malign/prolonged Ca^{2+} depletion promotes the accumulation of misfolded proteins, as the function of several chaperones is controlled by the concentration of luminal Ca^{2+}. It was shown that calreticulin and calnexin,[53] as well as GRP78/BiP activities are decreased by low luminal ER Ca^{2+} concentrations.[54] Furthermore, Ca^{2+} also modulate the interactions between different chaperones. For instance, the association between CRT and PDI is promoted at low luminal Ca^{2+} concentration,[55] thereby reducing the activity of PDI. The chaperoning function of PDI is thus inhibited under conditions of ER Ca^{2+} depletion. Conversely, CRT dissociates from PDI at higher Ca^{2+} concentrations and thus the activity of PDI is promoted when Ca^{2+} stores are full. Therefore, it is tempting to speculate that alterations of the ER Ca^{2+} concentration result in the accumulation of misfolded protein, and that this condition will trigger the UPR.

How do cells sense the accumulation of misfolded proteins in the lumen of the ER and transmit this information to the cytosol and nucleus? In yeast, the signal transduction pathway involves an ER transmembrane ser/thr kinase (Ire1p) that senses unfolded proteins in the ER and stimulates a downstream transcription factor. Upon ER stress, Ire1p oligomerizes and trans-phosphorylates, which activates its endonuclease activity. The target of the endonuclease is the mRNA of HAC. Once cleaved by Ire1p, HAC becomes active as a transcription factor and specifically bind to promoter regions containing an unfolded protein response element (UPRE). In mammals, even if many similarities were noticed, the pathways appear to be more complex and are so far not fully understood. Similar to yeast, the sensor is a transmembrane protein, Ire1α or Ire1β, that is able to trans-phosphorylate and possesses an endonuclease activity. In resting conditions, the chaperone GRP78/BiP binds to Ire1α, preventing its oligomerization. In case of unfolded protein accumulation, BiP binds misfolded proteins and releases Ire1α, which then becomes activated.[56] The target of Ire1α is almost certainly a mRNA, but no definitive target has been found so far. A likely candidate is ATF6, a basic leucine zipper (bZIP) transcription factor that belongs to the same family of ATF/CREB protein as HAC in yeast. No direct interaction between Ire1a and ATF6 has been described so far, but ATF6 is regulated by ER stress.[57] ATF6 recognizes specific sequence on promoter regions called ERSE (ER stress response element). Several chaperones, including GRP78/BiP, GRP94, and calreticulin contain this region within their promoter,[57] and the induction of these chaperones by tunacamycin or ER Ca^{2+} depletion required the presence of ERSE.

This mechanism might account for the increased expression of ER chaperones during cellular stress such as oxygen deprivation, glucose starvation, and treatments that inhibit protein glycosylation (tunicamycin) or induce a ER Ca^{2+} depletion (thapsigargin)[58,59] (reviewed in ref. 60). In addition, an ER stress caused by Ca^{2+} depletion or misfolded protein accumulation induces the expression of other proteins such as SERCA2b, the ubiquitous Ca^{2+}-ATPase of intracellular Ca^{2+} stores.[61] SERCA2b contains in its promoter region an ERSE element, suggesting that this Ca^{2+} transporter protein responds to an ER stress in a similar manner as ER chaperones. Interestingly, Ca^{2+} depletion does not seem to be the only initiator of the response, as tunicamycin, which does not change the ER Ca^{2+} levels, is also able to increase SERCA2b expression. This suggests that a generic response is induced regardless of the type of ER stress, which leads to the upregulation of several ER proteins. However, all the chaperones are not induced by an ER stress, and some chaperones respond only to specific stress stimuli. In WEHI7.2 mouse lymphoma cells, GRP78/GRP94 are not induced by treatment with thapsigargin, but increase in response to tunicamycin. Thus, an ER stress caused by ER Ca^{2+} depletion or unglycosylated proteins accumulation does not necessarily trigger the same signaling pathway.[62]

The increased expression of chaperones and of Ca^{2+}-regulatory proteins is clearly beneficial for the cell, and has been shown to protect cells against further stress-induced apoptosis. The protection conferred by specific chaperones, however, might be restricted to a specific stress condition. While GRP78 expression is enhanced by several ER stress stimuli, it seems to specifically protect cells against ER Ca^{2+} depletion-induced apoptosis, but not against tunicamycin-induced apoptosis.[63] A broader protection might be conferred by the general attenuation of protein synthesis, which occurs during an ER stress together with the specific increase in ER chaperones expression. In this case, the response occurs at the translational level. Another ER transmembrane protein, PERK, was shown to phosphorylate the eukaryotic translation initiation factor 2 (eIF2α), resulting in a reduction of the translation initiation.[64,65] The reduced protein synthesis diminishes the load of putative misfolded protein during ER stress, and thus serves as a protection against cell death. Accordingly, Harding[66] showed that Perk-/- cells are more susceptible to cell death during an ER stress. These cells have much higher levels of activated caspase-12 during treatment with thaspigargin or tunicamycin, which might explain their higher susceptibility to apoptotic stimuli (see below).

Although the cellular responses described above might allow cells to overcome an ER stress, an ER stress can also, in mammalian cells, lead to programmed cell death. Caspase 12 might be an important mediator in this process, as this caspase was recently found to be involved in ER mediated apoptosis.[50] Caspase-12 is an ER membrane bound protease that is located at the cytosolic side of the ER.[67] Treatment with thapsigargin, A23187, tunamycin, or brefeldin A cleaves procaspase-12, while other apoptotic stimuli that do not involve an ER stress (i.e., staurosporine) are ineffective.[67] In line with this finding, caspase-12 knock out mice are resistant to ER stress-induced apoptosis. m-calpain, a low-affinity Ca^{2+}-dependent protease distinct from the caspase family, was recently found to be involved in the activation of caspase-12. The cleavage of caspase-12 required millimolar Ca^{2+} concentrations, consistent with its activation by m-calpain. In addition, m-calpain also cleaves Bcl-xL, likely transforming this protein from an anti-apoptotic to an apoptotic agent.[68,69] Interestingly, in cells treated with etoposide (a topoisomerase II inhibitor), GRP 94 appeared to be cleaved by calpain and to generate a fragment of 80 KDa. This cleavage was selective for GRP94, as other chaperones did not get cleaved.[70] More recently, it was shown that tumor necrosis factor receptor-associated factor 2 (TRAF-2) is involved in the activation of caspase-12.[71] TRAF2 acts downstream from Ire1α and was shown to stimulate components of the c-Jun N-terminal kinase (JNK) pathway. In resting conditions, TRAF2 is associated with procaspase-12, but the complex dissociates during an ER stress, thus favoring the cleavage of procaspase-12 by proteins such as m-calpain. However, the precise Ca^{2+}-dependency of this process is not well established in vivo. Another pathway leading to apoptosis involves the transcription factor Gadd 153/CHOP, by an as yet unknown mechanism. Transcription of CHOP is induced by the UPR and follows a similar kinetic as GRP78/BiP (reviewed in ref. 51). In cells overexpressing CHOP the level of the anti-apoptotic protein Bcl-2 is dramatically reduced, possibly accounting for the susceptibility of these cells to apoptosis.[72] Similarly, in microglial MG5 cells, NO induced apoptosis, which is linked to an ER Ca^{2+} decrease, is also mediated by a stimulation of CHOP.[73]

In summary, the ER is emerging as a central player in apoptosis, being able to detect, transduce, and respond to a variety of stress signals. The ER stress response might enable yeast to survive under stress conditions, and, in mammalian cells, ensures that damaged cells are safely and efficiently removed by apoptosis. The ER stress response invariably interferes with the role of the ER as a protein factory and Ca^{2+} storage organelle, and is often caused by alterations in these two central ER functions. Changes in the Ca^{2+} concentration within the ER lumen, in particular, can both induce and execute the ER stress response, and have a direct impact on cellular function. Excessively high ER Ca^{2+} levels lead to apoptosis by activating Ca^{2+}-dependent targets located in the cytosol or in neighboring mitochondria. In contrast, low ER Ca^{2+} levels induce the ER stress response by promoting the accumulation of ER chaperones and of ER Ca^{2+} transport proteins. Calreticulin appears to play a critical role in sensing and correcting alterations in ER Ca^{2+} signals, and changes in calreticulin expression levels alters both the total and the free Ca^{2+} concentration within the ER lumen. The dual role of this Ca^{2+}-binding chaperone allows calreticulin to integrate variations in ER Ca^{2+} homeostasis and in protein folding, thereby linking Ca^{2+} signaling to apoptosis.

References

1. Kerr JF, Wyllie AH, Currie AR. Apoptosis: a basic biological phenomenon with wide-ranging implications in tissue kinetics. Br J Cancer 1972; 26(4):239-57.
2. Juin P, Pelletier M, Oliver L et al. Induction of a caspase-3-like activity by calcium in normal cytosolic extracts triggers nuclear apoptosis in a cell-free system. J Biol Chem 1998; 273(28):17559-64.
3. Kaiser N, Edelman IS. Calcium dependence of glucocorticoid-induced lymphocytolysis. Proc Natl Acad Sci USA 1977; 74(2):638-42.
4. Kaiser N, Edelman IS. Further studies on the role of calcium in glucocorticoid-induced lymphocytolysis. Endocrinology 1978; 103(3):936-42.
5. Kaiser N, Edelman IS. Calcium dependence of ionophore A23187-induced lymphocyte cytotoxicity. Cancer Res 1978; 38(11Pt 1):3599-603.
6. Dowd DR, MacDonald PN, Komm BS et al. Stable expression of the calbindin-D28K complementary DNA interferes with the apoptotic pathway in lymphocytes. Mol Endocrinol 1992; 6(11):1843-8.
7. McConkey DJ, Hartzell P, Nicotera P et al. Calcium-activated DNA fragmentation kills immature thymocytes. Faseb J 1989; 3(7):1843-9.
8. Aw TY, Nicotera P, Manzo L et al. Tributyltin stimulates apoptosis in rat thymocytes. Arch Biochem Biophys 1990; 283(1):46-50.
9. McConkey DJ, Chow SC, Orrenius S et al. NK cell-induced cytotoxicity is dependent on a Ca2+ increase in the target. Faseb J 1990; 4(9):2661-4.
10. Juntti-Berggren L, Larsson O, Rorsman P et al. Increased activity of L-type Ca2+ channels exposed to serum from patients with type I diabetes. Science 1993; 261(5117):86-90.
11. Jiang S, Chow SC, Nicotera P et al. Intracellular Ca2+ signals activate apoptosis in thymocytes: studies using the Ca(2+)-ATPase inhibitor thapsigargin. Exp Cell Res 1994; 212(1):84-92.
12. Kaneko Y, Tsukamoto A. Thapsigargin-induced persistent intracellular calcium pool depletion and apoptosis in human hepatoma cells. Cancer Lett 1994; 79(2):147-55.
13. Levick V, Coffey H, D'Mello SR. Opposing effects of thapsigargin on the survival of developing cerebellar granule neurons in culture. Brain Res 1995; 676(2):325-35.
14. Koike T, Martin DP, Johnson EM Jr. Role of Ca2+ channels in the ability of membrane depolarization to prevent neuronal death induced by trophic-factor deprivation: evidence that levels of internal Ca2+ determine nerve growth factor dependence of sympathetic ganglion cells. Proc Natl Acad Sci USA 1989; 86(16):6421-5.
15. Yano S, Tokumitsu H, Soderling TR. Calcium promotes cell survival through CaM-K kinase activation of the protein-kinase-B pathway. Nature 1998; 396(6711):584-7.
16. Berridge MJ, Lipp P, Bootman MD. The versatility and universality of calcium signalling. Nat Rev Mol Cell Biol 2000; 1(1):11-21.
17. Rizzuto R, Brini M, Murgia M et al. Microdomains with high Ca2+ close to IP3-sensitive channels that are sensed by neighboring mitochondria. Science 1993; 262(5134):744-7.
18. Szalai G, Csordas G, Hantash BM et al. Calcium signal transmission between ryanodine receptors and mitochondria. J Biol Chem 2000; 275(20):15305-13.
19. Pacher P, Thomas AP, Hajnoczky G. Ca2+ marks: miniature calcium signals in single mitochondria driven by ryanodine receptors. Proc Natl Acad Sci USA 2002; 99(4):2380-5.
20. Montero M, Alonso MT, Carnicero E et al. Chromaffin-cell stimulation triggers fast millimolar mitochondrial Ca2+ transients that modulate secretion. Nat Cell Biol 2000; 2(2):57-61.
21. Arnaudeau S, Kelley WL, Walsh JV Jr et al. Mitochondria recycle Ca(2+) to the endoplasmic reticulum and prevent the depletion of neighboring endoplasmic reticulum regions. J Biol Chem 2001; 276(31):29430-9.
22. Kroemer G. The proto-oncogene Bcl-2 and its role in regulating apoptosis. Nat Med 1997; 3(6):614-20.
23. Liu X, Kim CN, Yang J et al. Induction of apoptotic program in cell-free extracts: requirement for dATP and cytochrome c. Cell 1996; 86(1):147-57.
24. Kluck RM, Bossy-Wetzel E, Green DR et al. The release of cytochrome c from mitochondria: a primary site for Bcl-2 regulation of apoptosis. Science 1997; 275(5303):1132-6.
25. Mancini M, Nicholson DW, Roy S et al. The caspase-3 precursor has a cytosolic and mitochondrial distribution: implications for apoptotic signaling. J Cell Biol 1998; 140(6):1485-95.
26. Susin SA, Lorenzo HK, Zamzami N et al. Mitochondrial release of caspase-2 and -9 during the apoptotic process. J Exp Med 1999; 189(2):381-94.
27. Susin SA, Lorenzo HK, Zamzami N et al. Molecular characterization of mitochondrial apoptosis-inducing factor. Nature 1999; 397(6718):441-6.

28. Du C, Fang M, Li Y et al. Smac, a mitochondrial protein that promotes cytochrome c-dependent caspase activation by eliminating IAP inhibition. Cell 2000; 102(1):33-42.
29. Verhagen AM, Ekert PG, Pakusch M et al. Identification of DIABLO, a mammalian protein that promotes apoptosis by binding to and antagonizing IAP proteins. Cell 2000; 102(1):43-53.
30. Rizzuto R, Pinton P, Carrington W et al. Close contacts with the endoplasmic reticulum as determinants of mitochondrial Ca2+ responses. Science 1998; 280(5370):1763-6.
31. Khan AA, Soloski MJ, Sharp AH et al. Lymphocyte apoptosis: mediation by increased type 3 inositol 1,4,5-trisphosphate receptor. Science 1996; 273(5274):503-7.
32. Blackshaw S, Sawa A, Sharp AH et al. Type 3 inositol 1,4,5-trisphosphate receptor modulates cell death. Faseb J 2000; 14(10):1375-9.
33. Wang HG, Pathan N, Ethell IM et al. Ca2+-induced apoptosis through calcineurin dephosphorylation of BAD. Science 1999; 284(5412):339-43.
34. Jayaraman T, Marks AR. Calcineurin is downstream of the inositol 1,4,5-trisphosphate receptor in the apoptotic and cell growth pathways. J Biol Chem 2000; 275(9):6417-20.
35. Pinton P, Ferrari D, Rapizzi E et al. The Ca2+ concentration of the endoplasmic reticulum is a key determinant of ceramide-induced apoptosis: significance for the molecular mechanism of Bcl-2 action. Embo J 2001; 20(11):2690-701.
36. Nakamura K, Zuppini A, Arnaudeau S et al. Functional specialization of calreticulin domains. J Cell Biol 2001; 154(5):961-72.
37. Nakamura K, Bossy-Wetzel E, Burns K et al. Changes in endoplasmic reticulum luminal environment affect cell sensitivity to apoptosis. J Cell Biol 2000; 150(4):731-40.
38. Szalai G, Krishnamurthy R, Hajnoczky G. Apoptosis driven by IP(3)-linked mitochondrial calcium signals. Embo J 1999; 18(22):6349-61.
39. Hajnoczky G, Csordas G, Madesh M et al. Control of apoptosis by IP(3) and ryanodine receptor driven calcium signals. Cell Calcium 2000; 28(5-6):349-63.
40. Li P, Nijhawan D, Budihardjo I et al. Cytochrome c and dATP-dependent formation of Apaf-1/caspase-9 complex initiates an apoptotic protease cascade. Cell 1997; 91(4):479-89.
41. Zou H, Li Y, Liu X et al. An APAF-1.cytochrome c multimeric complex is a functional apoptosome that activates procaspase-9. J Biol Chem 1999; 274(17):11549-56.
42. Krajewski S, Tanaka S, Takayama S et al. Investigation of the subcellular distribution of the bcl-2 oncoprotein: residence in the nuclear envelope, endoplasmic reticulum, and outer mitochondrial membranes. Cancer Res 1993; 53(19):4701-14.
43. Pinton P, Ferrari D, Magalhaes P et al. Reduced loading of intracellular Ca(2+) stores and downregulation of capacitative Ca(2+) influx in Bcl-2-overexpressing cells. J Cell Biol 2000; 148(5):857-62.
44. Foyouzi-Youssefi R, Arnaudeau S, Borner C et al. Bcl-2 decreases the free Ca2+ concentration within the endoplasmic reticulum. Proc Natl Acad Sci USA 2000; 97(11):5723-8.
45. Muchmore SW, Sattler M, Liang H et al. X-ray and NMR structure of human Bcl-xL, an inhibitor of programmed cell death. Nature 1996; 381(6580):335-41.
46. Schendel SL, Montal M, Reed JC. Bcl-2 family proteins as ion-channels. Cell Death Differ 1998; 5(5):372-80.
47. Schendel SL, Xie Z, Montal MO et al. Channel formation by antiapoptotic protein Bcl-2. Proc Natl Acad Sci USA 1997; 94(10):5113-8.
48. Schlesinger PH, Gross A, Yin XM et al. Comparison of the ion channel characteristics of proapoptotic BAX and antiapoptotic BCL-2. Proc Natl Acad Sci USA 1997; 94(21):11357-62.
49. Welihinda AA, Tirasophon W, Kaufman RJ. The cellular response to protein misfolding in the endoplasmic reticulum. Gene Expr 1999; 7(4-6):293-300.
50. Nakagawa T, Zhu H, Morishima N et al. Caspase-12 mediates endoplasmic-reticulum-specific apoptosis and cytotoxicity by amyloid-beta. Nature 2000; 403(6765):98-103.
51. Kaufman RJ. Stress signaling from the lumen of the endoplasmic reticulum: coordination of gene transcriptional and translational controls. Genes Dev 1999; 13(10):1211-33.
52. Patil C, Walter P. Intracellular signaling from the endoplasmic reticulum to the nucleus: the unfolded protein response in yeast and mammals. Curr Opin Cell Biol 2001; 13(3):349-55.
53. Vassilakos A, Michalak M, Lehrman MA et al. Oligosaccharide binding characteristics of the molecular chaperones calnexin and calreticulin. Biochemistry 1998; 37(10):3480-90.
54. Ivessa NE, De Lemos-Chiarandini C, Gravotta D et al. The Brefeldin A-induced retrograde transport from the Golgi apparatus to the endoplasmic reticulum depends on calcium sequestered to intracellular stores. J Biol Chem 1995; 270(43):25960-7.
55. Corbett EF, Oikawa K, Francois P et al. Ca2+ regulation of interactions between endoplasmic reticulum chaperones. J Biol Chem 1999; 274(10):6203-11.

56. Bertolotti A, Zhang Y, Hendershot LM et al. Dynamic interaction of BiP and ER stress transducers in the unfolded-protein response. Nat Cell Biol 2000; 2(6):326-32.
57. Yoshida H, Haze K, Yanagi H et al. Identification of the cis-acting endoplasmic reticulum stress response element responsible for transcriptional induction of mammalian glucose-regulated proteins. Involvement of basic leucine zipper transcription factors. J Biol Chem 1998; 273(50):33741-9.
58. Lee AS. Mammalian stress response: induction of the glucose-regulated protein family. Curr Opin Cell Biol 1992; 4(2):267-73.
59. Kozutsumi Y, Segal M, Normington K et al. The presence of malfolded proteins in the endoplasmic reticulum signals the induction of glucose-regulated proteins. Nature 1988; 332(6163):462-4.
60. Lee AS. The glucose-regulated proteins: stress induction and clinical applications. Trends Biochem Sci 2001; 26(8):504-10.
61. Caspersen C, Pedersen PS, Treiman M. The sarco/endoplasmic reticulum calcium-ATPase 2b is an endoplasmic reticulum stress-inducible protein. J Biol Chem 2000; 275(29):22363-72.
62. McCormick TS, McColl KS, Distelhorst CW. Mouse lymphoma cells destined to undergo apoptosis in response to thapsigargin treatment fail to generate a calcium-mediated grp78/grp94 stress response. J Biol Chem 1997; 272(9):6087-92.
63. Miyake H, Hara I, Arakawa S et al. Stress protein GRP78 prevents apoptosis induced by calcium ionophore, ionomycin, but not by glycosylation inhibitor, tunicamycin, in human prostate cancer cells. J Cell Biochem 2000; 77(3):396-408.
64. Harding HP, Zhang Y, Ron D. Protein translation and folding are coupled by an endoplasmic-reticulum-resident kinase. Nature 1999; 397(6716):271-4.
65. Shi Y, Vattem KM, Sood R et al. Identification and characterization of pancreatic eukaryotic initiation factor 2 alpha-subunit kinase, PEK, involved in translational control. Mol Cell Biol 1998; 18(12):7499-509.
66. Harding HP, Zhang Y, Bertolotti A et al. Perk is essential for translational regulation and cell survival during the unfolded protein response. Mol Cell 2000; 5(5):897-904.
67. Nakagawa T, Yuan J. Cross-talk between two cysteine protease families. Activation of caspase-12 by calpain in apoptosis. J Cell Biol 2000; 150(4):887-94.
68. Fujita N, Nagahashi A, Nagashima K et al. Acceleration of apoptotic cell death after the cleavage of Bcl-XL protein by caspase-3-like proteases. Oncogene 1998; 17(10):1295-304.
69. Clem RJ, Cheng EH, Karp CL et al. Modulation of cell death by Bcl-XL through caspase interaction. Proc Natl Acad Sci USA 1998; 95(2):554-9.
70. Reddy RK, Lu J, Lee AS. The endoplasmic reticulum chaperone glycoprotein GRP94 with Ca(2+)-binding and antiapoptotic properties is a novel proteolytic target of calpain during etoposide-induced apoptosis. J Biol Chem 1999; 274(40):28476-83.
71. Yoneda T, Imaizumi K, Oono K et al. Activation of caspase-12, an endoplastic reticulum (ER) resident caspase, through tumor necrosis factor receptor-associated factor 2-dependent mechanism in response to the ER stress. J Biol Chem 2001; 276(17):13935-40.
72. McCullough KD, Martindale JL, Klotz LO et al. Gadd153 sensitizes cells to endoplasmic reticulum stress by down-regulating Bcl2 and perturbing the cellular redox state. Mol Cell Biol 2001; 21(4):1249-59.
73. Kawahara K, Oyadomari S, Gotoh T et al. Induction of CHOP and apoptosis by nitric oxide in p53-deficient microglial cells. FEBS Lett 2001; 506(2):135-9.

Chapter 14

Calreticulin in Cytotoxic Lymphocyte-Mediated Cytotoxicity

Dorothy Hudig and Reza Karimi

Abstract

New functions are implicated for calreticulin, based on its release from cytotoxic NK and T cells. Calreticulin is the only one of six "KDEL" (Lys-Asp-Glu-Leu carboxy terminal) endoplasmic reticulum chaperone proteins present in cytotoxic granules of these lymphocytes. Calreticulin is released when the lymphocytes release their granules to kill other cells, which suggests that it is likely to have an important role. In this chapter, we briefly review the contents of cytotoxic granules and how calreticulin might interact with these proteins. Then we discuss calreticulin inactivation of lysis mediated by perforin, a pore-forming protein which is essential for granule-mediated toxicity. Calreticulin is degraded by granzymes (proteases found within the cytotoxic granules), particularly by one granzyme, Chymase 1. Chymase 1 was previously identified as required for granule-mediated lysis, which could indicate that the calreticulin function(s) may be temporally regulated. We also introduce our "inactivation of the inactivator" hypothesis (granzyme-degradation of the inactivator calreticulin) as a focus for evaluation of the present information and as a focus to identify critical information to collect in the future. Finally, we provide the reader with additional functions of calreticulin that could be important for systemic immunity.

Introduction

The topic of this chapter is "What is calreticulin doing in cytotoxic lymphocyte granules?". The question arises because calreticulin is the only one of six "KDEL" (Lys-Asp-Glu-Leu carboxy terminal) endoplasmic reticulum (ER) chaperone proteins present in the lymphocyte granules.[1] Cytotoxic granules have been described as specialized lysosomes[2] that are designed to kill cells infected with pathogens, which means that the selective retention of calreticulin begs for a functional explanation. Currently, we are pursuing the hypothesis that calreticulin provides important control of granule-dependent killing. We first introduce the cytotoxic lymphocytes and the contents of their dangerous granules, including the pore-forming granule protein, perforin,[3,4] which is essential for granule-mediated death. Then we summarize information about the effects of calreticulin on perforin-mediated killing of cells. At the end, we consider other effects that could result when calreticulin is released from immune cells. At several points, we identify laboratories and their locations to emphasize that the progress in this field is truly from many countries and many disciplines.

Calreticulin, 2nd ed., edited by Paul Eggleton and Marek Michalak.

Cytotoxic Lymphocytes and the Contents of the Granules

What are cytotoxic lymphocytes and what is the purpose of their toxic granules?

Cytotoxic lymphocytes include natural killer (NK) and T (thymically-derived) killer lymphocytes. Both types of killer lymphocytes circulate in blood and through lymphoid organs of people and other mammals all the time, whether the individuals are healthy or diseased (reviewed, ref. 5). The NK and T lymphocytes control intracellular infections by viruses (with different viral species that can infect almost all cell types) or intracellular infections by species of bacteria which selectively live inside macrophages (such as tubercle bacilli). In humans, active NK lymphocytes, which lack T cell receptors for antigen, are ~3% of all the lymphocytes circulating in blood and constitutively contain cytotoxic granules.[6] NK cells are part of the innate immune system and immediately kill other cells after detecting their infection. NK cells also kill cells that are coated with specific antibodies[7] which could be to viral antigens in the plasma membranes of infected cells. In contrast, memory cytotoxic T cells lack abundant granules until granule formation is induced when individual antigen-specific T cells encounter their cognate antigens. The cytotoxic T cells and the antibody-directed killing by NK cells are part of adaptive immunity, acquired in response to specific infections. Both NK or T release (exocytose) their granules after the lymphocytes specifically bind to infected "target" cells. Granule-mediated killing affects only the target cells, while the NK or T killer cells survive without damage to move on to kill more infected cells. The target cell dies in one of two ways: direct lysis by disruption of its plasma membrane or by induced apoptosis during which its plasma membrane stays initially intact while damage begins inside the cell.

What is in the NK and T cytotoxic cell granules to make them lethal (and how would calreticulin interact with it)?

Toxicity is associated with proteins (rather than lipid mediators) found within the granules[8] Important granule contents are summarized in Table 14.1. Membrane damage is produced by two of the granule proteins, perforin and a saponin-like protein termed granulysin in humans.[9,10] Each of these proteins has direct cellular toxicity independent of the other; however, granulysin requires perforin to gain access to intracellular bacteria. Perforin, in extremely low quantities (estimated to be below 10^{-9} M and fewer than 500 molecules per cell), makes large pores in cells. The perforin pores cause rapid lysis (immediate necrosis) if they are stable in cell membranes. Even if they are cleared by the target cells, the pores can facilitate entry of granzymes into the targets. Once inside the target cells, several of the granzymes can mediate intracellular damage by cleaving pro-apoptotic protein substrates. In contrast, higher concentrations (~1 μM) of granulysin are needed to lyse cells or kill bacteria.[11] Granzymes, serine-dependent proteases of lymphocytes, are implicated in the control of perforin-dependent lysis.[12,13] Three granzymes, Gr A, Gr B and Gr K, are capable of inducing apoptosis after gaining intracellular access to target cell substrates. The granzymes are designated by letters which correspond to their genes and have different proteolytic activities (see the footnotes of Table 14.1). Granzymes alone lack toxicity as do granules isolated from perforin$^{-/-}$ mice (J. Holder and D. Hudig, unpublished results). The calreticulin that is found inside cytotoxic granules is also without toxicity as would be anticipated based on its known chaperone activities. Almost all of the granule proteins are NBterminally glycosylated and the initial, high mannose carbohydrates of the glycosylation could bind to calreticulin for transport[14] (which would be an early function of calreticulin). However, the chaperone functions fail to explain why calreticulin would be retained in the granules after the proteins are transported.

Table 14.1. Proteins of T and NK cytotoxic lymphocyte granules

Protein	Function	Interactions with Calreticulin	Comments
Perforin	Forms large, stable membrane pores in plasma membranes.	Binds in the absence of calcium and dissociates with mM calcium.	Essential for granule-mediated toxicity.
Granulysin	Lyses cells and bacteria	Unknown.	Porcine ortholog is NK-lysin; rodent ortholog unknown
Granzyme A	Initiates apoptosis upon intracellular access to target cell substrates.[36,37]	Unknown. Possible binding to untrimmed, NBlinked high mannose carbohydrates of Gr A.[42]	Tryptase enzyme activity (a).
Granzyme B	Initiates apoptosis upon intracellular access to target cell substrates.[38,39]	Unknown. Possible binding to untrimmed, NBlinked high mannose carbohydrates of Gr B.[42]	Asp-ase enzyme activity (b).
Granzyme H	Unknown.	Unknown.	Chymase enzyme activity (c).
Granzyme K	Initiates apoptosis upon intracellular access to unknown substrates.[40]	Unknown.	Tryptase enzyme activity.
Granzyme M	Unknown.	Unknown.	Met-ase (d). Found in NK cells but not T cells.[41]
Granzymes C, D, E, F, and G	Unknown.	Unknown.	Found only in rodents.
Sulfated proteoglycan	Unknown. May bind granzymes and perforin under physiological salt.	Unknown.	Very acidic chondroitin sulfate with serglycin core protein.
Calreticulin	Unknown.		Intragranule presence confirmed by exocytosis
Lysosomal glycosidases and cathepsins	Unknown.	Unknown. confirmed by exocytosis	Intragranule presence of several enzymes

(a) Trypsin-like specificity, cleaving after Lys and Arg residues.
(b) cleaving after Asp residues.
(c) Chymotrypsin-like specificity, cleaving after aromatic amino acid residues
(d) Cleaving preferentially after Met residues.

It is very important for these studies to firmly establish that calreticulin is really inside the cytotoxic granules.

This consideration is warranted because ER vesicles contaminate the granules during their isolation by density centrifugation.[15] The presence of calreticulin in granules was first noted by Jurg Tschopp"s laboratory in Switzerland. They discovered that calreticulin is a major contaminant when perforin is purified from granules by anion exchange chromatography.[16] Subsequently, Marek Michalak and Chris Bleackley and their colleagues in Canada reported that mRNA encoding calreticulin was expressed at ~ 10 fold higher levels in activated cytotoxic T lymphocytes than in other lymphocytes.[17] Since only cytotoxic lymphocytes have dense granules, the expression of message is consistent with sequestration of substantial amounts of calreticulin in the unique granules. The localization of calreticulin to cytotoxic granules was indicated by confocal microscopy[1] using antibodies to calreticulin and to granzyme B which indicated that these proteins were in common compartments. Calreticulin appeared in cytotoxic granules even in perforin$^{-/-}$ mice.[1] Finally, the extracellular release of calreticulin, after stimulation of cytotoxic T cells with antibodies to T cell antigen receptor-associated proteins,[1] confirms that calreticulin really is in secretory granules. The definitive localization of calreticulin encouraged us to search for a role for calreticulin in lysis.

The Role of Calreticulin in Perforin-Dependent Lysis

What information is available and how does it support our hypothesis that calreticulin controls perforin-dependent killing?

Our hypothesis is that calreticulin inactivates perforin lysis and that selected granzymes cleave calreticulin to "inactivate the inactivator" and let perforin be lytic (see Fig. 14.1). This hypothesis provides an intra-granule substrate for the granzymes in addition to the intra-target cell substrates for the granzymes A, B and K.

We found that calreticulin will inactivate perforin-mediated lysis.[18,19] Perforin is routinely assayed by hemolysis (hemoglobin release) of red blood cells (rbc"s) which are very sensitive to perforin. Alternately, perforin is assayed by lysis of other cells indicated by release of radioactive ^{51}Cr as a tracer. The radioactive method requires about 30-fold more perforin for cell lysis but represents the cells that are the physiological targets of the killer lymphocytes. Calreticulin blocks perforin lysis in both assays (for rbc"s[18,19] and unpublished results for K562 cell targets). Calreticulin blocks perforin-mediated lysis of rbc"s at 2×10^{-7} M concentrations of native calreticulin. Slightly greater (9×10^{-7} M) concentrations of recombinant calreticulin also block perforin lysis. When the three major domains of calreticulin are compared, the C-domain prevented lysis while the N and P (proline-rich) domains were lacking effects. The recombinant C domain was nearly as effective as whole recombinant calreticulin.[19] The inactivation of perforin persists for over five hours. This observation is consistent with either stable inactivation and/or with the unrelated loss of activity of perforin during the assay. It is striking that while calreticulin inactivates partially purified perforin it rarely inactivates lysis mediated by unfractionated granules.[19]

It was this observation that led us to discover that granzymes degrade calreticulin. First, we noted that the granzymes of unfractionated granules completely digest ^{125}I-labeled calreticulin (Hudig and Elliott, unpublished results). If we allow the assumption that the labeled and the endogenous calreticulin are equivalent, then the intra-granule calreticulin is also rapidly degraded by the endogenous granzymes. Later, we found that only one of four chymases (granzymes which cleave after aromatic amino acid residues, as does chymotrypsin) degrades calreticulin: chymase 1, which appears essential for lysis,[13,20] cleaves calreticulin at two sites within the P domain (Hudig and Schegg, unpublished results). The proteolytic sites probably reflect the specificity of chymase 1, particularly since the other granzyme chymases had no effect on calreticulin. The P domain cleavage sites are unusual since the C region of calreticulin is the most susceptible region to proteolysis.[21,22] A tryptase other than Gr A also cleaves calreticulin, at an as yet unidentified site. Furthermore, based on the ability of the C domain alone to block lysis, it may be that the tryptase cleavage is in the C domain and its activity participates in inactivation of calreticulin.

How could calreticulin inactivate perforin lysis?

We have considered and discounted several explanations. Calreticulin is unlikely to form soluble complexes with perforin that are able to prevent perforin from binding to cells even though the two proteins can interact in solution. Soluble perforin will bind to calreticulin in

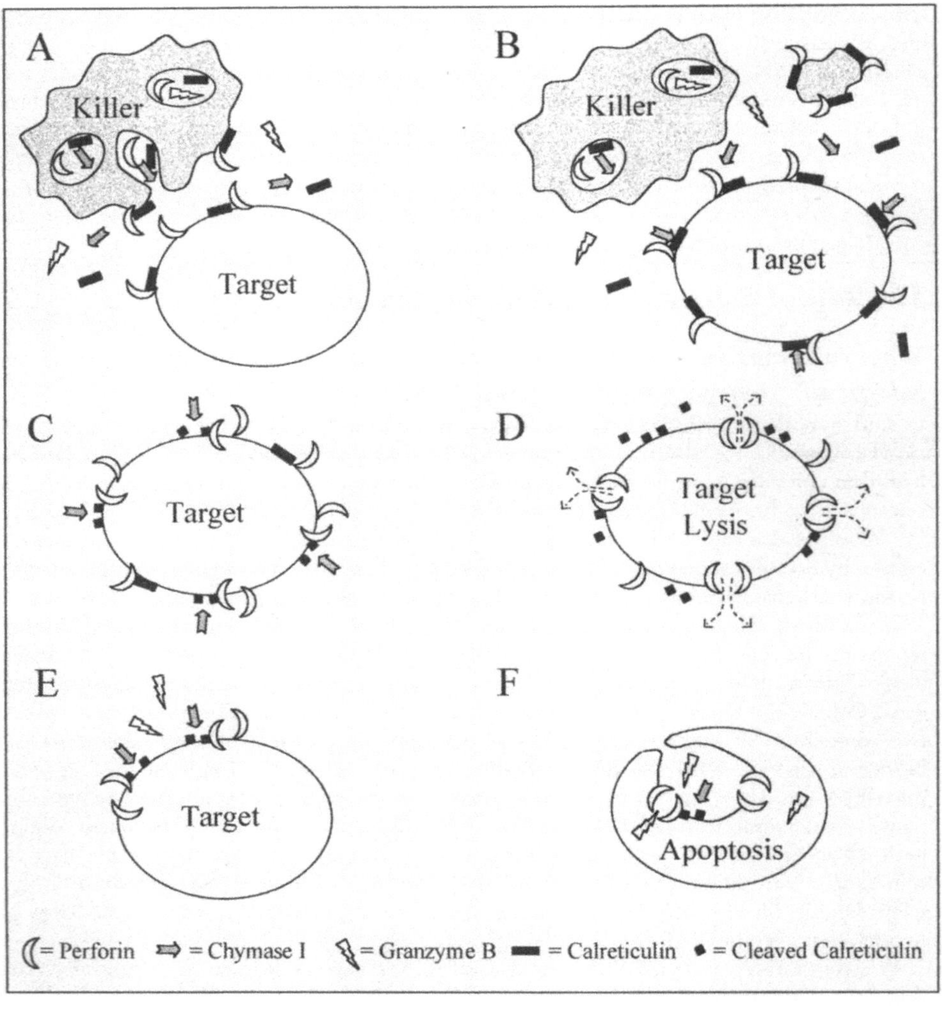

Figure 14.1. Model for regulation of perforin-mediated lysis by calreticulin and Chymase 1. In the scenario (A) calreticulin, perforin and the granzyme Chymase 1 are stored together in killer lymphocyte granules. After exocytosis, perforin and calreticulin will bind to the plasma membranes of the target cells and form complexes to prevent perforin from inserting into the membrane. Over time (B), the chymase will cleave membrane-bound calreticulin. Replacement of this calreticulin on the membrane in (B) will occur as long as the concentration of intact calreticulin is greater than its K_d for its specific binding site(s). After calreticulin is depleted by cleavage and/or diffusion, perforin can enter the membrane (C). Once perforin has formed pores, the cell may be directly damaged and leak to death (D). The regulation of perforin will also affect granzyme B-induced apoptosis. Perforin is also essential for granzyme B (Gr B, Asp-ase) to penetrate target cells where the protease cleaves several substrates to initiate apoptosis. Sublytic concentrations of perforin (E) stimulate cells to endocytose granzymes. Perforin-dependent injury that makes the endosome leak (F) is thought to be essential for granzyme B to gain access to intracellular substrates. Granzymes A and K (tryptases) also require perforin to gain intracellular access before they can initiate apoptosis.

the absence of calcium.[1] Perforin has the amino acid sequence KVFF (residues 439-442) which matches the first four amino acids of the KXFF[K/R]R sequence that supports calreticulin binding to the cytoplasmic domain of α-subunit of integrins[23] and to several steroid receptors.[24,25] However, the 0.1 mM or higher calcium concentrations that support perforin-dependent lysis dissociate soluble perforin-calreticulin complexes.[1] In other words, under lytic conditions, perforin and calreticulin are dissociated in solution; however, it should be noted that membrane-associated complexes in the presence of calcium are still possible. Competition of calreticulin with perforin for binding sites on membranes also seems unlikely. Perforin binds to phosphorylcholine[26] and calreticulin does not. Furthermore, there is insufficient calreticulin bound to red blood cells (at inactivating concentrations of calreticulin) to obscure substantial areas of membrane.[18] Calreticulin is also unlikely to prevent lysis by binding free calcium. The concentration of calreticulin that prevents lysis is insufficient to bind enough calcium to interfere with perforin lysis. Specifically, micro molar concentrations of calreticulin inactivate lysis but are insufficient to reduce 1 mM calcium to below 0.1 mM calcium, levels at which the lytic activity of perforin decreases. Our calculations have taken into account the 25-30 moles of calcium that can be bound per mole of calreticulin. Calreticulin also fails to inactivate the protease activities of the chymases and other granzymes that function in lysis.[19]

After consideration of the possibilities listed above, we have formed a hypothesis.

We propose that when calreticulin binds to target cell membranes, it acquires an altered conformation. In the new conformation we believe that it forms complexes with perforin to prevent perforin from entering into the membrane (Fig. 14.1A and B). In this model, Chymase 1 and/or another granzyme must degrade calreticulin to free perforin from the complex and allow it to enter the target cell membrane (Fig. 14.1C). Lysis will occur only after sufficient calreticulin has been cleaved (Fig. 14.1D) and/or the unbound calreticulin in equilibrium with the membrane has diffused away under in vivo conditions. Similarly, cleavage of calreticulin would be needed for perforin to form channels to allow granzyme B (or Gr A or Gr K) to enter cells and then cleave intracellular substrates to initiate apoptosis (Fig. 14.1E and F).

What on membranes binds calreticulin, to then promote localized calreticulin-perforin complex formation? Calreticulin has different affinities for the plasma membranes of different cells (and the concentrations that inactivate lysis are above these affinities). We have found that in the presence of calcium, calreticulin has a K_d of 2.7×10^{-7} M for rbc"s at unknown binding site(s).[18] Kuwabara et al., also in the USA, found a K_d " 7.4×10^{-9} M for calreticulin binding to endothelial cells, also at unknown binding site(s).[27] In addition, the cell surface protein heterodimer CD91 binds calreticulin,[28] as well as alpha-2 macroglobulin. Thus membrane interactions of calreticulin provide an alternate site for calreticulin-perforin interactions.

We don"t know why some cells, such as rbc"s, are much more susceptible to perforin than others, but it may correlate with the type and density of the calreticulin receptors. Other mechanisms may contribute to regulation of perforin damage. It should be added for perspective, that exocytic clearance of perforin lesions from resistant cells has compelling experimental support.[29,30] This mechanism may be completely independent of calreticulin.

What would be the biological advantages of " inactivation of the inactivator"?, you may ask.

The rates of pore formation would be slower, which would provide time for the killer cell to patch and release its perforin (indicated by the killer cell bleb with perforin and calreticulin in Fig. 14.1B). The scenario offers systemic protection: membranes of neighboring cells will remain protected from lysis by calreticulin if the chymase is inactivated (by plasma protease inhibitors). Furthermore, the presence of calreticulin on the cell surface of activated T lymphocytes,[31] as observed by Arosa et al. in Portugal, indicates that its location could also help protect the killer cells from the toxic agents they release.

Other Functions for Calreticulin in Immunity

What other effects could calreticulin have after it is released by immune cells?

When calreticulin binds to CD91 on macrophages, CD91 and calreticulin form a complex to promote uptake and clearance of apoptotic vesicles. For this occurrence, the apoptotic fragments must also be tagged with collectins such as complement C1q or mannose binding protein (see reference[32] and the chapter in this book by P.M. Henson and colleagues). It is the collectins that associate with calreticulin[33] in the CD91/CRT complexes and then the interaction triggers the cellular uptake of the collectin-coated vesicles. In a lymphocyte-mediated scenario, calreticulin would be present after induction of granzyme-induced apoptosis and be able to produce calreticulin-CD91 complexes on neighboring macrophages to promote clearance of the apoptotic fragments.

Another intriguing possibility is opened by the discoveries of Srivastava et al. in the USA. They found that extracellular complexes of calreticulin with foreign proteins can be ingested by cells to result in the intracellular degradation of the complexed protein. After degradation, the peptide fragments of the foreign protein are bound to MHC I (major histocompatibility complex) proteins. The MHC I proteins with the peptides are sent to the plasma membrane for antigen presentation to $CD8^+$ T cells,[34] providing a novel way to load peptides from extracellular proteins onto MHC I. The novel pathway allows MHC I (rather than MHC II) presentation of antigens that were synthesized outside the cytoplasm of the antigen-presenting cell. By this mechanism, viral and bacterial antigens from cells that died from NK cell-mediated perforin lytic disruption (rather than granzyme-induced apoptosis) could be salvaged. The salvaged antigens could then be recycled for presentation to $CD8^+$ T cells. This scenario would permit the immediate elimination of infections, mediated by the innate immunity of NK cells, to stimulate the long-lasting adaptive immunity of $CD8^+$ T cells that recognize antigens in the context of MHC I and not MHC II.

Neutrophils also release calreticulin,[35] as reported by Paul Eggleton and colleagues in the U.K., which indicates that there are additional sources of calreticulin for these immune functions which are likely to be unrelated to control of perforin.

Conclusions

There is a wide gap between experimentally reproducible data and physiologically valid conclusions. Descriptive information is needed to test our hypothesis, such as the amounts of calreticulin, calreticulin-degrading granzymes and perforin in granules. Critical in vitro experiments are also needed, using isolated perforin and calreticulin with differing concentrations of the granzyme calreticulinases. We could learn much from cellular experiments with lymphocytes lacking calreticulin (from $CRT^{-/-}$ mice) or expressing only calreticulin variants that would be refractory to granzyme degradation (from "knock-in calreticulin variants" of $CRT^{-/-}$ mice) . We hope to close this gap concerning the role of calreticulin in lysis, by our own discoveries and by integrating the rapid progress made by other researchers who are also addressing the roles of extracellular calreticulin.

Acknowledgments

This publication was supported in part by the NIH (USA) grant R01 CA38942. We thank Ms. Viki Elliott, Myra Godfrey, and Dr. W.H. Welch at the University of Nevada, Reno, for helpful criticism of the manuscript.

References

1. Andrin C, Pinkoski MJ, Burns K et al. Interaction between a Ca2+-binding protein calreticulin and perforin, a component of the cytotoxic T-cell granules. Biochemistry 1998; 37:10386-10394.
2. Page LJ, Darmon AJ, Uellner R et al. L is for lytic granules: lysosomes that kill. Biochim Biophys Acta 1998; 1401:146-156.
3. Liu C-C, Walsh CM, Young JDE. Perforin: structure and function. Immunol Today 1995; 16:194-201.
4. Tschopp J, Nabholz M. Perforin-mediated target cell lysis by cytolytic T lymphocytes. Ann Rev Immunol 1990; 8:279-302.
5. Henkart PA. Cytotoxic T lymphocytes. In: Paul WE Jr, ed. Fundamental Immunology. Philadelphia: Lippincott-Raven, 1999:1021-1050.
6. Bonavida B, Bradley TP, Grimm EA. The single cell assay in cell-mediated cytotoxicity. Im Today 1983; 4:196-198.
7. Trinchieri G. Biology of natural killer cells. Adv Immunol 1989; 47:187-376.
8. Smyth MJ, Kelly JM, Sutton VR et al. Unlocking the secrets of cytotoxic granule proteins. J Leukoc Biol 2001; 70:18-29.
9. Pena SV, Hanson DA, Carr BA et al. Processing, subcellular localization, and function of 519 (granulysin), a human late T cell activation molecule with homology to small lytic granule proteins. J Immunol 1997; 158: 2680-2688.
10. Pena SV, Krensky AM. Granulysin, a new human cytolytic granule-associated protein with possible involvement in cell-mediated cytotoxicity. Semin Immunol 1997; 9:117-125.
11. Stenger S, Hanson DA, Teitelbaum R et al. An antimicrobial activity of cytolytic T cells mediated by granulysin. Science 1998; 282:121-125.
12. Hudig D, Ewoldt GR, Woodard SL. Proteases and lymphocyte cytotoxic killing mechanisms. Curr Opin Immunol 1993; 5:90-96.
13. Woodard SL, Fraser SA, Winkler U et al. Purification and characterization of lymphocyte Chymase I, a granzyme implicated in perforin-mediated lysis. J Immunol 1998; 160:4988-4993.
14. Vassilakos A, Michalak M, Lehrman MA et al. Oligosaccharide binding characteristics of the molecular chaperones calnexin and calreticulin. Biochemistry 1998; 37:3480-3490.
15. Podack ER, Konigsberg PJ. 1984. Cytolytic T cell granules: Isolation, structural, biochemical, and functional characterization. J Exp Med 1984; 160:695-710.
16. Dupuis M, Schaerer E, Krause K-H et al. The calcium-binding protein calreticulin is a major constituent of lytic granules in cytolytic T lymphocytes. J Exp Med 1993; 177:1-7.
17. Burns K, Helgason CD, Bleackley RC et al. Calreticulin in T-lymphocytes. Identification of calreticulin in T-lymphocytes and demonstration that activation of T cells correlates with increased levels of calreticulin mRNA and protein. J Biol Chem 1992; 267:19039-19042.
18. Fraser SA, Michalak M, Welch WH et al. Calreticulin, a component of the endoplasmic reticulum and of cytotoxic lymphocyte granules, regulates perforin-mediated lysis in the hemolytic model system. Biochem Cell Biol 1998; 76:881-887.
19. Fraser SA, Karimi R, Michalak M et al. Perforin lytic activity is controlled by calreticulin. J Immunol 2000; 164:4150-4155.
20. Woodard SL, Jackson DS, Abuelyaman AS et al. Chymase-directed serine protease inhibitor that reacts with a single 30 kDa granzyme and blocks NK-mediated cytotoxicity. J Immunol 1994; 153:5016-5025.
21. Corbett EF, Michalak KM, Oikawa K et al. The conformation of calreticulin is influenced by the endoplasmic reticulum luminal environment. J Biol Chem 2000; 275:27177-27185.
22. Bouvier M, Stafford WF. Probing the three-dimensional structure of human calreticulin. Biochemistry 2000; 39:14950-14959.
23. Coppolino MG, Woodside MJ, Demaurex N et al. Calreticulin is essential for integrin-mediated calcium signalling and cell adhesion. Nature 1997; 386:843-847.
24. Burns K, Duggan B, Atkinson EA et al. Modulation of gene expression by calreticulin binding to the glucocorticoid receptor. Nature 1994; 367:476-480
25. Dedhar S, Rennie PS, Shago M et al. Inhibition of nuclear hormone receptor activity by calreticulin. Nature 1994; 367:480-483.
26. Tschopp J, Sch:afer S, Masson D et al. Phosphorylcholine acts as a Ca2+-dependent receptor molecule for lymphocyte perforin. Nature 1989; 337:272-274.

27. Kuwabara K, Pinsky DJ, Schmidt A et al. Calreticulin, an Antithrombotic Agent Which Binds to Vitamin K-dependent Coagulation Factors, Stimulates Endothelial Nitric oxide Production, and Limits Thrombosis in Canine Coronary Arteries. J Biol Chem 1995; 270:8179-8187.
28. Basu S, Binder RJ, Ramalingam T et al. CD91 is a common receptor for heat shock proteins gp96, hsp90, hsp70, and calreticulin. Immunity. 2001; 14:303-313.
29. Jones J, Morgan BP. Killing of cells by perforin. Resistance to killing is not due to diminished binding of perforin to cell membrane. Biochem J 1991; 280:199-204.
30. Jones J, Hallett MB, Morgan BP. Reversible cell damage by T-cell perforins. Calcium influx and propidium iodide uptake into K562 cells in the absence of lysis. Biochem J 1990; 267:303-307.
31. Arosa FA, De Jesus O, Porto G et al. Calreticulin is expressed on the cell surface of activated human peripheral blood T lymphocytes in association with major histocompatibility complex class I molecules. J Biol Chem 1999; 274:16917-16922.
32. Ogden CA, deCathelineau A, Hoffmann PR et al. C1q and mannose binding lectin engagement of cell surface calreticulin and CD91 initiates macropinocytosis and uptake of apoptotic cells. J Exp Med 2001; 194:781-795.
33. Sim RB, Moestrup SK, Stuart GR et al. Interaction of C1q and the collectins with the potential receptors calreticulin (cC1qR/collectin receptor) and megalin. Immunobiology 1998; 199:208-224.
34. Basu S, Srivastava PK. Calreticulin, a peptide-binding chaperone of the endoplasmic reticulum, elicits tumor- and peptide-specific immunity. J Exp Med 1999; 189:797-802.
35. Kishore U, Sontheimer RD, Sastry KN et al. Release of calreticulin from neutrophils may alter C1q-mediated immune functions. Biochem.J. 1997; 322:543-550.
36. Beresford PJ, Xia Z, Greenberg AH et al. Granzyme A loading induces rapid cytolysis and a novel form of DNA damage independently of caspase activation. Immunity 1999; 10:585-594.
37. Zhang D, Pasternack MS, Beresford PJ et al. Induction of rapid histone degradation by the cytotoxic T lymphocyte protease Granzyme A. J Biol Chem 2001; 276:3683-3690.
38. Kam C-M, Hudig D, Powers JC. Granzymes (lymphocyte serine proteases): characterization with natural and synthetic substrates and inhibitors. Biochim Biophys Acta 2000; 1477:307-323.
39. Sharif-Askari E, Alam A, Rheaume E et al. Direct cleavage of the human DNA fragmentation factor-45 by granzyme B induces caspase-activated DNase release and DNA fragmentation. EMBO J 2001; 20:3101-3113.
40. Shi L, Kam C-M, Powers JC et al. Purification of three lymphocyte granule serine proteases that induce apoptosis through distinct substrate and target interactions. J Exp Med 1992; 176:1521-1529.
41. Smyth MJ, Sayers TJ, Wiltrout T et al. Met-ase: cloning and distinct chromosomal location of a serine protease preferentially expressed in human natural killer cells. J Immunol 1993; 151:6195-6205.
42. Griffiths GM, Isaaz S. Granzymes A and B are targeted to the lytic granules of lymphocytes by the mannose-6-phosphate receptor. J Cell Biol 1993; 120:885-896.

Chapter 15

A Role for Calreticulin in the Clearance of Apoptotic Cells and in the Innate Immune System

Peter M. Henson

Abstract

Calreticulin has been shown to bind the collagenous tails of members of the collectin family of pattern recognition molecules. Its presence on cell surfaces then, implies a potential role for recognition of the collectins, and anything they may have bound via their globular head groups. In addition to foreign organisms, collectins bind to apoptotic cells and can mediate their uptake into phagocytic cells via this calreticulin-mediated recognition of the collagenous domains. As a non-transmembrane protein, calreticulin cannot transduce intracellular signals by itself but appears to use for this uptake and removal, a partner, CD91, which does exhibit appropriate intracellular signaling domains and function.

Introduction

The process of apoptosis, or programmed cell death, leads to numerous alterations in and on the cell that ultimately result in its clearance from the tissue. In simple terms, apoptosis involves steps that lead to destruction of the nuclear replicative apparatus as well as surface changes that allow the cell to be recognized and removed. While the signaling pathways that initiate these changes are complex, the end result under normal circumstances, is removal of the cell (Fig. 15.1). Studies over the last twenty years have shown this removal to represent a phagocytic event and have begun to define the receptors and mechanisms involved, as well as the usual consequences of the event. Some recent reviews on these subjects include.[1-7]

Tissues in metazoan animals do not remain static after development but are increasingly recognized to undergo turnover and remodeling, even in the adult. While this capacity may be more limited in mammals than, for example, amphibians, nevertheless, cell death, removal and replacement appear to be generally critical elements in normal tissue homeostasis. Key to these processes is the quiet, non-inflammatory and potentially regenerative nature of apoptosis and removal, which usually occurs in situ on a cell by cell basis. Thus, while professional phagocytes such as macrophages and immature dendritic cells are certainly most efficient at recognizing and internalizing apoptotic cells, the majority of cell types in the body seem capable of performing this function—including those from endodermal, epidermal or mesodermal origin. Removal is efficient and, usually, sporadic. This means that it has generally escaped notice and even now is very hard to identify when examination of tissues is limited to only snapshots in time. A marked example of the efficacy of this process is the normal removal of circulating neutrophils which has been estimated to encompass more than twice the total numbers within the circulation every 24 hours.[8] Massive remodeling, apoptosis and cell removal occurs during embryonic development, metamorphosis in amphibians, involution of the mammary gland or

Calreticulin, 2nd ed., edited by Paul Eggleton and Marek Michalak.

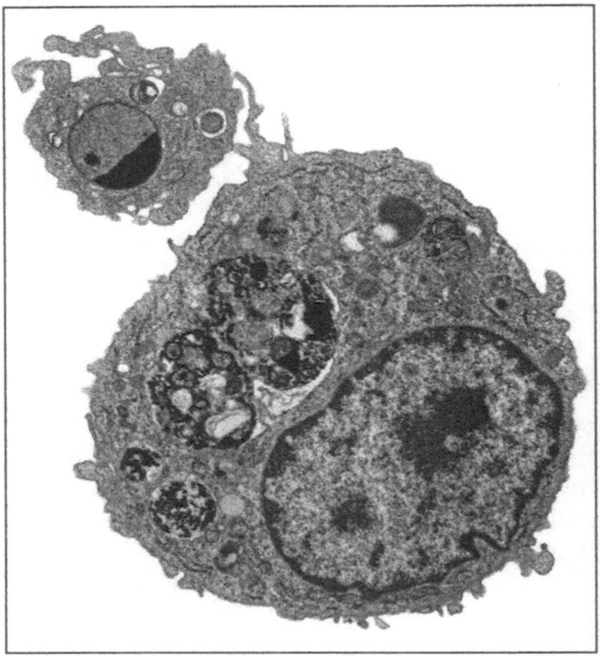

Figure 15.1. A macrophage ingesting apoprotic neutrophils. The electron micrograph was taken from a resolving pulmonary inflammatory response and depicts an apoptotic neutrophil in contact with the macrophage surface and apoptotic debris within phagosomes.

uterus after lactation or parturition in mammals, during the establishment and maintenance of the immune system and after every inflammatory response.

Clear, local examples of abnormal apoptotic cell removal are only beginning to be appreciated in mammals, likely because of significant redundancies in the removal processes. On the other hand, genetic studies in the nematode *Caenorhabditis elegans* have led to the identification of at least seven genes that contribute to apoptotic cell clearance. Most of these represent molecules now known to participate in signaling pathways involved in the ingestion of apoptotic cells. Only one of the identified genes has been implicated as a potential recognition molecule for apoptotic cells. This has led to speculation that even in this group of animals, there may be substantial redundancy in the receptors involved in specifically recognizing apoptotic cells with less redundancy in the signaling. It is this apoptotic cell recognition where calreticulin comes in, but, in an indirect fashion i.e., via recognition and binding of the collectin family of molecules to apoptotic cells, followed then by their ability to interact with cell surface calreticulin.

Recognition that calreticulin might play a role in apoptotic cell recognition and clearance comes from a concatenation of initially disparate observations. In 1997 Korb et al reported the binding of C1q to apoptotic keratinocytes and showed that this was localized to membrane blebs.[9] This led us to question the possible role of C1q and, indeed other members of the broadly defined collectin family of molecules, in apoptotic cell recognition. From a totally different direction, Botto, Walport and colleagues were investigating the potential links between C1q levels and autoimmune disease, particularly Systemic lupus erythematosus, SLE. They generated a C1qa$^{-/-}$ mouse and indeed showed it to have autoimmune abnormalities. In addition, they presented evidence for increased numbers of apoptotic cells in the kidneys of these animals[10]—a phenomenon that was entirely in keeping with defective apoptotic cell clearance. In support of this concept, deliberate instillation of apoptotic cells into the peritoneum of $C1q^{-/-}$ animals revealed a clearance defect that could be restored by re-addition of C1q.[11] Paradoxically, later studies have suggested that although C1q may participate in apoptotic cell clearance at a number of sites in the body, keratinocyte removal itself may not use this mechanism[12] since this latter was unaffected in the knockout mice.

From the other direction, a series of studies over many years had led to the recognition that one of the key receptors for C1q on cells was in fact calreticulin. Identification of cellular C1q receptors has had a long, complicated history and has generated numerous potential candidates. Ghebrehiwet and colleagues first characterized a binding molecule they called cC1qR,[13,14] because it appeared to recognize the collagenous tails of the C1q (see also ref. 15), in contrast to globular head binding "receptors" they called gC1qR. This cC1qR was later shown to be cell surface calreticulin.[16,17] The cC1qR (and thus, calreticulin) binds not only C1q, but also surfactant protein A (SP-A), mannose binding lectin (MBL) and conglutinin, i.e., members of the collectin family. It also interacts with another member of this family, surfactant protein D (SPD) (Eggleton personal communication and ref. 20). In fact a binding domain on calreticulin was identified that could be blocked with specific peptides.[18]

Putting this together led to experiments showing that C1q participated in the uptake of apoptotic cells by human monocyte-derived macrophages in vitro by a mechanism that could be blocked by anti-calreticulin antibodies.[19] This in turn directed attention to the possibility that other members of the collectin family could also mediate apoptotic cell uptake using the same mechanisms. Calreticulin-dependent uptake of apoptotic cells was shown in vitro for MBL[19] well as for SP-A (ref. 20 and see also ref. 21) and SP-D.[20] In addition, $SP\text{-}D^{-/-}$ mice exhibited an apoptotic cell clearance defect in the lungs.[20]

So much for the history, now let us examine the mechanisms involved.

The Collectin Family of Pattern Recognition, Innate Immune System, Molecules

Collectins are a family of multimeric, multifunctional, pattern-recognition molecules with a common three dimensional structure (see for example refs. 22-30). Globular head groups containing the recognition units are linked to long, remarkably similar, collagen-like tails by a hinge region. True collectins (SP-A, SP-D, MBL, conglutinin etc) exhibit C lectin activity in their heads—hence the name. C1q is similar in structure but has no lectin function. Unlike the others, the collagen tails here comprise three slightly different chains in contrast to the homo-trimers of the true collectins. C1q used to be thought different too because of its ability to interact with C1r and C1s to initiate the classical complement pathway. More recently a similar activity has been shown for MBL interacting with its own activation components.[22,28,31,32] For simplicity, therefore, we will consider these together as members of the collectin family.

C1q, MBL, SP-A and SP-D all have been shown to bind to the surface of apoptotic cells. However, the nature of this binding is not simple. Surprisingly, very little is known about the surface changes on cells that occur during apoptosis. Loss of the normal membrane phospholipid asymmetry is the most well characterized change and results in exposure of normally inner leaflet phosphatidylserine (PS).[33,34] The exact mechanisms are not understood in detail but the process seems to be driven by an increase in bidirectional phospholipid movement (scrambling) accompanied by a block of the normally rectifying aminophospholipid translocase so that any PS that reaches the outer leaflet is not returned to the inside.[35,36]

In addition to carbohydrates and proteins, each of these four collectin family proteins has been shown to bind phospholipids, including in some cases, phosphatidylserine itself (see reviews noted above). Whether PS is indeed the ligand for one or more of the collectin family remains to be determined. On the other hand, altered glycosylation of the apoptotic cell surface may also contribute. Uptake of apoptotic cells by the liver was shown to involve altered carbohydrates[37,38] and the MBL binding to apoptotic Jurkats was shown susceptible to mannosidase (unpublished observations).

In our hands, only MBL exhibited a marked increase in total binding to the surface of apoptotic cells compared with their viable counterparts[19,20]—the others bound to both viable and apoptotic cells. This should be contrasted with data reported by others for SP-A.[21] Whether the contrasting results reflect different cells or assays, the point is that the globular head groups do appear capable of interacting with ligands on normal cells and, likely have significant roles to play as a consequence. One example would be the increasing recognition that SP-A exhibits a potent anti-inflammatory effect in the lungs presumably as a function of stimulation by its globular head group of specific surface molecules ("receptors"). When initially encountered in our studies of C1q binding this lack of quantitative difference between binding to viable versus apoptotic cells posed a conceptual problem in trying to understand a potential role for C1q in apoptotic cell uptake. However, careful examination of the distribution of binding showed that the C1q attached to apoptotic cells in a highly localized, aggregated fashion. This fits the original description of C1q binding to keratinocyte blebs[9] and in later studies, by its globular heads, to apoptotic blebs on endothelial cells.[39] MBL, SP-A and SP-D all attach to apoptotic cells in a similar, highly localized fashion.[19,20]

Two important points emerge from these observations. For one, the data support primary interaction of the collectin family with apoptotic cells via their globular heads, thereby leaving their collagenous tails for engagement of the phagocyte. Secondly, the localized, aggregated, distribution may have important implications for modes of signaling to those phagocytes.

Collectin Interaction with Cell Surface Calreticulin

This subject is treated at length in refs. 16-17 and 40 in which the specific sites on the calreticulin that interact with the collagen-like tails of the collectins are defined and discussed. In Chapters 5 and 7 as well as in ref. 40 are considered the complex issues of calreticulin expression on the surface of cells. At this point in time, a wide variety of cells can be shown to exhibit this molecule on the surface. How the molecule gets there is a subject of some discussion. An increasing consensus seems to be emerging that it is actively transported to the membrane surface (see for example ref. 41) although the mechanisms involved are not yet clear. Calreticulin's KDEL, endoplasmic reticulum retrieval sequence, was thought to imply an alternative mechanism. For example, any cells dying in a culture or cell isolate might be expected to release some calreticulin from the ER which could then bind to putative surface ligands for this molecule on the surrounding cells. A number of arguments have been mounted against this possibility, including the observed uniformity of calreticulin amount from cell to cell. It is worth noting that gp96, another ER, KDEL-containing chaperone, is also becoming recognized as a widely distributed cell surface protein,[42] raising the same types of questions regarding how it gets there. The comparison is made even more relevant by the observation that both calreticulin and gp96 bind to CD91 (see below). An intriguing new possibility for mechanism of surface expression comes from the suggestion that the endoplasmic reticulum can exhibit transient connections with the plasma membrane, particularly during phagocytic events (DesJardins, M, personal communication). On the other hand, while the origin of cell-surface calreticulin is important to understand, how it got there may make no difference to its potential role in collectin binding and apoptotic cell removal.

Binding of collectin-coated particles or apoptotic cells was blocked by pretreatment with anti-calreticulin antibodies. In a further series of studies, the role of cell membrane calreticulin was demonstrated by using the receptor modulation approach pioneered by Silverstein.[43] Plating macrophages on ligands or antibody for a "receptor" (in this case calreticulin) leads to accumulation or sweeping of that receptor to the underneath of the cell leaving the upper surface selectively depleted of that receptor and any molecules it might be closely associated with.[19,20] In our case, we confirmed the efficacy of the modulation by probing the upper surface for remaining calreticulin with erythrocytes coated with anti-calreticulin antibodies. Plating macrophages onto any of the collectin family members led to selective depletion of calreticulin from the upper surface. Furthermore, plating on isolated collagenous tails from either C1q or SP-A had the same effect, supporting the orientation proposed, namely that the collectins bind the apoptotic cells by their globular heads and interact with calreticulin on the phagocytic cell via their tails. In all of these modulation experiments, removal of upper surface calreticulin was also accompanied by decreased uptake of apoptotic cells.

Interaction of Calreticulin with CD91/LRP As a Mechanism for Initiating Apoptotic Cell Internalization

As a non-transmembrane surface molecule, calreticulin presumably needs a partner for mediating intracellular signaling and internalization. A potential candidate for this has arisen in studies by Srivastava's group.[42,44,45] They have been investigating the role of heat shock proteins (chaperones) in the induction of immune responses and were also searching for a transmembrane signaling molecule (receptor) that would bind and internalize heat shock proteins and any peptides that may be attached to them. The prime candidate turned out to be CD91 (Fig. 15.2). Since calreticulin was one of the chaperones examined by them for peptide presentation, their demonstration that calreticulin bound to CD91 opened the potential that this molecule was responsible for collectin/calreticulin signaling and ultimately, for uptake of apoptotic cells.

CD91, also known as LDL receptor-related protein (LRP) and as one of the α-2 macroglobulin receptors (α2MR), is a large receptor comprising two non-covalently associated chains. The transmembrane chain exhibits signaling domains in its cytoplasmic region that seem to be

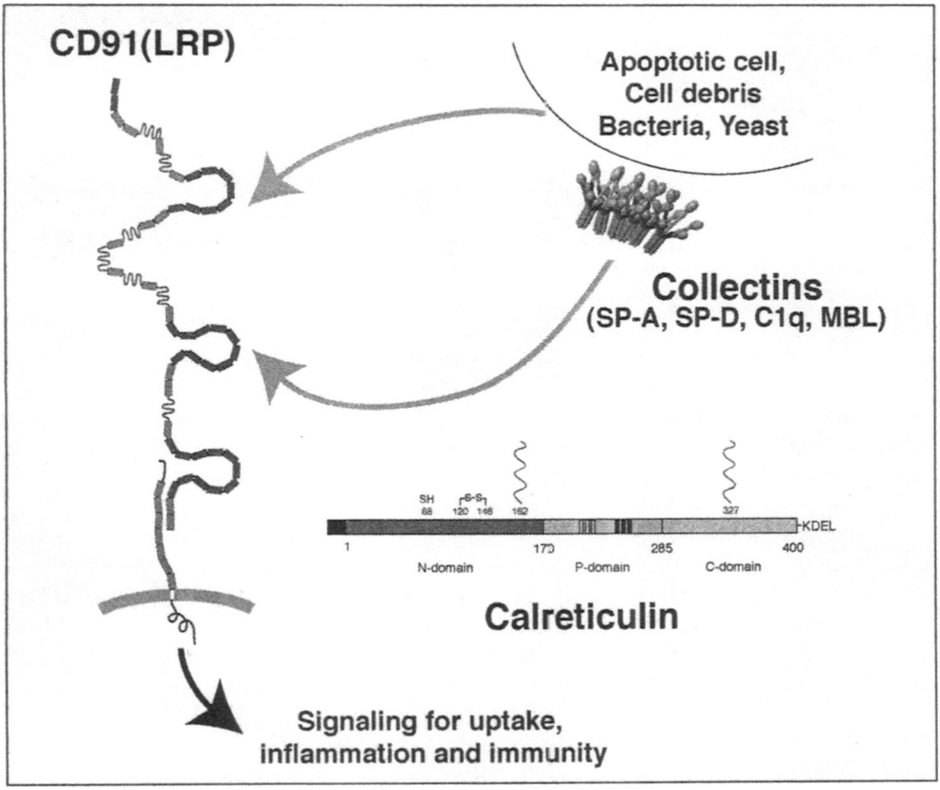

Figure 15.2. Scheme showing the potential orientation of collectins, calreticulin and CD91/LRP in the recognition and uptake of apoptotic cells.

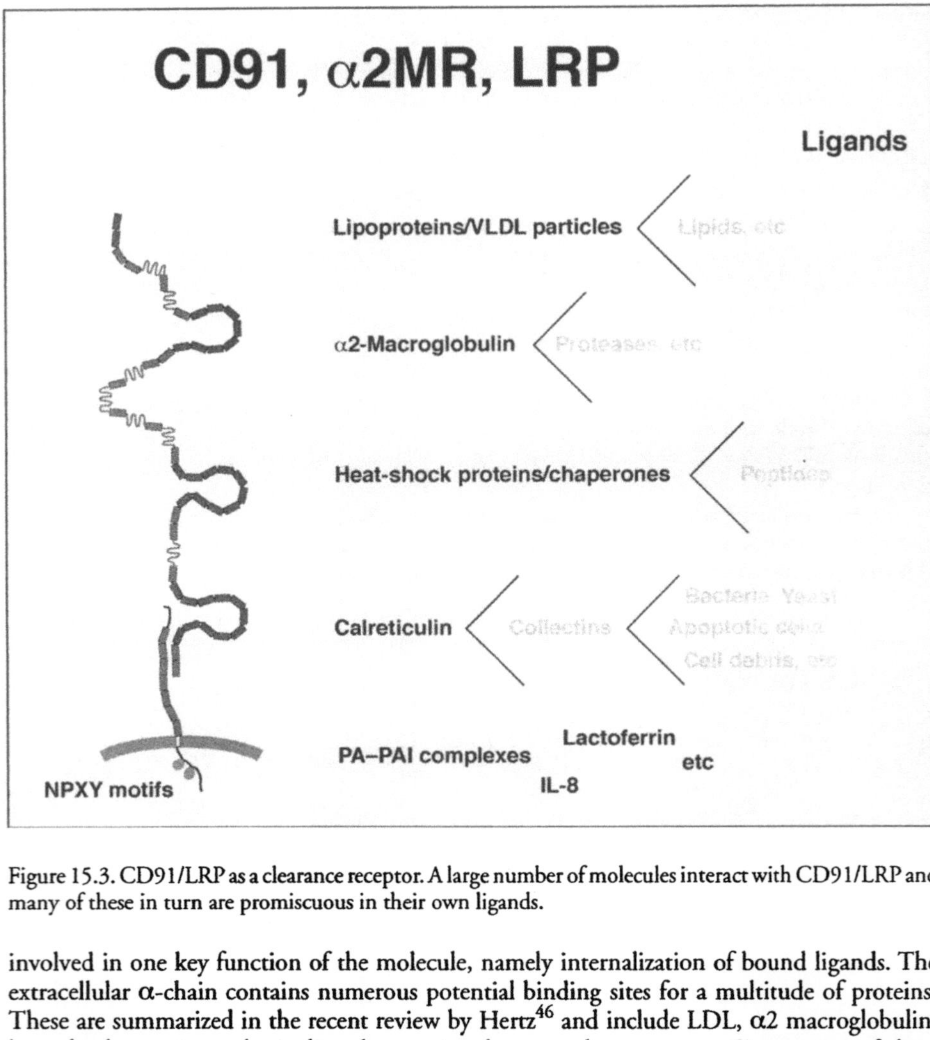

Figure 15.3. CD91/LRP as a clearance receptor. A large number of molecules interact with CD91/LRP and many of these in turn are promiscuous in their own ligands.

involved in one key function of the molecule, namely internalization of bound ligands. The extracellular α-chain contains numerous potential binding sites for a multitude of proteins. These are summarized in the recent review by Hertz[46] and include LDL, α2 macroglobulin, heat shock proteins, calreticulin, glycosaminoglycans and many more. Since many of these proteins themselves can bind other molecules (peptides, proteins, lipids etc) the whole system represents a potentially diverse, highly amplified mechanism for uptake and clearance—see Figure 15.3.

To address the involvement of CD91 in apoptotic cell removal, similar approaches were used[19,20] as previously outlined for calreticulin. Thus, for C1q, MBL, SP-A and SP-D attached to erythrocytes, anti-CD91 effectively blocked uptake into macrophages. Coating the erythrocytes with anti-calreticulin, calreticulin itself or α2-macroglobulin as an alternative CD91 ligand, also initiated uptake in an anti-CD91 inhibitable fashion. Plating macrophages onto surfaces coated with calreticulin, α2-macroglobulin, collectins or collectin collagenous tails all modulated CD91 from the upper surface of the cells as determined by diminished attachment and internalization of erythrocytes with attached anti-CD91 antibody. Finally calreticulin and CD91 were seen to co-localize on macrophage surfaces by immunofluorescence.

From these observations it is suggested that CD91 may act as a general candidate for internalization not only for lipoproteins, a2-macroglobulin, heat shock proteins and the like, but also for calreticulin and molecules that bind to this, including the collectins. This pathway would presumably apply not only to opsonization of apoptotic cells by the collectins but also to clearance of cell debris and foreign organisms. In this regard it is noteworthy that C1q (via its globular head groups) is an excellent ligand for mitochondria—interacting with not only mitochondrial proteins but also lipids such as cardiolipin.[47-49] When one remembers the likely evolutionary origin of mitochondria, this recognition becomes even more logical. We suspect in addition that the other collectins, through their extensive lipid and carbohydrate recognition properties will also be important in cell debris uptake. This potentially makes CD91 the ultimate in removal of dead cells and their components.

As mentioned above, much information on apoptotic cell clearance has come from genetic studies in *C. elegans*. This animal does contain a molecule homologous with mammalian calreticulin[50] and it has been suggested to play a role in certain forms of cell death, but more from its calcium sequestering activity than any implication of possible cell surface properties.[51] The original genetic screens did not show association of calreticulin with apoptotic cell removal and reports of calreticulin RNAi experiments in the worm were not accompanied by suggestions of abnormalities in cell clearance.[51]

CED-1 on the other hand may represent a more interesting story. Its recent cloning suggested that, unlike the other nematode molecules mentioned above as part of signal cascades, it might serve as a cell surface receptor.[52] Initial attempts to find a mammalian equivalent (using the whole gene as template) led to a suggested relationship with an endothelial cell scavenger receptor, SREC). Unfortunately the putative signaling domains did not match well. On the other hand, a search of the mammalian databases based on the intracellular domain of CED-1 led to demonstration of significant homology with the intracellular domain of CD91.[53] In fact the worm and mammalian domains serve equally to bind to CED6 (or its mammalian equivalent, GULP), an adapter molecule shown to be involved in apoptotic cell uptake in both *C. elegans* and mammalian systems.[54]

The dissimilarity between CD91 and CED-1 in their extracellular domains might suggest that, although the signaling domains are related, the organisms have developed quite different approaches to actually recognizing the apoptotic cells surface. In the case of CED-1 this may be direct binding but in mammalian systems, via a complex set of bridging molecules including calreticulin and the collectins. On the other hand one might ask whether bridging molecules linking apoptotic cells to CED-1 might exist in *C. elegans* and/or whether CD91 has an additional ability to directly recognize surfaces structures on apoptotic cells.

Mechanisms of Uptake and Signaling

This subject has been discussed at some length in recent papers and reviews (e.g., refs. 1,2,55,56). Current data are in keeping with a mechanism that is akin to stimulated macropinocytosis, not unlike processes involved in uptake of pathogenic Salmonella into epithelial cells[57] (Fig. 15.4). Stimulated macropinocytosis can be distinguished from phagocytosis induced through the Fc receptor which involves a "zipper" mechanism and results in close apposition between phagosomal membranes and the ingested particle. Macropinocytosis on the other hand leads to more spacious phagosomes and concurrent ingestion of extracellular fluid. It can also result in bystander uptake, i.e., stimulation of macropinocytosis by receptor ligation can mediate ingestion of any molecule or particle previously attached to the membrane in the local vicinity. Direct ligation of either calreticulin or CD91 on the cell can induce macropinocytosis and the ingestion of water soluble dye from the medium.[19]

Macropinocytosis also appears to be a key mechanism by which materials are taken up into immature dendritic cells, although here, the process seems to be constituitively turned on. Supply of calreticulin to such dendritic cells leads to uptake, processing and presentation of any associated peptides. Thus, addition of calreticulin isolated from different tumors to dendritic cells allowed them to induce tumor specific responses to appropriate T cells.[58] This type of experiment further supports the potential involvement of calreticulin-mediated uptake, not only in clearance of intracellular materials but also in participation in the interface between the innate and adaptive immune responses.

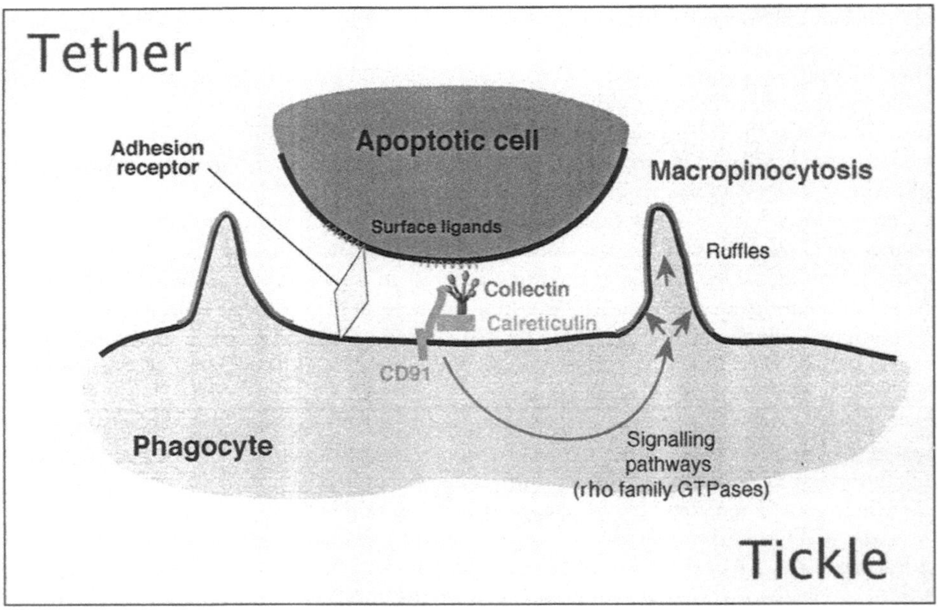

Figure 15.4. Proposed two-step recognition and uptake process for apoptotic cells. Tethering ligands cooperate with internalization receptors to optimize ingestion of these large particles.[55] The uptake mechanism is proposed to be a form of stimulated macropinocytosis.

Although we do not know in detail the signaling pathways involved in uptake induced by collectin/calreticulin/CD91, there does seem to be a common step in the requirement for *rac*-1 (equivalent to *C. elegans* CED-10) as seen in both apoptotic cell ingestion and macropinocytosis in general. As suggested above, GULP or CED-6 may be a necessary adaptor protein, but how specific this is to CD91 is not yet known. Since the collectins bind to apoptotic cells in a localized, aggregated fashion, and are ineffective when attached more diffusely to non-apoptotic cells, we have hypothesized that the collagenous tails also need to be aggregated in order to deliver an appropriate signal to the cell. This could mean the calreticulin and/or the CD91 must be crosslinked to initiate signaling or merely reflection of an avidity issue. Developing our understanding of the signaling pathways will be important and we suspect that they will be different from those involved in more classical phagocytic mechanisms. The very broad evolutionary distribution of these apoptotic cell and macropinocytotic uptake processes as well as their presence in most cell types suggest a strikingly "primitive" process.

Conclusions

The data support a crucial role for cell-surface calreticulin in recognition and removal of apoptotic cells, cell debris and intracellular constituents. Via the action of the collectin family of "defense collagens" the spectrum of recognized structures is broadened enormously. The collectin family of molecules has long been implicated in enhancing phagocytosis, in part by acting as opsonins (e.g., refs. 59,60). Since the time of Metchnikov, science approached phagocytosis from an infectious and protective bias. While in no way diminishing the importance in host defense, newer information on so-called "innate immune system molecules", including the collectins, suggests that a clear distinction between recognition of "foreign" and "self" in these systems may be significantly blurred. Removal of autologous damaged cells and debris is a critically important clean-up mechanism leading to restoration of normal structure and function. Beyond even this function, however, we are now coming to recognize that cells are

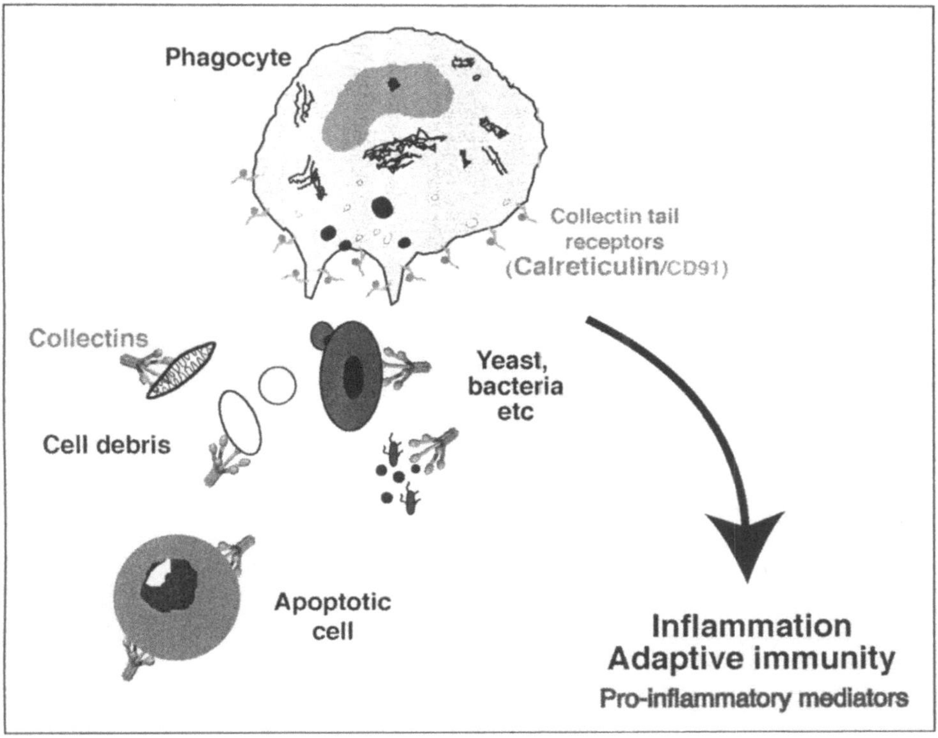

Figure 15.5. Collectins are innate immune system molecules that recognize and bind foreign materials and also apoptotic cells nd cell debris through their golbular heads. This presents their collagenous tails to calreticulin/CD91 complexes on phagocytes to initiate uptake and removal.

produced in the body in excess, both during development and in the adult. These excess cells, if not "used" or after they have performed their normal functions, are efficiently removed. As discussed above, we hypothesize an important role for cell surface calreticulin in this process. Thus, it seems reasonable to suggest that some of the profound developmental abnormalities seen in calreticulin$^{-/-}$ mice (e.g., ref. 61) may be due not only to its absence from the endoplasmic reticulum, but also because of abnormal apoptotic cell clearance. From an evolutionary perspective, parsimonious utilization of similar genes and processes for recognition and uptake of unwanted cells, dead cells and cell debris as well as for foreign organisms makes sense. The multiple roles in metazoa for calreticulin makes this a fascinating molecule, not only from the functional perspective but also in this evolutionary context. As already suggested, it would seem well worth while to question directly its role in direct or indirect recognition of dying cells and their products, even in organisms as diverse as invertebrates and plants.

References

1. Henson PM, Bratton DL, Fadok VA. Apoptotic cell removal. Curr Biol 2001; 11(19):R795-805.
2. Henson PM, Bratton DL, Fadok VA. The phosphatidylserine receptor: a crucial molecular switch? Nat Rev Mol Cell Biol 2001; 2(8):627-33.
3. Fadok VA, Bratton DL, Henson PM. Phagocyte receptors for apoptotic cells: recognition, uptake, and consequences. J Clin Invest 2001; 108(7):957-62.
4. Somersan S, Bhardwaj N. Tethering and tickling: a new role for the phosphatidylserine receptor. J Cell Biol 2001; 155(4):501-4.
5. Savill J, Fadok V. Corpse clearance defines the meaning of cell death. Nature 2000; 407(6805):784-8.
6. Chimini G. Engulfing by lipids: a matter of taste? Cell Death Differ 2001; 8(6):545-8.

7. Platt N, da Silva RP, Gordon S. Recognizing death: the phagocytosis of apoptotic cells. Trends Cell Biol 1998; 8(9):365-72.
8. Wintrobe M. Clinical Hematology. Philadelphia: Lea & Feibiger, 1981.
9. Korb LC, Ahearn JM. C1q binds directly and specifically to surface blebs of apoptotic human keratinocytes: complement deficiency and systemic lupus erythematosus revisited. J Immunol 1997; 158(10):4525-8.
10. Botto M, Dell'Agnola C, Bygrave AE et al. Homozygous C1q deficiency causes glomerulonephritis associated with multiple apoptotic bodies [see comments]. Nat Genet 1998; 19(1):56-9.
11. Taylor PR, Carugati A, Fadok VA et al. A hierarchical role for classical complement proteins in the clearance of apoptotic cells in vivo: a mechanism for protection from autoimmunity. J Exp Med 2000; in Press.
12. Pickering MC, Fischer S, Lewis MR et al. Ultraviolet-radiation-induced keratinocyte apoptosis in C1q-deficient mice. J Invest Dermatol 2001; 117(1):52-8.
13. Ghebrehiwet B, Kew RR, Gruber BL et al. Murine mast cells express two types of C1q receptors that are involved in the induction of chemotaxis and chemokinesis. J Immunol 1995; 155(5):2614-9.
14. Peerschke EI, Ghebrehiwet B. Platelet C1q receptor interactions with collagen- and C1q-coated surfaces. J Immunol 1990; 145(9):2984-8.
15. Malhotra R, Thiel S, Reid KB et al. Human leukocyte C1q receptor binds other soluble proteins with collagen domains. J Exp Med 1990; 172(3):955-9.
16. Stuart GR, Lynch NJ, Day AJ et al. The C1q and collectin binding site within C1q receptor (cell surface calreticulin). ImmunoPharmacol 1997; 38(1-2):73-80.
17. Sim RB, Moestrup SK, Stuart GR et al. Interaction of C1q and the collectins with the potential receptors calreticulin (cC1qR/collectin receptor) and megalin. Immunobiology 1998; 199(2):208-24.
18. Kovacs H, Campbell ID, Strong P et al. Evidence that C1q binds specifically to CH2-like immunoglobulin gamma motifs present in the autoantigen calreticulin and interferes with complement activation. Biochemistry 1998; 37(51):17865-74.
19. Ogden CA, deCathelineau A, Hoffmann PR et al. C1q and mannose binding lectin engagement of cell surface calreticulin and CD91 initiates macropinocytosis and uptake of apoptotic cells. J Exp Med 2001; 194(6):781-95.
20. Vandivier RW, Ogden CA, Fadok VA, Hoffman PR, Brown KK, Botto M, Walport MJ, Fisher JH, Henson PM, Greene KE. Role of surfactant proteins a, d and clq in the clearance of apoptotic cells in vivo and in vitro: Calreticulin and cd91 as a common collectin receptor complex. J Immunol 2002; 169:3978-3986.
21. Schagat TL, Wofford JA, Wright JR. Surfactant protein A enhances alveolar macrophage phagocytosis of apoptotic neutrophils. J Immunol 2001; 166(4):2727-33.
22. Holmskov UL. Collectins and collectin receptors in innate immunity. APMIS Suppl 2000; 100:1-59.
23. Clark HW, Reid KB, Sim RB. Collectins and innate immunity in the lung. Microbes Infect 2000; 2(3):273-8.
24. Hakansson K, Reid KB. Collectin structure: a review. Protein Sci 2000; 9(9):1607-17.
25. Crouch E, Hartshorn K, Ofek I. Collectins and pulmonary innate immunity. Immunol Rev 2000; 173:52-65.
26. van Rozendaal BA, van Golde LM, Haagsman HP. Localization and functions of SP-A and SP-D at mucosal surfaces. Pediatr Pathol Mol Med 2001; 20(4):319-39.
27. Kishore U, Reid KB. Structures and functions of mammalian collectins. Results Probl Cell Differ 2001; 33:225-48.
28. Hansen S, Holmskov U. Structural aspects of collectins and receptors for collectins. Immunobiology 1998; 199(2):165-89.
29. Tenner AJ. Membrane receptors for soluble defense collagens. Curr Opin Immunol 1999; 11(1):34-41.
30. Epstein J, Eichbaum Q, Sheriff S et al. The collectins in innate immunity. Curr Opin Immunol 1996; 8(1):29-35.
31. Turner MW, Hamvas RM. Mannose-binding lectin: structure, function, genetics and disease associations. Rev Immunogenet 2000; 2(3):305-22.
32. Petersen SV, Thiel S, Jensen L et al. Control of the classical and the MBL pathway of complement activation. Mol Immunol 2000; 37(14):803-11.
33. Fadok VA, Voelker DR, Campbell PA et al. Exposure of phosphatidylserine on the surface of apoptotic lymphocytes triggers specific recognition and removal by macrophages. J Immunol 1992; 148(7):2207-16.
34. Martin SJ, Reutelingsperger CP, McGahon AJ et al. Early redistribution of plasma membrane phosphatidylserine is a general feature of apoptosis regardless of the initiating stimulus: inhibition by overexpression of Bcl-2 and Abl. J Exp Med 1995; 182(5):1545-56.

35. Bratton DL, Fadok VA, Richter DA et al. Appearance of phosphatidylserine on apoptotic cells requires calcium- mediated nonspecific flip-flop and is enhanced by loss of the aminophospholipid translocase. J Biol Chem 1997; 272(42):26159-65.
36. Frasch SC, Henson PM, Kailey JM et al. Regulation of phospholipid scramblase activity during apoptosis and cell activation by protein kinase Cdelta. J Biol Chem 2000; 275(30):23065-73.
37. Ruzittu M, Carla EC, Montinari MR et al. Modulation of cell surface expression of liver carbohydrate receptors during in vivo induction of apoptosis with lead nitrate. Cell Tissue Res 1999; 298(1):105-12.
38. Dini L. Recognizing death: liver phagocytosis of apoptotic cells. Eur J Histochem 2000; 44(3):217-27.
39. Navratil JS, Watkins SC, Wisnieski JJ et al. The globular heads of C1q specifically recognize surface blebs of apoptotic vascular endothelial cells. J Immunol 2001; 166(5):3231-9.
40. Johnson S, Michalak M, Opas M et al. The ins and outs of calreticulin: from the ER lumen to the extracellular space. Trends Cell Biol 2001; 11(3):122-9.
41. Xiao G, Chung TF, Fine RE et al. Calreticulin is transported to the surface of NG108-15 cells where it forms surface patches and is partially degraded in an acidic compartment. J Neurosci Res 1999; 58(5):652-62.
42. Basu S, Binder RJ, Ramalingam T et al. CD91 is a common receptor for heat shock proteins gp96, hsp90, hsp70, and calreticulin. Immunity 2001; 14(3):303-13.
43. Michl J, Unkeless JC, Pieczonka MM et al. Modulation of Fc receptors of mononuclear phagocytes by immobilized antigen-antibody complexes. Quantitative analysis of the relationship between ligand number and Fc receptor response. J Exp Med 1983; 157(6):1746-57.
44. Binder RJ, Han DK, Srivastava PK. CD91: a receptor for heat shock protein gp96. Nat Immunol 2000; 1(2):151-5.
45. Binder RJ, Karimeddini D, Srivastava PK. Adjuvanticity of alpha 2-macroglobulin, an independent ligand for the heat shock protein receptor CD91. J Immunol 2001; 166(8):4968-72.
46. Herz J, Strickland DK. LRP: a multifunctional scavenger and signaling receptor. J Clin Invest 2001; 108(6):779-84.
47. Storrs SB, Kolb WP, Pinckard RN et al. Characterization of the binding of purified human C1q to heart mitochondrial membranes. J Biol Chem 1981; 256(21):10924-9.
48. Kovacsovics T, Tschopp J, Kress A et al. Antibody-independent activation of C1, the first component of complement, by cardiolipin. J Immunol 1985; 135(4):2695-700.
49. Ghebrehiwet B, Lim BL, Kumar R et al. gC1q-R/p33, a member of a new class of multifunctional and multicompartmental cellular proteins, is involved in inflammation and infection. Immunol Rev 2001; 180:65-77.
50. Smith MJ. A C. elegans gene encodes a protein homologous to mammalian calreticulin. DNA Seq 1992; 2(4):235-40.
51. Xu K, Tavernarakis N, Driscoll M. Necrotic cell death in C. elegans requires the function of calreticulin and regulators of Ca(2+) release from the endoplasmic reticulum. Neuron 2001; 31(6):957-71.
52. Zhou Z, Hartwieg E, Horvitz HR. CED-1 is a transmembrane receptor that mediates cell corpse engulfment in C. elegans. Cell 2001; 104(1):43-56.
53. Su HP, Nakada-Tsukui K, Tosello-Trampont AC et al. Interaction of CED-6/GULP, an adapter protein involved in engulfment of apoptotic cells, with CED-1 and CD91/LRP. J Biol Chem 2001; 29:29.
54. Liu QA, Hengartner MO. Human CED-6 encodes a functional homologue of the Caenorhabditis elegans engulfment protein CED-6. Curr Biol 1999; 9(22):1347-50.
55. Hoffmann PR, deCathelineau AM, Ogden CA et al. Phosphatidylserine (PS) induces PS receptor-mediated macropinocytosis and promotes clearance of apoptotic cells. J Cell Biol 2001; 155(4):649-60.
56. Leverrier Y, Ridley AJ. Requirement for Rho GTPases and PI 3-kinases during apoptotic cell phagocytosis by macrophages. Curr Biol 2001; 11(3):195-9.
57. Zhou D, Galan J. Salmonella entry into host cells: the work in concert of type III secreted effector proteins. Microbes Infect 2001; 3(14-15):1293-8.
58. Basu S, Srivastava PK. Calreticulin, a peptide-binding chaperone of the endoplasmic reticulum, elicits tumor- and peptide-specific immunity. J Exp Med 1999; 189(5):797-802.
59. Reid KB. Functional roles of the lung surfactant proteins SP-A and SP-D in innate immunity. Immunobiology 1998; 199(2):200-7.
60. Crouch EC. Surfactant protein-D and pulmonary host defense. Respir Res 2000; 1(2):93-108.
61. Mesaeli N, Nakamura K, Zvaritch E et al. Calreticulin is essential for cardiac development. J Cell Biol 1999; 144(5):857-68.

Chapter 16

Calreticulin and Tumor Suppression

Giovanna Tosato, Lei Yao and Sandra E. Pike

Abstract

Lymphoblastic cell lines derived by Epstein-Barr virus (EBV) immortalization of normal B lymphocytes can reproducibly inhibit tumor growth in nude mice. This anti-tumor activity is due, in part, to a mediator present in the culture supernatant of EBV-immortalized lymphoblastic cells identified as a fragment of calreticulin encompassing the NH_2 terminal domain. Recombinant calreticulin NH_2 domain (amino acids 1-180), but not calreticulin P domain (amino acids 181-189), can inhibit endothelial cell proliferation in vitro and angiogenesis in vivo. Calreticulin NH_2 domain can bind specifically to the extracellular matrix protein laminin and through this binding can interfere with endothelial cell attachment and spread on laminin-coated surfaces. It can also bind directly to endothelial cell membranes. When inoculated subcutanously into nude mice, recombinant calreticulin NH_2 domain prevents the development of subcutaneous tumors and reduces the growth of established tumors of various lineages. The capillary vascular network of tumor tissues from mice treated with calreticulin NH_2 domain is reduced compared to controls and scattered tumor cell apoptosis is noted, but tumor histology is otherwise similar. Thus, the NH_2 domain of calreticulin is detected in the culture supernatant of lymphoblastoid cells and can exert anti-tumor activities acting as an inhibitor of endothelial cell growth and neovascularization.

Introduction

Progressive growth and spread of tumors are dependent upon and adequate blood supply achieved through the co-option of existing blood vessels, neovascularization, and formation of blood channels (Fig. 16.1).[1-5] Tumors can promote new vessel formation through the release or induction of angiogenic factors, including members of the VEFG (vascular endothelial growth factor), FGF (fibroblast growth factor) and Ang (angiopoietin) protein family.[6-8] These molecules variously contribute to angiogenesis by promoting endothelial cell survival, mobilization, migration, growth, adhesion or matrix degradation.[9-11] Recently, a variety of experimental drugs that can reduce tumor blood supply have been reported to prevent or delay tumor development, and to suppress the growth or even promote the regression or dormancy of established tumors in experimental models.[9,12-20]

In the course of experiments designed to investigate the role of natural immunity in the regulation of Epstein-Barr virus infections, we noted that EBV-immortalized cell lines, which are not tumorigenic in the subcutaneous tissue of athymic mice, can delay the growth and even promote the regression of a variety of tumors established in these mice.[21,22] The anti-tumor effects exerted by the EBV-immortalized cells appeared to be vascular-based, and to require the expression of the EBV latency gene LMP-1.[23] Initial studies revealed that the murine chemokines IFN-γ inducible protein-10 (IP-10) and the monokine induced by IFN-γ (Mig) contributed to tumor regression in this model by exerting an angiotoxic effect directed at the established tumor vasculature.[22,24-28] However, IP-10 and Mig did not fully reproduce the anti-tumor activities exerted by EBV-immortalized cells. In particular, we noted that rate of tumor growth

Calreticulin, 2nd ed., edited by Paul Eggleton and Marek Michalak.

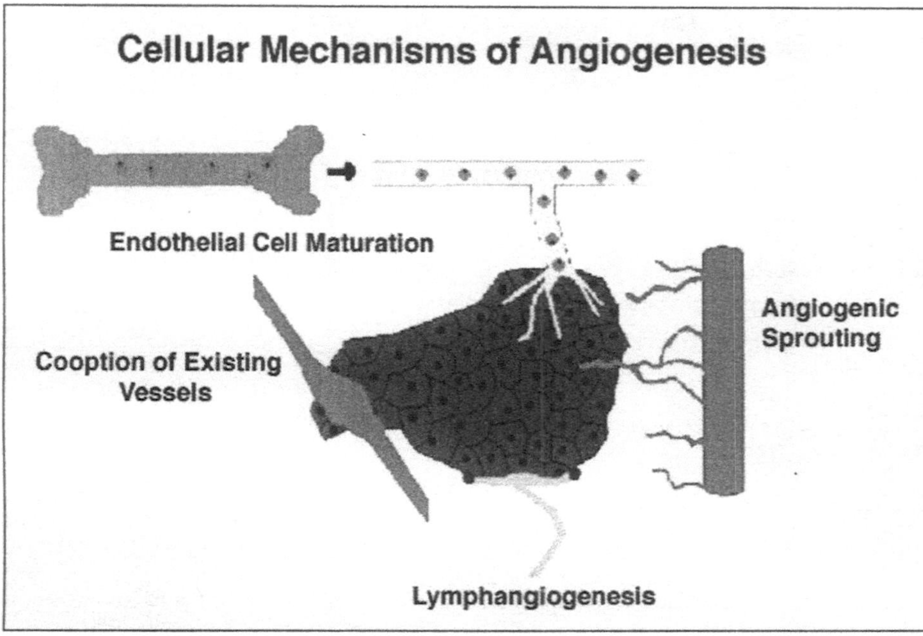

Figure 16.1. Schematic Representation of cellular mechanisms for tumor angiogenesis. Blood supply to progressively growing tumors is provided by angiogenic sprouting originating from the neighboring vessels, "cooption" of the existing vasculature, and maturation/migration of endothelial cell precursors/endothelial cells from the bone marrow.

was not initially reduced by IP-10 or Mig whereas it was typically reduced by the EBV-immortalized cells.[21] In addition, IP-10 and Mig failed to promote complete tumor regressions that were often induced by EBV-immortalized cells.[21] Since the conditioned medium from EBV-immortalized cell lines reduced somewhat tumor growth in mice, we raised the possibility that factors released by EBV-immortalized cells might inhibit angiogenesis and tumor growth.

Isolation of Calreticulin NH_2 Terminal Fragments and Calreticulin and Their Identification As Inhibitors of Endothelial Cells Proliferation

Culture supernatants from a variety of EBV-immortalized cell lines can inhibit to a varying degree the proliferation of endothelial cells derived from primary cultures of human umbilical cord endothelial cells (HUVEC) or bovine heart endothelial cells (FBHE). Due to its apparent high content of inhibitory activity, we selected conditioned medium from the VDS-O cell line[21] as a source of the endothelial cell inhibitory activity for purification. Using inhibition of bFGF (basic fibroblast growth factor)-induced endothelial cell proliferation as an assay to follow recovery of biological activity, we purified the inhibitory compounds from serum-free culture supernatants of the VDS-O cell line. Four sequential chromatographic steps were used, including metal chelating affinity, anion exchange, hydrophobic interaction, and anion exchange. In a representative purification, an initial loading of 340 mg total protein from 16 liters of supernatant yielded 4.5 μg of purified protein. The total biological activity of the starting supernatant was approximately 44×10^3 units (one unit being defined as the 1/2 maximal activity measured in a proliferation assay of FBHE cells), and the purified material contained approximately 9×10^2 units.

When analyzed by two-dimensional gel electrophoresis under reduced conditions and stained by Coomassie, two defined spots were identified with molecular masses of ~55 and ~20 kDa

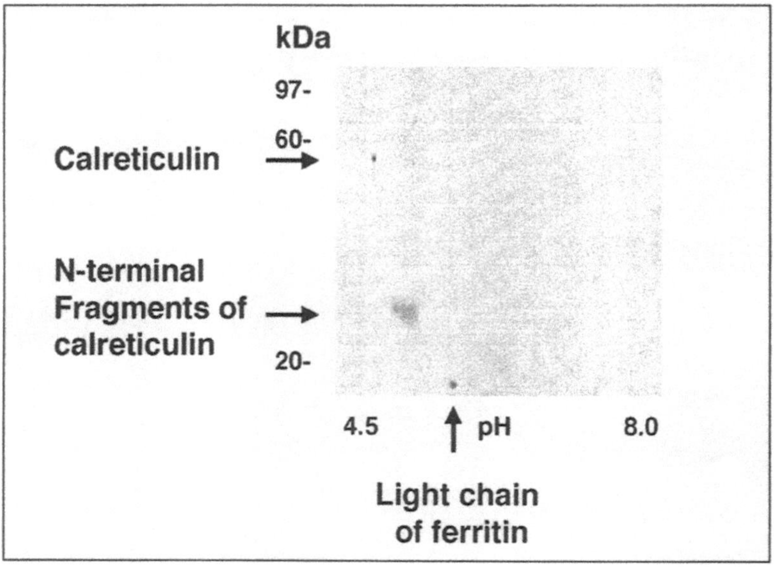

Figure 16.2. Two-dimensional gel electrophoresis of material purified from the culture supernatants of EBV-immortalized cells showing the isolation of three components identified as calreticulin, NH_2-terminal fragments of calreticulin, and the light chain of ferritin.

and apparent isoelectric point of 4.7 and 5.6, respectively (Fig. 16.2). In addition, a larger and poorly defined spot was identified with a relative mass of 25-30 kDa. By trypsin digestion and analysis of the fragments by ion trap mass spectrometry, the 55 kDa spot was identified as calreticulin whereas the 20 kDa spot was identified as the light chain of ferritin. Further evidence of the identity of the 55 kDa material as calreticulin was obtained by Western blotting of the purified material with a monospecific rabbit antiserum raised against recombinant human calreticulin. Interestingly, this antiserum also recognized the 25-30 kDa spot, suggesting that it may reflect fragments of calreticulin. Antibodies directed at calreticulin NH_2 (amino acids 6-19), but not at the COOH (amino acids 382-400) terminal domain recognized this 25-30 kDa material, suggesting that it reflects NH_2 terminal fragments of calreticulin. We therefore concluded that the biologically active, purified material contained calreticulin, NH_2-terminal fragments of calreticulin and the light chain of ferritin.

To assess whether the endothelial cell inhibitory activity derived from one of these purified components, we eluted separately from preparative gels material corresponding in size to the 55 and 25-35 kDa spots. Material eluted from a band of approximately 25-35 kDa and separately from a band of approximately 55kDa, but not from control parts of the gel, was found to inhibit endothelial cell proliferation. Thus, we had isolated calreticulin and NH_2 fragments of calreticulin from culture supernants of an EBV-immortalized cell line, and had obtained preliminary evidence that calreticulin and NH_2 fragments of calreticulin were active as inhibitors of endothelial cell proliferation. By contrast, recombinant purified light chain of human ferritin (1pg-1μg/ml) had minimal effect on the proliferation of endothelial cells.

As described above, we purified calreticulin and 25-30 kDa NH_2 terminal calreticulin fragments from culture supernatants of a lymphoblastoid cell line. The calreticulin fragments appeared more abundant than the 55 kDa intact calreticulin. One of the issues raised by these results is how calreticulin can reach the extracellular compartment in spite of its expression of the endoplasmic reticulum retrieval sequence KDEL at the C-terminous. It is possible that calreticulin is released from lymphoblastic cells as a consequence of cell death, much alike what was previously reported with cytotoxic cells engaged in cell killing of target cells.[29] In spite of optimal culture conditions we routinely observed 5-10% cell death in cultures used to generate conditioned medium for purification. Alternatively, it was suggested that calreticulin may follow the secretory pathway in spite of its ER retention signals.[30-33] An additional possibility is that calreticulin may undergo conformational change depending on the conditions of the ER luminal environment resulting in the formation of proteolytic fragments that may follow the secretory pathway. Experiments in vitro have shown that Ca^{2+} binding to the COOH-domain of calreticulin in the presence of ATP leads to conformational change of the protein allowing for trypsin cleavage of the C-domain and formation of a protease-resistant calreticulin fragment encompassing the N- and P-domains.[34] It will be important to further establish the regulation of calreticulin trafficking, potential conformational changes and proteolytic cleavage it may undergo both intracellularly and extracellularly.

Effects of Calreticulin and Calreticulin Fragments on Endothelial Cell Proliferation

To evaluate more closely whether calreticulin can inhibit endothelial cell proliferation, we have produced calreticulin and fragments of calreticulin in *E. coli*. Calreticulin (amino acids 1-400) was expressed as a fusion protein of either maltose-binding protein (MBP) or glutathione S-transferase (GST), and calreticulin NH_2 domain (amino acids 1-180) was expressed as a fusion protein of MBP. Additional recombinant proteins included MBP, GST, MBP-calreticulin P-domain, and a deletion mutant of calreticulin that lacks amino acids 1-120 (Δ120 calreticulin). The purified proteins were tested for their effects on endothelial cell proliferation induced by bFGF or VEGF.[35,36] At a concentration of 1 µg/ml, MBP-calreticulin and MBP-calreticulin N-domain inhibited significantly bFGF-induced proliferation of fetal bovine heart endothelial (FBHE) cells, whereas MBP did not (Fig. 16.3 A and B). Similarly, MBP-calreticulin and MBP-calreticulin N-domain inhibited significantly the proliferation of HUVEC induced by either bFGF (Table 16.1) or VEGF (not shown). A side-by-side comparison of full-length MBP-calreticulin, MBP-calreticulin N-domain and MBP-Δ120 calreticulin revealed that these proteins display similar endothelial cell growth inhibitory activity in vitro (Fig. 16.4). Additional experiments revealed that calreticulin N-domain inhibits endothelial cell cycle progression from the G_0/G_1 phase of the cell cycle to the S phase (unpublished). These results demonstrated that calreticulin can inhibit endothelial cell proliferation in vitro and that this activity likely resides in the N-terminal domain of the molecule.

In contrast to their inhibitory effects on bovine and human endothelial cell proliferation, calreticulin and calreticulin N-domain at concentrations of 0.5-10 µg/ml had minimal effect on the proliferation of a variety of other primary cells and cell lines. These included human peripheral blood mononuclear cells stimulated with phytohemagglutinin, B-cell enriched peripheral blood mononuclear cells stimulated with EBV, T-cell enriched mononuclear cells stimulated with pokeweed mitogen; human foreskin fibroblasts, Burkitt lymphoma cells lines, EBV-immortalized lymphoblastoid cell lines, a T-cell line (Molt-4), neuroblastoma cells (SK-N-MC), lung adenocarcinoma cells (NBA-MB-468), acute promyelocytic leukemia cells (HL-60), prostate adenocarcinoma cells (Tsu-Pr1, PC-3, Du145), Hodgkin's lymphoma cells (Hs445), colon adenocarcinoma cells (SW480), Wilms tumor cells (SK-NEP-1) and melanoma cells (A-375). Thus, calreticulin and calreticulin N-domain appear specific in their ability to inhibit endothelial cell growth.

Calreticulin and Calreticulin-N domain expressed in *E. coli* or purified from culture supernatants of lymphoblastoid cells inhibited endothelial cell proliferation in vitro. To exclude that low level endotoxin (<0.5 unit/microgram protein) contaminating E coli-derived proteins might contribute to endothelial cell growth inhibition, we examined the effects of protein denaturation on the biological activity. Boiling the protein for 2 minutes at 100 °C and storage of the protein for 14 days at 4 °C removed most of the biological activity, suggesting that an intact protein structure is required for activity. In addition, polymyxin B that is known to functionally inactivate endotoxin, had minimal effect on endothelial cell growth inhibition by calreticulin

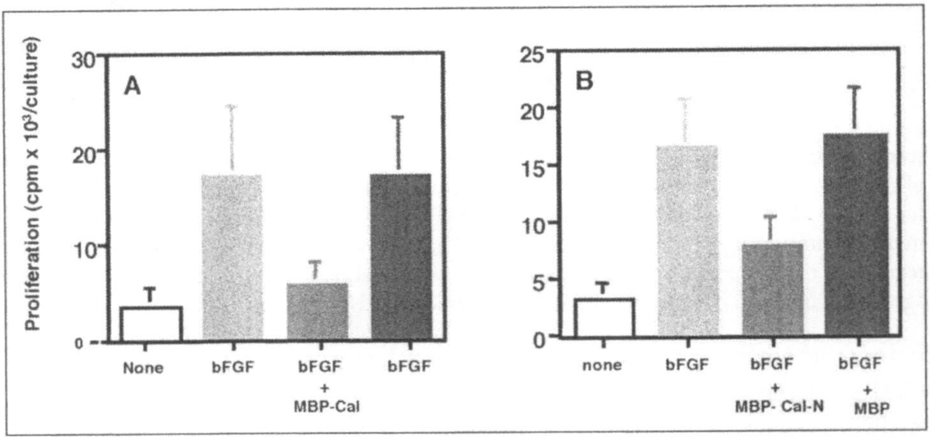

Figure 16.3. Inhibition of endothelial cell proliferation by calreticulin and the NH_2-domain of calreticulin. Fetal Bovine Heart Endothelial (FBHE) cells were cultured (800 cells/microwell in DMEM culture medium with 10% heat-inactivated fetal calf serum and 5μg/ml gentamicin) for 5 days with or without bFGF (25 ng/ml). DNA synthesis was measured by ^3H thymidine incorporation during the final 22 hr in culture. A) Effects of MBP or MBP-calreticulin (both at 1 μg/ml). The results reflect the means (±SD) of 16 experiments. Cal: calreticulin. B) Effects of MBP or MBP-NH_2domain of calreticulin (both at 1 μg/ml). The results reflect the means (±SD) of 9 experiments. Cal-N: NH_2-domain of calreticulin.

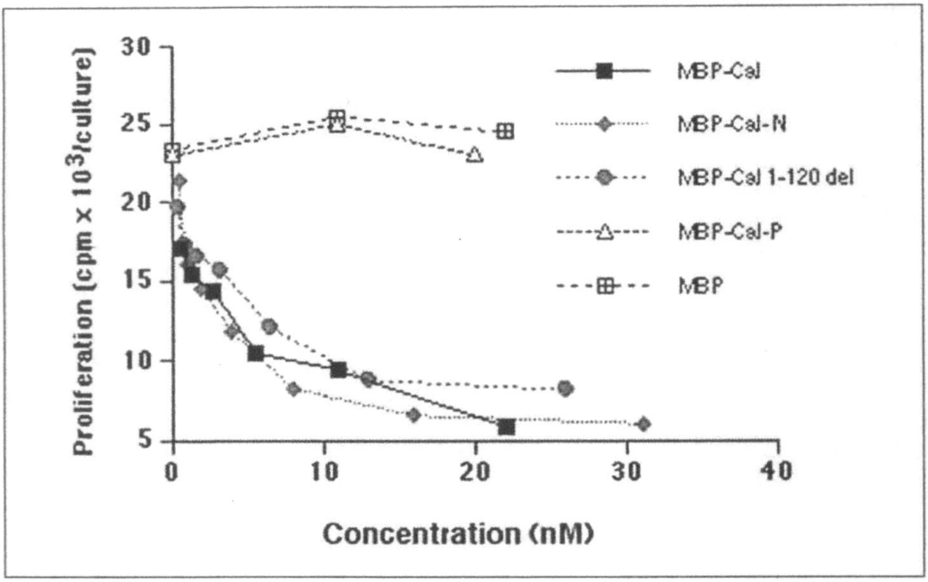

Figure 16.4. Comparative analysis of endothelial cell growth inhibition by MBP-calreticulin, MBP-calreticulin NH_2-domain, MBP-Δ1-120 calreticulin, MBP-calreticulin P-domain and MBP. Fetal Bovine Heart Endothelial (FBHE) cells were cultured (800 cells/microwell in DMEM culture medium with 10% heat-inactivated fetal calf serum and 5μg/ml gentamicin) for 5 days with or without bFGF (15 ng/ml). Recombinant purified proteins were added to culture at concentrations ranging between 0.4 and 32 nM. DNA synthesis was measured by ^3H thymidine incorporation during the final 22 hr in culture. MBP-Cal: MBP-calreticulin; MBP-Cal-N: MBP-calreticulin NH_2-domain; MBP-Cal1-120 del: MBP-Δ1-120 calreticulin; MBP-Cal-P: MBP-calreticulin P-domain; MBP:maltose-binding protein.

Table 16.1. Inhibition of endothelial cell proliferation by calreticulin and calreticulin NH_2 domain

Additions to Culture	Proliferation (mean cpm/culture)	Inhibition (%)
None	142	
bFGF	32,493	
MBP-calreticulin	13,231	59.3
MBP-calreticulin-NH_2	17,930	44.8
MBP	34,363	

Human umbilical vein endothelial cells ($3.5x10^3$/well) were cultured in medium alone or in medium supplemented with bFGF (25 ng/ml). Recombinant purified MBP, MBP-calreticulin or MBP-calreticulin NH_2 domain was added at 1 µg/ml. Proliferation was measured by ^{3}H thymidine incorporation during the final 20 hours of a 3-day culture. The results reflect the mean of triplicate cultures. Representative experiment.

N-domain when added to culture at concentrations of 0.5 µg/ml. Furthermore, the biological activity of different lots of protein purified from E coli was often unrelated to the degree of endotoxin contamination but depended upon protein concentration. Thus, it is unlikely that contamination with endotoxin is responsible for inhibition of endothelial cell growth by calreticulin.

Effects of Calreticulin on Endothelial Cell Attachment

Previous studies have described calreticulin binding to glycosylated laminin with an affinity of approximately 470 nM.[37] Laminins constitute a large family of high molecular weight glycoproteins composed of the assembly of three distinct chains, α, β and γ, held together by disulfite bonds.[38,39] They represent one of the components of the extracellular matrix and have been implicated in cell adhesion, migration, proliferation and differentialion.[40-42] In particular, laminins constitute the major component of vascular endothelium basement membrane,[43] and are produced by endothelial cells.[44,45] We examined the potential contribution of calreticulin binding to laminin for endothelial cell function. In particular, we wanted to establish whether the binding of calreticulin to laminin might underlie and explain endothelial cell inhibition by calreticulin and calreticulin N-domain.[46] In solid-phase assays, we confirmed that calreticulin can bind to laminin, and further determined that calreticulin N-domain can bind to laminin indistinguishably from the full-length molecule (Fig. 16.5). By contrast, calreticulin and calreticulin N-domain failed to bind to other extracellular matrix proteins, including collagen type IV, fibronectin, vitronectin, or tenascin-C.[46] Since endothelial cells are known to attach to laminin-coated surfaces but not to uncharged polystyrene surfaces, we examined the effects of calreticulin on endothelial cell attachment assays. Endothelial cells could attach to laminin, collagen and fibronectin-coated surfaces to a verying degree. However, calreticulin N-domain selectively inhibited endothelial cell attachment to laminin-coated surfaces and had minimal effect on cell binding to the other matrices (Fig. 16.6 A). This anti-adhesive effect of calreticulin N-domain was dose-dependent (Fig. 16.6 B), was competed by laminin (Fig. 16.6 C), and was also mimicked by antibodies against laminin.[46] These experiments demonstrate that calreticulin can specifically bind to laminin and through this binding can specifically reduce endothelial cell attachment to laminin-coated surfaces. Earlier studies have demonstrated that endothelial cells express several laminin-binding sites, including integrins, a high affinity 67-kDa receptor, and dystroglycan.[47-50] However, it is not clear whether calreticulin interferes with laminin binding to endothelial cells at one or more of these sites.

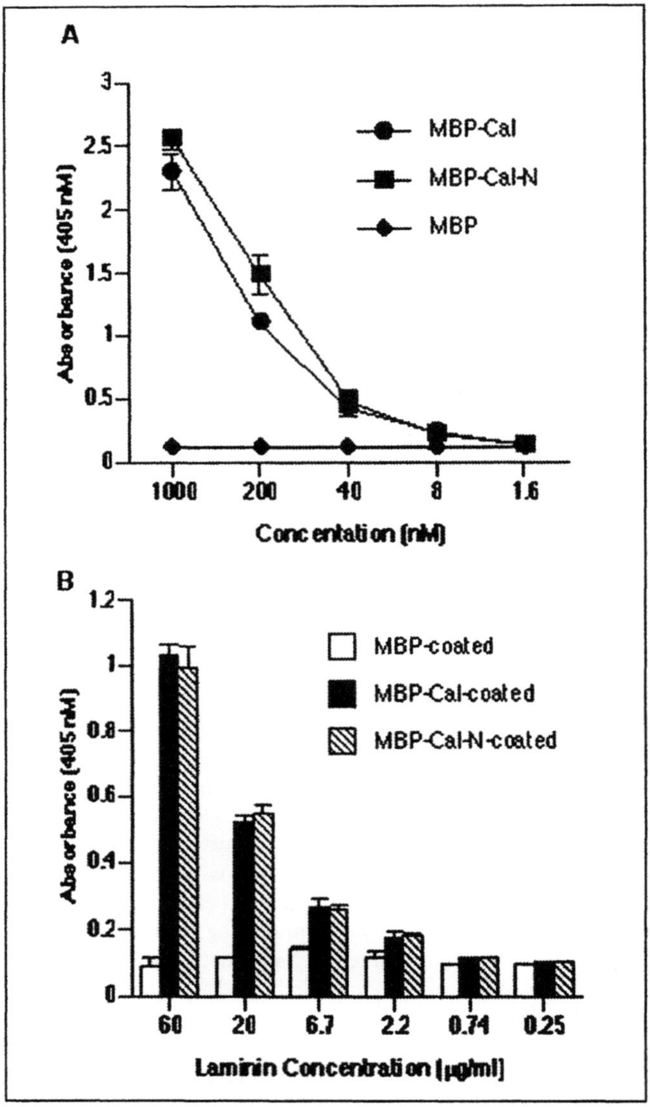

Figure 16.5. Calreticulin and calreticulin NH_2-domain can bind to murine laminin. A)Recombinant MBP-calreticulin, MBP-calreticulin NH_2-domain, and MBP (0.16-1000 nM) were tested for binding to laminin in solid-phase assays. Laminin (25 μg/ml) was immobilized onto microtiter wells. Protein binding to laminin was measured by use of a rabbit anti-MBP calreticulin NH_2-domain monospecific antiserum followed by a secondary affinity-purified goat anti-rabbit antiserum conjugated to alkaline phosphatase. The results reflect the mean absorbance at 405 nM. B)Laminin (0.25-60 μg/ml) was tested for binding to recombinant MBP-calreticulin, MBP-calreticulin NH_2-domain or MBP immobilized onto microtiter wells. Comparable coating by the proteins was ensured. Laminin binding to the wells coated with MBP-calreticulin, MBP-calreticulin NH_2-domain or MBP was detected by use of a rabbit anti-laminin antiserum and a secondary affinity-purified goat anti-rabbit serum conjugated to alkaline phosphatase.

One of the issues raised by these experiments is whether endothelial cell growth inhibition by calreticulin is a consequence of calreticulin binding to laminin and its anti-attachment effects. Endothelial cells unable to attach to a solid surface would be expected to stop proliferating and die. This possibility appears unlikely because calreticulin N-domain inhibited to a

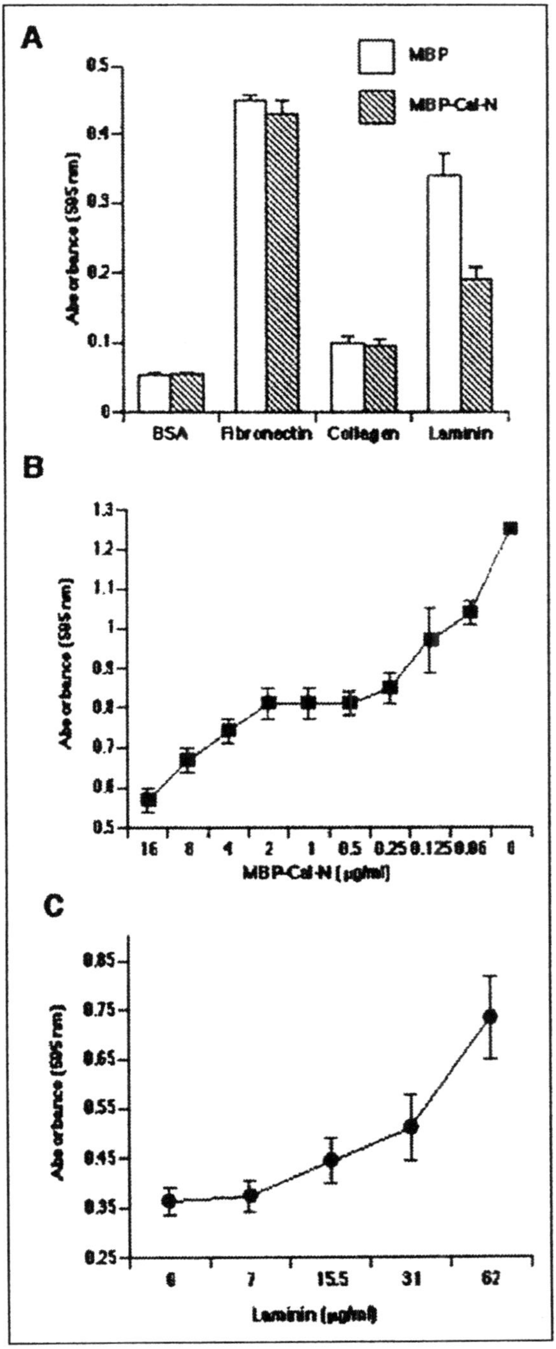

Figure 16.6. Effects of calreticulin NH_2-domain or laminin on endothelial cell attachment to laminin-coated wells. A. FBHE cells ($3x10^4$/well) were tested for their ability to bind to polystyrene wells that had been coated with 5 µg/ml bovine serum albumin (BSA), bovine fibronectin, murine collagen Type IV, or murine laminin. Cell binding to the plates was measured by absorbance after staining the cells bound to the plate. B. Dose-dependency of calreticulin NH_2-domain displacement of endothelial cell attachment to laminin-coated wells. C. Laminin displacement of endothelial cell attachment to laminin-coated wells.

similar degree endothelial cell proliferation irrespective of the surface the cells were cultured on, including charged uncoated polystyrene wells and wells coated with laminin, collagen type IV or vitronectin. This differs from the specificity for laminin of the anti-adhesive effect of calreticulin. In addition, calreticulin N-domain inhibited endothelial cell proliferation even when it was added to monolayers of endothelial cells. Furthermore, antibodies against laminin inhibited effectively and specifically endothelial cell attachment to laminin-coated wells but had minimal effect on endothelial cell proliferation once the cells were already attached. Thus, calreticulin N-domain can both interfere with endothelial cell attachment to laminin, the major component of vascular basement membranes, and can inhibit endothelial cell growth in response to growth factors. These appear to be two distinct properties of calreticulin, both residing within its N domain. Previously, binding studies demonstrated that radiolabeled calreticulin binds to monolayers of bovine aortic endothelial cells with a Kd of approximately 7.4 nM.[51] In vivo studies demonstrated that calreticulin infused intravenously into mice is rapidly cleared from the circulation and found in association with the vascular endothelium, particularly in the most vascular organs. Although the putative calreticulin binding site on endothelial cells has yet to be reported, it appeared to be distinct from that for factor IX.[51] Recently, CD91 was identified as a macrophage receptor for a number of heat shock proteins, including calreticulin, hsp70 and gp90, even though these proteins share little structural homology with each other.[52] In addition, α2 macroglobulin is a known ligand for CD91. However, CD91 expression was reported to be mostly restricted to macrophages and dendritic cells,[52] suggesting that it may not mediate calreticulin's effects on endothelial cells.

Calreticulin and Calreticulin N-Domain Inhibit Angiogenesis

A Matrigel-based assay was employed to evaluate the effects of calreticulin on angiogenesis in vivo (Fig. 16.7). Matrigel is a crude extract of extracellular matrix proteins including collagens and laminins. When it is impregneated with an endothelial cell growth factor and then injected in the subcutaneous tissues of mice, it induces a local angiogenic response. Endothelial cells will proliferate, invade the plug and form capillary structures at various stages of maturation.[53] The angiogenic response can be evaluated and quantified by microscopic analysis of the

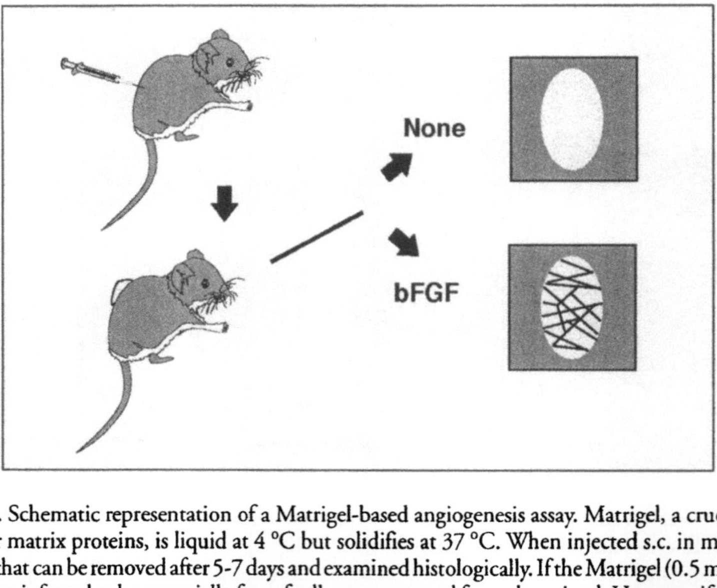

Figure 16.7. Schematic representation of a Matrigel-based angiogenesis assay. Matrigel, a crude extract of extracellular matrix proteins, is liquid at 4 °C but solidifies at 37 °C. When injected s.c. in mice, it forms a solid plug that can be removed after 5-7 days and examined histologically. If the Matrigel (0.5 ml) is injected alone, the pug is found to be essentially free of cells once removed from the animal. However, if the Matrigel is mixed with an endothelial cell growth factor (such as bFGF or VEGF) prior to s.c. injection into the mice, the plugs are found to contain numerous endothelial cells forming capillaries in various stages of maturation.

Table 16.2. Effects of calreticulin and calreticulin NH_2 domain on angiogenesis in vivo

Additions To Matrigel	Mean Surface Area Occupied by Cells ($mm^2/1.26x10^5mm^2$)	Inhibition (%)
None	671	
bFGF	16,732	
bFGF+GST-calreticulin	4,616	74
GST	17,890	
None	649	
bFGF	11,544	
bFGF+MBP-Calreticulin-N	5,286	54
bFGF+MBP	9,186	
None	487	
bFGF	14,472	
bFGF+MBP-calreticulin	5,112	65
bFGF+MBP-calreticulin-N	4,989	66
bFGF+MBP	13,472	

Mice were injected s.c. with Matrigel alone, Matrigel plus bFGF (150 ng.ml), Matrigel plus bFGF (150 ng/ml) plus GST-calreticulin, GST, MBP-calreticulin, MBP-calreticulin NH_2 domain, or MBP (all at 5 µg/ml). Plugs were removed after 7 days, and histological sections were stained with Masson's trichrome. The results reflect the mean surface area (expressed in mm^2) occupied by cells within a surface area of $1.26x10^5mm^2$; 12-15 non-overlapping fields were scanned in each plug using a semi-automated digital analyzer.

plugs removed from the animals after 5-8 days. When added to Matrigel plugs at concentrations of 5-10 µg/ml (Table 16.2), calreticulin and calreticulin N-domain inhibited significantly angiogenesis induced by bFGF.[35,36] It seems likely that calreticulin inhibits angiogenesis in this assays at least in part by exerting an anti-proliferative effect on growth factor-stimulated endothelial cells. Inhibition of endothelial cell attachment to laminin, a component of Matrigel, likely contributes to the anti-angiogenic effect. By contrast, we have no evidence that calreticulin inhibits endothelial cell migration or morphogenesis, mechanisms that appear critical to the antiangiogenic activities of a variety of other compounds such as angiostatin.[54]

Anti-Tumor Effects of Calreticulin and Calreticulin N-Domain

Calreticulin and calreticulin N-domain were first tested for their ability to prevent the growth of human Burkitt lymphoma in athymic mice (Fig. 16.8). Sublethally irradiated, 6-weeks old, female BALB/c nude mice were inoculated subcutaneously (s.c.) with 10^6 human Burkitt lymphoma cells (CA46 cell line). Beginning at the time of cell inoculation, the mice received daily s.c. inoculations (6 days/week for 14 days, 0.1 ml injection volume) of either control GST (20 µg/mouse) or GST-calreticulin (60 µg/mouse), adjacent to the site of cell injection (Fig. 16.8 A). By day 17, 12/12 mice injected with the control protein developed a tumor whereas only 4/13 mice injected with GST-calreticulin had developed a tumor (p=0.005). The tumor-bearing mice were sacrificed, and the remaining 9 mice were observed untreated. Tumors eventually developed in 8/9 mice that had originally received GST-calreticulin. The latest tumor appeared on day 36, 22 days after treatment had ended.

Calreticulin N-domain was tested in the same Burkitt lymphoma prevention model. Groups of mice received either control MBP (40 µg/mouse/day 6 days/week) or MBP-calreticulin

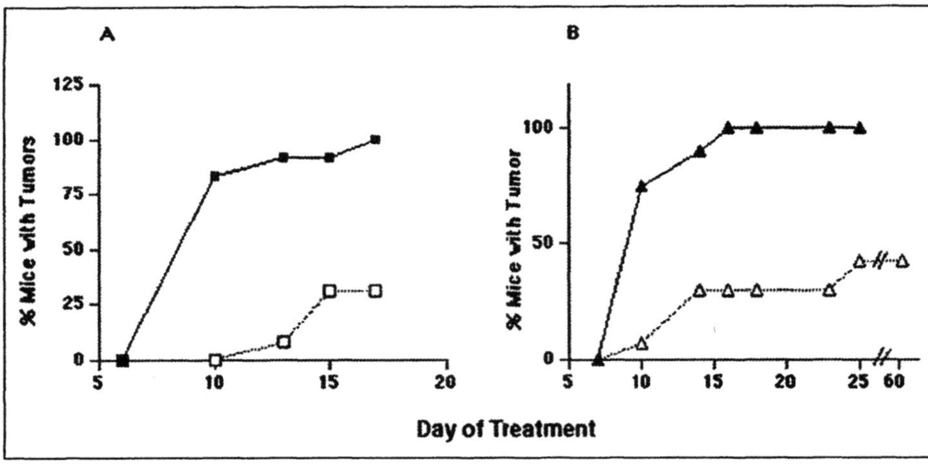

Figure 16.8. Calreticulin and calreticulin NH_2-domain prevent Burkitt tumor development in mice. Burkitt lymphoma cells (CA46 cell line, 10^7 cells/mouse) were inoculated s.c. into athymic BALB/c nude mice, 6 weeks of age. A) Beginning on the day of cell inoculation and continuing daily thereafter, 6 days/week, 12 mice were injected (s.c. proximal to the site of tumor cell injection) with control GST protein (20 µg/day x14 days; closed squares), and 13 mice were injected with GST-calreticulin (60 µg/day x14 days; open squares). B) Beginning on the day of cell inoculation and continuing daily thereafter, 6 days/week, 12 mice were injected (s.c. proximal to the site of tumor cell injection) with control MBP protein (40 µg/day x18 days; closed triangles), and 13 mice were injected with MBP-calreticulin NH_2-domain (60 µg/day x18 days; open triangles); all mice with tumor were killed on day 18. The remaining mice were observed and tumor cell development was monitored.

N-domain (60 µg/mouse, day 6 days/week) for at least 18 days, at which time all animals with tumor were sacrificed (Fig. 16.8 B). The remaining tumor-free animals continued treatment until tumors appeared. By day 18, 4/12 animals treated with MBP-calreticulin N domain as opposed to 12/12 control mice had developed a tumor (p=0.001). The mean (±SD) weight of tumors in the control group (0.43±0.2 g) was greater than the weight of tumors from MBP-calreticulin N domain treated animals (0.21±0.05). With continued treatment, an additional tumor appeared on day 23, but the remaining animals remained tumor free.

These experiments provided evidence that, over the short term, calreticulin could prevent the appearance of tumors in a proportion of animals. To test whether tumor prevention by calreticulin could be long lasting, we continued to inject mice that had not developed a tumor after 18 days of treatment. A group of 20 mice, initially injected with the Burkitt lymphoma cells (10^6 cell s.c.) and then treated from the day of tumor injection with calreticulin N-domain (60 µg/mouse, day 6 days/week), was continued on this treatment until tumor appearance. Most the mice (16/20) did not develop a tumor over 120 days of continued treatment. Two of the mice (2/20) developed a tumor on day 22 and 24 of treatment, but the subsequent rate of tumor growth (observed for additional 40 days) was reduced compared to the rate we usually observe in mice injected with the same tumor cells and left untreated. One other mouse (1/20) developed a late tumor on day 64 which grew slowly over the subsequent 20 days. The remaining mouse (1/20) developed a tumor on day 48 of treatment that grew rapidly over the subsequent week. All together, these experiments documented how calreticulin and calreticulin N-domain can prevent or delay Burkitt tumor growth in athymic mice.

In additional experiments, we tested for the anti-tumor activities of calreticulin and calreticulin N-domain against established Burkitt tumors (Fig. 16.9). In a representative experiment, the rate of tumor growth was significantly reduced in the group of mice (9 animals)

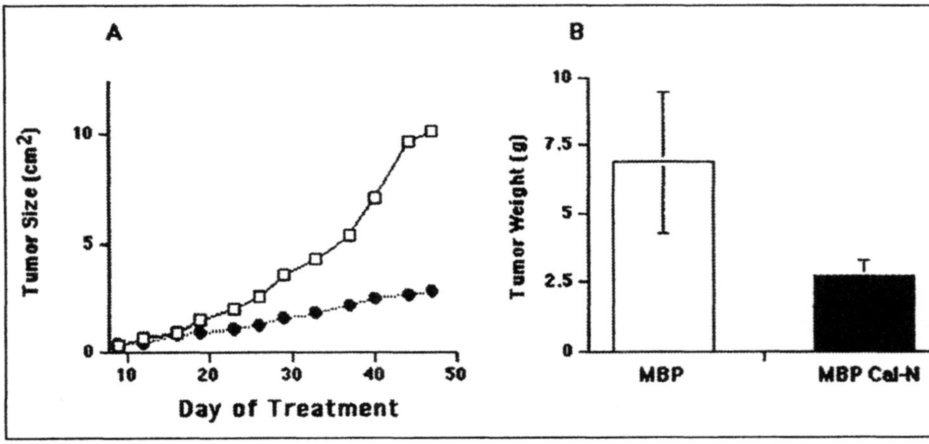

Figure 16.9. Calreticulin NH_2-domain inhibits the growth of established Burkitt tumors. Burkitt lymphoma cells (CA46 cell line, 10^7 cells/mouse) were inoculated s.c. into athymic BALB/c nude mice, 6 weeks of age. After a tumor appeared (at least 130 mm^2 in size), 9 mice were treated with MBP calreticulin NH_2-domain (200 μg/mouse/day, 0.1 ml injection volume, 6 days/week, x46 days) and 10 mice were treated with formulation buffer alone (0.1 ml x46 days). A) Tumor size in mice treated with buffer alone (close circles) or with MBP calreticulin NH_2-domain (open squares). Tumor size is expressed as the product of two-dimensional caliper measurements. B) Tumor weight expressed in g (±SD) measured after the tumors were removed in toto at completion of the experiment. The mice were treated for 46 days with buffer alone or with calreticulin NH_2-domain.

treated with calreticulin N domain (200 μg/mouse) compared with the controls (10 animals) treated with either buffer alone or MBP (p=0.003). Tumors were removed from the animals on day 28 after cell inoculation. The mean weight of tumors removed from the control group (6.9±2.6g) was significantly (p=0.0005) greater than the mean weight of tumors from the treatment group (2.7±0.6 g). In no case did established tumors regress with calreticulin treatment in this experiment and other subsequent confirmatory experiments. Rather, these results show that calreticulin N-domain can reduce the rate of growth of established Burkitt tumors. In vitro, calreticulin has minimal activity against growth or survival human Burkitt cells. Thus, one would not expect calreticulin to reduce the size an established tumor has already reached. Rather, by inhibiting endothelial cell division, one might expect calreticulin to reduce tumor neovascularization and slow down new tumor growth.

The histology of Burkitt tumors treated with MBP-calreticulin or MBP-calreticulin N domain was indistinguishable from that of controls with respect to the tissue architecture and the morphology of the tumor cells. However, specific staining for apoptotic cells using TUNEL revealed a larger number of apoptotic cells scattered thoughout the tumor in calreticulin as opposed to control treated tumors (Fig. 16.10 A). In contrast, the number of cell mitoses within the tumor tissue was indistinguishable between the treated and control groups. Consistent with this observation, staining for the nuclear antigen Ki67 revealed that the proportion of proliferating cells in tumors treated with calreticulin and control tumors was similar. Immunohistochemical staining of the tumor vasculature with anti-smooth muscle antibodies (Fig. 16.10 B) and with anti-CD31 antibodies (Fig. 16.10 C) showed that the number of vessels in calreticulin-treated tumors was significantly reduced compared with the control, particularly at the periphery of the tumor. Thus, the histological and immunohistochemical characteristics of calreticulin or calreticulin N-domain treated Burkitt tumors were remarkable for the paucity of distinguishing features from controls, these differences being limited to increased scattered cell death and reduced staining for endothelial and smooth muscle cells.

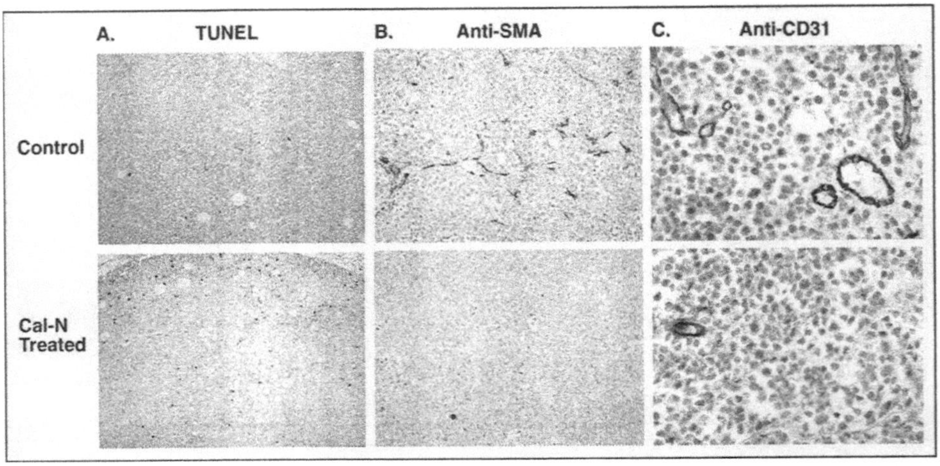

Figure 16.10. Immunohistochemical characterization of Burkitt tumors treated with calreticulin NH_2-domain. A) Evidence of increased cell apoptosis in Burkitt tumors treated with calreticulin NH_2-domain (bottom panel) as opposed to MBP alone (top panel). Apoptotic cells were visualized by TUNEL (original magnification 10x). B) Evidence for reduced smooth muscle cells in Burkitt tumors treated with calreticulin NH_2-domain (bottom panel) as opposed to MBP alone (top panel). Smooth muscle cells were detected by staining with an anti-smooth muscle actin antibody (original magnification 20 x). C) Evidence for reduced endothelial cells in Burkitt tumors treated with calreticulin NH_2-domain (bottom panel) as opposed to MBP alone (top panel). Endothelial cells were visualized by staining with an anti-CD31 antibody (original magnification 40 x).

In additional experiments, tumor cell lines of various lineages were inoculated s.c. in nude mice and the effects of calreticulin N-domain on tumor growth tested (Fig. 16.11). Similar to the results with Burkitt lymphoma cells, calreticulin N-domain reduced significantly the growth of colon carcinoma (SW-480 cell line), Wilms tumor (SK-NEP-1 cell line), myelomonocytic lymphoma (HL-60 cell line), rhabdomyosarcoma (A-204 cell line), melanoma (A-375 cell line), and breast adenocarcinoma (MDA-MB-468 cell line) established s.c. in nude mice. In addition, calreticulin N-domain reduced the growth of a syngeneic murine lymphoma induced by inoculation of the Raw 8.1 cell line into athymic BALB/c mice. In no case complete tumor regressions were observed with calreticulin N-domain treatment. These experiments demonstrated that calreticulin N-domain exerts its anti-tumor activities regardless of tumor lineage. Based on the anti-angiogenic activities of calreticulin and the general dependence of tumors on adequate neovascularization, it is not surprising that the anti-tumor effects on calreticulin are observed regardless the tumor lineage.

As discussed above, one would not expect a pure inhibitor of angiogenesis to cause regression of established tumors. Rather, such drug would be expected to reduce or slow down tumor growth. In this context, angiogenesis inhibitors might be beneficial in conjunction with other therapeutic modalities directed at the tumor cells themselves or at the established tumor vasculature. We elected to combine calreticulin N-domain with interleukin-12 (IL-12), a multifunctional cytokine with direct and indirect anti-tumor effects (Fig. 16.12). When combined with IL-12, calreticulin N-domain was significantly more effective at reducing tumor growth in nude mice than when administered alone. In many cases, Burkitt lymphomas and colon carcinomas were reduced to microscopic tumors by the combined effects of calreticulin N-domain and IL-12.[55]

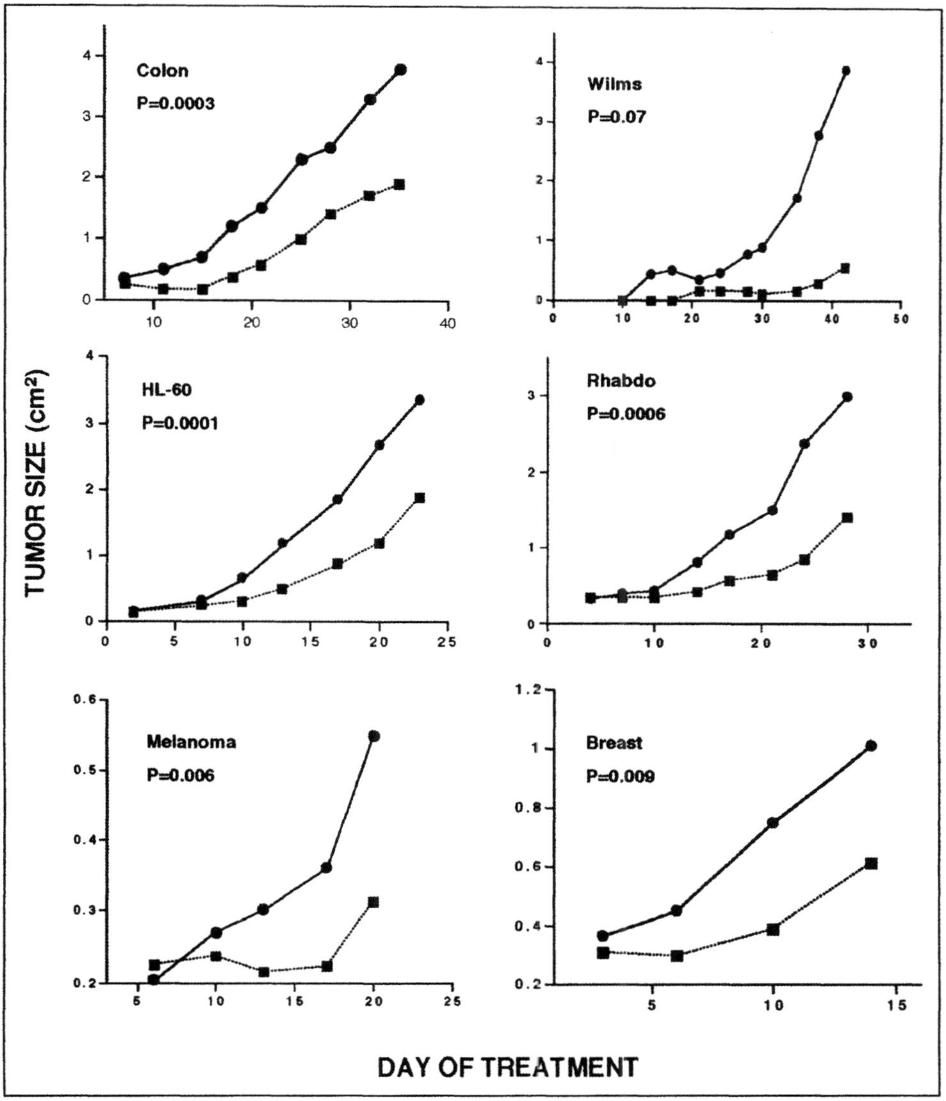

Figure 16.11. Inhibition of tumor growth by calreticulin NH_2-domain. Groups of thymic mice (BALB/c nude mice, 6 weeks of age), 8-12 mice/group, were injected s.c. with human tumor cell lines of various lineages. The cell lines included SW-480 (colon carcinoma); SK-NEP-1 (Wilms tumor); HL-60 (myelomonocytic leukemia/lymphoma); A-204 (rhabdomyosarcoma); A-375 (melanoma); and MDA-MB-468 (breast adenocarcinoma). The number of cells injected ($4\text{-}8 \times 10^6$) was chosen to produce 100% tumor take within 3 weeks. Beginning the day after tumor cell inoculation and continuing daily thereafter, 6 days/week, the mice were injected with either formulation buffer alone (0.1 ml/mouse, adjacent to the site of tumor cell injection) or with MBP-calreticulin NH_2-domain (100-200 μg/mouse/day). The significance of group differences in the rate of tumor growth (expressed as P value) was determined by Wilcoxon Rank Sums test. Closed circles: buffer alone; closed squares: MBP-calreticulin NH_2-domain.

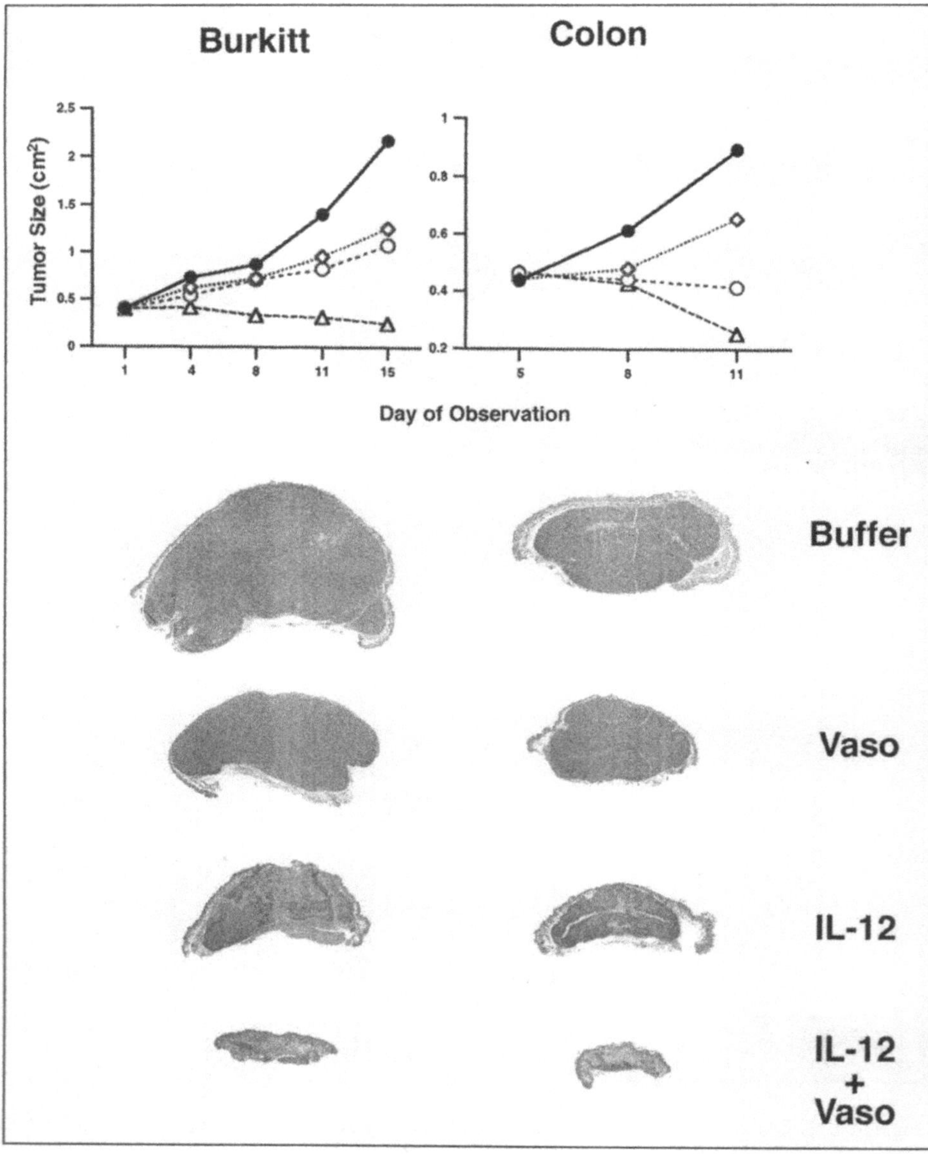

Figure 16.12. Treatment of experimental Burkitt lymphomas and colon carcinoma with calreticulin NH_2-domain alone, IL-12 alone, or the combination of calreticulin NH_2-domain and IL-12. Burkitt lymphoma cells (CA46 cells line, $8x10^6$ cells/mouse) or colon carcinoma cells (SW-620 cell line, $8x10^6$ cells/ mouse) were inoculated s.c. into athymic mice (BALB/c nude mice, 6 weeks of age) and 1-5 days later treatment was started. There were 11-18 mice/group. The treatment groups included buffer alone (closed circles); MBP-calreticulin NH_2-domain (open diamonds, 100 µg/mouse/day, 6 days/week); murine IL-12 (open circles, 100 ng/mouse/day, 6 days week), or the combination (open triangles) of MBP-calreticulin NH_2-domain (100 µg/mouse/day, 6 days/week) plus IL-12 (100 ng/mouse/day, 6 days week). Tumor size was measured at the indicated time-points as the two-dimensional caliper measurement. At completion of the experiment, all tumors were removed, fixed and processed for histology. Representative gross morphology of tumors from each group is shown.

Recently, another group has confirmed the anti-angiogenic and anti-tumor activities of calreticulin.[56] Using a different approach, lung metastatic tumor was induced by intravenous inoculation of tumor cells (TC-1 cell line), and treatment with calreticulin was given by gene gun inoculation of calreticulin DNA (beginning 2 days after injection of the tumor cells, every week for a total of 4 shots or 64 μg DNA). In this model, calreticulin expressed endogenously by the mice, reduced significantly the number of lung metastasis and caused a reduction of tumor vascularization. Confirming its direct anti-antiangiogenic activity, this group also found that the anti-tumor activity of calreticulin in this model did not require either the presence of T cells or NK cells.[56] These results are consistent with those from our experiments where athymic mice were used in tumor prevention or tumor treatment models. It is important to note that young athymic mice, particularly after total body irradiation with 400 rad, have very few, if any, circulating T lymphocytes. Thus, it is unlikely that in our experiments as in some of the experiments reported by Cheng et al,[56] T cell immunity contributed to the anti-tumor activities of calreticulin. However, calreticulin has been shown to associate with endogenous peptides delivered to the ER by TAP-1 and TAP-2 transporters and by MHC-class I/β2 microglobulin molecules.[57,58] It has also been shown to associate with exogenous peptides, including tumor-derived peptides, and to favor peptide-specific T cells responses in vitro.[52,59] Therefore, it was possible that in immunocompetent mice bearing certain immunogenic tumors, calreticulin might serve the dual role of an angiogenesis inhibitor and a promoter of specific tumor immunity. Indeed, when calreticulin was fused to a model tumor antigen (HPV-16 E7) and the chimeric DNA molecule was used as an immunogen in immunocompetent mice, it effectively protected the mice against a tumor challenge with E7 expressing cells, more effectively than calreticulin or E7 DNA alone.[56] In addition, the chimeric calreticulin/E7 DNA was significantly more effective than E7 alone and somewhat more effective than calreticulin DNA alone at protecting mice from metastatic pulmonary disease.[56] These experiments provide evidence that calreticulin may be useful as a cancer therapeutic both based on its anti-angiogenic properties and its immune potentiating effects.

Concluding Remarks

With the decoding of the human genome we have come to appreciate that the number of genes it contains is smaller than expected. This has suggested that many proteins must play multiple functional roles to satisfy the complexities of mammalian organisms. Calreticulin is an example of multi-functional protein possessing different biological activities residing in separable or overlapping domains. Here we have attempted to summarize the results of experiments that have unveiled a previously unrecognized property of calreticulin as a regulator of endothelial cell growth, angiogenesis and tumor growth. As our understanding of endothelial cell development, maturation and morphogenesis advances, a more clear characterization of the role of calreticulin as a regulators of vasculogenesis/angiogenesis will emerge. At present many questions remain open. Can we better define the domains of calreticulin that are important to its endothelial cell regulator functions? In the target endothelial cells, what mediates the inhibitory effects of calreticulin? Does calreticulin play a role in vasculogenesis during development? Does it play a role in vascular remodeling after birth? Is it involved in the pathogenesis of certain vascular pathologies? Can calreticulin be developed into a viable therapeutic drug with anti-angiogenic activities? Much excitement has surrounded the discovery of angiogenesis inhibitors and the realization that such compounds hold promise as anti-cancer agents. Future careful work will be needed to fully evaluate calreticulin functions as a regulator of the vascular system.

References

1. Folkman J. Anti-angiogenesis: new concept for therapy of solid tumors. Ann Surg 1972; 175(3):409-416.
2. Hanahan D, Folkman J. Patterns and emerging mechanisms of the angiogenic switch during tumorigenesis. Cell 1996; 86(3):353-364.
3. Asahara T, Murohara T, Sullivan A et al. Isolation of putative progenitor endothelial cells for angiogenesis. Science 1997; 275(5302):964-967.
4. Holash J, Maisonpierre PC, Compton D et al. Vessel cooption, regression, and growth in tumors mediated by angiopoietins and VEGF. Science 1999; 284(5422):1994-1998.
5. Lyden D, Young AZ, Zagzag D et al. Id1 and Id3 are required for neurogenesis, angiogenesis and vascularization of tumour xenografts [see comments]. Nature 1999; 401(6754):670-677.
6. Gullino PM. Angiogenesis and oncogenesis. J Natl Cancer Inst 1978; 61(3):639-643.
7. Fukumura D, Xavier R, Sugiura T et al. Tumor induction of VEGF promoter activity in stromal cells. Cell 1998; 94(6):715-725.
8. Carmeliet P. Mechanisms of angiogenesis and arteriogenesis. Nat Med 2000; 6(4):389-395.
9. Carmeliet P, Lampugnani MG, Moons L et al. Targeted deficiency or cytosolic truncation of the VE-cadherin gene in mice impairs VEGF-mediated endothelial survival and angiogenesis. Cell 1999; 98(2):147-157.
10. Carmeliet P, Jain RK. Angiogenesis in cancer and other diseases. Nature 2000; 407(6801):249-257.
11. Yancopoulos GD, Davis S, Gale NW et al. Vascular-specific growth factors and blood vessel formation. Nature 2000; 407(6801):242-248.
12. Kim KJ, Li B, Winer J et al. Inhibition of vascular endothelial growth factor-induced angiogenesis suppresses tumour growth in vivo. Nature 1993; 362(6423):841-844.
13. Ferrara N, Alitalo K. Clinical applications of angiogenic growth factors and their inhibitors. Nat Med 1999; 5(12):1359-1364.
14. Rastinejad F, Polverini PJ, Bouck NP. Regulation of the activity of a new inhibitor of angiogenesis by a cancer suppressor gene. Cell 1989; 56(3):345-355.
15. Dameron KM, Volpert OV, Tainsky MA et al. Control of angiogenesis in fibroblasts by p53 regulation of thrombospondin-1. Science 1994; 265(5178):1582-1584.
16. O'Reilly MS, Holmgren L, Shing Y et al. Angiostatin: a novel angiogenesis inhibitor that mediates the suppression of metastases by a Lewis lung carcinoma [see comments]. Cell 1994; 79(2):315-328.
17. O'Reilly MS, Boehm T, Shing Y et al. Endostatin: an endogenous inhibitor of angiogenesis and tumor growth. Cell 1997; 88(2):277-285.
18. Dong Z, Greene G, Pettaway C et al. Suppression of angiogenesis, tumorigenicity, and metastasis by human prostate cancer cells engineered to produce interferon-beta. Cancer Res 1999; 59(4):872-879.
19. Gohongi T, Fukumura D, Boucher Y et al. Tumor-host interactions in the gallbladder suppress distal angiogenesis and tumor growth: involvement of transforming growth factor beta1. Nat Med 1999; 5(10):1203-1208.
20. Zhang M, Volpert O, Shi YH et al. Maspin is an angiogenesis inhibitor. Nat Med 2000; 6(2):196-199.
21. Tosato G, Sgadari C, Taga K et al. Regression of experimental Burkitt's lymphoma induced by Epstein-Barr virus-immortalized human B cells. Blood 1994; 83(3):776-784.
22. Angiolillo AL, Sgadari C, Sheikh N et al. Regression of experimental human leukemias and solid tumors induced by Epstein-Barr virus-immortalized B cells. Leuk Lymphoma 1995; 19(3-4):267-276.
23. Cherney BW, Sgadari C, Kanegane C et al. Expression of the Epstein-Barr virus protein LMP1 mediates tumor regression in vivo. Blood 1998; 91(7):2491-2500.
24. Angiolillo AL, Sgadari C, Taub DD et al. Human interferon-inducible protein 10 is a potent inhibitor of angiogenesis in vivo. J Exp Med 1995; 182(1):155-162.
25. Angiolillo AL, Sgadari C, Tosato G. A role for the interferon-inducible protein 10 in inhibition of angiogenesis by interleukin-12. Ann N Y Acad Sci 1996; 795:158-167.
26. Sgadari C, Angiolillo AL, Cherney BW et al. Interferon-inducible protein-10 identified as a mediator of tumor necrosis in vivo. Proc Natl Acad Sci USA 1996; 93(24):13791-13796.
27. Sgadari C, Angiolillo AL, Tosato G. Inhibition of angiogenesis by interleukin-12 is mediated by the interferon-inducible protein 10. Blood 1996; 87(9):3877-3882.
28. Sgadari C, Farber JM, Angiolillo AL et al. Mig, the monokine induced by interferon-gamma, promotes tumor necrosis in vivo. Blood 1997; 89(8):2635-2643.
29. Urbe S, Page LJ, Tooze SA. Homotypic fusion of immature secretory granules during maturation in a cell-free assay. J Cell Biol 1998; 143(7):1831-1844.
30. Zhu Q, Zelinka P, White T et al. Calreticulin-integrin bidirectional signaling complex. Biochem Biophys Res Commun 1997; 232(2):354-358.

31. Borisjuk N, Sitailo L, Adler K et al. Calreticulin expression in plant cells: developmental regulation, tissue specificity and intracellular distribution. Planta 1998; 206(4):504-514.
32. Zuber C, Spiro MJ, Guhl B et al. Golgi apparatus immunolocalization of endomannosidase suggests post-endoplasmic reticulum glucose trimming: implications for quality control. Mol Biol Cell 2000; 11(12):4227-4240.
33. Day PJ, Owens SR, Wesche J et al. An interaction between ricin and calreticulin that may have implications for toxin trafficking. J Biol Chem 2001; 276(10):7202-7208.
34. Corbett EF, Michalak KM, Oikawa K et al. The conformation of calreticulin is influenced by the endoplasmic reticulum luminal environment. J Biol Chem 2000; 275(35):27177-27185.
35. Pike SE, Yao L, Jones KD et al. Vasostatin, a calreticulin fragment, inhibits angiogenesis and suppresses tumor growth. J Exp Med 1998; 188(12):2349-2356.
36. Pike SE, Yao L, Setsuda J et al. Calreticulin and calreticulin fragments are endothelial cell inhibitors that suppress tumor growth. Blood 1999; 94(7):2461-2468.
37. McDonnell JM, Jones GE, White TK et al. Calreticulin binding affinity for glycosylated laminin. J Biol Chem 1996; 271(14):7891-7894.
38. Burgeson RE, Chiquet M, Deutzmann R et al. A new nomenclature for the laminins. Matrix Biol 1994; 14(3):209-211.
39. Timpl R, Brown JC. The laminins. Matrix Biol. Aug 1994; 14(4):275-281.
40. Kleinman HK, Cannon FB, Laurie GW et al. Biological activities of laminin. J Cell Biochem 1985; 27(4):317-325.
41. Grant DS, Tashiro K, Segui-Real B et al. Two different laminin domains mediate the differentiation of human endothelial cells into capillary-like structures in vitro. Cell 1989; 58(5):933-943.
42. Kanemoto T, Reich R, Royce L et al. Identification of an amino acid sequence from the laminin A chain that stimulates metastasis and collagenase IV production. Proc Natl Acad Sci USA. Mar 1990; 87(6):2279-2283.
43. Sanes JR, Engvall E, Butkowski R et al. Molecular heterogeneity of basal laminae: isoforms of laminin and collagen IV at the neuromuscular junction and elsewhere. J Cell Biol 1990; 111(4):1685-1699.
44. Alitalo K. Production of both interstitial and basement membrane procollagens by fibroblastic WI-38 cells from human embryonic lung. Biochem Biophys Res Commun 1980; 93(3):873-880.
45. Tokida Y, Aratani Y, Morita A et al. Production of two variant laminin forms by endothelial cells and shift of their relative levels by angiostatic steroids. J Biol Chem 1990; 265(30):18123-18129.
46. Yao Lei PSE, Tosato G. Laminin Binding to the Calreticulin Fragment Vasostatin Regulates Endothelial cell Functin. Journal of Leukocyte Biology 2002; 71(1):47-53.
47. Hynes RO. Integrins: versatility, modulation, and signaling in cell adhesion. Cell 1992; 69(1):11-25.
48. Malinoff HL, Wicha MS. Isolation of a cell surface receptor protein for laminin from murine fibrosarcoma cells. J Cell Biol 1983; 96(5):1475-1479.
49. Mecham RP. Receptors for laminin on mammalian cells. Faseb J 1991; 5(11):2538-2546.
50. Montanaro F, Lindenbaum M, Carbonetto S. alpha-Dystroglycan is a laminin receptor involved in extracellular matrix assembly on myotubes and muscle cell viability. J Cell Biol 1999; 145(6):1325-1340.
51. Kuwabara K, Pinsky DJ, Schmidt AM et al. Calreticulin, an antithrombotic agent which binds to vitamin K-dependent coagulation factors, stimulates endothelial nitric oxide production, and limits thrombosis in canine coronary arteries. J Biol Chem 1995; 270(14):8179-8187.
52. Basu S, Binder RJ, Ramalingam T et al. CD91 is a common receptor for heat shock proteins gp96, hsp90, hsp70, and calreticulin. Immunity 2001; 14(3):303-313.
53. Yao L, Sgadari C, Furuke K et al. Contribution of natural killer cells to inhibition of angiogenesis by interleukin-12. Blood 1999; 93(5):1612-1621.
54. Troyanovsky B, Levchenko T, Mansson G et al. Angiomotin: an angiostatin binding protein that regulates endothelial cell migration and tube formation. J Cell Biol 2001; 152(6):1247-1254.
55. Yao L, Pike SE, Setsuda J et al. Effective Targeting of Tumor Vasculature by the Angiogenesis Inhibitors Vasostatin and Interleukin-12. BLOOD 2000; 96(5):1900-1905.
56. Cheng WF, Hung CF, Chai CY et al. Tumor-specific immunity and antiangiogenesis generated by a DNA vaccine encoding calreticulin linked to a tumor antigen. J Clin Invest 2001; 108(5):669-678.
57. Sadasivan B, Lehner PJ, Ortmann B et al. Roles for calreticulin and a novel glycoprotein, tapasin, in the interaction of MHC class I molecules with TAP. Immunity 1996; 5(2):103-114.
58. Spee P, Neefjes J. TAP-translocated peptides specifically bind proteins in the endoplasmic reticulum, including gp96, protein disulfide isomerase and calreticulin. Eur J Immunol 1997; 27(9):2441-2449.
59. Basu S, Srivastava PK. Calreticulin, a peptide-binding chaperone of the endoplasmic reticulum, elicits tumor- and peptide-specific immunity. J Exp Med 1999; 189(5):797-802.

CHAPTER 17

Calreticulin's Role(s) in Autoimmune Disorders

Richard D. Sontheimer, Doina Racila, Emil Racila, Paul Eggleton and Suzanne Donnelly

Abstract

For over ten years autoantibodies (Aab) against calreticulin (CRT) have been reported in a number of autoimmune disorders including rheumatoid arthritis, Sjögren's syndrome, celiac disease and complete congenital heart block. The most studied group is patients with systemic lupus erythematosus (SLE), where Aab against CRT have been detected in 40% of all patients. A number of studies have sought to assess the pathogenicity of such antibodies. The production of CRT Aab shows similarities to classical heat shock stress responses such as hsp90, 70 and 60. Aab to CRT and hsp may penetrate living cells and interfere with cell function. Moreover since CRT plays an immunoregulatory role with antigen presentation through the MHC class I pathway, Aab may interfere with this important process. Furthermore, there is extensive evidence to show CRT binds to the first component of complement—C1q. This leads to a number of consequences; first, C1q binding to CRT prevents Aab against CRT binding to CRT. It has been observed in individuals with low levels of C1q, a higher rate of photosensitive lupus occurs. Second, C1q may bind to CRT on the cell surface of apoptotic neutrophils enhancing their clearance. In lupus patients, C1q does not appear to bind to these cells and therefore apoptotic mechanisms may be impaired in lupus patients. Third, CRT interaction with C1q leads to inhibition of classical complement activation that might control the inflammatory response to immune complex accumulation. This chapter reviews the immune related functions of CRT and the multistage pathogenic considerations of CRT autoantibody production.

Introduction

CRT is an abundant intracellular protein found in most nucleated cells. It is localized primarily in the lumen of the endoplasmic reticulum (ER) where it is considered to have roles in the regulation of calcium storage and signaling and in protein processing. It is now recognized that CRT may be expressed on the surface of circulating leucocytes, and under certain conditions such as infection and inflammation, released from these cells into the extracellular milieu. The production of autoantibodies to CRT has been described in a wide variety of human autoimmune diseases from the organ specific primary biliary cirrhosis to the multi-system condition systemic lupus erythematosus. In this chapter we review the immune related biology of CRT and current knowledge pertaining to its role as an autoantigen in autoimmune disease.

Cellular Localization of CRT

CRT possesses a C-terminal KDEL endoplasmic reticulum (ER)1 retrieval sequence, and as such is located primarily in the ER. Recent studies demonstrate that CRT may also be found in other intracellular compartments such as within the secretory granules of cytotoxic lymphocytes, and on the surface of several cell types including peripheral blood lymphocytes and neutrophils. Arosa et al report that CRT is expressed on the cell surface of activated human

Calreticulin, 2nd ed., edited by Paul Eggleton and Marek Michalak.

peripheral blood T lymphocytes, where it is physically associated with a pool of unfolded MHC class I molecules.[1] How and under what conditions CRT escapes ER retention mechanisms and moves into the cytoplasm, nucleus, or plasma membrane is unknown at present. Similarly, functions associated with CRT outside the ER are the subject of ongoing research.

Up regulation of CRT expression has been reported under various cellular conditions including as a result of oxidative stress, viral infection (T cells) or cellular activation (T cells). Several lines of investigation have suggested that CRT expression is altered in mammalian cells by UV radiation.[2] Prior studies by several groups including our own have demonstrated that CRT is up regulated and expressed on the surface of viable cells including transformed and non-transformed human epidermal keratinocytes following clinically relevant doses of UVB irradiation.[3] Finally, in disease states such as SLE elevated CRT has been directly demonstrated in patient sera.[4] This may represent the shedding of apoptotic blebs containing CRT from circulating apoptotic cells, which in turn are targeted by antiphospholipid antibodies and cell mediated attack.

Immune Related Functions of CRT

In the ER, CRT contributes to several important immune related cellular activities in addition to its role in calcium storage and signaling. Additionally, new evidence suggests that CRT on the cell surface may have immune related functions:

CRT Is a Molecular Chaperone Involved in Antigen Presentation

Antigen presentation by MHC class I molecules is dependent on the integrity of cellular processes of assembly, folding and stabilization which take place in the ER. In the ER, class I MHC molecules are assembled with antigenic peptide while bound to a number of protein co-factors or chaperones, one of which is CRT. ER chaperones are required for the formation of appropriate disulfide bonds in the class I heavy chain, which in turn are necessary for class I MHC molecules to bind peptide antigen securely. After peptide loading, CRT dissociates from the MHC class I dimers (see chapter 9), allowing the final transport of mature MHC class I molecules to the cell surface, providing continuous reporting of cellular events to cytotoxic T lymphocytes. The functional importance of CRT in this process is supported by studies on CRT gene deletion mice which show a decrease in cell surface mouse MHC class I (H2-Db and H2-Kb) expression of up to 75% and an associated 50-100% abrogation in Kb and Db restricted presentation of viral peptide epitopes (see Elliott, chapter 9)

CRT Is a Signaling Molecule in Fibroblasts

Evidence for a role for CRT in integrin mediated signaling pathways is derived from gene deletion/transfection cell line and animal studies. Integrins are important mediators of cell adhesion to extracellular ligands. They can transduce biochemical signals both into and out of cells and are involved in the regulation of cell shape, motility, growth and differentiation. Coppolino et al developed CRT-deficient embryonic stem cells and fibroblasts from CRT mutant mice and found CRT to be an essential modulator both of integrin adhesive functions and integrin-initiated signaling.[5] Fadel et al show that stable over-expression of full-length ER-targeted CRT correlates with an increased adhesiveness in transformed mouse L fibroblasts.[6,7] Changes in CRT expression were also noted to affect the tyrosine phosphorylation of cellular proteins in these cells, suggesting that CRT may play a role in protein-tyrosine kinases and/or phosphatase signaling pathways from the ER.

Surface CRT on Activated T Lymphocytes

In 1999, Arosa et al demonstrated cell surface expression of CRT following activation of both CD8+ and CD4+ T lymphocytes. Resting T cells, by contrast, were found to express low amounts of cell surface CRT.[1] These investigators noted co-precipitation with a 46-kDa protein which they identify as unfolded MHC I in activated T cells. The authors point out the similarities between the behavior of the CRT-MHC class I association in the ER and the T cell surface of activated human T cells and suggest that this could be related to an important but uncharacterized biological function related to cell proliferation, perhaps as an intermediary in the process of iron uptake and transfer to intracellular compartments in dividing T lymphocytes. Alternatively, T cell surface CRT may perform a chaperone function by mediating the refolding of unfolded MHC class I molecules that arise as a result of T cell division. Evidence in support of either of these hypotheses has yet to be established.

CRT Interacts with the First Component of Complement, C1q

CRT is known to bind via specific binding sites to C1q, a subunit of the first component of complement, which provides the trigger for activation of the classical complement cascade.[8-10] The C terminal globular head region of C1q binds to the CH2 domains of immune complex fixed immunoglobulin and initiates the classical complement pathway, resulting in solubilization and clearance of these precipitating complexes from plasma. Failure of clearance results in immune complex precipitation in target organs such as the kidney in SLE, leading to tissue inflammation and damage.

Both the globular heads and the collagen tail region of C1q have been shown to bind in vitro to the N terminal half of CRT which contains a number of short regions of 7-10 amino acids that show sequence similarity to the putative C1q binding region in the CH2 domain of IgG. A number of C1q binding sites on the CRT molecule have now been identified, including several containing a CH2-like motif similar to the ExKxKx C1q binding motif found in the CH2 domain of IgG. A number of these peptides were shown by Kovacs et al to inhibit binding of C1q to IgG.[11] Moreover, several of the peptides were capable of inhibiting the classical pathway of complement activation in vitro. Hence, in autoimmune disease characterized by immune complex formation in vivo, the interaction of CRT with C1q may interfere with the ability of C1q to associate with immune complexes.

While studies of the role of CRT in complement mediated immune complex clearance at the level of the organism are lacking, Kishore et al[12] have confirmed the inhibitory effect of CRT released from activated ex vivo neutrophils on complement dependent hemolysis.

CRT Is an Apoptotic Bridging Molecule

CRT and other rheumatic disease-associated autoantigens including Ro have been localized to the plasma membrane blebs of epidermal keratinocytes undergoing UV triggered apoptosis.[13,14] The complement component C1q has been shown to bind to these apoptotic blebs on both epidermal keratinocytes and endothelial cells.[15,16] Evidence is accumulating that interactions between C1q and CRT on these blebs play an important physiologic role in the clearance of apoptotic cells, promoting effective removal of the potentially autoantigenic contents of these blebs. In a model proposed by Hanson (see chapter 15) and confirmed by studies by Ogden and ourselves, C1q is proposed to bind directly to an as yet unidentified receptor on the apoptotic cell. Interaction is then proposed to occur between free extracellular CRT and CD91 which functions as a signaling partner for CRT on the surface of the engulfing phagocyte. Initial evidence from our laboratory further suggests that CRT located in apoptotic blebs is the C1q receptor on apoptotic neutrophils. By employing a number of immunochemical techniques, it has been possible to study the interaction of C1q with fresh and apoptotic neutrophils from healthy subjects and patients with SLE. Confocal analysis has confirmed that the ER blebs of apoptotic neutrophils are targeted by C1q and that the C1q may interact directly with cell surface CRT as confirmed by phage display analysis. It is also clear that C1q co-localizes with CD91 on the surface of normal monocytes. However what is striking in the SLE patients is that C1q does not bind to apoptotic cells from non-selected lupus patients with active disease, despite both CRT and CD91 being present on the cells. The possible role of autoantibodies to CRT in blocking this interaction is discussed later.

Of related interest is the proposition that CRT is important in protection against nitric oxide induced apoptotic cell death of pancreatic cells. Oyomodara et al report that NO depletes ER Ca^{2+} and causes ER stress leading to apoptosis.[17] Over-expression of CRT increased the Ca2+ content of the ER and afforded protection to cells against NO-mediated apoptosis-a process implicated in pancreatic cell disruption in the autoimmune disease diabetes mellitus. Again, a possible pathogenic role for anti-CRT antibodies in the disease may be speculated.

CRT As Autoantigen

Clinical Associations of Anti-CRT Autoantibody Production

Since the original description of anti-CRT autoantibody production in diseases known to be associated with anti-Ro (SSA) antibodies by one of the authors,[18,19] a variety of clinically and pathologically unrelated diseases have been reported to be associated with anti-CRT Aab formation. These range from the classic systemic autoimmune diseases such as SLE and Sjögrens syndrome to those not usually associated with autoantibody production such as halothane hepatitis and parasitic infection. Anti-CRT antibodies have been reported in conditions as detailed in Table 17.1. Why and how autoantibodies should be generated to a predominantly intracellular protein remains uncertain, as does the pathophysiological role of such antibodies in autoimmune disease.

Historical Perspective

RDS and colleagues initially reported anti-CRT autoantibody production to occur in human diseases which were recognized to be associated with an autoimmune response to the Ro/SS-A (Ro) ribonucleoprotein (RNP) particle system such as Sjögren's syndrome and SLE (also the non-systemic subset subacute cutaneous lupus erythematosus, SCLE). This observation led to a series of studies and some degree of controversy regarding the possibility of a molecular association between CRT and Ro RNP. The basis for this controversy has previously been discussed[3,20] and will be only summarized here.

Cellular Association of Ro RNP and CRT

Human CRT was purified in 1988 from Wil-2 cell extract by virtue of its ability to be precipitated in the Ouchterlony double immunodiffusion technique by monospecific Ro autoimmune sera.[21] It was assumed at that time that CRT was a polypeptide component of the Ro RNP complex. The following year, it was noted that approximately 70% of Ro sera reacted by ELISA with a synthetic peptide corresponding to the N-terminal amino acid sequence of human CRT.[22] The human CRT gene was cloned and sequenced in 1990 and localized to the short arm of chromosome 19.[23] Additional studies indicated that human CRT was highly homologous with the Onchocercal immunodominant antigen, Ral-1, and with an Aplysia memory molecule".[24] In 1992, the genomic organization of human CRT was described and its promoter was found to be homologous to the other resident endoplasmic luminal proteins.[25]

Subsequently, other investigators had difficulty demonstrating that Ro sera reacted with various forms of CRT and that CRT was a Ro RNP-associated protein. In addition, CRT autoantibodies began to be reported in diseases that were not conventionally associated with Ro autoantibody production such as rheumatoid arthritis and celiac disease.

In 1996, a full length recombinant form of human CRT expressed in *E. coli* was shown to bind to hYRNA (the RNA moiety of the Ro RNP) by electromobility gel shift analysis.[26] These same studies indicated that CRT and the 52 kDa Ro protein interacted by the yeast di-hybrid technique. In 1997, native human Wil-2 cell CRT was shown to associate with a subpopulation of Ro RNP.[27] Murine CRT was subsequently independently shown to be physically associated with 60 kDa Ro & 52 kDa Ro polypeptides in-vivo via epitope spreading experiments.[28] In 1999, C1q was reported to inhibit the binding of CRT autoantibodies to CRT.[10]

Table 17.1. Clinical associations of anti-CRT antibodies

Disease Associated with Anti-CRT Production	%	Cross Reactivity of Anti-CRT with Other Disease Associated Autoantibody
AutoImmune		
Systemic lupus erythematosus	~40%	Anti Ro RNP
Subacute Cutaneous LE		(not all anti-CRT show this cross specificity)
Sjogrens Syndrome	~15%	
Rheumatoid Arthritis		
Mixed /Undifferentiated		
Connective Tissue Disease		
Primary Biliary Cirrhosis	60%	Ig A
Auto Immune Hepatitis	60%	Ig G
Coeliac	~30%	Anti-gliaden ab
Halothane Hepatitis		
Parasitic Infection	up to 60%	Ig A predominantly
Onchocerca volvulus		
Schistosoma mansoni		
Trypanosoma cruz		
Leishmania donovani		
Necator americanus		
Ixodes scapularis		
Yersinia sp.		
Healthy controls	7%	

Reasons for differing observations relating to the relationship between human CRT and the Ro RNP particle system may relate to the differing methodologies applied in the study of this protein. Recombinant forms of antigenic CRT may fail to undergo critical post-translational modifications involved in the creation of conformational epitopes necessary for maximal autoantibody binding to native CRT. There is some experimental support for this hypothesis. Studies by Lieu, Sontheimer and colleagues employing a highly purified native form of human Wil-2 cell CRT in a non-denaturing counterimmunoelectrophoresis assay technique indicated that all 28 SCLE patients studied produced anti-CRT autoantibodies.[20] However, subsequent studies from the same group of investigators have indicated that only 38% of a cohort of SCLE patients were found to produce CRT autoantibodies when assayed by an enzyme-linked immunosorbent assay (ELISA) technique employing a full length recombinant form of human CRT expressed in *E. coli.*[29]

The effects of autoantibody assay conditions on the structural properties of the antigenic form of CRT employed could also be an important variable. Lieu, Sontheimer and colleagues found much lower rates of CRT autoantibody detection by Western blot analysis (a denaturing assay) compared to counterimmunoelectrophoresis (a non-denaturing assay) when examining the same Ro autoantibody-containing serum specimens.[20]

Another source of variation in CRT antigen structure could be differences in the native configurations of CRT purified from different laboratory cell lines. The original studies by Lieu, Sontheimer and colleagues employed Wil-2 cells for this purpose while others used HeLa and other cell lines. Particularly, Wil-2 is an Epstein-Barr (EB) virus-transformed human B cell lymphoblastoid line and, while there has been no systematic examination of the effects of the EB virus on CRT expression, one report has suggested that the expression of CRT may be dysregulated in EB transformed cells.[30] Other viruses such a the cytomegalovirus have been shown to alter the cellular expression of CRT.[31] Inapparent infection of cell lines with such organisms or with other subtle infectious agents (e.g., mycoplasma) could possibly play a role in variant CRT autoantigenic structure.

In summary, several lines of evidence continue to suggest that native forms of mammalian CRT directly interact with a subpopulation of Ro RNP particles. Curiously, this subpopulation does not appear to contain the La/SS-B polypeptide. In addition to autoantibodies to linear sequence epitopes on CRT, Aab to conformational epitopes appear to be present in patient sera that immunoprecipitate CRT:Ro RNP complexes. The existence of such conformational epitopes could account for some of the variant observations in this area. Technical/methodological considerations may also contribute to the different observations reported regarding the association of CRT and the Ro RNP system. Future studies of the clinical significance of CRT autoantibody production should focus more on the use of native forms of CRT and/or eukaryotically-expressed recombinant forms of human CRT that best represent the post-translational modified configurations of native human CRT. In addition, autoantibody assay conditions should take into account observations related to the structural configuration changes can CRT (e.g., phosphorylation, glycosylation, calcium and zinc ion binding capacity, hYRNA binding capacity).

How Does CRT Become Accessible to the Adaptive Immune System?

Autoimmune diseases such as SLE are characterized by the production of autoantibodies to a number of normally intracellular antigens such as double stranded DNA, histone DNA complexes and ribonucleoproteins in addition to CRT. There are a number of mechanisms by which these intracellular constituents are thought to relocate to the cell surface where they become targeted as autoantigens by the adaptive immune system leading to a break in self tolerance. They may become accessible during the process of regulated cell death or apoptosis. Blebs are formed as the cells round up and these contain high concentrations of the normally intracellular antigens Ro/La/DNA and CRT. However, this process is physiological and tightly regulated to ensure swift and effective, non-inflammatory clearance of apoptotic cells by the phagocytic system. Defective clearance of apoptotic blebs would result in their persistence, packed full of potential autoantigens, allowing peripheral tolerance to be broken.

As noted, although CRT is a predominantly intracellular protein, it is now known to be expressed on the surface of a variety of cell types—usually up regulated as a result of oxidative stress, UV irradiation (fibroblasts), viral infection (T cells) or cellular activation (T cells)—where it is accessible to the adaptive immune system.

Why CRT Might Be Targeted As Nonself

CRT and Epitope Spreading

In addition to its role in physiological aspects of antigen presentation, CRT is also involved in a potentially pathological aspect of antigen recognition—that of epitope spreading, a process by which an antigenic response to a single antigen spreads to involve other antigens in association with it. As CRT is frequently associated with the autoantigen Ro in apoptotic blebs, it is possible that anti CRT antibodies are generated by this phenomenon. Evidence exists to support this in animal studies; epitope spreading has been demonstrated to play a role in the autoimmune response to CRT that can be generated in mice by active immunization with Ro autoantigens.[32]

Parasite Molecular Mimicry

CRT protein homologues have been identified in a number of human parasites including the filarial parasite and cause of river blindness *Onchocerca*, the blood fluke *Schistosoma*, the trypanosome leishmania and the malarial parasite *Plasmodium*; however evidence for the release of parasitic CRT in acute parasitic infection is not, as yet, forthcoming. Interestingly, in one parasite, *Necator americanum*[33](see Protozoal Chapter) cell surface CRT may be a defense mechanism used by the nematode to avoid attack by the host innate immune mechanisms. Since Necator CRT is known to block classical complement activation, this may render the worm highly resistant to attack by complement.

Association of Cellular CRT with Viral Envelope Proteins and RNA

CRT interacts with nascent viral glycoproteins of a number of viruses as they transit through the mammalian cell ER during their replication cycle. Examples include glycoproteins expressed by hepatitis C,[34] HIV,[35] foot and mouth disease,[36] and rubella.[37]

Additionally, CRT has been demonstrated to interact with the measles virus H glycoprotein in the ER, with resulting up regulation of cellular and cell surface CRT expression.[38] Other viruses such as cytomegalovirus (CMV) have also been shown to alter the cellular expression of CRT[31] and dysregulation of CRT expression in EB transformed cells has been suggested.[30] Such interactions between virus glycoproteins and chaperone proteins may be significant in the targeting of CRT as non self. A phosphorylated form of human CRT binds specifically to a rubella viral 3'(+) stem-loop RNA structures and thereby serves as a mammalian host cell replication factor for the rubella virus.[39]

Can the CRT Autoimmune Response Be Viewed As a Heat Shock Response?

CRT been confirmed to be a heat shock (hsp)/stress response gene. A variety of clinical conditions associated with the CRT autoimmune response are also associated with antibodies to classical hsp such as hsp90, hsp70, hsp60. In general, such Aab have low disease specificity, and are found in a range of autoimmune connective tissue diseases similar to those associated with anti-CRT antibody production as well as with diseases not conventionally associated with autoimmunity including atherosclerotic cardiovascular disease, schizophrenia, and heat stress illness. In patients who produce hsp, the associated clinical features are similar to those observed in association with anti-CRT, raising the possibility that Aab directed against linear sequence epitopes on CRT might in fact represent a heat shock/stress response pattern of autoimmunity.

Preliminary studies have been carried out attempting to correlate anti-CRT and anti-hsp70/hsp-90 Aab in serum specimens from the same patients. Such studies have revealed no significant positive correlation between CRT Aab levels and either levels of Aab against hsp 90 or hsp70.[29] However, it should be noted that Aab to one conventional hsp family (eg hsp90) can be seen in one disease state to the exclusion of Aab to other conventional hsp (e.g., hsp70). A similar exclusivity might exist for anti-CRT and other types of hsp autoimmune responses.

As a heat shock response gene, environmental cellular stressors (nutritional depravation, exposure to heavy metals) might be important in the induction of autoimmunity to CRT. Recent work has suggested that necrotic cell death as opposed to apoptotic cell death results in the release of CRT and other hsp into the extracellular space and that such extracellular hsp can serve as immunological adjuvants by delivering tumor related peptides to CD91 receptor–bearing dendritic antigen presenting cells.[40-43] (Fig. 17.1). The possibility exists that traditional hsp might be able to deliver CRT autoantigenic peptides to T cells under conditions that could result in an autoimmune response to such peptides.

Observed Immunochemical Characteristics of the CRT Aab Response

Another approach to understanding the origin and significance of the human CRT autoimmune response is to examine its specific immunochemical features. All three major immunoglobulin classes (IgG, IgA, IgM) appear to be represented in the CRT Aab repertoire. This observation alone argues against a "natural autoimmunity" role for CRT Aab, as natural Aab that play a normal physiological function are often restricted to the IgM class. In addition, IgG CRT Aab have indirectly been observed to be restricted to the complement-fixing, T

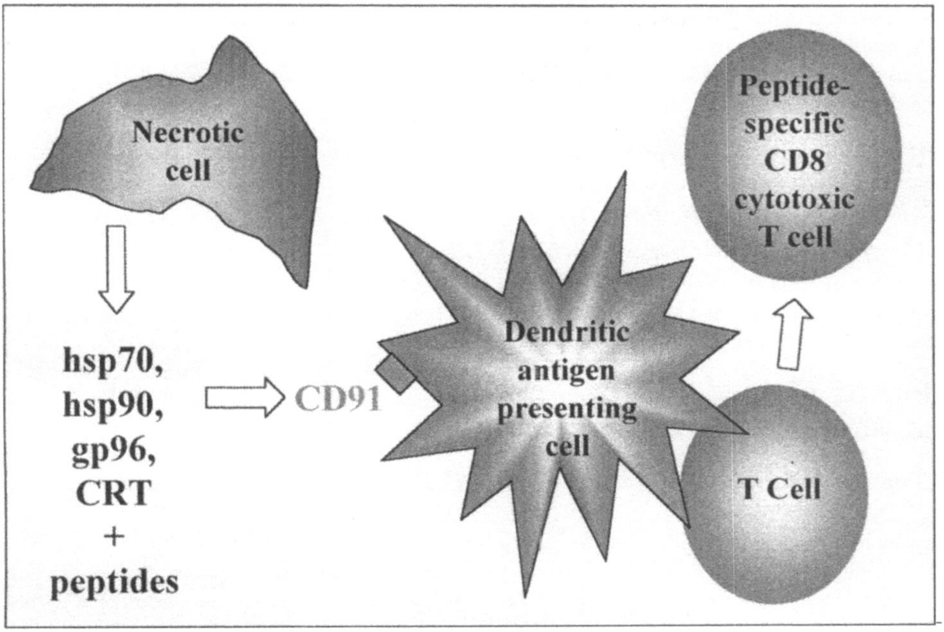

Figure 17.1. CRT and the cross-antigen presentation pathway (syn. cross-priming, indirect antigen presentation).

cell-dependent IgG subclasses, IgG1 and IgG3.* [44] These immunochemical features would suggest that CRT Aab production is under direct CD4 T cell control. However, little work directed toward identifying CRT-specific T cells has been reported to date. One preliminary report identified CRT-specific T cells in the blood of rheumatoid arthritis patients.[45] The existence of CRT-specific T cells is implied in an experimental murine model of epitope spreading between CRT and other polypeptide components of the Ro/SS-A ribonucleoprotein particle.[28] In addition, individuals that produce IgG Aab to CRT expressed by infecting parasites such as *Onchocerca volvulus* or *Schistosoma mansoni* could also be predicted to harbor helper-inducer CRT specific T cells.

As with many other rheumatic disease-related autoantigens, CRT is a highly-conserved cellular molecule. Aab to rheumatic disease autoantigens as a rule are directed at functionally-active sites on the corresponding autoantigens. Most work indicates that CRT Aab target epitopes on the N-terminal domain of CRT[22,46] (Fig. 17.3). In conjunction with this, several functional attributes of CRT have been linked to the amino terminus (e.g., RNA binding). It is not clear how these functions might relate to Aab production. However, it is tempting to speculate that viral RNA binding to the amino terminus of CRT as has been demonstrated for the rubella virus[47] could be a stimulus for CRT Aab production.

CRT Specific Cell Mediated Immune Responses

CRT-specific T-cells have been noted in patients with rheumatoid arthritis.[45] This observation has been supported by the identification of a peptide from the CRT sequence in the DR4Dw4/DR53 class II molecules isolated from rheumatoid arthritis patients.[48]

---

*The IgG subclass composition of the autoantibody response to a native form of human CRT encountered in patients with subacute cutaneous LE has been examined indirectly and reported (44;51). The antigenic form of native human CRT employed in these studies was originally identified via its association with a subpopulation of Ro/SS-A RNP particles.

Direct cellular cytotoxicity mediated by antigen-specific T cells is dependent on the effector molecule, perforin, stored in cytotoxic T-cell granules. Interestingly, CRT has been show to be present in cytotoxic granules where it appears to play a stabilizing function with respect to perforin actions.[49]

Pathogenetic Significance of the CRT Autoimmune Response

An Etio-Pathogenetic Paradigm to View the Relationship of CRT with a Complex Immune Disease Such As SLE

One instructive paradigm for organizing the diverse etio-pathogenetic considerations relating to polygenic human autoimmune disease such as systemic lupus erythematosus (SLE) is a sequential, multistage conceptual approach. Figure 17.2 illustrates such a paradigm. This approach can be adapted to organize the diverse observations related to the immune response to CRT as individual autoantigen and the interpretation of different roles for CRT both physiologic/protective and pathologic.

The Role of Anti-CRT Aab As Effectors of Cell and Tissue Damage

By binding with antigen, anti-CRT antibodies could subvert the role of CRT in important cellular regulatory processes and so contribute to inflammation in autoimmne disease. The intracellular location of CRT has previously been considered to argue against a pathologic role for anti CRT antibodies in the disease states with which they are associated, and indeed to date no evidence exists for their pathogenicity. While it is true that most of the physiological functions of CRT relate to its intracellular function, recent evidence is persuasive that surface and extracellular CRT have important functional roles in vital processes such as the recognition and clearance of apoptotic cells. Hence, anti-CRT Aab may abrogate these functions if they block important recognition and binding sites on the CRT molecule. In support of this possibility has been the recent observation that CRT Aab purified from SLE serum can reverse the inhibitory action of CRT on C1q-mediated hemolytic function.[50] Such properties require further elucidation in clinically based studies and recent work in our laboratory has begun to address this. As previously noted, we have demonstrated a failure of C1q to bind to apoptotic neutrophils from un-selected SLE patients, a condition characterized by anti-CRT Aab production. One can speculate that such antibodies may interfere with the C1q/CRT/CD91 interaction necessary for effective clearance of apoptotic cells. Three of five major antigenic sites within CRT as assessed by Aab mapping of the sites with anti-CRT antisera from lupus patients are positioned in the same regions that C1q binds (Fig. 17.3). Alternatively, there may be other specific cell surface molecules that interact with C1q that are lacking on the apoptotic cells of SLE patients. Both these scenarios are currently being investigated.

Final Thoughts on the Role of CRT in Autoimmune Disease

It is clear from this review of current knowledge of the immune related functions of CRT that its influence may be protective or pathological depending on the cell type and context. Aetiopathological models can be proposed based on current knowledge, however these are of necessity speculative at the present time. Further carefully designed studies which directly compare the effect of CRT in disease and the healthy state are now required in order clarify exactly which of these influences predominates and how this may contribute to specific disease pathogenesis. Similarly, as understanding of the functions of CRT in the immune system increases, reciprocal studies of the effects of disease associated anti-CRT antibodies will become imperative if we are to establish the dominant effects of CRT and anti-CRT antibodies in the pathogenesis of autoimmune disease.

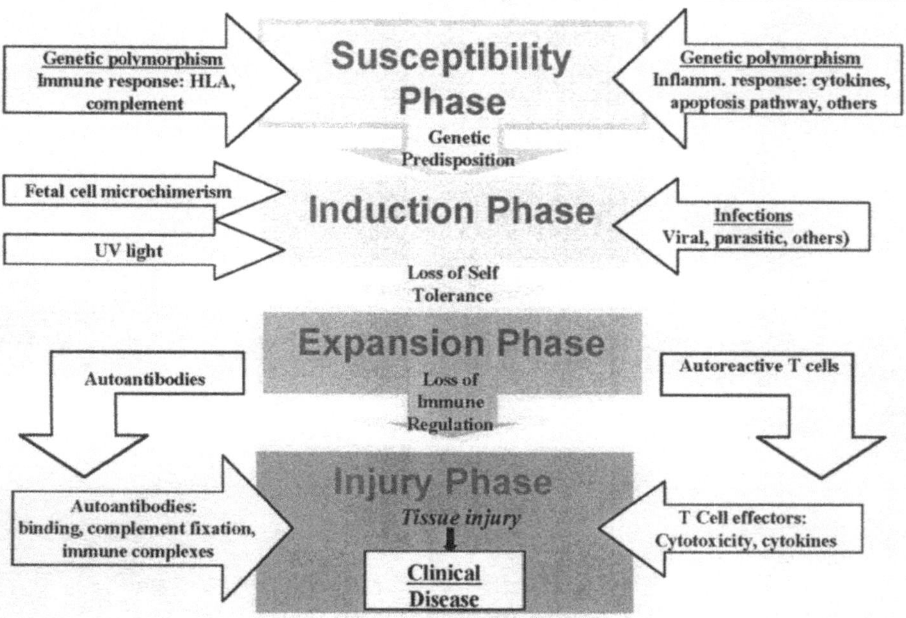

Susceptibility Phase. Human autoantibody responses are characteristically genetically restricted. For example, individuals displaying a specific autoantibody response are often found to possess a particular set of immune responsiveness gene alleles. The genetic background upon which the human CRT autoantibody response develops has not yet been systematically analyzed, but CRT has been shown to be a constituent of a subpopulation of human Ro RNP particles suggesting the possibility that the CRT AAB response could also be linked to the 8.1 ancestral haplotype.

Induction Phase. The factors responsible for the induction of the human CRT autoimmune response are not yet known. Environmental and inflammatory stimuli such as UV light and viral infection could be important in initiating a process that leads to loss of immunological tolerance to autoantigens such as CRT.

Expansion Phase. Once immunological tolerance has been breached, the resulting autoimmune responses must be expanded before tissue injury and clinical disease occurs. During this phase, high levels of AAB and autoreactive T cells are generated. Mechanisms by which autoimmune responses are expanded include affinity maturation and epitope spreading.

Injury Phase. During this phase, high levels of AAB and autoantigen specific T-cells conspire through a number of immunological effector mechanisms to produce tissue injury and clinical disease. Such conventional AAB-dependent effector mechanisms include antibody-mediated cytolysis, Fc-mediated opsonization, immune complex formation, and antibody-dependent, cell-mediated cytotoxicity.

There could be a rather unique mechanism by which CRT AAB might create problems. C1q binds to CRT in a defined region and the interaction between these two polypeptides is thought to be functional. It has been reported that C1q can block the binding of CRT AAB to CRT. This would imply the possibility that CRT AAB could interfere with the binding of C1q to CRT.

Autoreactive T cells can cause cellular injury by direct cell-mediated cytotoxicity. Evidence has been presented indicating the presence of CRT-specific T-cells in patients with rheumatoid arthritis. This observation has been supported by the identification of a peptide from the CRT sequence in the DR4Dw4/DR53 class II molecules isolated from rheumatoid arthritis patients.

Direct cellular cytotoxicity mediated by antigen-specific T cells is dependent on the effector molecule, perforin, stored in cytotoxc T-cell granules. Interestingly, CRT has been show to be present in cytotoxic granules where it appears to play a stabilizing function with respect to perforin actions.

Figure 17.2. An etio-pathogenic paradigm for polygenic autoimmune-mediated inflammatory diseases.

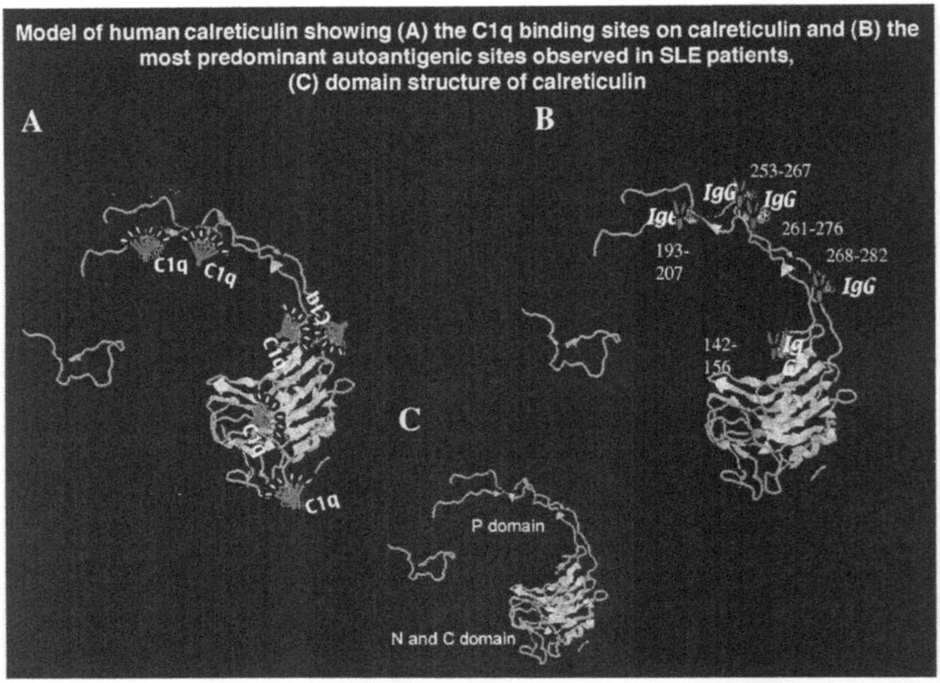

Figure 17.3. Position of some CLq and autoantibody binding sites identified on CRT.

Acknowledgments

Dr. R. D. Sontheimer holds the John S. Strauss Endowed Chair in Dermatology at the University of Iowa College of Medicine/University of Iowa Hospitals and Clinics (University of Iowa Health Care). This publication was made possible by Grant Number AR19101 from the NIH (NIAMS). Its contents are solely the responsibility of the authors and do not necessarily represent the official views of NIAMS. This work was also supported by a Dermatology Department endowment from the Herzog Foundation, the generous support of the Joseph Marshall Family of Morningside, Iowa, and Arthritis Research Campaign (Grant EO543-PE).

References

1. Arosa FA, de Jesus O, Porto G et al. Calreticulin is expressed on the cell surface of activated human peripheral blood T lymphocytes in association with major histocompatibility complex class I molecules. J Biol Chem 1999; 274(24):16917-16922.
2. Kawashima T, Zappi EG, Lieu T-S et al. Impact of ultraviolet radiation on the cellular expression of Ro/SS-A-autoantigenic polypeptides. Dermatology 1994; 189(Suppl. 1):6-10.
3. Nguyen TQ, Racila D, Sontheimer RD. Role of calreticulin autoimmunity in the pathogenesis of photosensitive cutaneous lupus erythematosus. In: Tsokos GC, Kammer G, eds. Lupus: Molecular and Cellular Pathogenesis. Totowa: The Humana Press, 1999:514-528.
4. Kishore U, Sontheimer RD, Sastry KN et al. The systemic lupus erythematosus (SLE) disease autoantigen—calreticulin can inhibit C1q-association with immune complexes. Clin Exp Immunol 1997; 108:181-190.
5. Coppolino MG, Woodside MJ, Demaurex N et al. Calreticulin is essential for integrin-mediated calcium signalling and cell adhesion. Nature 1997; 386(6627):843-847.
6. Fadel MP, Dziak E, Lo CM et al. Calreticulin affects focal contact-dependent but not close contact-dependent cell-substratum adhesion. J Biol Chem 1999; 274(21):15085-15094.
7. Fadel MP, Szewczenko-Pawlikowski M, Leclerc P et al. Calreticulin affects beta-catenin-associated pathways. J Biol Chem 2001; 276(29):27083-27089.

8. Eggleton P, Lieu TS, Zappi EG et al. Calreticulin is released from activated neutrophils and binds to C1q and mannan-binding protein. Clin Immunol Immunopathol 1994; 72:405-409.
9. Stuart GR, Lynch NJ, Lu J et al. Localization of the C1q binding site within C1q receptor/calreticulin. FEBS Letters 1996; 397(2-3):245-249.
10. Racila D, Sontheimer RD. C1q inhibits autoantibody binding to calreticulin. Lupus 1999; 8:300-304.
11. Kovacs H, Campbell ID, Strong P et al. Evidence that C1q binds specifically to CH2-like immunoglobulin gamma motifs present in the autoantigen calreticulin and interferes with complement activation. Biochemistry 1998; 37(51):17865-17874.
12. Kishore U, Sontheimer RD, Sastry KN et al. Release of calreticulin from neutrophils may alter C1q-mediated immune functions. Biochem J 1997; 322:543-550.
13. Casciola-Rosen LA, Anhalt G, Rosen A. Autoantigens targeted in systemic lupus erythematosus are clustered in two populations of surface structures on apoptotic keratinocytes. J Exp Med 1994; 179:1317-1330.
14. Rosen A, Casciola-Rosen L, Ahearn J. Viral antigens are concentrated and clustered with specific autoantigens in the surface blebs on apoptotic cells. Arthritis Rheum 1994; 37(9 Suppl):S368.
15. Korb LC, Ahearn JM. C1q binds directly and specifically to surface blebs of apoptotic human keratinocytes—complement deficiency and systemic lupus erythematosus revisited. J Immunol 1997; 158(10):4525-4528.
16. Navratil JS, Watkins SC, Wisnieski JJ et al. The globular heads of C1q specifically recognize surface blebs of apoptotic vascular endothelial cells. J Immunol 2001; 166(5):3231-3239.
17. Oyadomari S, Takeda K, Takiguchi M et al. Nitric oxide-induced apoptosis in pancreatic beta cells is mediated by the endoplasmic reticulum stress pathway. Proc Natl Acad Sci USA 2001; 98(19):10845-10850.
18. Lieu T-S, Newkirk M, Capra JD et al. Molecular characterization of human Ro/SS-A antigen: amino terminal sequence of the protein moiety of human Ro/SS-A antigen and immunological activity to a corresponding synthetic peptide. J Clin Invest 1988; 82:96-101.
19. Lieu TS, Newkirk MM, Arnett FC et al. A major autoepitope is present on the amino terminus of a human SS-A/Ro polypeptide. J Autoimmunity 1989; 2:367-374.
20. Sontheimer RD, Nguyen TQ, Cheng ST et al. Calreticulin and autoimmunity. In: Michalak M, ed. Calreticulin. Austin: Landes Bioscience, 1996:117-139.
21. Lieu TS, Newkirk MM, Capra JD et al. Molecular characterization of human Ro/SS-A antigen. Amino terminal sequence of the protein moiety of human Ro/SS-A antigen and immunological activity of a corresponding synthetic peptide. J Clin Invest 1988; 82:96-101.
22. Lieu T-S, McCauliffe DP, Newkirk MM et al. A major autoepitope is present on the amino terminus of the human Ro/SS-A polypeptide. J Autoimmunity 1989; 2:367-374.
23. McCauliffe DP, Lux FA, Lieu T-S et al. Molecular cloning, expression, and chromosome 19 localization of a human Ro/SS-A autoantigen. J Clin Invest 1990; 85:1379-1391.
24. McCauliffe DP, Zappi E, Lieu T-S et al. A human Ro/SS-A autoantigen is the homologue of calreticulin and is highly homologous with onchocercal RAL-1 antigen and an aplysia "memory molecule". J Clin Invest 1990; 86:332-335.
25. McCauliffe DP, Yang Y-S, Wilson J et al. The 5'-flanking region of the human calreticulin gene shares homology with the human GRP78, GRP94, and protein disulfide isomerase promoters. J Biol Chem 1992; 267:2557-2562.
26. Cheng ST, Nguyen TQ, Yang YS et al. Calreticulin binds hYRNA and the 52 KD polypeptide component of the Ro/SS-A ribonucleoprotein autoantigen. J Immunol 1996; 156:4484-4491.
27. Lieu TS, Sontheimer RD. A subpopulation of Wil-2 cell calreticulin molecules is associated with Ro/SS-A ribonucleoprotein particles. Lupus 1997; 6(1):40-47.
28. Kinoshita G, Keech CL, Sontheimer RD et al. Spreading of the immune response from 52kDa Ro and 60kDa Ro to calreticulin in experimental autoimmunity. Lupus 1998; 7:7-11.
29. Racila DM, Sontheimer RD. The autoantibody response to human calreticulin displays features of a non-classical heat shock protein autoimmune response. J Invest Dermatol 2002; in press.
30. Pike SE, Yao L, Jones KD et al. Vasostatin, a calreticulin fragment, inhibits angiogenesis and suppresses tumor growth. J Exp Med 1998; 188(12):2349-2356.
31. Zhu J. Ultraviolet B irradiation and cytomegalovirus infection synergize to induce the cell surface expression of 52-kD/Ro antigen. Clin Exp Immunol 1996; 103(1):47-53.
32. McCluskey J, Farris AD, Keech CL et al. Determinant spreading: lessons from animal models and human disease.
33. Kasper G, Brown A, Eberl M et al. A calreticulin-like molecule from the human hookworm Necator americanus interacts with C1q and the cytoplasmic signalling domains of some integrins. Parasite Immunology 2001; 23(3):141-152.

34. Merola M, Brazzoli M, Cocchiarella F et al. Folding of hepatitis C virus E1 glycoprotein in a cell-free system. J Virol 2001; 75(22):11205-11217.
35. Land A, Braakman I. Folding of the human immunodeficiency virus type 1 envelope glycoprotein in the endoplasmic reticulum. Biochimie 2001; 83(8):783-790.
36. O'Donnell VK, Pacheco JM, Henry TM et al. Subcellular distribution of the foot-and-mouth disease virus 3A protein in cells infected with viruses encoding wild-type and bovine-attenuated forms of 3A. Virology 2001; 287(1):151-162.
37. Nakhasi HL, Ramanujam M, Atreya CD et al. Rubella virus glycoprotein interaction with the endoplasmic reticulum calreticulin and calnexin. Arch Virol 2001; 146(1):1-14.
38. Bolt G. The measles virus (MV.) glycoproteins interact with cellular chaperones in the endoplasmic reticulum and MV infection upregulates chaperone expression. Arch Virol 2001; 146(11):2055-2068.
39. Singh NK, Atreya CD, Nakhasi HL. Identification of calreticulin as a rubella virus RNA binding protein. Proc Natl Acad Sci USA 1994; 91:12770-12774.
40. Basu S, Binder RJ, Suto R et al. Necrotic but not apoptotic cell death releases heat shock proteins, which deliver a partial maturation signal to dendritic cells and activate the NF-kappa B pathway. Int Immunol 2000; 12(11):1539-1546.
41. Przepiorka D, Srivastava PK. Heat shock protein--peptide complexes as immunotherapy for human cancer. Mol Med Today 1998; 4(11):478-484.
42. Srivastava P. Interaction of heat shock proteins with peptides and antigen presenting cells: chaperoning of the innate and adaptive immune responses. Annu Rev Immunol 2002; 20:395-425.
43. Binder RJ, Karimeddini D, Srivastava PK. Adjuvanticity of alpha 2-macroglobulin, an independent ligand for the heat shock protein receptor CD91. J Immunol 2001; 166(8):4968-4972.
44. Lieu TS, Reimer CB, Sontheimer RD. Immunoglobulin class and subclass profile of the Ro/SS-A autoantibody response. J Invest Dermatol 1988; 90:158-164.
45. Jacob L, Bartosz-Roznowski A, Houen G et al. Evidence for human calreticulin as an autoantigen specifically recognized by peripheral blood T cells of non-immunosuppressed patients with rheumatoid arthritis and other arthritic diseases. Arthritis Rheumat 1997; 40(9 Suppl):S197-S-197.
46. Eggleton P, Ward FJ, Johnson S et al. Fine specificity of autoantibodies to calreticulin: epitope mapping and characterization. Clin Exp Immunol 2000; 120(2):384-391.
47. Atreya CD, Pogue GP, Singh NK et al. Role of calreticulin in rubella virus replication. In: Michalak M, ed. Calreticulin. Austin: Landes Bioscience, 1996:89-115.
48. Verreck FA, Elferink D, Vermeulen CJ et al. DR4Dw4/DR53 molecules contain a peptide from the autoantigen calreticulin. Tissue Antigens 1995; 45(4):270-275.
49. Fraser SA, Karimi R, Michalak M et al. Perforin lytic activity is controlled by calreticulin. J Immunol 2000; 164(8):4150-4155.
50. van den Berg RH, Siegert CE, Faber-Krol MC et al. Anti-C1q receptor/calreticulin autoantibodies in patients with systemic lupus erythematosus (SLE). Clin Exp Immunol 1998; 111(2):359-364.
51. Bennion SD, Ferris C, Lieu T-S et al. IgG subclasses in the serum and skin in subacute cutaneous lupus erythematosus and neonatal lupus erythematosus. J Invest Dermatol 1990; 95:643-646.

Chapter 18

Cell Surface Calreticulin: Role in Signaling Thrombospondin Anti-Adhesive Activity

Silvia M. Goicoechea and J.E. Murphy-Ullrich

Abstract

Cell adhesion is a multi-step process initiated by receptor binding to extracellular matrix (ECM) components. This triggers intracellular signaling cascades leading to spreading and organization of the cytoskeleton with focal adhesions. Essential cellular processes such as locomotion, gene expression and survival are regulated by both signaling and mechanical forces generated by the organized cytoskeleton. The anti-adhesive protein, thrombospondin (TSP), induces changes in the cytoskeleton characterized by a loss of focal adhesions and stress fibers. This de-adhesive activity of TSP is localized to amino acids 17-35 in the N-terminal heparin binding domain and can be replicated by a peptide (hep I) with this sequence. Our laboratory recently showed that calreticulin (CRT) on the surface of bovine aortic endothelial (BAE) cells binds to TSP and mediates its ability to stimulate the loss of focal adhesions and stress fibers in several cell types. Wild type but not *crt–/–* cells respond to TSP/hep I. Responsiveness to TSP/hep I can be restored by incubation of cells with exogenous CRT. The TSP binding site in CRT is located in the N-terminal domain (amino acid 19-36) and TSP binding to CRT is enhanced by divalent ions Ca^{2+} and Zn^{2+}. TSP binding to CRT on the cell surface activates signaling cascades that involve a pertussis toxin-sensitive G protein signal that stimulates activation of phosphoinositide 3 kinase (PI3K) and ERK, culminating in disassembly of focal adhesions. Recent data show that LDL-receptor-related protein (LRP) acts as a co-receptor with CRT to mediate TSP/hep I-stimulated focal adhesion disassembly. These studies identify a novel function for the cell surface form of calreticulin as a mediator of de-adhesive events and they show that cell surface calreticulin is capable of co-operatively triggering intracellular signals.

Introduction—Calreticulin: A Ubiquitous Protein with Diverse Functions

CRT is a widely-expressed calcium-binding protein found mainly in the endoplasmic reticulum, but also in other cellular compartments.[1] In the lumen of the ER, CRT acts as a molecular chaperone and regulates calcium homeostasis.[2-4] The localization of CRT to other cellular compartments, including the cell surface, as secreted forms, and possibly, in the cytoplasm and nucleus, is prompting reconsideration of CRT as a mediator of a broader array of cellular functions. CRT apparently plays a critical role in the development of the myocardium as CRT knockout animals exhibit severe cardiac deformities.[5,6] Evidence of the physiologic importance of CRT is shown by recent data reporting that a fragment of the N-terminus (amino acids 120-180) termed vasostatin inhibits angiogenesis by blocking endothelial cell proliferation, thereby suppressing tumor cell growth.[7] CRT may also regulate protein transcription by altering β-catenin phosphorylation through a unique ER signaling pathway and altering Wnt signaling.[8]

Calreticulin, 2nd ed., edited by Paul Eggleton and Marek Michalak.

In this chapter, we report a novel function for CRT at the cell surface—its role as a receptor for the de-adhesive matricellular protein TSP and mediator of focal adhesion disassembly. We will describe the evidence for CRT expression at the cell surface, its interactions with TSP and its ability to signal focal adhesion disassembly, and finally, we will discuss the potential implications of cell surface CRT on cell function, particularly during cellular stress conditions.

Calreticulin Is a Cell Surface Protein

Expression of CRT has been reported on the surface of several types of cells.[9-12] CRT has an N-terminal signal sequence and potentially can be transported to the cell surface.[2] Gray et al showed that cell surface CRT on fibroblasts binds to the β chain of fibrinogen mediating its mitogenic activity.[10] White et al found that CRT is expressed on the external cell surface as a putative mannoside lectin, which triggers mouse melanoma cell spreading.[11] Recently, Arosa et al reported finding CRT expressed on the cell surface of activated human peripheral blood T lymphocytes, where it is physically associated with a pool of unfolded MHC class I molecules.[9] In addition, it has been reported that CRT is also localized on the surface of neutrophils and participates in the pertussis toxin-sensitive signal transduction pathway stimulated by L5, an anti-microbial peptide.[13] Consistent with these findings, we have also documented CRT on the cell surface.[14] CRT can also be released from neutrophils during inflammation. Numerous studies have reported the presence of CRT on endothelial cells.[12,15,16] CRT on the surface of endothelial cells appears to be capable of modulating cell function as it has been shown that CRT is involved in the production of IL-8 by human umbilical vein endothelial cells.[16] CRT is readily detectable on the surface of resting bone marrow vascular endothelial cells and its expression is up regulated in response to inflammatory mediators.[12] Finally, CRT binds specifically and reversibly to bovine aortic endothelial cells in vitro (Kd approximately 7.4 nM), suggesting that secreted CRT can also modulate endothelial cell behavior.[15]

It is not clear how CRT is able to leave the ER and be transported to the cell surface. However, the expression of other KDEL containing proteins at the cell surface has recently been shown.[17] In fact, newly synthesized CRT appears to be preferentially transported to the cell surface where it has a half-life of approximately 12-hr.[18] CRT expressed on the surface of neuronal cells is turned over via a lysosomal degradation pathway.[18] Interestingly, the amino-terminal heparin-binding domain of TSP, which contains the sequence that mediates focal adhesion disassembly (hep I) also mediates rapid, lysosomal degradation of TSP via interactions with heparan sulfate proteoglycans and LRP.[19,20] It is possible that CRT binding to the hep I sequence may facilitate degradation of CRT.

We have initial data suggesting that under conditions of cellular stress, such as serum-deprivation of BAE cells, the expression of CRT at the cell surface is increased. This is consistent with reports suggesting that ER proteins can "escape" to the cell surface during cellular stress. Correspondingly, TSP expression is also up regulated by cell stress. It is possible that TSP engagement of CRT signaling at the cell surface represents an adaptive mechanism for cell responses to injury or repair.[21]

TSP-Mediated Focal Adhesion Disassembly

TSP was first identified and characterized as a protein released from the alpha granules of platelets upon activation by thrombin.[22] It is synthesized by and deposited into the ECM of cultured endothelial cells, fibroblasts and smooth muscle cells among others and is widely distributed in numerous tissues. TSPs are a family of 5 related glycoproteins. The best-characterized family member, TSP1, is a large homotrimer (450 kDa) consisting of multiple domains that interact with a number of macromolecules and growth factors (Fig. 18.1). A number of receptors have been identified for these various domains, including syndecan, LRP, integrins, CD36, and CD47.[20,23] TSP has been shown to be a major endogenous inhibitor of angiogenesis, a physiologic regulator of TGF-β activation, as well as a modulator of cell-substrate and cell-cell adhesion, motility, cell proliferation, and phagocytosis.[21,24-27]

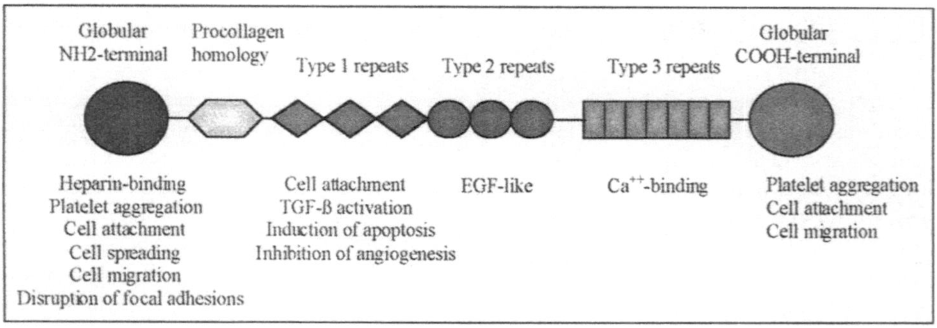

Figure 18.1. Schematic model of a single subunit of TSP-1 molecule. This cartoon represents a monomeric subunit of the TSP1 trimer. Like other matrix proteins, TSP is composed of multiple repeats of modular domains. Putative functions and binding interactions for these domains are indicated.

Cell Adhesion and De-adhesion

Cell adhesion is a continuous and reversible process necessary for normal cell survival, growth and control of specific gene expression. It results in the linkage of the cytoskeleton to the extracellular matrix (ECM).[28] Most ECM glycoproteins promote organization of the cytoskeleton and the formation of focal adhesions. Focal adhesions are submembranous scaffolds of structural, signaling, and adapter molecules which, link the actin cytoskeleton to the ECM, resulting in a firm adhesive state characteristic of stationary cells.[29] Focal adhesions not only physically link a cell to its matrix, but they also regulate the balance of tensional forces in the cell. Thus, focal adhesions regulate cell function through both mechanical and biochemical signals. Cell adhesion and signaling through these substrates have been studied extensively. However, there exists another class of ECM proteins, termed "matricellular" proteins, which function as adapters and modulators of cell-matrix interactions.[30,31] These proteins include TSP 1 and 2, tenascins, and SPARC. As substrates, these proteins are only capable of supporting the initial and intermediate stages of cell adhesion: attachment and spreading.[30,32-34] Focal adhesions and stress fiber formation, characteristic of strong cell adhesion, are rarely observed. In contrast, when the soluble form of these matricellular proteins is presented to cells in a strong adhesive state, they have primarily de-adhesive affects.

As a matricellular protein, TSP has a complex role in regulation of cell adhesion. As a substrate, TSP1 has both adhesive and anti-adhesive characteristics depending on the experimental conditions used to immobilize TSP and cells. As a soluble protein TSP1 stimulates the loss of focal adhesions and stress fibers in spread, adherent bovine aortic endothelial cells plated in fibronectin substrates.[32,35] TSP1 also stimulates the disassembly of focal adhesions in fibroblasts and smooth muscle cells as well and this activity is independent of the substrate used to support strong adherence. Treatment of adherent cells with TSP1 results in focal adhesion restructuring and alterations in the stress fibers but has no effect on cell spreading and integrin clustering.[35,36] The loss of both α-actinin and vinculin also occurs at focal adhesion sites. However, almost no change is observed in $\alpha_v\beta_3$ integrin or talin clustering.[32,37] The focal adhesion disassembling activity of TSP1 is localized to the amino-terminal heparin-binding domain, consistent with earlier data showing that TSP1 activity could be blocked by either heparin or a monoclonal antibody specific for the amino-terminal heparin-binding domain. Specifically, de-adhesive activity of TSP1 is localized to amino acids 17-35.[36] Synthetic peptides containing this sequence (hep I) induce loss of stress fibers and focal adhesions identical to that of TSP1.

Cell Surface CRT As a Receptor for TSP-Mediated Focal Adhesion Disassembly

Experiments were performed to identify the receptor for TSP that mediates disassembly of focal adhesions in bovine aortic endothelial (BAE) cells. A 60 kDa hep I-binding protein was isolated from detergent extracts of BAE cells. This protein has N-terminal sequence homology to CRT and is immunoreactive with anti-CRT antibodies. These studies also demonstrated that the 60 kDa protein competitively blocked focal adhesion disassembly by hep I and TSP.[14]

The ability of antibodies to CRT incubated with BAE cells to block focal adhesion disassembly by TSP suggested that CRT is expressed on the surface of BAE cells where it can mediate focal adhesion disassembly. Biotinylation studies using a membrane-impermeable form of biotin confirmed the surface localization of CRT.[14] Immunolocalization by confocal microscopy and by fluorescence-activated cell sorting (FACS) of intact cells further confirmed the cell surface localization of CRT (Fig. 18.2).[14]

CRT on the cell surface mediates TSP-stimulated focal adhesion disassembly. As mentioned, incubation of BAE cells with antibodies to CRT blocked focal adhesion disassembly by TSP. In addition, GST-CRT incubated with either TSP or hep I similarly blocked the ability of TSP/hep I to stimulate focal adhesion disassembly (Fig. 18.3). The ability of cell surface CRT to mediate focal adhesion disassembly is specific for TSP since the anti-CRT antibodies had no effect on either SPARC or tenascin C-mediated focal adhesion disassembly.[14] TSP-mediated focal adhesion disassembly through CRT is not limited to BAE cells, since anti-CRT antibodies also blocks hep I-mediated focal adhesion disassembly in a uterine smooth muscle cell line (ELT-3) and in bovine embryonic fibroblasts. These cells also stain for cell surface CRT.[14]

The role of CRT as a receptor for TSP/hep I-induced focal adhesion disassembly was further demonstrated using mouse embryonic fibroblasts (MEF) from *crt*$^{-/-}$ and wild type embryos.[38] Since CRT knockout cells do not express CRT, they provide an excellent tool to investigate the role of CRT on TSP and hep I-induced focal adhesion disassembly. These studies showed that as expected, wild type cells were responsive to TSP and hep I-mediated focal adhesion disassembly. In contrast, TSP or hep I did not induce-focal adhesion disassembly in CRT null cells, although they still responded to tenascin-C and SPARC stimulated focal adhesion disassembly. Responsiveness to TSP/hep I was restored by stable transfection of CRT null cells with CRT. In addition, simple incubation of CRT null cells with GST-CRT rescued responsiveness to TSP/hep I.[57] These latter experiments indicate that CRT bound to the cell surface is apparently sufficient to mediate TSP signaling of focal adhesion disassembly, since under these conditions CRT is only restored to the cell surface and not to the ER. Interestingly, CRT null fibroblasts have somewhat larger focal adhesions as viewed by interference reflection microscopy than do wild type fibroblasts, perhaps indicating that endogenous TSP-CRT signal some basal level of adhesive modulation.

Studies were done to show that TSP interacts with purified CRT protein. Immunoprecipitation experiments using recombinantly expressed glutathione S-transferase-CRT (GST-CRT) and purified TSP from platelets showed that there is a specific interaction between CRT and TSP. CRT recognizes the hep I sequence of TSP, since TSP/CRT complex formation can be blocked specifically by hep I peptide.[14] TSP/CRT complex formation is enhanced in buffers-containing physiologic concentrations of calcium and zinc, suggesting that this interaction is modulated by divalent ions.[57]

In recent studies, we identified the binding site for TSP in CRT. The N-domain is required for TSP binding and focal adhesion disassembly. Using a series of N domain truncation mutants and overlapping peptides, we identified an 18 amino acid sequence as the putative TSP binding site in CRT: amino acids 19-36 (RWIESKHKSDFGKFVLSS). Synthetic peptides containing this sequence block TSP and hep I-mediated focal adhesion disassembly identical to that of CRT. We also observed that TSP binding to the N-domain is enhanced in the presence of physiologic levels of calcium (2mM), but still detectable at subphysiologic levels. Zinc similarly enhanced binding to the N-domain.[57] However, the 18 amino acid sequence in CRT

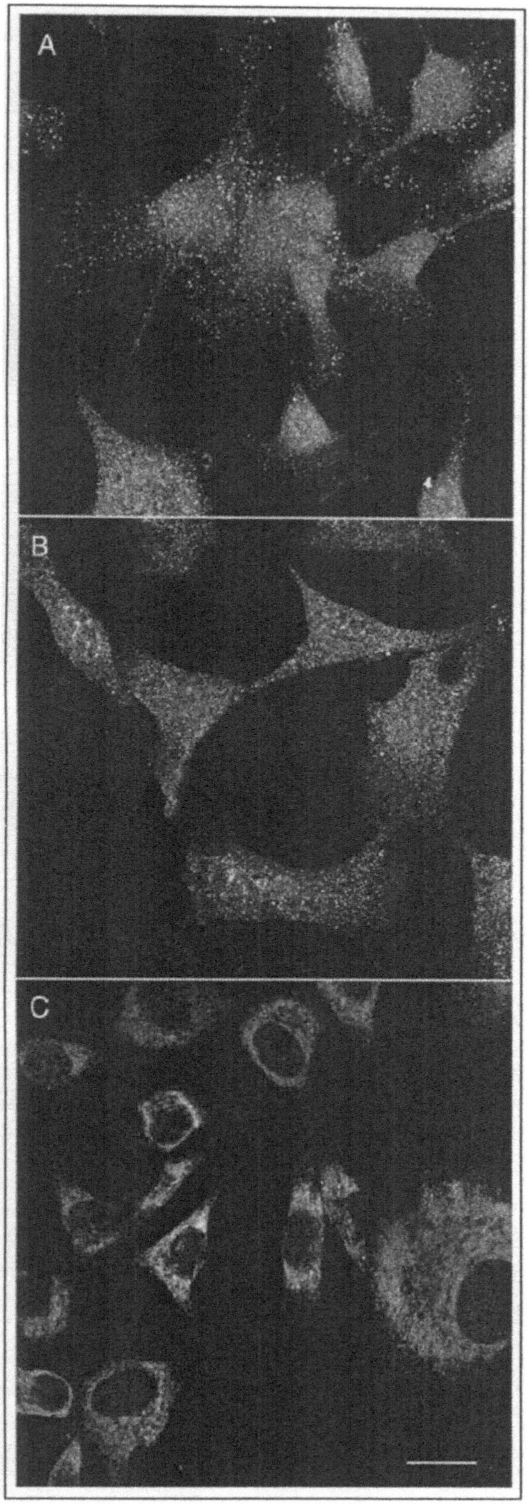

Figure 18.2. Immunofluorescence of cell surface CRT by confocal microscopy. Non-permeabilized BAE cells were fixed in 3% paraformaldehyde for 10 minutes and processed for immunofluorescence with anti-calreticulin antiserum (A) or rabbit anti-N-terminal calreticulin IgG (B). Staining of non-permeabilized cells was compared with staining of fixed cells permeabilized with 0.1% Triton X-100 with rabbit anti-calreticulin (C). Cells were viewed using laser confocal microscopy. Bar = 20 μm. Reprinted with permission from: Goicoechea S, Orr AW, Pallero MA, Eggleton P, Murphy-Ullrich JE 2000; 275:36358-36368. " 2000 The American Society for Biochemistry and Molecular Biology, Inc.

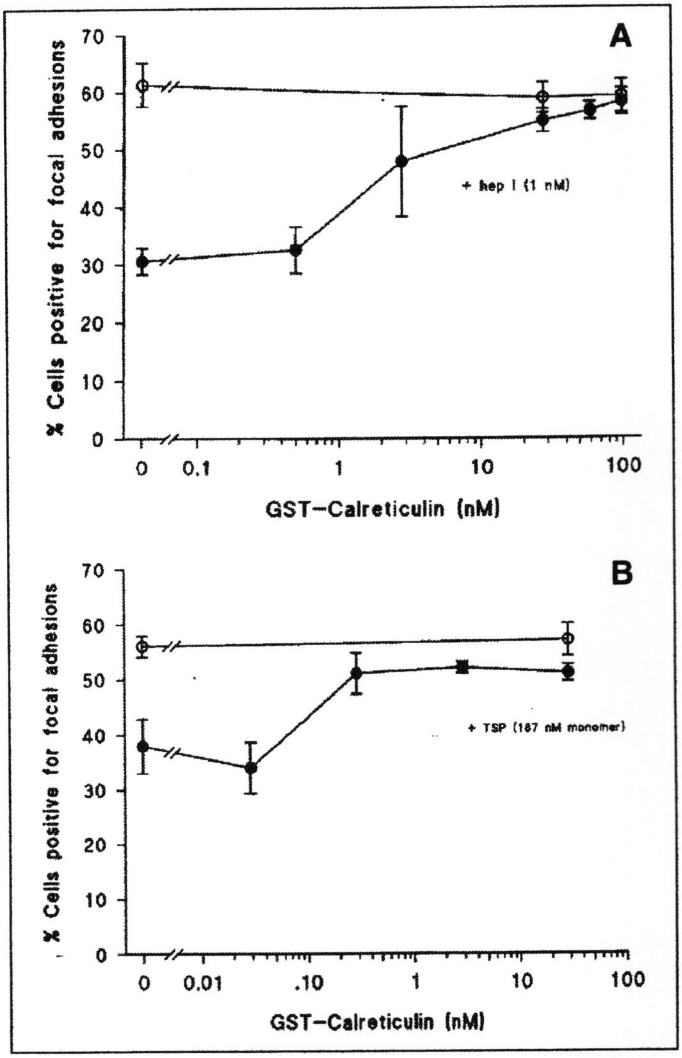

Figure 18.3. Soluble GST-CRT blocks TSP and hep I-mediated focal adhesion disassembly. BAE cells were grown on coverslips until near confluence. Hep I (1 nM) (A) or TSP (167 nM) (B) were incubated in the absence or the presence of increasing concentrations of GST-CRT (closed symbols) for 20 minutes before addition to washed BAE cells. Cells were then incubated with these proteins for 1 hour at 37°C, fixed and scored for the number of cells positive for focal adhesions as determined by IRM. Replicate coverslips were also incubated with increasing concentrations of GST-CRT (open symbols) in the absence of hep I (A) or TSP (B). Results are expressed as the mean percent of cells positive for focal adhesions ± S.D., n = 3. Reprinted with permission from: Goicoechea S, Orr AW, Pallero MA, Eggleton P, Murphy-Ullrich JE 2000; 275:36358-36368. " 2000 The American Society for Biochemistry and Molecular Biology, Inc.

involved in TSP binding has no apparent Ca^{2+} binding sites. Although a Ca^{2+}-binding site has not been identified in the N-terminal domain of CRT, it is possible that Ca^{2+} somehow affects the conformation/accessibility of the TSP binding sequence.

Signaling of CRT/TSP Focal Adhesion Disassembly

Initial studies showed that basal levels of cyclic GMP-dependent protein kinase (PKG) activity are necessary for focal adhesion disassembly as selective inhibitors of PKG block this process and late passage rat aortic smooth muscle cells that are deficient in PKG do not respond to hepI/CRT stimulation.[39] Further studies showed that TSP/hep I binding to CRT activates phosphoinositide 3 kinase (PI3K) and that activation of PI3K is necessary for focal adhesion disassembly.[14,37] CRT is important for transmission of this signal as *crt*$^{-/-}$ mouse embryonic fibroblasts are unable to activate PI3K in response to hep I stimulation. We also observed that TSP and hep I-induced focal adhesion disassembly involves ERK phosphorylation.[58] Hep I stimulates a transient increase in ERK phosphorylation and focal adhesion disassembly is blocked by a MEK inhibitor. Again, CRT is necessary for this signaling as CRT null fibroblasts do not phosphorylated ERK in response to TSP/hep I stimulation.

CRT Signaling Through Heterotrimeric G Proteins

Hep I-mediated focal adhesion disassembly is sensitive to inhibition by pertussis toxin, suggesting the involvement of heterotrimeric $G_{i\alpha}$ proteins. Recent data suggest that $G_{i\alpha 2}$ and $G_{\beta\gamma}$ mediate focal adhesion disassembly through TSP/CRT signaling.[58] Pertussis toxin blocks downstream signaling events such as PI3K activation and ERK phosphorylation that are induced by TSP/hep I binding. Similarly, CRT, which is peripherally associated with membranes of activated neutrophils, has been reported to signal in a pertussis toxin-sensitive manner, raising the possibility that CRT-heterotrimeric G protein signaling represents a general mechanism for cell surface CRT signaling.[13]

Our data show that CRT signals TSP/hep I-mediated focal adhesion disassembly from the peripheral membrane to the inside of the cell. However, it is still unclear as to how a peripheral protein can transmit intracellular signals. We know that in the absence of stimulation with TSP/hep I, CRT binding to either CRT-expressing BAE cells or to CRT-null mouse embryonic fibroblasts is not sufficient to signal focal adhesion disassembly. This suggests that expression of CRT at the cell surface is not in itself sufficient to signal. Furthermore, these data support the idea that interactions with TSP are necessary to "activate" CRT so that it can signal. This "activation" might involve TSP-induced changes in CRT conformation, resulting in CRT association with a transmembrane protein that then can act as a co-receptor and transmit signals (Fig. 18.4).

We now have evidence that the transmembrane protein, LDL-receptor-related protein (LRP), can act as a co-receptor with CRT to mediate TSP/hep I stimulated focal adhesion disassembly.[59] LRP is a known ligand of both CRT and TSP. Previous reports show that LRP (CD91) acts as a receptor for heat shock proteins, including CRT.[40] More recently it was shown that CRT on the surface of macrophages signals apoptotic cell ingestion through association with CD91.[41] Our data show that LRP on BAE cells co-precipitates with CRT, that CRT binds LRP, and that LRP antagonists block hep I stimulated focal adhesion disassembly. TSP/hep I binding to CRT might trigger a ligand-dependent association with LRP or activation of a pre-existing complex of CRT and LRP to signal focal adhesion disassembly (Fig. 18.4). It is also worth mentioning that LRP has also been shown to be involved in pertussis toxin-sensitive G protein signaling and association with G protein subunits.[42,43] In addition, α_2-macroglobulin binding to LRP on peritoneal macrophages activate PI3K and a 2-3-fold increase in PIP_3 levels.[44] These data are consistent with our hypothesis that TSP-CRT-binding to LRP can trigger both pertussis toxin-sensitive G protein-mediated signaling of focal adhesion disassembly.

Physiologic Significance of Cell Surface Calreticulin

Calreticulin and Cell Adhesion

Our work identifying a role for cell surface CRT as mediator of de-adhesive changes differs from data showing that ER or cytoplasmic CRT promotes stable cell adhesion.[45-50] CRT in the

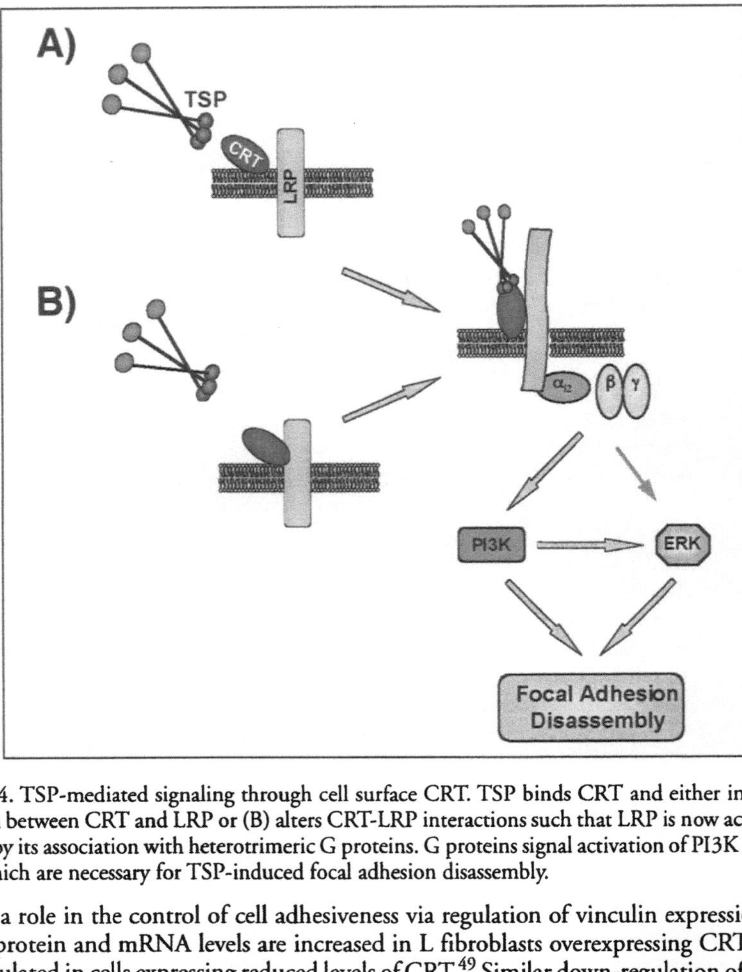

Figure 18.4. TSP-mediated signaling through cell surface CRT. TSP binds CRT and either induces (A) association between CRT and LRP or (B) alters CRT-LRP interactions such that LRP is now activated, as indicated by its association with heterotrimeric G proteins. G proteins signal activation of PI3K and ERK, both of which are necessary for TSP-induced focal adhesion disassembly.

ER plays a role in the control of cell adhesiveness via regulation of vinculin expression. Both vinculin protein and mRNA levels are increased in L fibroblasts overexpressing CRT and are down-regulated in cells expressing reduced levels of CRT.[49] Similar down-regulation of vinculin expression is observed in epithelial cells with a diminished level of CRT.[46] This coincides with an increase of total cellular phosphotyrosine, suggesting that the effects of CRT on cell adhesiveness may involve modulation of protein tyrosine kinase or phosphates activities, which can affect the stability of focal contacts.[46] It has also been shown that overexpression of CRT correlates with increased adhesiveness in transformed fibroblasts and decreased tyrosine phosphorylation of cellular proteins including β-catenin, a component of cadherin-mediated adhesion complexes and part of the Wnt signaling pathway. This signaling occurs via activation of gene expression and alteration of tyrosine phosphorylation.[8] On the other hand, CRT associates transiently with the cytoplasmic domains of integrin α subunits during spreading and this interaction can influence integrin-mediated cell adhesion to extracellular matrix.[47,50] CRT-deficient embryonic stem cells have impaired integrin-mediated adhesion to fibronectin, although integrin expression is unaltered. Taken together, these results suggest that intracellular CRT promotes stable cell adhesion. This is in contrast to our studies in which CRT on the cell surface is involved in destabilizing cytoskeletal organization and cell adhesion. However, there are examples of signaling mediators or cell adhesion molecules having differential effects on cell adhesion depending on cell type, the initial adhesive state of the cell, and the particular milieu of receptors and matrix molecules.[32] Nonetheless, our work is the first to show a destabilizing effect of CRT on cell adhesion. It remains to be determined as to how the relative adhesive and anti-adhesive functions of this protein are regulated under various physiologic circumstances.

Calreticulin and Apoptosis

CRT also modulates cell survival. CRT-deficient cardiac myocytes show increased apoptosis suggesting that CRT protects against apoptotic stimuli, although CRT failed to protect embryonic fibroblasts under endoplasmic reticulum stress conditions.[6] In several cell types, including prostate cancer cells and neurolastoma-glioma hybrids, calreticulin also protects against pro-apoptotic factors, including calcium toxicity induced by ionophore treatment.[51-54] Induction of apoptosis by geranylgeraniol in HL-60 cells is preceded by inhibition of CRT expression, suggesting that down-regulation of the protective effect of CRT is necessary for engagement of the apoptotic machinery.[55] In contrast, overexpression of CRT increased the sensitivity of HeLa cells to both thapsigargin- and staurosporine-induced apoptosis: this is consistent with the finding that CRT-deficient cells were resistant to apoptosis, having decreased caspase 3 and cytochrome c release.[38] These latter results suggest that CRT is part of the cell death pathway. These conflicting observations may reflect differing mechanisms of triggering cell death. In the ER, CRT may attenuate calcium toxicity by buffering luminal calcium levels. Whether CRT expressed at the cell surface also modulates cell death pathways remains to be determined: it is interesting to speculate that its enhanced expression during cell stress is consistent with a role in preventing cell death.

Calreticulin and Cell Migration

The heparin-binding domain of TSP stimulates endothelial cell migration.[25,26,56] Consistent with these findings, recent data from our lab show that TSP and hep I also stimulate cell migration.[60] Although little is known about involvement of CRT in cell migration, cell surface CRT, acting as a mediator of TSP signaling, can mediate cell migration. Increased cell locomotion is consistent with the nature of the cytoskeletal changes induced by TSP signaling through CRT. Further studies will continue to examine the role of CRT in modulating cell motility.

Summary and Significance

TSP binding to a cell surface form of CRT triggers reorganization of the cytoskeleton with restructuring of focal adhesion plaques and unbundling of actin stress fibers while maintaining the cells in a spread, extended morphology. This adhesive state we have termed "intermediate" cell adhesion. TSP-CRT interactions stimulate activation of PI3K, ERK phosphorylation and focal adhesion disassembly. These signals require the action of pertussis toxin-sensitive heterotrimeric G proteins. The biological significance of TSP stimulation of intermediate cell adhesion through CRT signaling is not known. There is little information in the literature regarding this adhesive state; yet, cell adhesiveness is a dynamic and reversible process. Both TSP and CRT are proteins that are up regulated in response to cellular stress and injury. It is interesting to speculate that TSP signaling of the intermediate adhesive state through binding to CRT enhances cell survival during times when a cell down-regulates its adhesiveness in order to facilitate motility. In contrast, loss of cell adhesiveness by disrupting integrin-matrix interactions results in apoptotic cell death. Furthermore, cells could be prompted by alterations in cytoskeletal forces on the nucleus to increase expression of proteins that are involved in responses to injury. The intermediate adhesive state could thus be a fundamental mechanism whereby a cell can do what it must to engage in repair processes without it becoming part of a suicide mission. This has significance not only for tissue remodeling during morphogenesis and wound healing processes, but also for chronic disease processes such as atherosclerosis and arthritis where the endothelium must respond to chronic stress. It will be interesting to assess the role of CRT signals from the cell surface in these physiologic processes and disease states.

Acknowledgments

The essential contributions of Manual Antonio Pallero, Wayne Orr, and Claudio Pedraza to these studies are gratefully acknowledged. The authors wish to thank Drs. Paul Eggleton, (University of Oxford) and Marek Michalak, (University of Alberta) for the generous supply of reagents, helpful advice, and supportive collaborations. This work was supported by NIH grant HL44575 to JMU, an Established Investigator Award from the AHA, Special Awardee in Thrombosis from Genentech, to JMU and an AHA post-doctoral fellowship (Grant # 0020534B) to SMG.

References

1. Johnson S, Michalak M, Opas M et al. The ins and outs of calreticulin: from the ER lumen to the extracellular space. Trends Cell Biol 2001; 11(3):122-9.
2. Michalak M, Corbett EF, Mesaeli N et al. Calreticulin: one protein, one gene, many functions. Biochem J 1999; 344(Pt 2):281-92.
3. Camacho P, Lechleiter JD. Calreticulin inhibits repetitive intracellular Ca2+ waves. Cell 1995; 82(5):765-71.
4. Bastianutto C, Clementi E, Codazzi F et al. Overexpression of calreticulin increases the Ca2+ capacity of rapidly exchanging Ca2+ stores and reveals aspects of their lumenal microenvironment and function. J Cell Biol 1995; 130(4):847-55.
5. Mesaeli N, Nakamura K, Zvaritch E et al. Calreticulin is essential for cardiac development. J Cell Biol 1999; 144(5):857-68.
6. Rauch F, Prud'homme J, Arabian A et al. Heart, brain, and body wall defects in mice lacking calreticulin. Exp Cell Res 2000; 256(1):105-11.
7. Pike SE, Yao L, Jones KD et al. Vasostatin, a calreticulin fragment, inhibits angiogenesis and suppresses tumor growth. J Exp Med 1998; 188(12):2349-56.
8. Fadel MP, Szewczenko-Pawlikowski M, Leclerc P et al. Calreticulin affects beta-catenin-associated pathways. J Biol Chem 2001; 276(29):27083-9.
9. Arosa FA, de Jesus O, Porto G et al. Calreticulin is expressed on the cell surface of activated human peripheral blood T lymphocytes in association with major histocompatibility complex class I molecules. J Biol Chem 1999; 274(24):16917-22.
10. Gray AJ, Park PW, Broekelmann TJ et al. The mitogenic effects of the B beta chain of fibrinogen are mediated through cell surface calreticulin. J Biol Chem 1995; 270(44):26602-6.
11. White TK, Zhu Q, Tanzer ML. Cell surface calreticulin is a putative mannoside lectin which triggers mouse melanoma cell spreading. J Biol Chem 1995; 270(27):15926-9.
12. Guo WX, Ghebrehiwet B, Weksler B et al. Up-regulation of endothelial cell binding proteins/ receptors for complement component C1q by inflammatory cytokines. J Lab Clin Med 1999; 133(6):541-50.
13. Cho JH, Homma K, Kanegasaki S et al. Activation of human neutrophils by a synthetic anti-microbial peptide, KLKLLLLLKLK-NH2, via cell surface calreticulin. Eur J Biochem 1999; 266(3):878-85.
14. Goicoechea S, Orr AW, Pallero MA et al. Thrombospondin mediates focal adhesion disassembly through interactions with cell surface calreticulin. J Biol Chem 2000; 275(46):36358-68.
15. Kuwabara K, Pinsky DJ, Schmidt AM et al. Calreticulin, an antithrombotic agent which binds to vitamin K- dependent coagulation factors, stimulates endothelial nitric oxide production, and limits thrombosis in canine coronary arteries. J Biol Chem 1995; 270(14):8179-87.
16. van den Berg RH, Faber-Krol M, van Es LA et al. Regulation of the function of the first component of complement by human C1q receptor. Eur J Immunol 1995; 25:2206-2210.
17. Xiao G, Chung TF, Pyun HY et al. KDEL proteins are found on the surface of NG108-15 cells. Brain Res Mol Brain Res 1999; 72(2):121-8.
18. Xiao G, Chung TF, Fine RE et al. Calreticulin is transported to the surface of NG108-15 cells where it forms surface patches and is partially degraded in an acidic compartment. J Neurosci Res 1999; 58(5):652-62.
19. Murphy-Ullrich JE, Mosher DF. Interactions of thrombospondin with endothelial cells: receptor-mediated binding and degradation. J Cell Biol 1987; 105(4):1603-11.
20. Mikhailenko I, Krylov D, Argraves KM et al. Cellular internalization and degradation of thrombospondin-1 is mediated by the amino-terminal heparin binding domain (HBD). High affinity interaction of dimeric HBD with the low density lipoprotein receptor-related protein. J Biol Chem 1997; 272(10):6784-91.

21. Bornstein P. Thrombospondins as matricellular modulators of cell function. J Clin Invest 2001; 107(8):929-34.
22. Baenzinger NL, Brodie GN, Majerus PW. A thrombin-sensitive protein of human platelet membranes. Proc Natl Acad Sci USA 1978; 68:240-249.
23. Chandrasekaran L, He CZ, Al-Barazi H et al. Cell contact-dependent activation of alpha3beta1 integrin modulates endothelial cell responses to thrombospondin-1. Mol Biol Cell 2000; 11(9):2885-900.
24. Murphy-Ullrich JE, Poczatek M. Activation of latent TGF-beta by thrombospondin-1: mechanisms and physiology. Cytokine Growth Factor Rev 2000; 11(1-2):59-69.
25. Taraboletti G, Roberts DD, Liotta LA. Thrombospondin-induced tumor cell migration: haptotaxis and chemotaxis are mediated by different molecular domains. J Cell Biol 1987; 105(5):2409-15.
26. Taraboletti G, Morbidelli L, Donnini S et al. The heparin binding 25 kDa fragment of thrombospondin-1 promotes angiogenesis and modulates gelatinase and TIMP-2 production in endothelial cells. Faseb J 2000; 14(12):1674-6.
27. Jimenez B, Volpert OV, Crawford SE et al. Signals leading to apoptosis-dependent inhibition of neovascularization by thrombospondin-1. Nat Med 2000; 6(1):41-8.
28. Teti A. Regulation of cellular functions by extracellular matrix. J Am Soc Nephrol 1992; 2(10 Suppl):S83-7.
29. Couchman JR, Rees DA. The behaviour of fibroblasts migrating from chick heart explants: changes in adhesion, locomotion and growth, and in the distribution of actomyosin and fibronectin. J Cell Sci 1979; 39:149-65.
30. Sage EH, Bornstein P. Extracellular proteins that modulate cell-matrix interactions. SPARC, tenascin, and thrombospondin. J Biol Chem 1991; 266(23):14831-4.
31. Bornstein P. Diversity of function is inherent in matricellular proteins: an appraisal of thrombospondin 1. J Cell Biol 1995; 130(3):503-6.
32. Greenwood JA, Murphy-Ullrich JE. Signaling of de-adhesion in cellular regulation and motility. Microsc Res Tech 1998; 43(5):420-32.
33. Yan Q, Sage EH. SPARC, a matricellular glycoprotein with important biological functions. J Histochem Cytochem 1999; 47(12):1495-506.
34. Adams JC, Lawler J. Diverse mechanisms for cell attachment to platelet thrombospondin. J Cell Sci 1993; 104(Pt 4):1061-71.
35. Murphy-Ullrich JE, Hook M. Thrombospondin modulates focal adhesions in endothelial cells. J Cell Biol 1989; 109(3):1309-19.
36. Murphy-Ullrich JE, Gurusiddappa S, Frazier WA et al. Heparin-binding peptides from thrombospondins 1 and 2 contain focal adhesion-labilizing activity. J Biol Chem 1993; 268(35):26784-9.
37. Greenwood JA, Pallero MA, Theibert AB et al. Thrombospondin signaling of focal adhesion disassembly requires activation of phosphoinositide 3-kinase. J Biol Chem 1998; 273(3):1755-63.
38. Nakamura K, Bossy-Wetzel E, Burns K et al. Changes in endoplasmic reticulum luminal environment affect cell sensitivity to apoptosis. J Cell Biol 2000; 150(4):731-40.
39. Murphy-Ullrich JE, Pallero MA, Boerth N et al. Cyclic GMP-dependent protein kinase is required for thrombospondin and tenascin mediated focal adhesion disassembly. J Cell Sci 1996; 109(Pt 10):2499-508.
40. Basu S, Binder RJ, Ramalingam T et al. CD91 is a common receptor for heat shock proteins gp96, hsp90, hsp70, and calreticulin. Immunity 2001; 14(3):303-13.
41. Ogden CA, deCathelineau A, Hoffmann PR et al. C1q and mannose binding lectin engagement of cell surface calreticulin and CD91 initiates macropinocytosis and uptake of apoptotic cells. J Exp Med 2001; 194(6):781-95.
42. Goretzki L, Mueller BM. Low-density-lipoprotein-receptor-related protein (LRP) interacts with a GTP-binding protein. Biochem J 1998; 336(Pt 2):381-6.
43. Wang XS, Gruenstein E. Rapid elevation of neuronal cytoplasmic calcium by apolipoprotein E peptide. J Cell Physiol 1997; 173(1):73-83.
44. Misra UK, Pizzo SV. Binding of receptor-recognized forms of alpha2-macroglobulin to the alpha2-macroglobulin signaling receptor activates phosphatidylinositol 3-kinase. J Biol Chem 1998; 273(22):13399-402.
45. Zhu Q, Zelinka P, White T et al. Calreticulin-integrin bidirectional signaling complex. Biochem Biophys Res Commun 1997; 232(2):354-8.
46. Fadel MP, Dziak E, Lo CM et al. Calreticulin affects focal contact-dependent but not close contact- dependent cell-substratum adhesion. J Biol Chem 1999; 274(21):15085-94.
47. Coppolino MG, Woodside MJ, Demaurex N et al. Calreticulin is essential for integrin-mediated calcium signalling and cell adhesion. Nature 1997; 386(6627):843-7.

48. Rojiani MV, Finlay BB, Gray V et al. In vitro interaction of a polypeptide homologous to human Ro/SS-A antigen (calreticulin) with a highly conserved amino acid sequence in the cytoplasmic domain of integrin alpha subunits. Biochemistry 1991; 30(41):9859-66.
49. Opas M, Szewczenko-Pawlikowski M, Jass GK et al. Calreticulin modulates cell adhesiveness via regulation of vinculin expression. J Cell Biol 1996; 135(6 Pt 2):1913-23.
50. Leung-Hagesteijn CY, Milankov K, Michalak M et al. Cell attachment to extracellular matrix substrates is inhibited upon downregulation of expression of calreticulin, an intracellular integrin alpha-subunit-binding protein. J Cell Sci 1994; 107(Pt 3):589-600.
51. Liu N, Fine RE, Simons E et al. Decreasing calreticulin expression lowers the Ca2+ response to bradykinin and increases sensitivity to ionomycin in NG-108-15 cells. J Biol Chem 1994; 269(46):28635-9.
52. Zhu N, Wang Z. Calreticulin expression is associated with androgen regulation of the sensitivity to calcium ionophore-induced apoptosis in LNCaP prostate cancer cells. Cancer Res 1999; 59(8):1896-902.
53. Liu H, Miller E, van de Water B et al. Endoplasmic reticulum stress proteins block oxidant-induced Ca2+ increases and cell death. J Biol Chem 1998; 273(21):12858-62.
54. Johnson RJ, Liu N, Shanmugaratnam J et al. Increased calreticulin stability in differentiated NG-108-15 cells correlates with resistance to apoptosis induced by antisense treatment. Brain Res Mol Brain Res 1998; 53(1-2):104-11.
55. Nakajo S, Okamoto M, Masuda Y et al. Geranylgeraniol causes a decrease in levels of calreticulin and tyrosine phosphorylation of a 36-kDa protein prior to the appearance of apoptotic features in HL-60 cells. Biochem Biophys Res Commun 1996; 226(3):741-5.
56. Taraboletti G, Roberts D, Liotta LA et al. Platelet thrombospondin modulates endothelial cell adhesion, motility, and growth: a potential angiogenesis regulatory factor. J Cell Biol 1990; 111(2):765-72.
57. Goicoechea S, Pallero MA, Eggleton P, Michalak M, Murhpy-Ullrich JE. The anti-adhesive activity of thrombospondin is mediated by the N-terminal domain of cell surface calreticulin. J Biol Chem 2002 Oct 4;277(40):37219-28.
58. Orr AW, Pallero MA, Murphy-Ullrich JE. Thrombospondin stimulates focal adhesion disassembly through Gi- and phosphoinositide 3-kinase-dependent ERK activation. J Biol Chem 2002 June 7;277(23):20453-60.
59. Orr AW, Pedraza CE, Palleo MA, Elzie CA, Goicoechea S, Strickland DK, Murphy-Ullrich JE. Low density lipoprotein receptor-related protein is a calreticulin co-receptor that signals focal adhesion disassembly. J Cell Bio., in press.
60. Orr AW, Elzie CA, Kucik DF, Murphy-Ullrich JE. Thrombospondin signaling through the calreticulin/LDL receptor-related protin co-complex stimulates random and directed cell migration. J Cell Science, in press.

CHAPTER 19

Calreticulin Regulation of Lung Endothelial NOS Activity

Jawaharlal M. Patel, Jianliang Zhang, Yong D. Li and Edward R. Block

Abstract

Increased synthesis of a multifunctional calcium binding protein calreticulin has been reported under diverse physiologic and pathophysiologic conditions in various tissues including stimulation of vascular endothelium by angiotensin IV (Ang-IV), a metabolic product of Ang-II. Ang-IV-mediated early and sustained activation of lung endothelial cell nitric oxide synthase (eNOS) is mediated through increased mobilization of intracellular calcium and by increased expression of calreticulin. Immunoprecipitation and confocal imaging studies revealed that eNOS and calreticulin are co-localized in Ang-IV-stimulated lung endothelial cells. Catalytic activity of purified eNOS in the absence of calmodulin was increased in a concentration-dependent fashion by calreticulin. The studies monitoring the effect of calreticulin on the rate of electron transfer from the reductase to the oxygenase domain of eNOS revealed that the calreticulin/eNOS interaction promotes electron transfer and mimics eNOS activation in the absence of exogenous calmodulin and enhances electron transfer and the catalytic activity of eNOS in the presence of calmodulin. Thus, calreticulin/eNOS protein:protein interaction enhances the rate of electron transfer, a critical event in the regulation of the catalytic activity of eNOS.

Introduction

Vascular endothelium is a metabolically active tissue that regulates pulmonary and systemic vascular tone through metabolism and/or generation and release of vasoactive mediators such as angiotensin peptides and nitric oxide (NO).[1,2] For example, lung endothelial cells metabolize angiotensin I (Ang- I) → Ang-II → Ang-III → Ang-IV by catalytic action of Ang-converting enzyme, aminopeptidase A, and aminopeptidase M, respectively.[3] A number of recent studies have identified and characterized the presence of Ang-IV-specific receptors in a variety of tissues including lung.[4-8] Receptor-mediated activation of signal transduction pathways are know to result in direct and/or G protein-coupled activation of phospholipases C and D (PLC/PLD), leading to a rapid increase in intracellular calcium release. Increased intracellular calcium is regulated by multiple mechanisms. One of the most ubiquitous of these mechanisms involves the PLC-phosphatidylinositol 3,4,5-triphosphate (PIP3) pathway, which releases calcium from intracellular stores, namely the endoplasmic reticulum (ER).[9-11] These signaling mechanisms are associated with activation of several calcium-dependent enzymes including eNOS.[12,13] Agonist-mediated intracellular calcium mobilization can also play a critical role in modulation of cell function such as the regulation of ER resident calcium binding proteins including calreticulin.[14-16] Metabolic pathways of Ang and Ang-IV receptor-mediated activation of signaling events leading to activation of eNOS and expression of calreticulin are shown in Figure 19.1.

Calreticulin, 2nd ed., edited by Paul Eggleton and Marek Michalak.

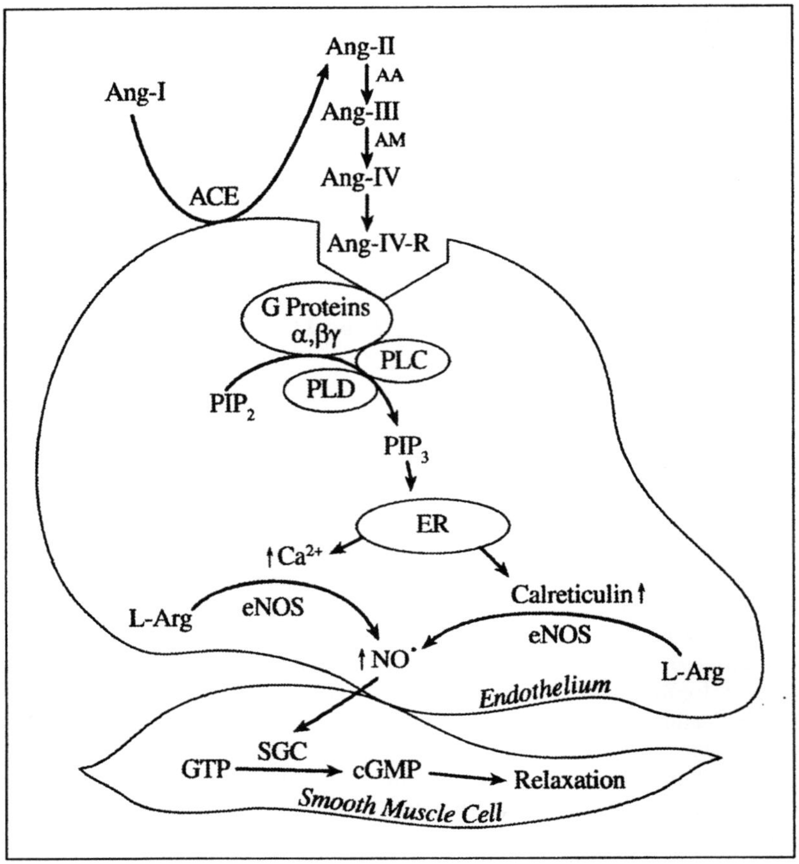

Figure 19.1. Angiotensin (Ang) peptide metabolism and Ang-IV receptor-mediated signaling pathway leading to activation of lung endothelial cell eNOS, calreticulin expression, and nitric oxide (NO)/guanosine 3'5'-cyclic monophosphate (cGMP)-linked vasorelaxation. Phospholipase c, PLC; phospholipase D PLD; phosphatidylinositol 4,5 diphosphate (PIP2) or 3,4,5 triphosphate (PIP3); endoplasmic reticulum, ER; guanosine 5'triphosphate, GTP; soluble guanylate cyclase, SGC, L-arginine, L-Arg.

The catalytic activity of eNOS is calcium- and calmodulin-dependent and is transiently activated by agonist-mediated signaling pathways that increase intracellular calcium mobilization.[17-20] The catalytic activity of eNOS is regulated by multiple post-transcriptional mechanisms including protein:protein interaction. The present review focuses on molecular events involved in Ang-IV-mediated regulation of the catalytic activity of eNOS in context with intracellular calcium mobilization, calreticulin expression, and calreticulin:eNOS protein:protein interaction in lung endothelial cells.

Biochemistry and Physiology of Ang-IV

The vascular endothelium processes a variety of biologically active substances in the circulation, including Ang-IV, a metabolic product of the renin-angiotensin system as shown below in animals and humans.[21-23]

Angiotensinogen NH_2-Asp-Arg-Val-Tyr-Ile-His-Pro-Phe-His-Leu-Val-Ile-His-COOH
_-- Renin
Ang-I NH_2-Asp-Arg-Val-Tyr-Ile-His-Pro-Phe-His-Leu-COOH
_-- ACE
Ang-II NH_2-Asp-Arg-Val-Tyr-Ile-His-Pro-Phe-COOH
_-------------------- Aminopeptidase A
Ang-III NH_2-Arg-Val-Tyr-Ile-His-Pro-Phe-COOH
_-------------------- Aminopeptidase M
Ang-IV NH_2-Val-Tyr-Ile-His-Pro-Phe-COOH

The presence of Ang-IV-specific receptor sites has been identified in several extra-pulmonary tissues including brain, heart, and kidney[24-26] as well as in cultured coronary microvascular, aortic, and lung endothelial cells.[3,7,8] Although the functional role of Ang-IV is not fully defined, it has been reported that Ang-IV increases blood flow in brain and kidney[27,28] and causes endothelium-dependent vasodilation of porcine pulmonary artery and rabbit brain arterioles.[3,29] In fact, Ang-IV has been shown to activate the lung endothelial cell isoform of nitric oxide (NO) synthase (eNOS) by a receptor-mediated pathway, leading to increases in NO release, production of guanosine 3',5'-cyclic monophosphate (cGMP), and NO-cGMP-mediated pulmonary artery vasodilation.[3,11] Agonist-mediated activation of signaling pathways that increase mobilization of intracellular calcium is known to transiently activate eNOS.[1] However, Ang-IV-stimulation of lung endothelial cells results in transient and sustained activation of eNOS.[3] Agonist-mediated intracellular calcium mobilization is also known to play a critical role as a second messenger in the regulation of a variety of functions, including cell proliferation, gene expression, protein expression, and protein:protein interaction.[14-16] Depletion of calcium from the ER under diverse physiologic and pathophysiologic conditions can facilitate a process that results in upregulation of a group of calcium binding proteins, including calreticulin.[11,30-37] Ang-IV-mediated transient and sustained activation of eNOS appear to be associated with intracellular calcium mobilization as well as with increased expression of the calcium binding protein calreticulin.

Calreticulin Expression and Function: Role of Cell Stimulation/Injury

Calreticulin is a 60-kDa, highly conserved ubiquitous calcium binding protein of the endoplasmic reticulum (ER) membrane with three distinct structural and functional domains (N-/P-/C-domains). The C- and P-domains consist of low and high affinity calcium binding sites, respectively[38-40] (Fig. 19.2). The N-terminal domain of calreticulin does not bind calcium, but it is believed to be one of the major sites of protein:protein interaction in diverse cell types that results in functional modulation of the binding proteins.[41-45] C-terminal KDEL (Lys-Asp-Glu-Leu) ER retention signal is responsible for retention of calreticulin in the ER lumen.[1,2] Despite the presence of the KDEL ER retention signal in the C-terminal, calreticulin is also localized in various subcellular compartments, including the cytoplasm, the nucleus, and the cell surface membrane,[46-49] and is also know to be secreted from various mammalian cells including vascular endothelium.[50-52] Agonist or chemical-induced depletion of calcium from the ER or sarcoplasmic reticulum, known intracellular calcium storage sites, can facilitate a process that results in increased synthesis of a group of ER resident proteins including calreticulin. For example, it is increasingly clear from a number of studies that up-regulation of calreticulin expression occurs in response to a variety of cellular stress/injury or stimuli including amino acid deprivation,[30,31] heat shock,[32,33] exposure to heavy metals[34], irradiation and viral infection,[35] cytotoxic T-cell stimulation,[36] inhibition of N-linked glucosylation by tunicamycin,[37] and depletion of ER calcium stores due to stimulation by agonists such as Ang-IV.[11,52] As such, the wide distribution and varying degree of expression of calreticulin under diverse pathophysiologic conditions provide unique features enabiling calreticulin to be associated with a broad array of cellular functions.

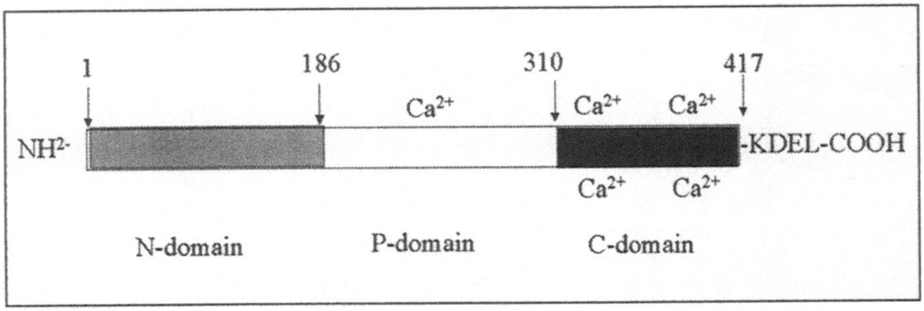

Figure 19.2. Structure of human calreticulin and its N-/P-/C-domains.

In the last 10 years, support for important physiological roles for calreticulin has been documented by numerous studies. For example, calreticulin has been recognized as a multifunctional protein involved in gene expression, receptor regulation, cell adhesion, autoimmunity, apoptosis, vascular remodeling and angiogenesis, endothelial cell NO release, viral RNA binding and replication, cytotoxic T cell as well as neutrophil activation, and modulation of signaling events leading to altered calcium homeostasis.[38,39,46,53-61] The role of calreticulin in the regulation of vascular endothelial-dependent responses such as NO release, remodeling, and angiogenesis are of particular interest. In this context, we focused on Ang-IV-mediated responses involving intracellular calcium mobilization, calreticulin expression, and NOS activation in lung vascular endothelial cells.

Structure, Function, and Regulation of eNOS Activity

NO is synthesized by mammalian cells by three major isoforms of NOS namely, constitutive endothelial and neuronal isoforms eNOS and nNOS, respectively, and an inducible isoform iNOS.[17,18] These isomers are members of a novel family of enzymes containing heme oxygenase and cytochrome P-450 reductase domains that catalyze the oxidative metabolism of cationic amino acid L-arginine to generate L-citrulline, NO, and H_2O.[19,20] The amino terminal oxygenase domain of eNOS protein contains bindings sites for heme, tetrahydrobiopterin, zinc, and L-arginine and the carboxy-terminal reductase domain consists of sites for NADPH, flavin adenine dinucleotide (FAD), flavin mononucleotide (FMN), and electron transfer control element (ETE) (Fig. 19.3).[17-20,62,63] These two domains are linked by a calcium/calmodulin binding region. The catalytic activity of eNOS is dependent on the rate of electron transfer from NADPH in the reductase domain to the oxygenase domain, a required process for the oxidative metabolism of L-arginine to generate NO.[62,63]

The catalytic activity of eNOS is regulated by transcriptional and post-transcriptional mechanisms. For example, shear stress as well as bioactive agents such as cyclosporine A, vascular endothelial growth factor, insulin, oxidized LDL, lysophosphatidylcholine, and transforming growth factor-_ are known to upregulate eNOS expression through modulation of putative cis-elements including GATA and Sp1 motifs and/or specific regulatory elements such as activator protein-1, nuclear factor-1 element, and sterol regulatory element in the eNOS gene.[13,65-69]

The catalytic activity of eNOS is also regulated by multiple post-transcriptional mechanisms involving a variety of factors including phosphorylation/dephosphorylation state, active site cysteine modulation in oxygenase and reductase domains, myristylation/palmitolation and subcellular localization, and by protein:protein interaction with calmodulin, caveolin, heat shock protein 90, and components of the cytoskeleton.[70-76,79] All of these known mechanisms involve a limited or transient, but not sustained, activation of eNOS through modulation of phosphorylation of critical amino acid residues or by translocation of eNOS within intracellular domains, as the Golgi, ER, and plasmalemmal caveolae, through calcium mobilization that promotes eNOS caveolin dissociation and calmodulin association.[76-79] Although calmodulin is an essential regulator of inter- and intra-domain electron transfer, the catalytic activity of

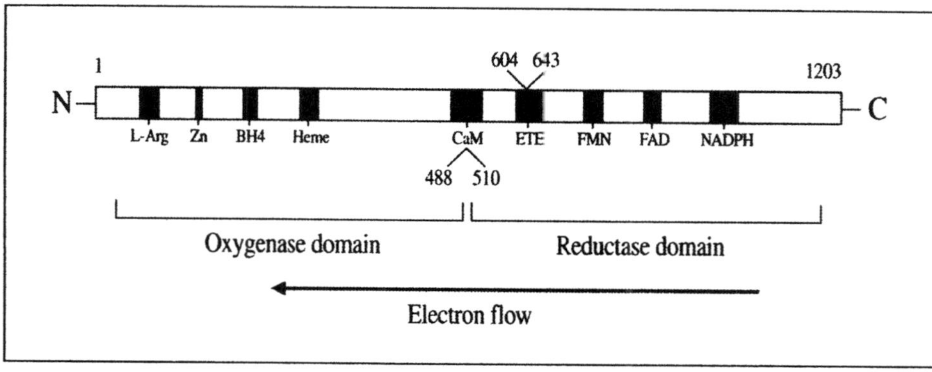

Figure 19.3. Oxygenase and reductase domains of eNOS: binding site or distinct region for calmodulin (CaM), electron transfer control element (ETE), flavin mononucleotide (FMN), flavin adenine dinucleotide (FAD), and NADPH in the reductase domain, and heme, tertahydrobiopterin (BH_4), zinc (Zn), and L-arginine (L-Arg) in the oxygenase domain.

eNOS can be transiently facilitated through increased intracellular calcium release (calcium-dependent) or at basal levels of calcium (calcium-independent pathways) by various stimuli including agonists, fluid shear stress, or growth factors.[69-71] A recent report indicates that mutation of serine 1179 to the negatively charged aspartate increases eNOS acitivity by increasing electron transfer at basal calcium levels in purified bovine

eNOS.[80] Thus, it is clear from these reports that the catalytic activity of eNOS is greatly influenced by critical events such as calcium modulation, phosphorylation of target sites, and calmodulin binding characteristics which in turn appear to regulate the rate of electron transfer which is controlled by ETE.

A number of recent reports using molecular biology approaches and purified NOS preparations have characterized the role of ETE present in the reductase domain of eNOS and nNOS but absent in iNOS for its putative autoinhibitory role in electron transfer and its control of catalytic activity of NOS isoforms.[62,63] The fact that deletion of ETE from human or bovine eNOS does not alter the structure or active center of eNOS but drastically changes its dependence on calcium or calmodulin binding with enhanced electron transfer and catalytic activity provides experimental evidence that ETE plays a critical role in the regulation of eNOS activity.[62,63] However, nothing is known about the mechanisms by which ETE regulates electron transfer and whether endogenously generated factors such as increased expression of calreticulin and modulation of ETE's function affect eNOS activity.

Ang-IV eNOS Activation: Link to Cellular Calcium and Calreticulin

As stated above the catalytic activity of eNOS is regulated by multiple mechanisms involving transcriptional and post-transcriptional processes. Ang-IV-mediated activation of eNOS is regulated by a post-transcriptional mechanism since the expression of eNOS protein in 4 to 12 hr Ang-IV-stimulated lung endothelial cells was comparable to that in control cells.[3] Endothelial cell release of NO is enhanced by receptor-mediated agonists such as acetylcholine, bradykinin, histamine, serotonin, and substance P via signal transduction-mediated transient activation of eNOS.[1] However, unlike other agonists described to date, Ang-IV-receptor-mediated activation of eNOS persists for up to 12 hr (Fig. 19.4A). Ang-IV-stimulation in the absence of extracellular calcium causes a rapid increase in intracellular calcium release in a dose-dependent manner which is significantly greater than the basal level of calcium.[11] This Ang-IV-mediated activation of eNOS is clearly dependent on intracellular calcium release since pre-incubation of cells with the calcium chelator BAPTA-AM blocked Ang-IV-mediated activation of eNOS (Fig. 19.4B).

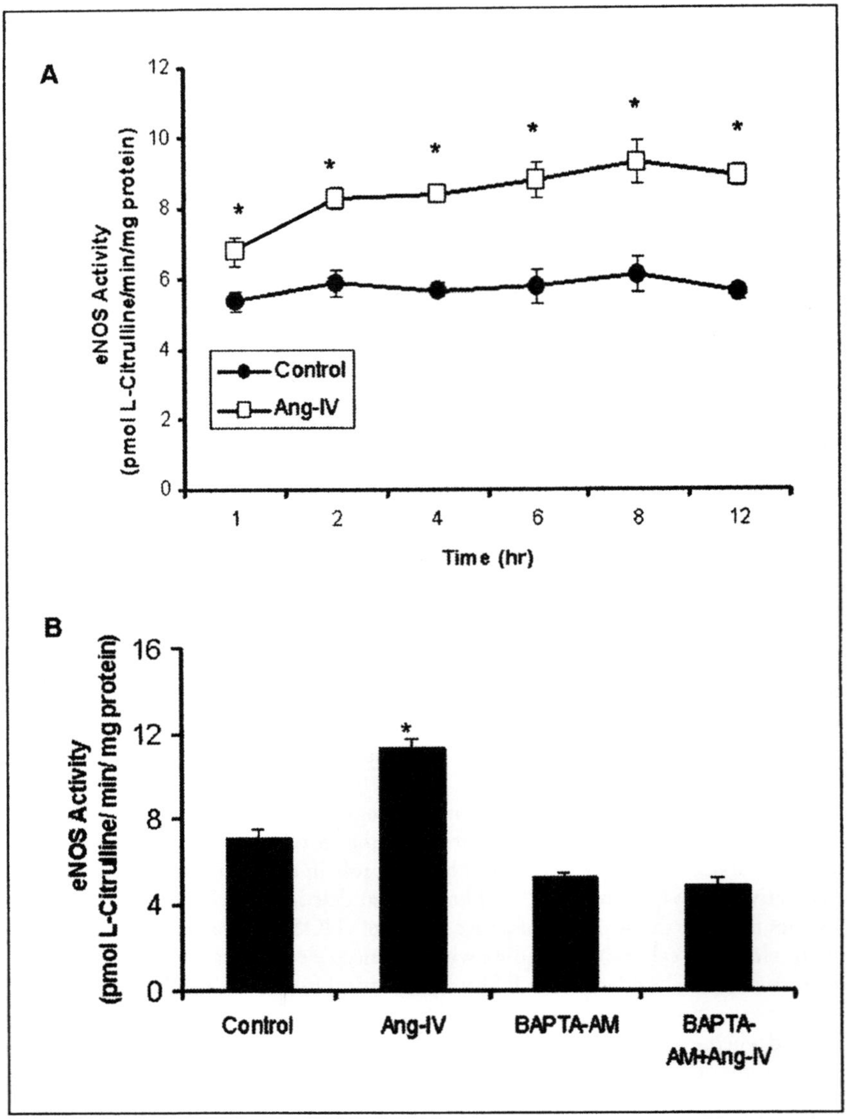

Figure 19.4. Time-dependent activation of eNOS by Ang-IV (panel A) and the effect of the intracellular calcium chelator BAPTA-AM on Ang-IV activated eNOS (panel B) in lung endothelial cells. Cell monolayers were incubated in RPMI 1640 medium alone (control) and with Ang-IV (1 μM) for 1 to 12 hr at 37° C. To examine the effects of the calcium chelator BAPTA-AM, cells were incubated in Tyrode solution without calcium with or without BAPTA-AM (50 μM) for 30 min and then incubated 2 hr at 37° C with or without Ang-IV (μM). Respective controls were incubated in Tyrode solution alone or Tyrode olution containg 50 μM BAPTA-AM under identical conditions. After incubation, total membrane fraction eNOS activity was measured by monitoring the metabolism of L-[^{3}H]-L-arginine to L-citrulline. Data represent mean ± SE; n=4-6 for each data point in both panels. * $p < 0.05$ vs control in each panel.

Although Ang-IV-stimulated transient activation of eNOS through intracellular mobilization of calcium is consistent with other agonist-mediated responses, Ang-IV-mediated sustained activation of eNOS is unique in the context of known post-transcriptional mechanisms regulating the catalytic activity of eNOS. Agonist and/or chemical-mediated depletion of ER

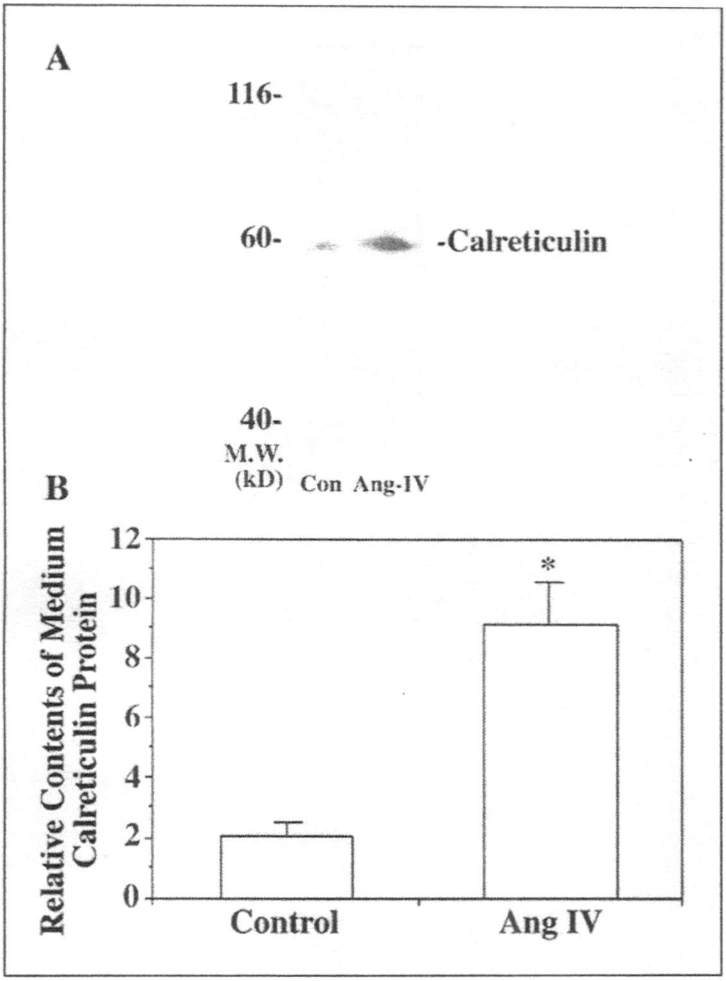

Figure 19.5. Ang-IV-stimulated expression of calreticulin increases its secretion into the extracellular medium. Lung endothelial cell monolayers in RPMI 1640 containing 1 μM Ang-IV or RPMI 1640 only (Con) were incubated for 2 hr at 37° C. After incubation, 60 ml of media were collected and concentrated, and calreticulin was indentified by immunoblot analysis. A: representative data drom 1 of 3 independent experiments. B: results of densitometric analysis of blots from 3 independent experiments (mean ± SE). * $p < 0.05$ vs control (Reprinted from reference #52 with permission from the American Physiological Society).

calcium stores have been linked with increased expression of ER resident calcium binding proteins.[30-37,52] In addition, Ang-IV has been reported to induce protein synthesis of plasminogen activator inhibitor 1 in bovine aortic endothelial cells[81] and calreticulin expression in porcine pulmonary artery endothelial cells.[52] Ang-IV has also been reported to increase DNA and RNA synthesis in mammalian cells including lung endothelial cells.[82-84] Ang-IV-mediated expression of calreticulin was significantly increased as early as 2 hr and remained elevated for 6 hr in lung endothelial cells.[52] As shown in Figure 19.5, increased levels of calreticulin were observed in the medium of cells stimulated with Ang-IV. Ang-IV-mediated intracellular calcium release is critical to the increased calreticulin expression as the calcium chelator BAPTA-AM blocked Ang-IV-induced expression of calreticulin in lung endothelial cells (Fig. 19.6). Thus,

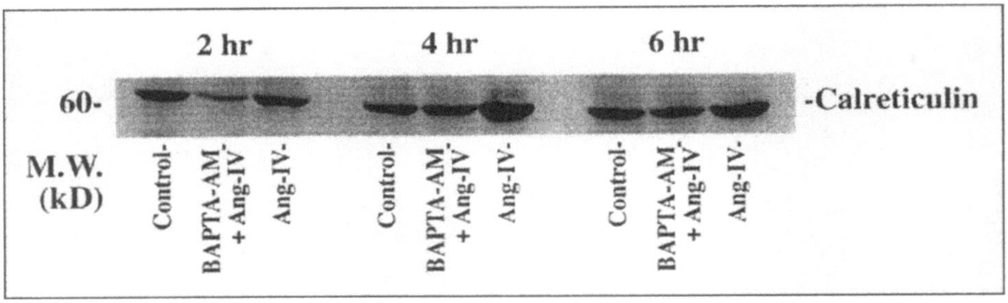

Figure 19.6. Effect of the intracellular calcium chelator BAPTA-AM on Ang-IV-mediated calreticulin expression. Lung endothelial cell monolayers were incubated in Tyrode solution without calcium with or without 50 _M BAPTA-AM for 30 min and were then incubated for 2, 4, or 6 hr at 37_ C with or without 1 _M Ang-IV. After incubation, cell lysate proteins (50 _g) were fractionated on a 7.5% SDS polyacrylamide gel, blotted onto PVDF membranes, and then hybridized with human anti-calreticulin monoclonal antibody (Reprinted from reference #52 with permission from the American Physiological Society).

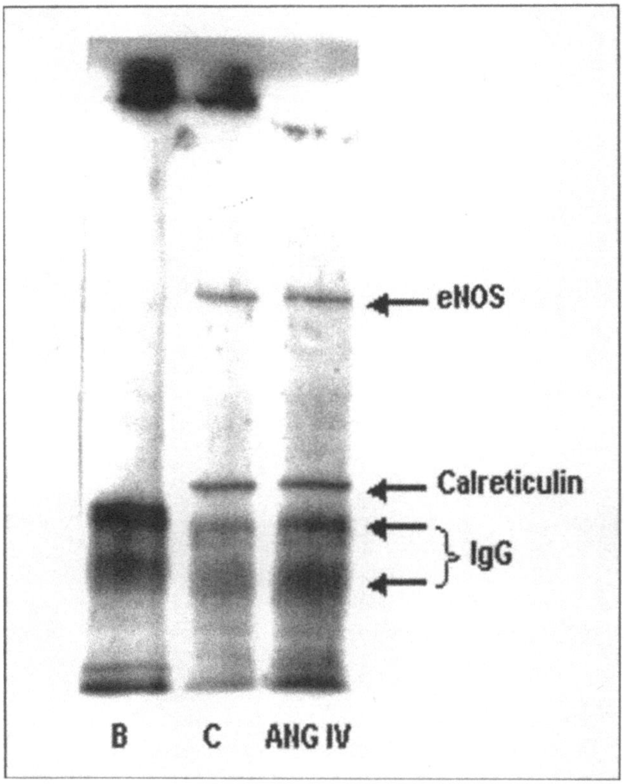

Figure 19.7. Interaction between eNOS and calreticulin. Lung endothelial cell monolayers in RPMI 1640 containing 1 μM Ang-IV or RPMI 1640 alone (control, C) were incubated for 4 hr at 37°C. After incubation, cell lysate protein (50 μg) or buffer only (blank, B) were immunoprecipitated with anti-eNOS monoclonal antibody. Immunoprecipitated complexes were denatured by boiling and then fractionated on PVDF membranes and hybridized with monoclonal eNOS and calreticulin antibodies. Data are representative of 2 independent experiments with similar results (Reprinted from reference #52 with permission from the American Physiological Society).

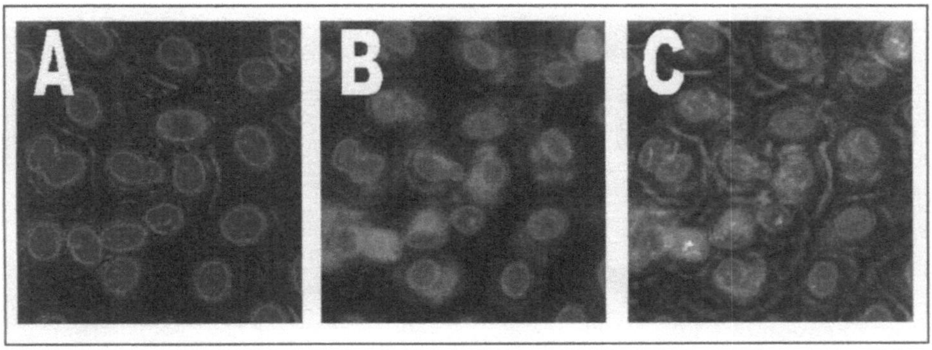

Figure 19.8. Confocal image of eNOS (panel A, red labeling), calreticulin (panel B, green labeling), and eNOS/calreticulin (panel C, yellow/orange color represents co-localization of eNOS/calreticulin) in Ang-IV-stimulated lung endothelial cells. Blue color represents nuclear labeling in all panels. Confluent cell monolayers cultured on Costar 96-well clear bottom plates were incubated with Ang-IV (1 μM) in RPMI 1640 medium for 4 hr. After incubation cells were washed, fixed/permeabilized overnight in Histochoice, and blocked for 30 min with 5% normal donkey serum in phosphate buffered saline. After 40 min incubation at room temperature with rabbit polyclonal anti-calreticulin antibody and monoclonal anti-eNOS antibody in 5% donkey serum, the cells were washed and incubated for an additional 40 min with Cyanine-conjugated donkey anti-mouse antibody and fluorescein isothiocynate-conjugated donkey anti-rabbit antibody. The secondary antibodies were affinity purified with rabbit or mouse serum, respectively. The nuclei were labeled with 300 nM DAPI (component A) blue fluorescent counter stain for 5 min. After incubation cells were washed and confocal microscopy was performed using a Ziess LSM 510 laser scanning microscope.

Ang-IV-mediated release of calcium from intracellular stores plays a major role in the regulation of eNOS activity and expression of calreticulin.

Some of the distinct characteristics of calreticulin include in vitro and in vivo peptide-binding and/or protein:protein interaction as well as its ability to bind specifically to vascular endothelium with significant implications for the regulation of vascular remodeling and angiogenesis.[42-44,52,55,58,59,85,86] We postulated that these responses are likely are mediated, at least in part, through modulation of eNOS function and NO release. To test this, we examined the potential role of calreticulin in the regulation of the catalytic activity of eNOS, focusing on protein:protein interaction and the comparative effects of calmodulin and calreticulin on the rate of electron transfer from the reductase to the oxygenase domain of eNOS. To determine whether a protein:protein interaction exists between calreticulin and eNOS in intact lung endothelial cells, eNOS from control and Ang-IV-stimulated cells was immunoprecipitated using anti-eNOS monoclonal antibody, and the immunoprecipitates were analyzed for the presence of eNOS and calreticulin proteins. As shown in Figure 19.7, the presence of eNOS and calreticulin proteins in the immunoprecipitates indicates that eNOS is present in an immunoprecipitable complex with calreticulin in lung endothelial cells. To support this, a confocal imaging technique was used to determine whether calreticulin co-localizes with eNOS in Ang-IV-stimulated cells. As shown in Figure 19.8, the presence of eNOS (panel A) and calreticulin (panel B) were identified by independent labeling with respective antibodies. Panel C shows co-localized eNOS/calreticulin following double labeling with calreticulin and eNOS antibodies. Thus, interaction of calreticulin with eNOS protein may be responsible for modulation of the catalytic activity of eNOS.

Previous work by several groups has clearly demonstrated that a protein:protein interaction with calmodulin, a calcium-binding protein, is absolutely critical for the regulation of eNOS activity in the presence of elevated levels of intracellular calcium (calcium-dependent) and in the presence of basal levels of calcium (calcium-independent pathway).[13,70,71] In addition, calmodulin is also required to shuttle the electrons transferred under the control of the electron

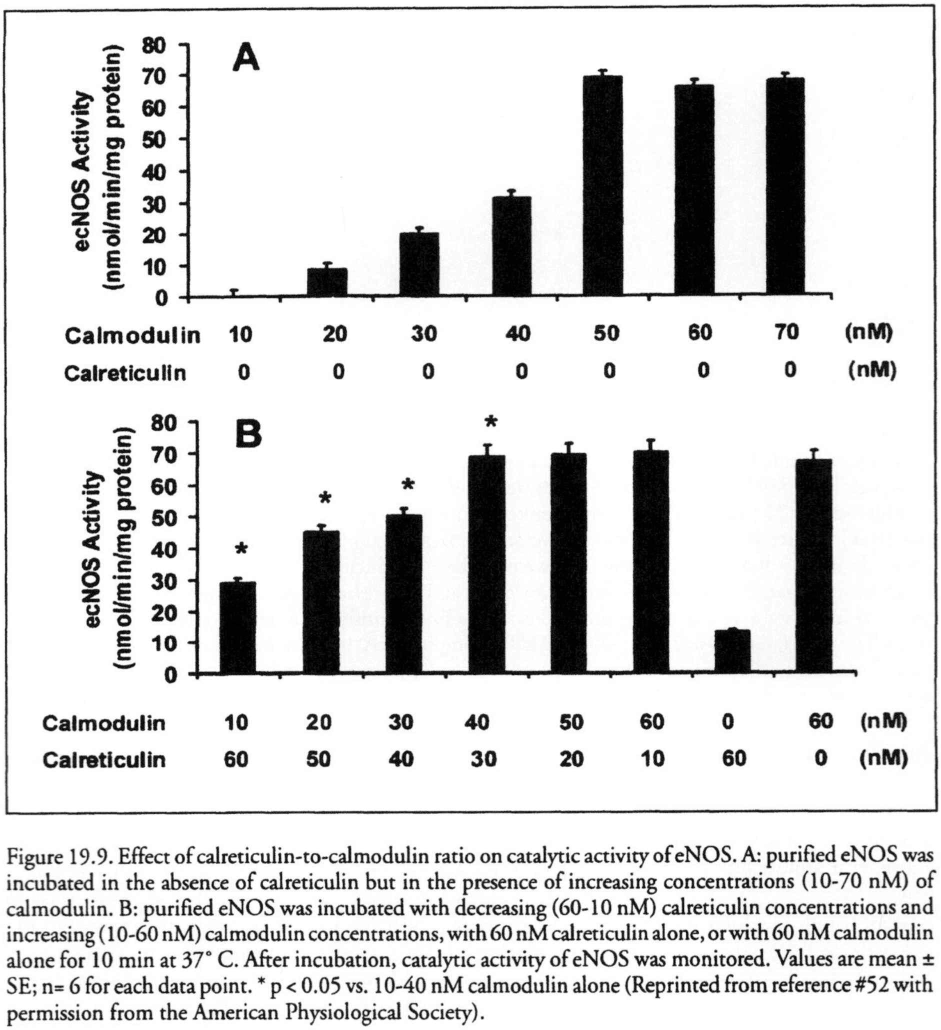

Figure 19.9. Effect of calreticulin-to-calmodulin ratio on catalytic activity of eNOS. A: purified eNOS was incubated in the absence of calreticulin but in the presence of increasing concentrations (10-70 nM) of calmodulin. B: purified eNOS was incubated with decreasing (60-10 nM) calreticulin concentrations and increasing (10-60 nM) calmodulin concentrations, with 60 nM calreticulin alone, or with 60 nM calmodulin alone for 10 min at 37° C. After incubation, catalytic activity of eNOS was monitored. Values are mean ± SE; n= 6 for each data point. * $p < 0.05$ vs. 10-40 nM calmodulin alone (Reprinted from reference #52 with permission from the American Physiological Society).

transfer element from the reductase to the oxygenase domain. For example, in the absence of calmodulin the electron transfer process is abolished resulting in near total inhibition of the catalytic activity of eNOS.[62-64] To identify possible competitive effects between calreticulin and calmodulin on eNOS activity, the catalytic activity of purified eNOS was determined in the presence of increasing or decreasing concentrations of calmodulin and calreticulin. As shown in Figure 19.9A, in the absence of calreticulin, the catalytic activity of eNOS was increased by calmodulin in a concentration-dependent fashion with maximal activation observed at 50 nM calmodulin. In the absence of calmodulin, 60 nM calreticulin resulted in only limited activation of eNOS (Fig. 19.9B), whereas calreticulin-to-calmodulin ratios of 60:10, 50:20, 40:30, and 30:40 significantly increased the catalytic activity of eNOS in comparison to eNOS activities at 10, 20, 30, and 40 nM calmodulin alone. This indicates that calreticulin enhances the catalytic activity of eNOS in the presence of calmodulin. However, the more interesting observation is that calreticulin alone at 60 nM increases the catalytic activity of eNOS. This suggests that a protein:protein interaction of calreticulin and eNOS, specifically at the calmodulin binding site on eNOS, may be responsible for modulation of the catalytic activity of eNOS. Although the precise nature of the interaction between calreticulin and eNOS remains to be determined,

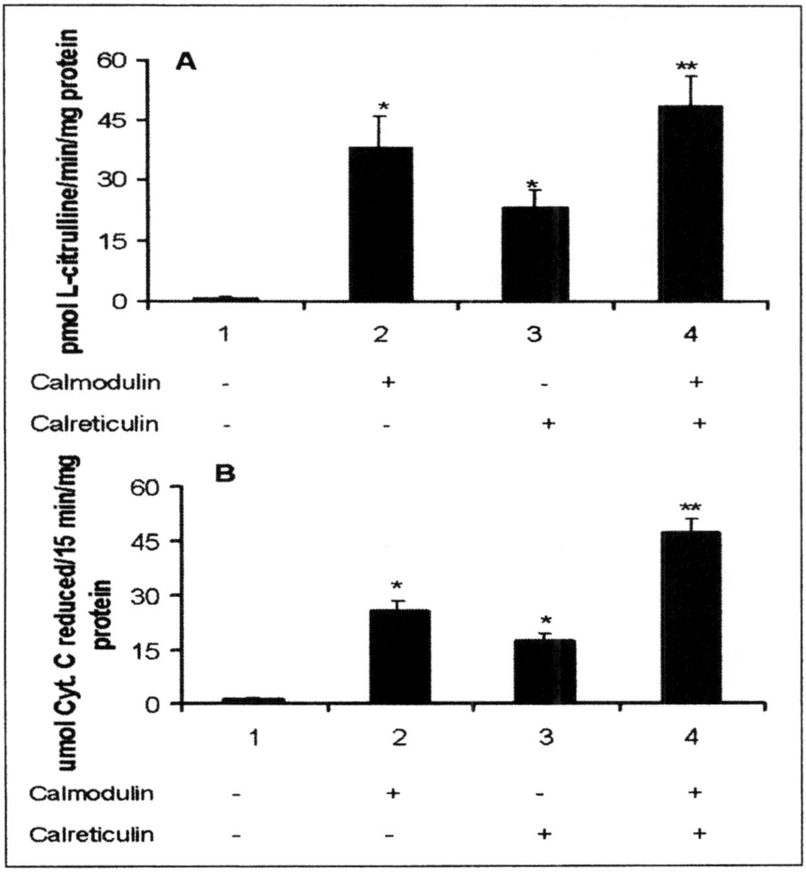

Figure 19.10. Calmodulin/calreticulin-mediated activities of eNOS (panel A) and NADPH cytochrome c reductase (panel B). For eNOS activity, purified eNOS (5 μg protein) was incubated in Tris-HCl buffer (50 mM, pH 7.4) containing calmodulin or calreticulin (30 nM each) or Tris-HCl buffer only for 10 min at 37° C. After incubation the catalytic activity of eNOS was monitored by measuring formation of L-citrulline from L-arginine. NADPH cytochrome c reductase activity was measured by continuous monitoring of absorption at 550 nm against an enzyme deficient blank sample, and the rate of cytochrome c reduction was calculated. Values are mean ± SE; n= 4 for each data point in both panels A and B. * $p < 0.001$ vs # 1 and ** $p < 0.01$ vs # 2 and 3 in both panels A and B.

it is possible that calreticulin can also bind at the site of the electron transfer control element in the reducase domain of eNOS to influence the rate of electron transfer and eNOS activity.

To identify the involvement of the electron transfer control element in the regulation of eNOS activity in context with calreticulin/eNOS interaction and whether such interaction promotes electron transfer and enhances the catalytic activity of eNOS, the catalytic activity of NADPH cytochrome c reductase was monitored as an index of electron transfer from reductase domain to the oxygenase domain with and without the presence of calmodulin, calreticulin, or calmodulin + calreticulin. Figure 19.10 A and B show the effects of calmodulin, calreticulin, and calmodulin + calreticulin on the catalytic activities of eNOS and NADPH cytochrome c reductase, respectively. In the absence of calmodulin or calreticulin, the catalytic activities of eNOS and NADPH cytochrome c reductase are nearly zero. However, in the presence of calmodulin or calreticulin the activities of eNOS and NADPH cytochrome c reductase are significantly enhanced. Moreover, the presence of both calmodulin and calreticulin resulted in further enhancement of these activities. These results indicate that the calreticulin/eNOS interaction promotes electron transfer and mimics eNOS activation in the absence of exogenous calmodulin. The interaction of calreticulin and eNOS also enhances electron transfer and the catalytic activity of eNOS in the presence of calmodulin.

Concluding Remarks

Vascular endothelial eNOS-catalyzed release of NO plays a critical role mediating diverse physiologic and pathophysiologic events in animals and humans.[87,88] Endothelium-derived NO regulates pulmonary and extra-pulmonary vascular function under physiologic conditions. The amount and the rate of NO release under various pathophysiologic conditions are believed to be major factors that determine NO-mediated physiologic responses such as vasorelaxation, angiogenesis, apoptosis, and cell proliferation and migration.[87-91] Multiple factors are responsible for the regulation of the catalytic activity of eNOS including post-transcriptional mechanisms such as protein:protein interaction and the regulation of electron transfer by the intra-molecular electron transfer control element in the reductase domain. The physiologic significance of Ang-IV-mediated intracellular calcium release, calreticulin expression, and eNOS is important in relation to NO biology and endothelium-dependent vascular function. Similarly, increased expression of calreticulin under diverse pathophysiologic conditions and calreticulin-mediated activation of eNOS may help restore impaired NO production due to diminished or reduced catalytic activity of eNOS associated with a variety of δ pulmonary diseases and/or injuries.

References

1. Davis MG, Fulton GJ, Hagen PO. Clinical biology of nitric oxide. Br J Surg 1995; 82:1598-1610.
2. Ignarro LJ, Buga GM, Wood KS. Endothelium-derived relaxing factor from pulmonary artery and vein possesses pharmacologic and chemical properties identical to those of nitric oxide radical. Circ Res 1987; 61:866-879.
3. Patel JM, Martenes JR Li YD et al. Angiotensin IV receptor-mediated activation of lung endothelial NOS is associated with vasorelaxation. Am J Physiol Lung Cell Mol Physiol 1998; 275:L1061-L1068.
4. Wright JW, Harding JW. Brain angiotensin receptors AT1, AT2, AT4 and their functions. Regul Pept 1995; 59:269-295.
5. Hanesworth JM, Sardinia MF, Krebs LT et al. Elucidation of a specific binding site for angiotensin II (3-8), angiotensin IV in mammalian heart membranes. J Pharmacol Exp Therp 1993; 266:1036-1042.
6. Handa RK, Krebs LT, Harding JW et al. Angiotensin IV AT4 receptor system in the rat kidney. Am J Physiol Renal Physiol 1998; 274:F290-F299.
7. Hall KL, Venkateswaran S. Hanesworth JM et al. Characterization of a functional angiotensin IV receptor on coronary microvascular endothelial cells. Regul Pept 1995; 58:107-115.
8. Riva L, Galzin AM. Pharmacologic characterization of a specific binding site for angiotensin IV in cultured porcine aortic endothelial cells. Eur J Pharmacol 1996; 305:193-199.
9. Berridge MJ. Inositol triphosphate and calcium signaling. Nature 1994; 361:315-325.
10. Sekar MC, Hokin LE. The role of phosphoinositides in signal transduction. J Memb Biol 1986; 89:193-210.
11. Chen S, Patel JM, Block ER. Angiotensin IV-mediated pulmonary artery vasorelaxation is due to endothelial intracellular calcium release. Am J Physiol Lung Cell Physiol 2000; 279:L849-L856.
12. Michel T, Feron O. Nitric oxide synthases: which, where, how, and why? J Clin Invest 1997; 100:2146-2152.
13. Govers R, Rabelink TJ. Cellular regulation of endothelial nitric oxide synthase. Am J Physiol Renal Physiol 2001; 280:F193-F206.
14. Booth C, Koch LE. Perturbation of cellular calcium induces secretion of luminal proteins. Cell 1989; 59:729-737.
15. Kohn EC, Alessandro R, Spooster J et al. Angiogenesis: role of calcium-mediated signal transduction. Proc Natl Acad Sci USA 1995; 92:1307-1311.
16. Marsen DR, Simonson MS, Dunn MJ. Thrombin-mediated ET-1 gene regulation involves CaM kinase and calreticulin in human endothelial cells. J Cardiovasc Pharmacol 1995; 26:S1-S4.
17. Marletta MA. Nitric oxide synthase:aspects concerning structure and catalysis. Cell 1994; 78:927-930.

18. Forstermann U, Closs EI, Pollock JS et al. Nitric oxide synthase isoforms:Characterization, purification, molecular cloning, and functions. Hypertension 1994; 23:1121-1131.
19. Ignaro LJ. Biosynthesis and metabolism of endothelium-derived nitric oxide. Annu Rev Pharmacol Toxicol 1990; 30:535-560.
20. Nathan C, Xie QW. Regulation of biosynthesis of nitric oxide. J Biol Chem 1994; 269:13725-13728.
21. Wright JW, Keebs LT, Stobb JW et al. The angiotensin IV system: functional implications. Front Neuroendocrinol 1995; 16:23-52.
22. Campbell DJ, Kladis A. Simultaneous radioimmunoassay of six angiotensin peptides in arterial and venous plasma of man. J Hypertension 1990; 8:165-172.
23. Shibasaki Y, Mori TY, Tsutumi Y et al. Differential kinetics of circulating angiotensin IV and II after treatment with angiotensin II type 1 receptor antagonist and their plasma levels in patients with chronic renal failure. Clin Neph 1999; 51:83-91.
24. Wright JW, Harding JW. Brain angiotensin receptors AT1, AT2, AT4 and their functions. Regul Pept 1995; 59:269-295.
25. Handa RK. Characterization and signaling of the AT(4) receptor in human proximal tubule epithelial (HK-2) cells. J Amer Soc Neph 2001; 12:440-409.
26. Chai SY, Bastias MA, Clune EF et al. Distribution of angiotensin IV binding sites (AT(4) receptor) in the human forebrain, midbrain and pons as visualized by in vitro receptor autoradiography. J Chem Neuroanatomy 2000; 20:339-348.
27. Kramar EA, Harding JW, Wright JW. Angiotensin II- and IV-induced changes in cerebral blood flow roles of AT1, AT2, and AT4 receptor subtypes. Regul. Pept. 1997; 68:131-138.
28. Harding JW, Wright JW, Swanson GN et al. AT4 receptors: specificity and distribution. Kidney Int 1994; 46:1510-1512.
29. Haberl RL, Decker PJ, Einhaupl KM. Angiotensin degradation products mediate endothelium-dependent dilation of rabbit brain arterioles. Circ Res 1991; 68:1621-1627.
30. Heal R, McGivan JD. Induction of calreticulin expression in response to amino acid deprivation in Chinese hamster ovary cells. Biochem J 1998; 329:389-394.
31. Plakidoudymock S, McGivan JD. Calreticulin-a stress protein induced in a renal epithelial cell line Nb-1 by amino acid deprivation. Cell Calcium 1994; 16:1-8.
32. Jethmalani SM, Henle KJ. Calreticulin association with stress proteins:implications for chaperone function during heat stress. J Cell Biochem 1998; 69:30-43.
33. Conway EM, Liu LL, Nowakowski B et al. Heat shock sensitive expression of calreticulin:in vitro and in vivo up-regulation. J Biol Chem 1995; 270:17011-17016.
34. Nguyen TQ, Capra D, Sotheimer RD. Calreticulin is transcriptionally upregulated by heat shock, calcium, and heavy metals. Mol Immunol 1996; 33:379-386.
35. Zhu J. Ultraviolet B irradiation and cytomegalovirus infection synergize to induce the cell surface expression of 52-kD/Ro antigen. Clin Exp Immunol 1996; 103:47-53.
36. Dupuis M, Schaerer E, Krause KH et al. The calcium binding protein calreticulin is a major constituent of lytic granules in cytolytic T lymphocytes. J Exp Med 1993; 177:1-7.
37. Llewellyn DH, Kendall JM, Sheikh FN et al. Induction of calreticulin expression in HeLa cells by depletion of endoplasmic reticulum calcium store and inhibition of N-linked glycosylation. Biochem J 1996; 318:555-560.
38. Michalak M, Milner RE, Burns K et al. Calreticulin. Biochem J 1992; 285:681-692.
39. Michalak M, Corbett EF, Mesaeli N et al. Calreticulin: one protein, one gene, many functions. Biochem J 1999; 344:281-292.
40. White TK, Zhu Q, Tanzer ML. Cell surface calreticulin is putative mannoside lactin which triggers mouse melanoma cell spreading. J Biol Chem 1995; 270:15926-15929.
41. Burns K, Duggan B, Atkinson EA et al. Modulation of gene expression by calreticulin binding to glucocorticoid receptor. Nature 1994; 367:476-480.
42. Baksh S, Burns K, Andrin C et al. Interaction of calreticulin with protein disulfide isomerase. J Biol Chem 1995; 270:31338-31344.
43. Niar S, Wearsch PA, Mitchell DA et al. Calreticulin displays in vivo peptide binding activity and can elicit CTL responses against bound peptides. J Immunol 1999; 162:6426-6432.
44. Cunningham TJ, Jing H, Wang Y et al. Calreticulin binding and other biological activities of survival peptide Y-P30 including effects of systemic treatment of rats. Expt Neurology 2000; 163:457-468.
45. Dedhar S. Novel functions of calreticulin:interaction with integrins and modulation of gene expression. Trends Biochem Sci 1994; 19:269-271.
46. Gray AJ, Park PW, Brokelmann TJ et al. The mitogenic effects of the B chain of fibrinogen are mediated through cell surface calreticulin. J Biol Chem 1995; 270:26602-26606.

47. Darby PJ, Kwan CY, Daniel EE. Caveolae from canine airway smooth contain the necessary components for a role in calcium handling. Am J Physiol Lung Cell Mol Physiol 2000; 279:L1226-L1235.
48. Ramsamooj P, Notario V, Dristchilo A. Enhanced expression of calreticulin in the nucleus of radioresistant squamous carcinoma cells in response to ionizing radiation. Cancer Res 1995; 55:3016-3021.
49. Roderick HL, Cambell AK, Llewellyn DH. Nuclear localization of calreticulin in vivo is enhanced by its interaction with glucocorticoid receptors. FEBS Lett 1997; 405:181-185.
50. Booth C, Koch LE. Perturbation of cellular calcium induces secretion of luminal proteins. Cell 1989; 59:729-737.
51. Panon W. Erdjument H, Marks AH et al. Components of the protein synthesis and folding machinery are induced in vascular smooth muscle cells by hypertrophic and hyperplastic agents. J Biol Chem 1995; 270:21404-21410.
52. Patel JM, Li YD, Zhang JL et al. Increased expression of calreticulin is linked to Ang-IV-mediated activation of lung endothelial NOS. Am J Physiol Lung Cell Mol Physiol 1999; 277:L794-L801.
53. Eggleton P, Llewellyn DH. Pathophysiological roles of calreticulin in autoimmune disease. Scand J Immunol 1999; 49:466-473.
54. Johnson S, Michalak M, Opas M et al. The ins and outs of calreticulin:from the ER lumen to the extracellular space. Trends Cell Biol 2001; 11:122-129.
55. Goicoecha S, Orr AW, Pallero MA et al. Thrombospondin mediates focal adhesion disassembly through interactions with cell surface calreticulin. J Biol Chem 2000; 275:36358-36368.
56. Singh NK, Atreya CD, Nakhasi HL. Identification of calreticulin as a rubella virus RNA binding protein. Proc Natl Acad Sci USA 1994; 91:12770-12774.
57. Nakamura K, Bossy-Wetzel E, Burns K et al. Changes in endoplasmic reticulum luminal environment affect cell sensitivity to apoptosis. J Cell Biol 2000; 150:731-740.
58. Pike SE, Yao L, Setsuda J et al. Calreticulin and calreticulin fragment are endothelial cell inhibitors that suppress tumor growth. Blood 1999; 94:2461-2468.
59. Kuwabara K, Pinsky DJ, Schmidt AM et al. Calreticulin, an antithrombotic agent which binds to vitamin K-dependent coagulation factors, stimulates endothelial nitric oxide production, and limits thrombosis in canine coronary arteries. J Biol Chem 1995; 270:8179-8187.
60. Xu W, Longo FJ, Wintermantel MR et al. Calreticulin modulates capacitative calcium influx by controlling the extent of inositol 1,4,5-triphosphate-induced calcium store depletion. J Biol Chem 2000; 275:36676-36682.
61. Nakamura K, Zuppini A, Arnaudeau S et al. Functional specialization of calreticulin domains. J Cell Biol 2001; 154:961-972.
62. Salerno JC, Harris DE, Irizarry K et al. An autoinhibitory control element defines calcium -regulated isoforms of nitric oxide synthase. J Biol Chem 1997; 272:29769-29777.
63. Nashida CR, Ortiz de Montellano PR. Autoinhibition of endothelial nitric oxide synthase:identification of an electron transfer control element. J Biol Chem 1999; 274:14692-14698.
64. Chen PF, Wu KK. Characterization of the roles of the 594-645 region in human endothelial nitric oxide synthase in regulating calmodulin binding and electron transfer. J Biol Chem 2000; 275: 13155-13163.
65. Noris M, Morigi M, Donadelli R et al. Nitric oxide synthesis by cultured endothelial cells is modulated by flow conditions. Circ Res 1995; 76:536-543.
66. Navarro-Antolin J, Rey-Campos J, Lamas S. Transcriptional induction of endothelial nitric oxide gene by cyclosporine A: a role for activator protein-1. J Biol Chem 2000; 275:3075-3080.
67. Kuboki K, Jiang ZY, Takahara N et al. Regulation of endothelial constitutive nitric oxide synthase gene expression in endothelial cells and in vivo:a specific vascular action of insulin. Circulation 2000; 101:676-681.
68. Kroll J, Waltenberger J. VEGF-A induces expression of eNOS and iNOS in endothelial cells via VEGF receptor-2 (KDR). Biochem Biophys Res Commun 1998; 252:743-746.
69. Hirata K, Miki N, Kuroda Y et al. Low concentrations of oxidized low-density lipoprotein and lysophosphatidylcholine upregulate constitutive nitric oxide synthase mRNA expression in bovine aortic endothelial cells. Circ Res 1995; 76:958-962.
70. Dimmeler S, Fleming I, Fisslthaler B et al. Activation of nitric oxide synthase in endothelial cells by AKT-dependent phosphorylation. Nature 1999; 399:601-605.
71. Garcia-Cardena G, Fan R. Shah V et al. Dynamic activation of endothelial nitric oxide synthase by HSP 90. Nature 1998; 392:821-824.
72. Chen PF, Tsai AL, Wu KK. Cysteine 184 of endothelial cell nitric oxide synthase is involved in heme coordination and catalytic activity. J Biol Chem 1994; 269:25060-25066.

73. Zhang JL, Patel JM, Li YD et al. Reductase domain cysteines 1048 and 1114 are critical for catalytic activity of human endothelial cell nitric oxide synthase as probed by site-directed mutagenesis. Biochem Biophys Res Commun 1996; 226:293-300.
74. Garcia-Cardena G, Oh P, Liu J et al. Targeting of nitric oxide synthase to endothelial cell caveolae via palmitoylation:implications of nitric oxide signaling. Proc Natl Acad Sci USA 1996; 93:6448-6455.
75. McDonald KK, Zharikov S, Block ER et al. A caveolar complex between cationic amino acid transporter 1 and endothelial nitric oxide synthase may explain the "arginine paradox". J Biol Chem 1997; 272:31213-31216.
76. Shaul PW. Regulation of endothelial nitric oxide synthase:location, location, location. Annu Rev Physiol 2002; 64:749-774.
77. Prabhaker P, Thatte HS, Goetz RM et al. Receptor-regulated translocation of endothelial nitric oxide synthase. J Biol Chem 1998; 273:27483-27488.
78. Goetz RM, Thatte HS, Prabhaker P et al. Estradiol induces the calcium-dependent translocation of endothelial nitric oxide synthase. Proc Natl Acad Sci USA 1999; 96:2788-2793.
79. Zharikov SI, Sigova AA, Chen S et al. Cytoskeletal regulation of the L-arginine/NO pathway in pulmonary artery endothelial cells. Am J Physiol Lung Cell Mol Physiol 2001; 280:L465-L473.
80. McCabe TJ, Fulton D. Roman LJ et al. Enhanced electron flux and reduced calmodulin dissociation may explain calcium-independent eNOS activation by phosphorylation. J Biol Chem 2000; 275:6123-6128.
81. Kerins DM, Hao Q, Vanghan DE. Angiotensin induction of PAI-1 expression in endothelial cells is mediated by the hexapeptide angiotensin IV. J Clin Invest 1996; 96:2515-2520.
82. Pawlikowski M, Kunert-Radek J. Angiotensin IV-stimulates the proliferation of rat anterior pituitary cells in vitro. Biochem Biophys Res Commun 1997; 232:292-293.
83. Wang L, Eberhard M, Erne P. Stimulation of DNA and RNA synthesis in cultured rabbit cardiac fibroblasts by angiotensin IV. Clin Sci 1995; 88:557-562.
84. Li YD, Block ER, Patel JM. Activation of multiple signaling modules are critical in angiotensin IV-induced lung endothelial cell proliferation. Am J Physiol Lung Cell Mol Physiol 2002; 283:L707-L716.
85. Dai E, Stewart M, Ritchie B et al. Calreticulin, a potential vascular regulatory protein, reduces intimal hyperplasia after arterial injury. Arterioscler Thromb Vasc Biol 1997; 17:2359-2368.
86. Pike SE, Yao L, Jones KD et al. Vasostatin, a calreticulin fragment, inhibits angiogenesis and suppresses tumor growth. J Exp Med 1998; 188:2349-2356.
87. Barnes J, Belvisi MG. Nitric oxide and lung disease. Thorax 1993; 48:1034-1043.
88. Moncada S, Palmer RMJ, Higgs EA. Nitric oxide: physiology, pathophysiology, and pharmacolohy. Pharcol Rev 1991; 43:109-142.
89. DeMeester SL, Qiu Y, Buchaman TG et al. Nitric oxide inhibits stress-induced endothelial cell apoptosis. Crit Care Med 1998; 26:1500-1509.
90. Mebmer UK, Ankarcrom M, Nicotera P et al. p53 expression in nitric oxide-induced apoptosis FESEB Lett 1994; 355:23-26.
91. Shimizu SM, Natio S, Yamamoto T et al. Stimulation of nitric oxide synthase during oxidative endothelial cell injury. Biochem Pharmacol 1998; 55:77-83.

Chapter 20

Role of Calreticulin in *Leishmania* Parasite Secretory Pathway and Pathogenesis

Alain Debrabant, Nancy Lee, Dennis M. Dwyer and Hira L. Nakhasi

Abstract

The trypanosomatid parasites *Leishmania* and *Trypanosoma* are the causative agents of human diseases such as leishmaniasis, Chagas disease or African sleeping sickness. Some proteins secreted by these protozoan parasites represent virulence factors and contribute to the survival of these pathogen in the human hosts. Therefore, alteration in the secretion of these proteins could result in attenuation of parasite virulence. Calreticulin is an endoplasmic reticulum chaperone protein involved in the quality control of glycoprotein folding in higher eukaryotes. Previously, we isolated and characterized a calreticulin homologue (LdCR) from *Leishmania donovani*, the causative agent of visceral leishmaniasis. To assess whether modulation of LdCR level in the parasite could affect the release of secretory proteins by *Leishmania*, we established *L. donovani* cell lines overexpressing LdCR or its putative N-, P-, or C-domains. In this report, we show that the secretion of secretory acid phosphatase and possibly other proteins trafficking through the secretory pathway of the parasite were affected as a result of overexpression of LdCR P- or C-domain. In addition, parasites expressing either the LdCR N- or P-domain showed significant decrease in survival inside macrophages in vitro. Taken together, these results suggest that disruption of the functions of calreticulin in *Leishmania* can result in an alteration of the parasite secretory pathway and also reduce its virulence in vitro.

Leishmania Biology

Leishmania are a group of trypanosomatid parasites which infect about 12 million people worldwide, with ~ half million new clinical cases reported annually.[1,2] There is a spectrum of diseases caused by the different species of *Leishmania*, which range from mild cutaneous lesions to often lethal visceral leishmaniasis. The later involves liver, spleen and bone marrow. The visceral disease is also known as kala-azar and is a significant infectious disease in the developing world and recently in developed world because of the business travelers, US military personnel and their family stationed in endemic areas and co-infection of human immunodeficiency virus infection.[3] At present there is no vaccine against leishmaniasis. Further, only few drugs are available to treat leishmaniasis patients, however, they are very toxic and poorly effective in immunocompromized individuals. In addition, *Leishmania* parasites have developed resistance to these drugs.

Leishmania has a digenetic life cycle (Fig. 20.1). The motile flagellated extracellular forms, called promastigotes, reside in the alimentary canal of the sandfly vector. The parasites mature into the infectious metacyclic forms and migrate toward the anterior part of the gut of the sandfly. They are transmitted into a vertebrate host by the bite of the sandfly during the process of obtaining the blood meal. Host macrophages take up the parasites by phagocytosis into

Calreticulin, 2nd ed., edited by Paul Eggleton and Marek Michalak.

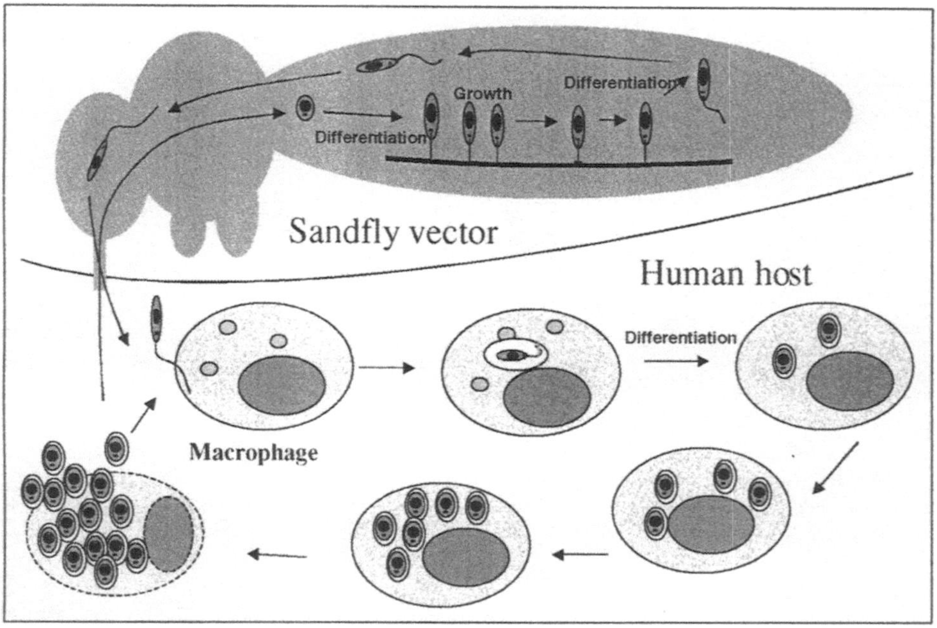

Figure 20.1. *Leishmania* life cycle. *Leishmania* has a digenetic life cycle. The sandfly vector becomes infected when feeding on the blood of an infected individual or an animal reservoir host. Inside the human host, the *Leishmania* parasites live in the macrophages as round, non-motile amastigotes (A). The macrophages are ingested by the fly during the blood meal and the amastigotes are released into the gut of the insect. Almost immediately the amastigotes transform into the motile, elongated, flagellate promastigote form (P). The promastigotes then colonize the alimentary tract of the fly, where they live extracellularly and multiply asexually by binary fission. During the first four to five days after feeding, the promastigotes differentiate into the infectious metacyclic forms and migrate to the anterior region of the gut of the insect. When the sandfly next feeds on a mammalian host, metacyclic promastigotes are transferred to the host along with the saliva. Once in the host, the metacyclic promastigotes are phagocytozed by the macrophages where they rapidly revert to the amastigote form. The *Leishmania* amastigotes are able to resist the microbicidal action of the lysosome acid hydrolases and so survive and multiply inside phagolysosomes of macrophages, eventually leading to the lysis of the macrophages. The released amastigotes are taken up by naive macrophages and so the cycle continues. Ultimately all the organs containing macrophages and phagocytes are infected, especially the spleen, liver and bone marrow.

phagolysosomes. Inside the phagolysosomes, the parasites differentiate into non-motile amastigote forms which are able to survive in this hydrolytic environment.[4] The change from promastigotes to amastigotes and vice versa is associated with dramatic biochemical and molecular changes.[5-7] These changes have been correlated with changes in gene expression.[8-17] Although not thoroughly understood, these biochemical and molecular changes may help the parasite to withstand the harsh environment of macrophages inside the vertebrate host. In addition to the above mentioned changes, it is postulated that molecules present on the parasite cell surface or secreted by *Leishmania* may be involved in mechanisms for the parasite survival inside the host.[18]

Cell surface of *Leishmania* is composed of many complex molecules which range from lipophosphoglycans to various glycoproteins such as enzymes and transporters[19,20] (Fig. 20.2). In addition a significant number of proteins are secreted into the outside environment such as secretory acid phosphatase,[21] chitinase,[22] or thiol-specific antioxidant.[23] Both the secreted and the membrane anchored molecules are thought to be putative virulence factors which can determine the *Leishmania* pathogenesis. It is believed that membrane bound and secreted

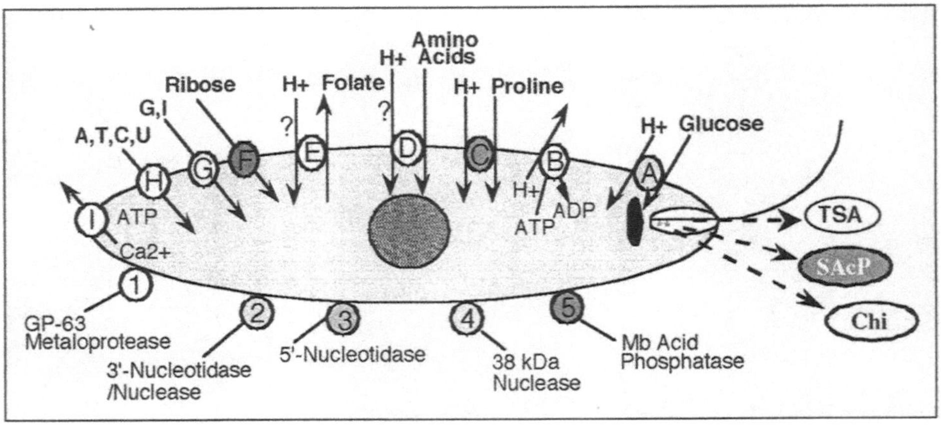

Figure 20.2. *Leishmania* cell surface and secreted proteins. Schematic diagram of a *Leishmania* promastigote showing some better characterized cell surface transporters (A-I) or enzymes (1-5) (see Schneider et al. for review)[19] Only a few *Leishmania* secreted proteins have been characterized to date. These include the secretory acid phosphatase family[21] (SAcP), a chitinase[22] (Chi) and thiol-specific antioxidant[23] (TSA).

proteins traffic in *Leishmania*, as well as in other trypanosomatids such as *Trypanosoma cruzi*[24] and *Trypanosoma brucei*,[25] via a typical eukaryotic secretion pathway. The proteins are folded in the ER and then are transported via Golgi to the flagellar reservoir on to either the cell surface or outside the cell.[26,27] Therefore, the processing of putative virulence factors in the ER and their transport via Golgi is essential for the pathogenesis of this class of parasites.

Secretory Pathway in Trypanosomatids

Little is known about the process of protein secretion or trafficking let alone the factors that are involved in these processes in trypanosomatids, which are primitive eukaryotes.[28] Among the trypanosomatids, the secretion of soluble proteins is best understood in *Leishmania donovani*,[21,29] and *Trypanosoma cruzi*,[30] a causative agent of Chagas disease than *Trypanosoma brucei*, a causative agent of African sleeping sickness. The schematic diagram (Fig. 20.3) depicts a possible pathway by which the proteins are transported via ER and Golgi into flagellar pocket to be either secreted outside the cell or presented on the cell surface.

There are differences between trypanosomatids and higher eukaryotes in various features of protein trafficking including quality control of proteins such as N-glycosylation, oligosaccharide processing and folding.[24] For example, there are significant differences in the pathway leading up to the formation of N-glycosylation in trypanosomatids. There are 11-12 isoprene residues in the dolichol moiety compared to 19-21 isoprene residues in mammalian cells.[31-33] This group of parasites does not synthesize dolichol-P-Glc.[34] In these parasites, the oligosaccharide ligands such as $Man_6GlcNAc_2$, $Man_7GlcNAc_2$ or $Man_9GlcNAc_2$ which are transferred on to the protein moiety lack the three glucose molecules as normally found in the oligosaccharides of higher eukaryotic cells.[31,35,36] The activity and specificity of trypanosomatid oligosaccharidyl-transferases is significantly lower than the mammalian transferases.[37] Both glucosyltransferase, which adds the one glucose molecule to the oligosaccharide chain, and glucosidase II, which removes the lone glucose molecule during the quality control process, have been detected in trypanosomatid cells.[37] However, these cells lack glucosidase I,[37-39] which is present in higher eukaryotic cells, and functions in removal of two glucose molecules from the triglucosylated oligosaccharide ligand so as to allow the monoglucosylated form interact with the ER chaperones.[40]

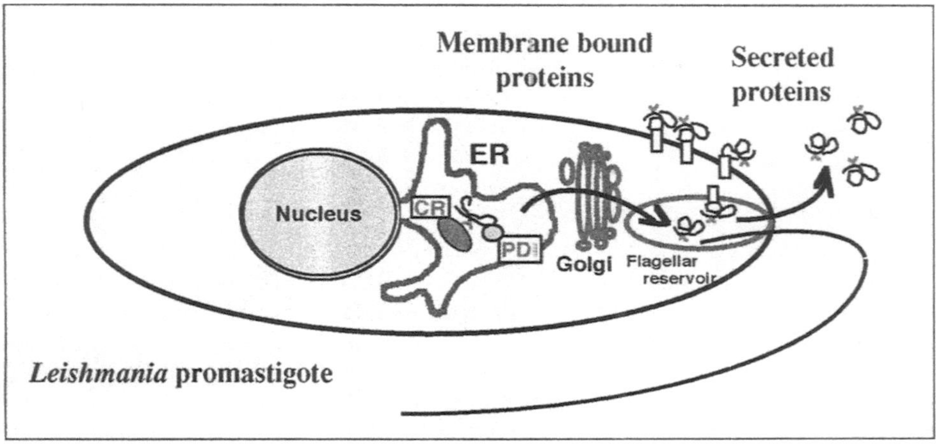

Figure 20.3. Secretory pathway of *Leishmania. Leishmania* are polarized secretory cells. Nascent proteins translocated in the endoplasmic reticulum (ER) are N-linked glycosylated in the ER, then traffic through the Golgi apparatus where they are further modified (O-linked glycosylated, phosphorylated etc.). These glycoproteins are subsequently transferred, probably via vesicular trafficking, to the parasite flagellar reservoir prior to being targeted either to the cell surface or to be released/secreted outside the cell. There is evidence that parasite nascent glycoproteins interact with chaperone proteins in the ER and undergo folding similar to that described for higher eukaryotes. Two recently identified *Leishmania* ER chaperones are indicated: calreticulin (CRT) and protein disulfide isomerase (PDI).

Characterization of ER Chaperones in Trypanosomatids

In addition to transferases and glucosidase,[41] known quality control components of the trypanosomatid cell include chaperone proteins such as immunoglobulin-heavy-chain-binding protein (BiP) and putative protein disulfide isomerase (PDI) homologue in *T. brucei*,[42,43] calreticulin in *T. cruzi*[24] and in *L. donovani*.[44] Sequence comparisons of *T. brucei* BiP and PDI with similar proteins from other organisms showed a significantly high similarity in amino acid and had other hallmarks of the ER chaperone proteins such as signal sequence, the ER retention signal and immuno localization in the ER.[42,43] Similarly comparison of both *Leishmania* and *T. cruzi* CRT sequences with CRT from other organisms revealed that there is both significant amino acid sequence conservation as well as the conservation of putative functional domains (Fig. 20.4). These domains include both high and low affinity Ca^{2+} binding domains (Fig. 20.4). The conservation of trypanosomatid chaperone protein sequences with higher eukaryotic cells suggests a common function even though the two are evolutionarily distinct and diverged millions of years ago.

Besides sequence similarity, the functional conservation of trypanosomatid chaperone proteins has also been established. For example, functional analysis of both the *T. brucei* BiP and *T. cruzi* calreticulin revealed that they function as chaperones as in higher eukaryotic cells. Transient interactions between either *T. brucei* BiP and the variant surface glycoproteins, or *T. cruzi* CRT and the lysosomal proteinase cruzipain, have been demonstrated.[24,45] The *T. cruzi* CRT specifically recognized free monoglucosylated high-mannose type oligosaccharides present on the cruzipain. This interaction was further confirmed by abrogating the interaction between CRT and cruzipain upon enzymatic removal of the glucose moiety from the cruzipain.[24] Mutation analysis of *T. brucei* BiP basic signals responsible for the ER retention and transport of variant surface glycoproteins through the ER revealed that they are conserved through the evolution.[45] Thus even though the above mentioned studies are limited in scope, they suggest that chaperone functions have been evolutionarily conserved in the eukaryotic systems.

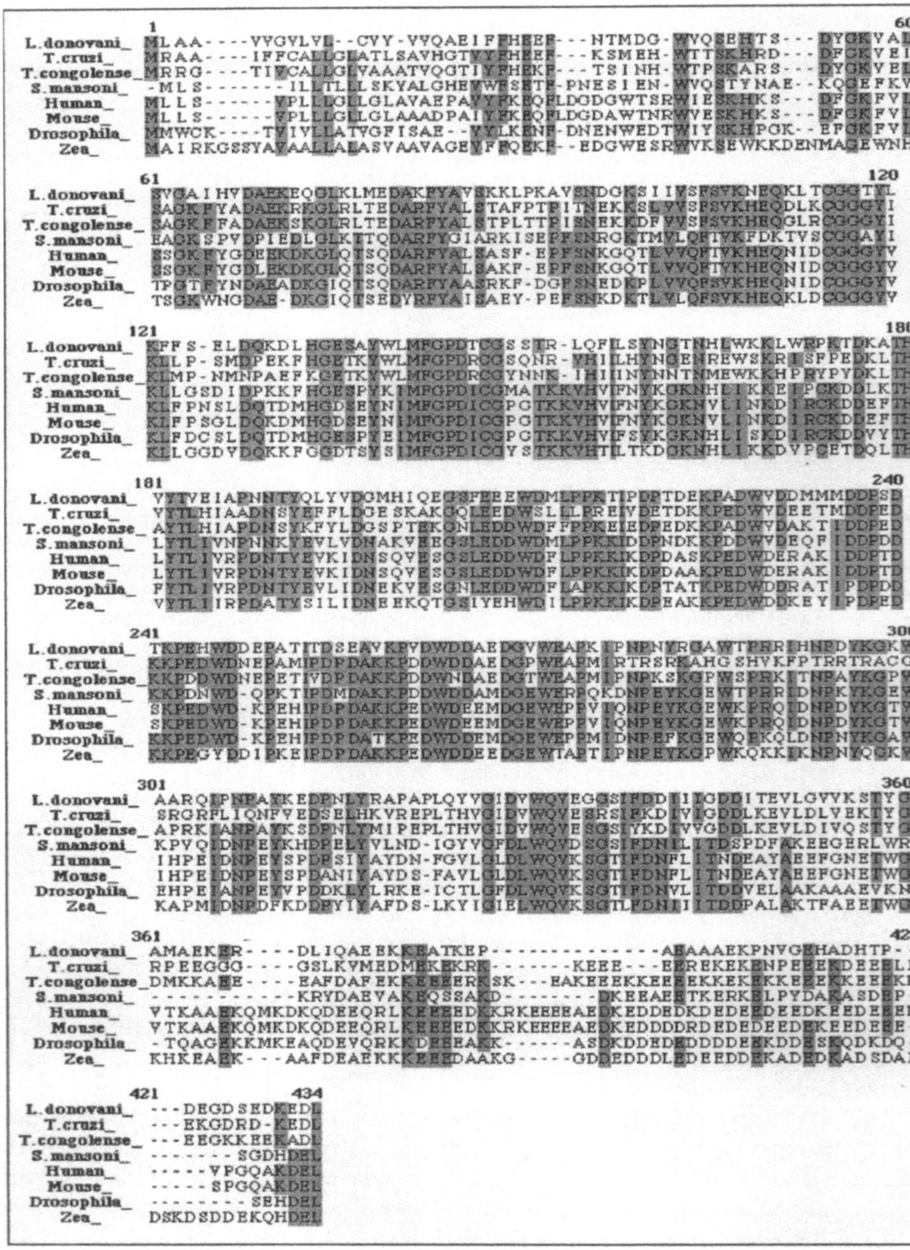

Figure 20.4. Clustal alignment of calreticulin protein sequences. Multiple alignment of calreticulin sequences from various organisms: *L. donovani* (AAB17728); *Trypanosoma cruzi* (AAD22175); *Trypanosoma congolense* (AAK52926); *Schistosoma mansoni* (AAA29854); human (AAA51916); mouse (CAA33053); *Drosophila melanogaster* (CAA45791) and *Zea mays* (CAA86728). Amino acids are listed in standard one-letter code and identical amino acids are indicated by shaded areas. Spaces introduced for sequence alignment are represented by dashes.

Role of Calreticulin in *Leishmania* Secretory Pathway

Characteristics of the L. donovani *Calreticulin*

The gene encoding the *L. donovani* CRT (LdCR) has been reported previously.[44] The deduced LdCR protein has a 3-domain structure similar to the structure of CRT from other organisms.[46] The LdCR putative N-, P- and C-domains are flanked by an N-terminal signal peptide for the translocation of the protein into the ER, and by a C-terminal KEDL ER retention signal.[44] Further, an LdCR recombinant protein made in *E. coli* was shown to bind Ca^{2+} in vitro suggesting that LdCR can bind Ca^{2+} in vivo which is one of the hallmarks of this family of chaperones. In addition, LdCR was also shown to be glycosylated and to have autokinase activity in vitro. More recently, the LdCR protein was shown to be localized in the ER of *L. donovani* by immunofluorescence (Fig. 20.5). The reticular pattern of fluorescence obtained using an anti-LdCR specific antibody (Fig. 20.5A) was very similar to that obtained with an anti-BiP antibody (Fig. 20.5B) which localized BiP in the ER compartment of the parasite. These results also indicate that the C-terminal KEDL sequence of LdCR is a functional ER retention signal in *Leishmania.*

Alteration of LdCR Expression by Gene Disruption

Since CRT is a known ER chaperone and plays a role in quality control of both membrane bound and secreted glycoproteins in higher eukaryotes, we explored the function of LdCR in the *Leishmania* secretory pathway. We used several approaches in order to assess the role LdCR in *Leishmania.* The first approach was to disrupt the single copy gene encoding LdCR from the parasite genome. Such a gene knock strategy has been successfully used previously to delete or disrupt several genes from the *Leishmania.*[47-49] *Leishmania* being a diploid organism, the two alleles of a single copy gene need to be sequentially deleted in order to generate a null mutant for a specific gene. However, gene rearrangements are likely to occur when attempting to knock out essential genes form the *Leishmania* genome.[48] This approach proved to be unsuccessful with LdCR gene (Debrabant, unpublished). Multiple gene rearrangement were observed in double drug resistant parasite clones selected after the disruption of the second LdCR allele. Despite multiple attempts, LdCR null mutants have not been obtained, suggesting that LdCR must play a critical role in the parasite development. This observation is supported by the fact that CRT knock out is embryonically lethal in mice.[50] Furthermore, we have not been able to detect a gene for calnexin in *Leishmania* (Debrabant, unpublished) which plays a role functionally redundant to CRT in higher eukaryotes. Thereby further suggesting that the presence of CRT is essential in *Leishmania.*

Dominant Negative Expression of LdCR-Domains in L. donovani

Episomal Expression of LdCR-Domains in *L. donovani*

A second approach used to disrupt the function of LdCR in *Leishmania* was to overexpress either the full length protein or each of its putative domains separately in the parasite using a leishmanial expression system. To that end, *L. donovani* promastigotes were transfected with plasmid constructs encoding either the entire LdCR protein or solely its N-, P-, or C-domains. Each of these expressed proteins contained an N-terminal LdCR signal peptide and a C-terminal hemagglutinin (HA)-tag fused to the KEDL sequence (Fig. 20.6A). The expression of the HA-tagged proteins by stably transfected parasites was confirmed by Western-blots using an anti-HA antibody (Fig. 20.6B). This antibody reacted with a protein doublet at ~ 58-60 kDa in lysates of parasites transfected with plasmid construct encoding the full length of LdCR (Fig. 20.6B, lane 8). The anti HA antibody also reacted with ~30 kDa (Fig. 20.6B, lane 9), ~32 kDa (Fig. 20.6B, lane 10) and ~18 kDa (Fig. 20.6B, lane 11) proteins in lysates of transfected parasites with plasmids containing the N-, P- and C- domains of LdCR respectively. The apparent high molecular mass of the expressed N- and P-domains observed in these experiments

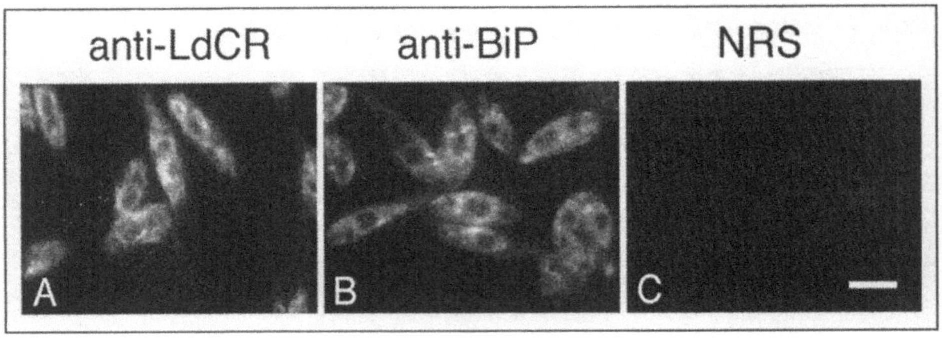

Figure 20.5. Immunolocalization of calreticulin in *Leishmania. L. donovani* wild type promastigotes were fixed in suspension in 4% paraformaldehyde in phosphate buffer (PBS) for 20 min on ice, washed in PBS and air dried onto glass slides. Cells were permeabilized in absolute methanol at –20°C for 6 min, rinsed in PBS and incubated for 30 min in 5% bovine serum albumin (BSA) in PBS. Subsequently, cells were incubated for 1 h with either the anti-LdCR (A), the anti-BiP (B), or a normal rabbit serum control (NRS, C) at appropriate dilutions in 1% BSA in PBS. Following three washes in PBS, cells were incubated for 1 h with fluorescein goat-anti rabbit antibodies (1/200 dilution) diluted in 1% BSA in PBS. Cells were further washed in PBS, mounted in Vectashield and observed using an epifluorescence microscope. Bar in (C) represents 10 micrometers.

(Fig. 20.6B, lanes 9 and 10) probably reflects the glycosylation of the N-domain and the proline rich content of the P-domain respectively.

These Western blots were also reacted with an anti-LdCR antibody (Fig. 20.6B). This antibody reacted with the endogenous ~60 kDa LdCR in all lysates of transfected parasites (Fig. 20.6B, lanes 2-6) including the control which contains the expression plasmid alone (Fig. 20.6B, lane 2). Results showed that the endogenous level of expression of the LdCR was similar in all transfected parasites. The anti-LdCR antibody also reacted with a ~32 kDa and a ~ 28 kDa protein in lysates of cells expressing the LdCR P-domain (Fig. 20.6B, lane 5). The ~32 kDa protein corresponds to the expressed P-domain and the lower ~28 kDa protein probably corresponds to a degradation product of ~32 kDa protein or could also reflect a specific proteolytic cleavage product of the P-domain. However, the anti-LdCR antibody reacted with neither the N- nor with the C-domain expressed proteins (Fig. 20.6B, lanes 4 and 6), suggesting that most of the antigenic epitopes of the LdCR are localized within its central P-domain. Further, immunofluorescence analyses of transfected parasites using both anti-HA and anti BiP antibodies showed colocalization of the HA-tagged proteins with BiP in all transfectants, therefore suggesting proper targeting and retention of the expressed proteins in the ER of these parasites (data not shown).

Effect of Expression of LdCR-Domains on Acid Phosphatase Secretion

Having determined that the HA-tagged expressed proteins were properly expressed by and targeted to the ER of the transfected parasites, we assessed whether such expression had an effect on the secretory pathway of these parasites. The secretory acid phosphatases (SAcPs) were used as marker proteins in these assays since they represent the major secreted proteins by *Leishmania*[21] and their trafficking through the parasite secretory pathway has been previously established.[21,51,52] Results of pulse/chase experiments followed by immunoprecipitation of radiolabeled SAcPs showed that 35-40% of the radiolabeled SAcPs, that were associated with the LdCR-KS, LdCR-F or LdCR-N cells after 10 min of pulse labeling, were found in the culture supernatants of these cells after 1h chase (Fig. 20.7A), and that 45-55% remained associated with the cells after the same period (Fig. 20.7B). In contrast, only 10-15% of the radiolabeled SAcPs, that were associated with the LdCR-P or LdCR-C cells after 10 min of pulse labeling, were found in the culture supernatants of these cells after 1h chase (Fig. 20.7A), and 75-90%

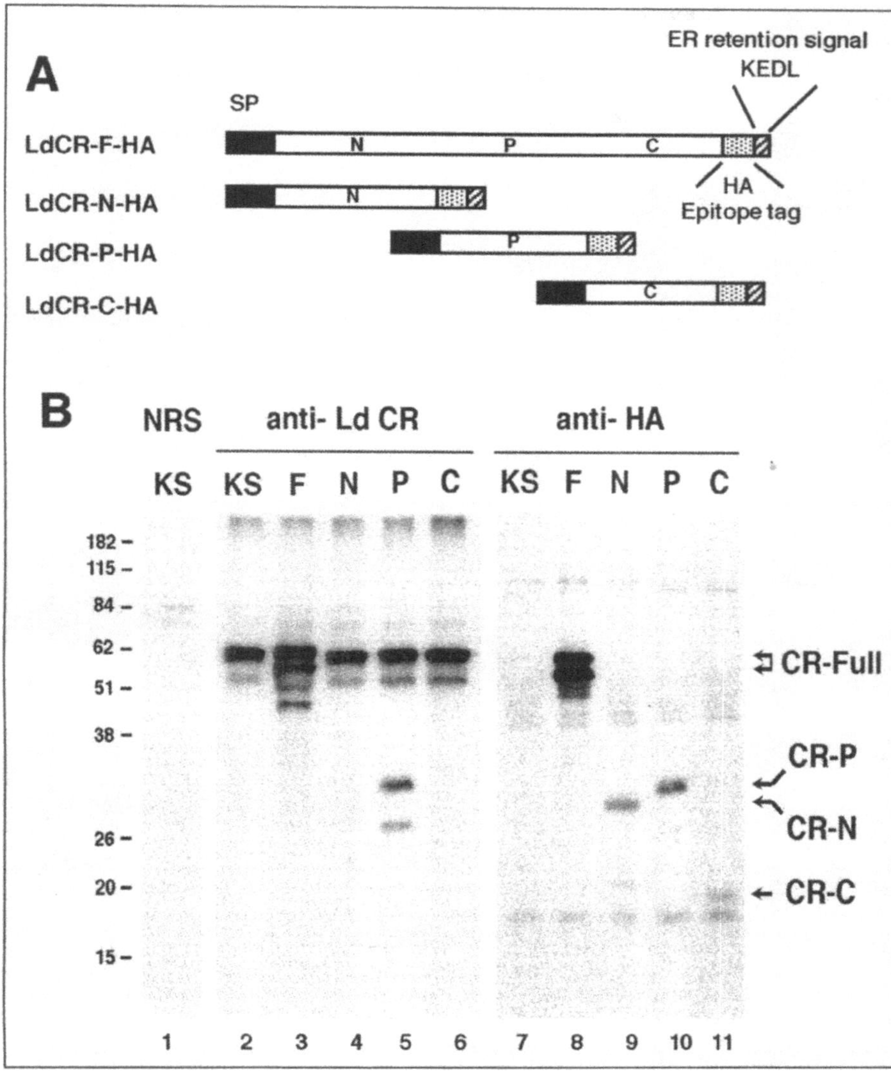

Figure 20.6. Expression of LdCR and its putative domains in *Leishmania*. A) Map of the episomaly expressed full length (LdCR-F-HA), N-domain (LdCR-N-HA), P-domain (LdCR-P-HA), or C-domain (LdCR-C-HA) proteins. The four expressed proteins contain the same N-terminal LdCR signal peptide (SP, solid black box) and C-terminal hemagglutinin epitope tag (HA) followed by the LdCR KEDL ER retention signal. Nucleotide sequences encoding these HA-tagged proteins were generated by PCR using the wild type LdCR gene as a template and appropriate primers and were cloned into the *[pKS NEO] Leishmania* expression plasmid. *L. donovani* promastigotes were subsequently transfected with the resulting plasmid constructs and selected for growth in the presence of geneticin (G418). B) Western blot of lysates from *L. donovani* promastigotes transfected with either the expression plasmid alone (KS, control) or plasmids encoding LdCR-F-HA (F), LdCR-N-HA (N), LdCR-P-HA (P), or LdCR-C-HA (C) proteins and reacted with either an anti-HA or anti-LdCR or control rabbit (NRS) antibodies. Molecular mass in kDa of protein standards is shown on the left. Arrows on the right indicate either the full length LdCR (CR-Full) or each of the expressed proteins.

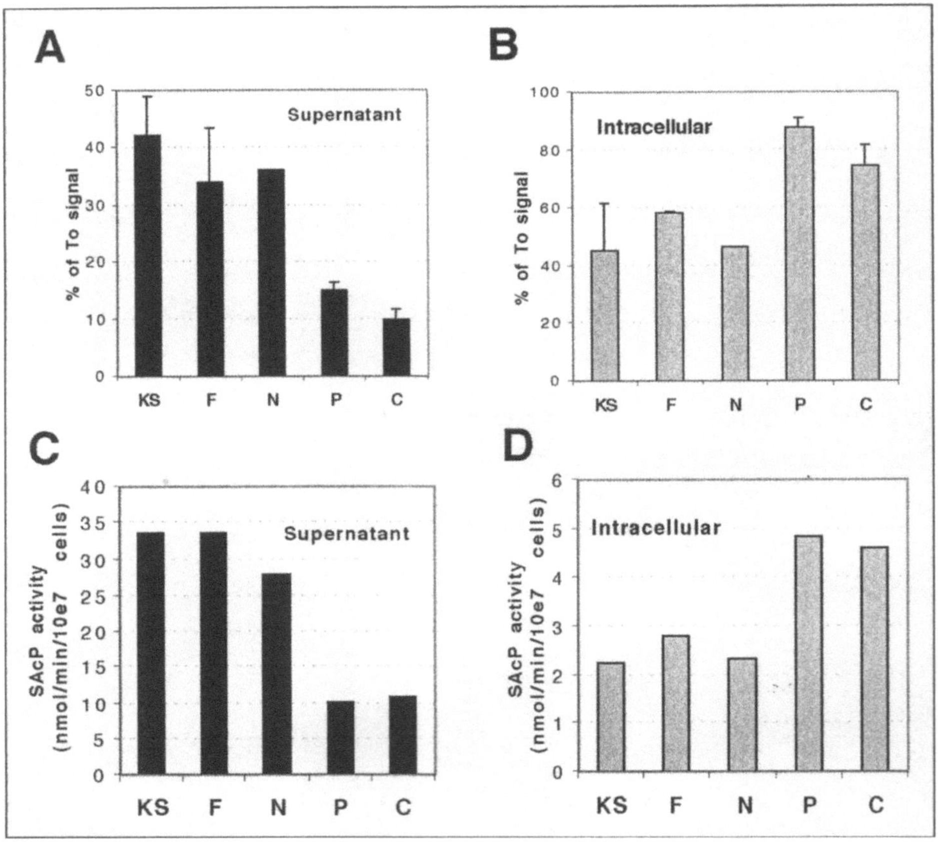

Figure 20.7. Secretion of SAcP by *Leishmania* transfectants. Panels A and B: Quantification of either intracellular (A) or secreted (B) radiolabeled SAcP proteins from parasites transfected with either the expression plasmid alone (KS, control) or plasmid encoding LdCR-F-HA (F), LdCR-N-HA (N), LdCR-P-HA (P), or LdCR-C-HA (C) proteins. Log-phase transfected promastigotes were pulse labeled with [^{35}S]methionine for 10 min and chased in complete culture medium for 1 h at 26°C. Both 10 min pulse labeled and 1 h chased cell pellets and culture supernatants were collected. The cell pellets were lysed (10^8 cells/ml) in NET lysis buffer (50 mM Tris, 150 mM NaCl, 1% Nonidet P40, 5 mM EDTA, 0.5 % deoxycholic acid, 0.1% , SDS, 10 µg/ml leupeptin, 4 µg/ml aprotinin, pH 7.5) for 30 min on ice and subsequently centrifuged at 10,000 x g for 30 min at 4°C. The labeled cell-free culture supernatants were centrifuged at 10,000 x g for 15 min to eliminate remaining cell debris. Subsequently, 50 µl aliquots of labeled cell lysates (either from 10 min pulse or after 60 min chase) or 100 µl aliquots of 1 h chased culture supernatants were diluted to 500 µl with NET lysis buffer and subjected to an immunoprecipitation using an anti-SAcP specific antibody. Affinity purified proteins were analyzed by SDS-PAGE and phosphorimaging. The amount of radiolabeled SAcP proteins from each sample was quantified using ImageQuant software. Results show the radiolabeled SAcP proteins either in the cell pellets (A) or in the culture supernatants (B) of each transfectant, after 1 h chase, and expressed as percent of the amount of labeled SAcP proteins measured inside cells after 10 min pulse. Panels C and D: Quantification of either intracellular (C) or secreted (D) SAcP enzymatic activity from transfected parasites described in panels A and B. After 24 h incubation in fresh culture medium at 26°C, cells and culture supernatants from these parasite cultures were harvested and assayed for acid phosphatase activity using paranitrophenyl phosphate as substrate as previously described.[64] The tartrate sensitive acid phosphatase enzyme activity (SAcP activity) is expressed as nmol of substrate hydrolyzed per min per 10^7 cells (nmol/min/10^7 cells).

remained associated with the cells after the same period (Fig. 20.7B). These results suggest that the episomal expression of either P- or C-domains of LdCR in *Leishmania* can alter the trafficking/processing of the SAcP proteins resulting in their apparent intracellular accumulation.

In order to determine whether the SAcP proteins were retained inside these cells because they were misfolded, the steady state levels of SAcP enzymatic activity were measured in lysates of all transfectants. In addition, the amount of SAcP enzymatic activity was also measured in supernatants of these cell after 24 h of culture in fresh medium. Results showed that SAcP activity was detected in culture supernatants of all transfected parasites (Fig. 20.7C). In agreement with our pulse/chase results above, supernatants of LdCR-P and LdCR-C transfectants had significantly less (~70%) SAcP activity than supernatants of either control cells LdCR-KS or LdCR-F transfectants (Fig. 20.7C). The expression of the N-domain of LdCR had limited effect on the secretion of SAcP since culture supernatant of LdCR-N contained only ~15% less SAcP activity than that of control cells (Fig. 20.7C). Also in agreement with our pulse/chase results above, the steady state levels of SAcP activity measured in lysates of transfected LdCR-P and LdCR-C cells were ~ two-fold higher than that of control LdCR-KS cells (Fig. 20.7D) whereas LdCR-F and LdCR-N transfectants had similar intracellular levels of SAcP activity as control cells (Fig. 20.7D).

Taken together, these results suggest that the overexpression of either the P- or C-domains of LdCR in *Leishmania* results in an intracellular accumulation of active, and therefore probably properly folded, SAcPs in these transfected parasites. Further, such overexpression was also associated with a significant reduction of SAcP secretion by these two transfected parasites.

Effect of LdCR-Domains Expression on SAcP-Chaperone Interaction

The above results suggest that the SAcP proteins are retained inside the transfected LdCR-P and LdCR-C parasites. One of the mechanisms by which proteins can be retained inside cells is via interaction with ER chaperones such as CRT or BiP.[53,54] These two chaperones have been shown to help to retain misfolded or incompletely assembled proteins in the ER that are further targeted to proteasome degradation.[53,54] Such retention of misfolded proteins by CRT and BiP is a part of the quality control of protein folding in the ER.[41,55] Therefore, in order to determine whether the apparent intracellular accumulation of SAcP proteins inside LdCR-P and LdCR-C transfectants was the result of its retention by the ER chaperone CRT or BiP, these parasites were pulse labeled for 10 min with ^{35}S-methionine and chased for up to 30 min in complete culture medium. Subsequently, lysates of these radiolabeled cells were subjected to immunoprecipitation (IP) using anti-LdCR or anti-BiP antibodies. Affinity bound proteins were subsequently subjected to a second IP using an anti-SAcP antibody to determined whether SAcP proteins were co-immunoprecipitated with either LdCR or BiP during the first IP. Protein samples were further analyzed by SDS-PAGE and fluorography. Results in Figure 20.8 showed co-immunoprecipitation of SAcP with BiP from lysates of control LdCR-KS , LdCR-P and LdCR-C transfectants, after 10 min of radiolabel (Fig. 20.A, lane 1, 6, and 11 respectively). Further, the kinetics of interaction between SAcP and BiP during the 30 min chase was very similar between the three transfectants (Fig. 20.B). These results suggest that SAcP proteins do not interact with BiP for longer periods of time in LdCR-P and LdCR-C transfectants and therefore suggest that BiP/SAcP interaction is probably not involved in the intracellular accumulation and thus has no role in the reduced secretion of SAcP in these two transfectants.

In these experiments, results of co-immunoprecipitations of LdCR and SAcP using an anti-LdCR antibody were inconclusive and failed to provide evidence of direct interaction between these two proteins. These results suggest that SAcPs may not interact directly with LdCR in vivo. This possibility is in agreement with Molinari and Helenius[56] who showed that an interaction between CRT and a glycoprotein is favored if multiple N-linked glycans are present in the first 50 N-terminal amino acid residues of the protein. Since SAcPs have only one N-linked glycan in the first 50 amino acid residues, at position 44,[57] it is possible that they

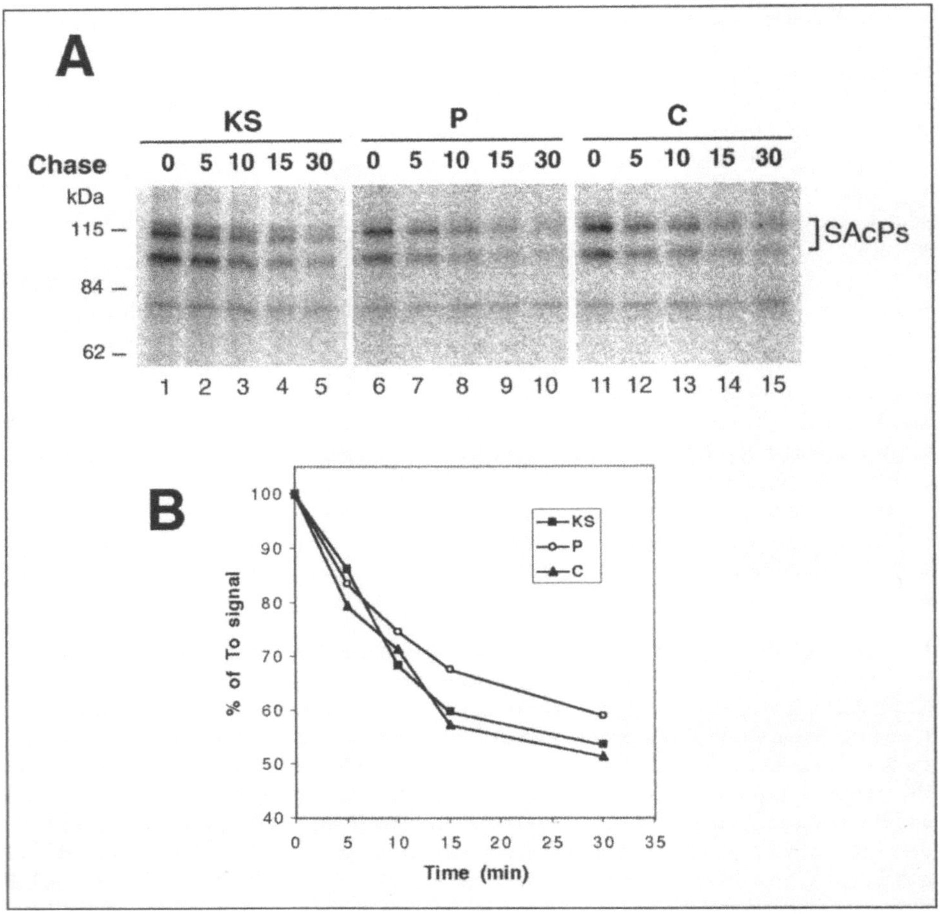

Figure 20.8. Interaction of SAcP and BiP in *Leishmania* transfectants. Control parasites (KS) or parasites expressing LdCR-P-HA (P), or LdCR-C-HA (C) were pulse labeled with [^{35}S]methionine for 10 min and chased in complete culture medium for 0, 5, 10, 15, and 30 min at 26°C. For each time point, labeled cells were collected, lysed and processed for immunoprecipitations (IP) as above (Fig. 20.7) using an anti-BiP specific antibody. Affinity bound proteins were subsequently subjected to a second IP using an anti-SAcP antibody. Protein samples were subsequently analyzed by SDS-PAGE and fluorography (A). The molecular mass of protein standards are shown on the left. The radiolabeled SAcP proteins are indicated on the right. The SAcP proteins in each lane were quantitated by phosphorimaging analysis and expressed as percent of the labeled SAcP measured in cell lysates after 10 min pulse labeling (To) (B).

do not interact directly with LdCR or with sufficient affinity to have been detected in our immunoprecipitation experiments. In addition, our results are also in agreement with Molinari and Helenius who showed that a protein is more likely to interact with BiP than with CRT if it does not contain any N-glycan within the first 50 amino acid residues.[56]

Taken together, our results showed that the intracellular accumulation and reduced secretion of SAcPs observed in parasites expressing either the P- or C-domain of LdCR is not the result of their retention by the ER chaperone LdCR or BiP. The exact mechanisms involved in such altered trafficking of SAcP proteins in the two transfected parasites remain to be determined.

Dominant-Negative Effect of Expression of Putative Domains of LdCR on the Parasite Survival in Macrophages in Vitro

Infection of Macrophages with LdCR Transfectants

The above results suggest that the episomal expression of either the P-domain or the C-domain of LdCR in *Leishmania* can affect the release/secretion of proteins trafficking through of the parasite secretory pathway. Since some of these proteins (including the SAcPs) are thought to be involved in the parasite survival inside the host macrophages,[18,21] we examined the viability of these transfected parasites inside macrophages in vitro. To that end, peritoneal macrophages isolated from BALB/c mice were infected in vitro with the control (LdCR-KS) and LdCR-F, LdCR-N, LdCR-P, and LdCR-C transfected parasites. Macrophages were infected with promastigotes from the stationary phase of cultures, which contain the infectious metacyclic forms of the parasite[58] and were maintained in culture for 72 h. The percent of infected macrophages in these cultures was counted at 6, 24, 48, and 72 h (Fig. 20.9).

Results showed that after 6 h of infection, 80-90% of the macrophages were infected in all cultures (Fig. 20.9). The number of infected macrophages decreased gradually over time to reach ~30% in cultures infected with the control parasites (LdCR-KS). This decrease probably corresponds to the elimination of non metacyclic parasites that are not fit to establish an infection in macrophages. A similar decrease of the number of infected macrophages was observed in macrophage cultures infected with LdCR-C transfected parasites (Fig. 20.9). In contrast, the number of infected macrophages decreased significantly faster in macrophage cultures infected with either LdCR-F, LdCR-N, or LdCR-P and reached only 5-10% after 72 h (Fig. 20.9). Similar infection experiments performed using human derived macrophages also showed a significant decrease of LdCR-P and LdCR-F parasite survival inside macrophages compared to control LdCR-KS parasites (not shown). These results indicate that overexpressing LdCR or the P- or N-domain of LdCR in *Leishmania* can result in a significant reduction of parasite survival in macrophages in vitro. In addition, since mice and hamsters are good model systems to study leishmaniasis in vivo,[59,60] the pathogenesis of the above transfectants is currently being assessed in these animals.

From Dominant Negative Expression of LdCR Domains to Reduced Parasite Virulence

The schematic diagram (Scheme-1) illustrates the possible mechanism of dominant negative expression of P-domain or N-domain of LdCR, observed in transfected parasites, on the secretion/release of a parasite glycoproteins and resulting in the decrease parasite survival in macrophages. In normal cells or control transfected parasites (i.e., LdCR-KS), nascent N-linked glycosylated proteins in the ER, containing a terminal glucose, would interact with the P-domain of the endogenous parasite calreticulin (Scheme-1, route #1). It was shown in higher eukaryotes that ERP57, a member of the PDI family, is part of the folding complex via interaction with the N- domain of CRT and contributes to the folding of the glycoprotein by catalyzing the formation of its disulfide bonds.[61,62] After their interaction with similar folding complexes (LdCR + ERP57) in *Leishmania* , the properly folded parasite glycoproteins would be secreted into the extracellular milieu, or be targeted to the parasite cell surface and assume their biological functions, possibly contributing to the overall parasite survival inside the macrophage. However, in parasites expressing the P-domain of LdCR (LdCR-P transfectants), nascent glycoproteins will interact either with the expressed P-domain which is present and freely available in the ER of these transfected parasites or with the endogenous LdCR (Scheme-1, route #2). The interaction with the expressed P-domain will be "sterile" and the glycoproteins will have to further interact with the endogenous, fully active, LdCR in order to be properly folded. Such P-domain expression will render the folding machinery in LdCR-P cells less efficient and

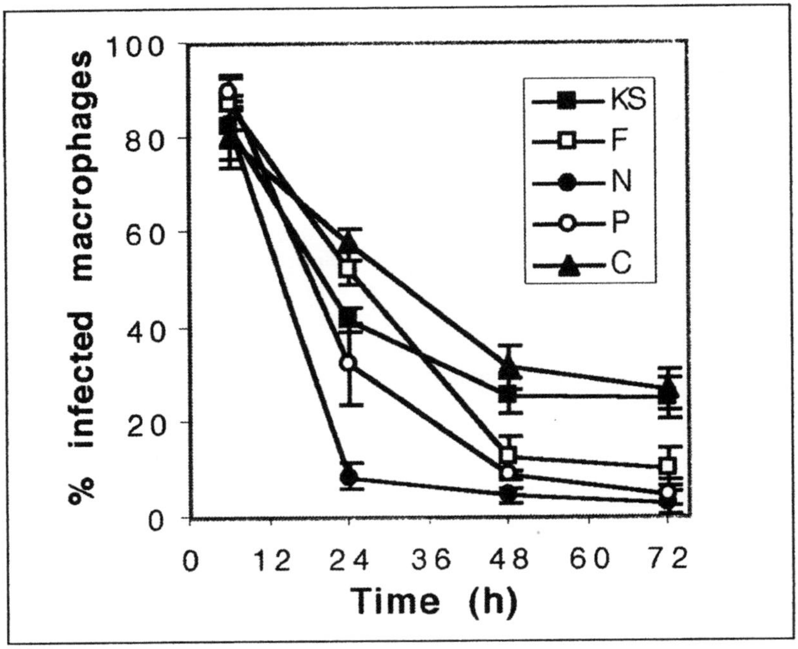

Figure 20.9. Infection of macrophages with *Leishmania* transfectants. Peritoneal macrophages from BALB/c mice were resuspended at 2 x 10^5 cells/ml in RPMI containing 25 mM Hepes, 10% fetal bovine serum, 2 mM L-glutamine, 100 IU/ml penicillin, 50 µg/ml streptomycin, 5 µg/ml gentamycin, plated in eight chamber Lab-Tek tissue culture slides (0.5 ml/well) and incubated overnight at 37°C in 5% CO_2. Non adherent cells were eliminated by extensive washing in RPMI. Promastigotes from stationary phase cultures of LdCR-KS, LdCR-F, LdCR-N, LdCR-P and LdCR-C were added to the mouse macrophages cultures (10:1, parasite to macrophage ratio) for 5 hours at 37°C in 5% CO2. After incubation, free extracellular parasites were removed by repeated washing in RPMI, and the cultures were incubated in macrophage culture medium for another 72 hours. At 6, 24, 48 and 72 hours post infection, the culture medium was removed from the culture slides, the slides were air-dried for 5 min at room temperature, fixed by immersion in absolute methanol for 10 min and stained using Diff-Quick Stain set. For each culture, a minimum of 300 macrophages were counted. Values are expressed as percentage of macrophages that were infected by *Leishmania*.

could result in insufficient amount of properly folded glycoproteins being released by these transfected parasites. If such glycoproteins in fact represent parasite virulence factors, the survival of P-domain expressing parasites within macrophages may be affected. Further, it was shown recently that ERP57 can interact with the P-domain of CRT in mammalian cells.[63] If such interaction also occurs in *Leishmania*, it may be affected by the episomal expression of the LdCR P-domain in LdCR-P transfectants and thus affect the ER folding machinery of these parasites.

Similarly, the folding of glycoproteins could also be affected in parasites overexpressing the N-domain of LdCR. The expressed N-domain present in the ER of these transfected parasites would compete with the endogenous LdCR for the binding of ERP57 (Scheme-1, route #3). If ERP57 is not present in the complex LdCR/glycoprotein, the folding of the later glycoproteins will not occur properly (i.e., poor disulfide bonds formation) and such glycoproteins will remain in the ER until they encounter active folding complexes containing ERP57. Such transfected parasites may be less efficient at processing glycoprotein virulence factors trafficking through the ER resulting in their reduced survival in macrophages.

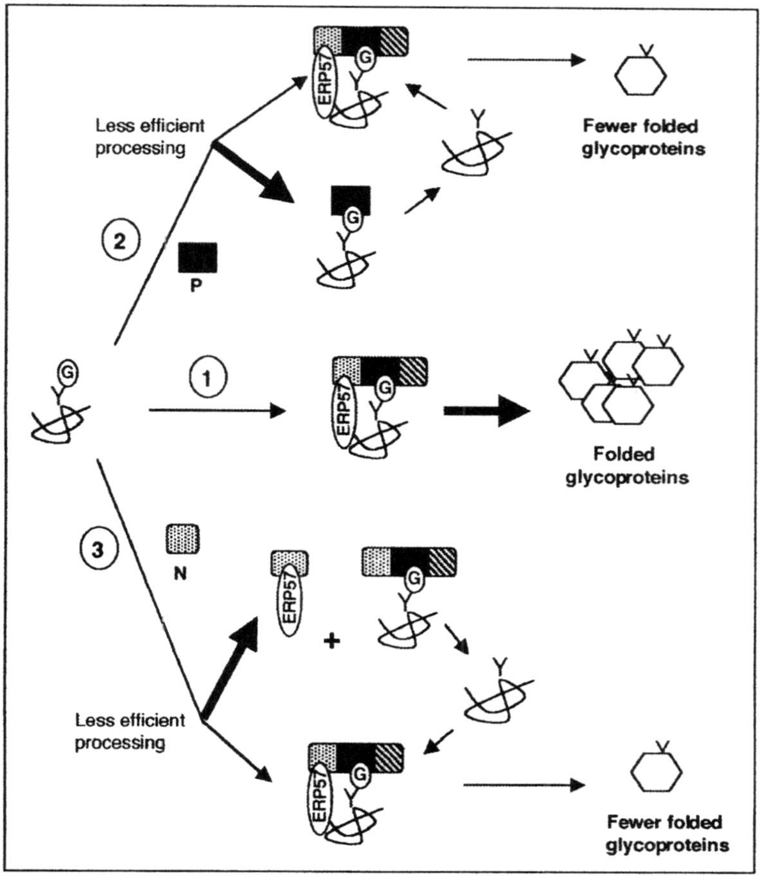

Scheme-1. Model of dominant negative interaction between glycoproteins and N- and P-domains of LdCR. The schematic diagram illustrates the possible mechanism of dominant negative expression of P-domain or N-domain of LdCR on the parasite secretion/release of glycoproteins. Route #1 represents the normal situation where nascent N-linked glycosylated proteins in the ER, containing a terminal glucose, interact with the P-domain of the endogenous parasite calreticulin. A member of the PDI family (ERP57), part of the protein complex, catalyzes the formation of disulfide bonds in the bound glycoproteins. The properly folded glycoproteins are released from LdCR, continue on the secretory pathway through the ER and Golgi, and are either secreted into the extracellular milieu, or targeted to the parasite cell surface, where they assume their biological functions (e.g., possibly contributing to the overall parasite survival inside macrophage). Route #2 represents the conditions in parasites expressing the P-domain of LdCR. In this pathway, nascent glycoproteins will interact with the expressed P-domain which is present and freely available in the ER of these transfected parasites thus interfering in the normal interaction with the endogenous LdCR. The interaction between glycoproteins and expressed P-domain will be "sterile" and the glycoproteins will have to further interact with the endogenous, fully active, LdCR in order to be properly folded. Such P-domain overexpression will render the folding machinery in LdCR-P cells less efficient, and could result in an insufficient amount of properly folded glycoproteins being released by these transfected parasites. If these glycoproteins represent parasite virulence factors, the survival of these transfected parasites inside macrophages might be affected. Route #3 represents parasites expressing the N-domain of LdCR. In this pathway. The expressed N-domain present in the ER of these transfected parasites will compete with the endogenous LdCR for the binding of ERP57. If ERP57 is not present in the LdCR/glycoprotein complex, the folding of the later glycoproteins will not occur properly (i.e., no disulfide bonds will form) and such glycoproteins will remain in the ER until they encounter active folding complexes containing ERP57. These transfected parasites may also be less efficient at processing glycoprotein virulence factors trafficking through the ER resulting in their reduced survival phenotype in macrophages.

Conclusion

Results presented in this paper provide a basis for understanding the secretory pathway of *Leishmania* parasites specifically with regard to the role of calreticulin in that pathway. We showed that overexpressing either the P- or the C-domain of LdCR in *Leishmania* resulted in a significant decrease in SAcP secretion by the parasites. SAcP were used as marker proteins in these experiments, however, other proteins trafficking through the secretory pathway of the parasite could also be affected. This might affect many mechanisms in which these proteins play a role, such as mechanisms involved in parasite survival inside its human host and insect vector (e.g., resistance to the host immune system or the salvage of essential nutrients). This hypothesis was supported by results of our macrophage infection studies which showed that parasites expressing the P-domain of LdCR have decreased survival inside macrophages compared to control cells. The SAcP proteins are probably not involved in that stage of parasite development since parasites expressing the C-domain of LdCR, which also show a significant decrease of SAcP secretion, showed no decrease in survival in these experiments. These results also suggest that LdCR-P and LdCR-C overexpression down regulate SAcP secretion via different mechanisms, and that other proteins besides SAcPs are being affected as a result of LdCR-P overexpression. We propose a model of dominant negative interaction between the expressed P-domain and nascent glycoproteins in the ER in these transfected parasites that would result in a less efficient folding machinery and a reduction in the secretion of virulence factors.

Similarly, a model of dominant negative expression of the N-domain of LdCR in the ER of transfected parasites could explain the reduction of parasite survival in macrophages. In that model, the expressed LdCR N-domain in the ER would bind ERP57 resulting in an increase number of inactive folding complexes containing nascent glycoproteins and LdCR but lacking ERP57. Such dominant negative expression of LdCR-N would also result in an inefficient ER folding machinery and reduced protein secretion. Further, this study also showed that parasites overexpressing the full length LdCR also showed reduced parasite survival inside macrophages even though no significant reduction of SAcP secretion was observed in these parasites. However, this does not exclude the possibility that secretion of virulence factors other than SAcPs is affected in LdCR-F transfectants. Further studies are needed to describe this phenomenon. In addition, since LdCR is a calcium binding protein, a change in calcium homeostasis in these transfected parasites could result in a general negative effect on the parasite secretory pathway. Such calcium imbalance in the ER could also contribute to the phenotypes observed for LdCR-P and LdCR-C transfectants, since both of these domains contain calcium binding sites, however this remains to be demonstrated.

Finally, this study showed that altering the function of the ER chaperone calreticulin in *Leishmania* can affect the secretory pathway of this parasite and also reduce the parasite survival/virulence inside macrophages in vitro. This study suggests that this approach could be further exploited to develop a live attenuated vaccine against human leishmaniasis.

Acknowledgments

We thank Dr. Greg Matlashewski (McGill University, Montreal, Canada) for providing the expression plasmid *[pKS NEO]* and Dr. James Bangs (University of Wisconsin-Madison, Madison, WI) for providing the anti-BiP antibody. We thank Drs. S. Mendez and S. Bertholet for their help with the macrophage infections. We also thank Dr. R. Duncan for his critical review of this manuscript.

References

1. Desjeux P. Human leishmaniases: epidemiology and public health aspects. World Health Stat Q 1992; 45:267-275.
2. Handman E. Leishmaniasis: current status of vaccine development. Clin Microbiol Rev 2001; 14:229-243.
3. Desjeux P, Piot B, O'Neill K et al. Co-infections of leishmania/HIV in south Europe. Med Trop 2001; 61:187-193.
4. Molyneux D, Killick-Kendrick R. Morphology, ultrastructure and life cycles. In: Peters W, Killick-Kendrick R, eds. The leishmaniases in biology and medicine. London: Academic Press, 1987:121-176.
5. Alexander J, Satoskar AR, Russell DG. Leishmania species: models of intracellular parasitism. J Cell Sci 1999; 112:2993-3002.
6. McConville MJ, Blackwell JM. Developmental changes in the glycosylated phosphatidylinositols of Leishmania donovani. Characterization of the promastigote and amastigote glycolipids. J Biol Chem 1991; 266:15170-15179.
7. Mukkada AJ, Meade JC, Glaser TA et al. Enhanced metabolism of Leishmania donovani amastigotes at acid pH: an adaptation for intracellular growth. Science 1985; 229:1099-1101.
8. Bahr V, Stierhof YD, Ilg T et al. Expression of lipophosphoglycan, high-molecular weight phosphoglycan and glycoprotein 63 in promastigotes and amastigotes of Leishmania mexicana. Mol Biochem Parasitol 1993; 58:107-121.
9. Cairns BR, Collard MW, Landfear SM. Developmentally regulated gene from Leishmania encodes a putative membrane transport protein. Proc Natl Acad Sci USA 1989; 86:7682-7686.
10. Charest H, Matlashewski G. Developmental gene expression in Leishmania donovani: differential cloning and analysis of an amastigote-stage-specific gene. Mol Cell Biol 1994; 14:2975-2984.
11. Joshi M, Dwyer DM, Nakhasi HL. Cloning and characterization of differentially expressed genes from in vitro-grown 'amastigotes' of Leishmania donovani. Mol Biochem Parasitol 1993; 58:345-354.
12. Kidane GZ, Samaras N, Spithill TW. Cloning of developmentally regulated genes from Leishmania major and expression following heat induction. J Biol Chem 1989; 264:4244-4250.
13. Shapira M, McEwen JG, Jaffe CL. Temperature effects on molecular processes which lead to stage differentiation in Leishmania. Embo J 1988; 7:2895-2901.
14. Pogue GP, Lee NS, Koul S et al. Identification of differentially expressed Leishmania donovani genes using arbitrarily primed polymerase chain reactions. Gene 1995; 165:31-38.
15. Streit JA, Donelson JE, Agey MW et al. Developmental changes in the expression of Leishmania chagasi gp63 and heat shock protein in a human macrophage cell line. Infect Immun 1996; 64:1810-1818.
16. Turco SJ, Sacks DL. Expression of a stage-specific lipophosphoglycan in Leishmania major amastigotes. Mol Biochem Parasitol 1991; 45:91-99.
17. Duncan R, Alvarez R, Jaffe CL et al. Early response gene expression during differentiation of cultured Leishmania donovani. Parasitol Res 2001; 87:897-906.
18. Dwyer DM, Gottlieb M. The surface membrane chemistry of Leishmania: its possible role in parasite sequestration and survival. J Cell Biochem 1983; 23:35-45.
19. Schneider P, Bordier C, Etges R. Membrane proteins and enzymes of Leishmania. Subcell Biochem 1992; 18:39-72.
20. Ferguson MA. The surface glycoconjugates of trypanosomatid parasites. Philos Trans R Soc Lond B Biol Sci 1997; 352:1295-1302.
21. Bates PA, Dwyer DM. Biosynthesis and secretion of acid phosphatase by Leishmania donovani promastigotes. Mol Biochem Parasitol 1987; 26:289-296.
22. Shakarian AM, Dwyer DM. The Ld Cht1 gene encodes the secretory chitinase of the human pathogen Leishmania donovani. Gene 1998; 208:315-322.
23. Webb JR, Campos-Neto A, Ovendale PJ et al. Human and murine immune responses to a novel Leishmania major recombinant protein encoded by members of a multicopy gene family. Infect Immun 1998; 66:3279-3289.
24. Labriola C, Cazzulo JJ, Parodi AJ. Trypanosoma cruzi calreticulin is a lectin that binds monoglucosylated oligosaccharides but not protein moieties of glycoproteins. Mol Biol Cell 1999; 10:1381-1394.
25. Bangs JD. Surface coats and secretory trafficking in African trypanosomes. Curr Opin Microbiol 1998; 1:448-454.

26. Overath P, Stierhof Y, Wiese M. Endocytosis and secretion in trypanosomatid parasites—tumultuous traffic in a pocket. Trends Cell Biol 1997; 7:27-33.
27. Landfear SM, Ignatushchenko M. The flagellum and flagellar pocket of trypanosomatids. Mol Biochem Parasitol 2001; 115:1-17.
28. Sogin M. History assignment: when was the mitochondrion founded? Curr Opin Genet Dev 1997; 7:792-799.
29. Ilg T, Menz B, Winter G et al. Monoclonal antibodies to Leishmania mexicana promastigote antigens. I. Secreted acid phosphatase and other proteins share epitopes with lipophosphoglycan. J Cell Sci 1991; 99:175-180.
30. Andrews NW, Whitlow MB. Secretion by Trypanosoma cruzi of a hemolysin active at low pH. Mol Biochem Parasitol 1989; 33:249-256.
31. Parodi AJ, Quesada-Allue LA. Protein glycosylation in Trypanosoma cruzi. I. Characterization of dolichol-bound monosaccharides and oligosaccharides synthesized "in vivo". J Biol Chem 1982; 257:7637-7640.
32. Quesada-Allue LA, Parodi AJ. Novel mannose carrier in the trypanosomatid Crithidia fasciculata behaving as a short alpha-saturated polyprenyl phosphate. Biochem J 1983; 212:123-128.
33. Low P, Dallner G, Mayor S et al. The mevalonate pathway in the bloodstream form of Trypanosoma brucei. Identification of dolichols containing 11 and 12 isoprene residues. J Biol Chem 1991; 266:19250-19257.
34. de la Canal L, Parodi AJ. Synthesis of dolichol derivatives in trypanosomatids. Characterization of enzymatic patterns. J Biol Chem 1987; 262:11128-11133.
35. Parodi AJ, Quesada Allue LA, Cazzulo JJ. Pathway of protein glycosylation in the trypanosomatid Crithidia fasciculata. Proc Natl Acad Sci USA 1981; 78:6201-6205.
36. Mendelzon DH, Previato JO, Parodi AJ. Characterization of protein-linked oligosaccharides in trypanosomatid flagellates. Mol Biochem Parasitol 1986; 18:355-367.
37. Bosch M, Trombetta S, Engstrom U et al. Characterization of dolichol diphosphate oligosaccharide: protein oligosaccharyltransferase and glycoprotein-processing glucosidases occurring in trypanosomatid protozoa. J Biol Chem 1988; 263:17360-17365.
38. Parodi AJ, Lederkremer GZ, Mendelzon DH. Protein glycosylation in Trypanosoma cruzi. The mechanism of glycosylation and structure of protein-bound oligosaccharides. J Biol Chem 1983; 258:5589-5595.
39. Trombetta SE, Bosch M, Parodi AJ. Glucosylation of glycoproteins by mammalian, plant, fungal, and trypanosomatid protozoa microsomal membranes. Biochemistry 1989; 28:8108-8116.
40. Ellgaard L, Molinari M, Helenius A. Setting the standards: quality control in the secretory pathway. Science 1999; 286:1882-1888.
41. Parodi AJ. Role of N-oligosaccharide endoplasmic reticulum processing reactions in glycoprotein folding and degradation. Biochem J 2000; 348:1-13.
42. Bangs JD, Uyetake L, Brickman MJ et al. Molecular cloning and cellular localization of a BiP homologue in Trypanosoma brucei. Divergent ER retention signals in a lower eukaryote. J Cell Sci 1993; 105:1101-1113.
43. Hsu MP, Muhich ML, Boothroyd JC. A developmentally regulated gene of trypanosomes encodes a homologue of rat protein-disulfide isomerase and phosphoinositol-phospholipase C. Biochemistry 1989; 28:6440-6446.
44. Joshi M, Pogue GP, Duncan R et al. Isolation and characterization of Leishmania donovani calreticulin gene and its conservation of the RNA binding activity. Mol. Biochem. Parasitol 1996; 81:53-64.
45. Bangs JD, Brouch EM, Ransom DM et al. A soluble secretory reporter system in Trypanosoma brucei. Studies on endoplasmic reticulum targeting. J Biol Chem 1996; 271:18387-18393.
46. Nakhasi HL, Pogue GP, Duncan RD et al. Implications of calreticulin function in parasite biology. Parasitology Today 1998; 14:157-160.
47. Cruz A, Coburn CM, Beverley SM. Double targeted gene replacement for creating null mutants. Proc Natl Acad Sci USA 1991; 88:7170-7174.
48. Dumas C, Ouellette M, Tovar J et al. Disruption of the trypanothione reductase gene of Leishmania decreases its ability to survive oxidative stress in macrophages. Embo J 1997; 16:2590-2598.
49. Cunningham ML, Titus RG, Turco SJ et al. Regulation of differentiation to the infective stage of the protozoan parasite Leishmania major by tetrahydrobiopterin. Science 2001; 292:285-287.
50. Mesaeli N, Nakamura K, Zvaritch E et al. Calreticulin is essential for cardiac development. J Cell Biol 1999; 144:857-868.

51. Bates PA, Hermes I, Dwyer DM. Leishmania donovani: immunochemical localization and secretory mechanism of soluble acid phosphatase. Exp Parasitol 1989; 68:335-346.
52. Bates PA, Hermes I, Dwyer DM. Golgi-mediated post-translational processing of secretory acid phosphatase by Leishmania donovani promastigotes. Mol Biochem Parasitol 1990; 39:247-255.
53. Hurtley SM, Bole DG, Hoover-Litty H et al. Interactions of misfolded influenza virus hemagglutinin with binding protein (BiP). J Cell Biol 1989; 108:2117-2126.
54. Klausner RD, Sitia R. Protein degradation in the endoplasmic reticulum. Cell 1990; 62:611-614.
55. Zhang JX, Braakman I, Matlack KE et al. Quality control in the secretory pathway: the role of calreticulin, calnexin and BiP in the retention of glycoproteins with C-terminal truncations. Mol Biol Cell 1997; 8:1943-1954.
56. Molinari M, Helenius A. Chaperone selection during glycoprotein translocation into the endoplasmic reticulum. Science 2000; 288:331-333.
57. Shakarian AM, Ellis SL, Mallinson DJ et al. Two tandemly arrayed genes encode the (histidine) secretory acid phosphatases of Leishmania donovani. Gene 1997; 196:127-137.
58. Sacks DL. Metacyclogenesis in Leishmania promastigotes. Exp Parasitol 1989; 69:100-103.
59. Goyal N, Guru PY, Rastogi AK. Status of glutathione in lymphoid tissues of golden hamster during Leishmania donovani infection. Indian J Biochem Biophys 1994; 31:211-213.
60. Ghedin E, Charest H, Zhang WW et al. Inducible expression of suicide genes in Leishmania donovani amastigotes. J Biol Chem 1998; 273:22997-23003.
61. Michalak M, Corbett EF, Mesaeli N et al. Calreticulin: one protein, one gene, many functions. Biochem J 1999; 344:281-292.
62. High S, Lecomte FJ, Russell SJ et al. Glycoprotein folding in the endoplasmic reticulum: a tale of three chaperones? FEBS Lett 2000; 476:38-41.
63. Frickel EM, Riek R, Jelesarov I et al. TROSY-NMR reveals interaction between ERp57 and the tip of the calreticulin P-domain. Proc Natl Acad Sci USA 2002; 99:1954-1959.
64. Gottlieb M, Dwyer DM. Identification and partial characterization of an extracellular acid phosphatase activity of Leishmania donovani promastigotes. Mol Cell Biol 1982;2:76-81.

Chapter 21

The Hookworm Calreticulin Conundrum

D.I. Pritchard, N. Girod, A. Brown, R. Caddick, D.S.W. Hooi, R.J. Quinnell, S.J. Johnson and P. Eggleton

Abstract

Hookworm parasites possess effective defence mechanisms against host innate and adaptive immune responses such as complement activation, eosinophilia and the respiratory burst in granulocytes. However there does appear to be a strong relationship between helminthic infection and the allergic phenotype in infected individuals.

One candidate parasite antigen implicated in each of the above phenomena is calreticulin. It was originally identified by screening parasite cDNA expression libraries with IgE antibody, and once cloned to homogeneity was shown to modulate complement activation and bind to the signalling domains of integrins involved in haemostasis and inflammation. Hence the term the hookworm calreticulin conundrum. In hookworms, calreticulin is acting as a potential pro-inflammatory allergen, yet has the ability to moderate inflammation. However, to fulfil these functions, calreticulin should be secreted by the parasite, in a manner already described for fleas and ticks.

The work described in the present chapter summarizes work published in this field to date, addresses the issue of calreticulin secretion, describes experiments with human basophils to examine the true allergenicity of calreticulin, and discusses the likelihood that the response in humans is driven by an autoimmune component.

Introduction

Hookworm calreticulin was cloned serendipitously during a systematic search for hookworm allergens.[1] In this study, IgE-rich plasma from hookworm infected individuals from rural Papua New Guinea (Madang Province) was used to screen a cDNA library from adult *Necator americanus.* Using a highly specific anti-IgE as a second antibody, a number of allergens were cloned, including calreticulin. This may be an important finding, given the importance of T-helper 2 responses to protective immunity to this parasite[2] and the apparent protective effect of hookworm infections on the development of allergy in the tropics.[3] However, the finding does in turn present what we term the hookworm '*calreticulin conundrum*'.

Hookworm Calreticulin May Be Secreted to Perform Important Biological Functions at the Host Parasite Interface

Calreticulin was first identified as a calcium binding protein of the muscle sarcoplasmic reticulum.[4] Subsequently, calreticulin was shown to be highly abundant in non-muscle tissues, and it is now known to be one of the major calcium binding proteins of the endoplasmic reticulum,[5] where it acts as a molecular chaperone. The C domain of calreticulin terminates with the ER retrieval sequence, KDEL leading to the commonly held belief that calreticulin is retrieved to, and retained within, the endoplasmic reticulum.[6]

Calreticulin, 2nd ed., edited by Paul Eggleton and Marek Michalak.

Nevertheless, evidence is increasing to support the appearance of KDEL proteins at the cell surface,[7] and calreticulin has been found at the cell surface,[8,9,10] associated with secretory granules and perforin in T cells,[11] and in the secretions of ectoparasites.[12,13,14] Calreticulin is also expressed at the surface of apoptotic cells.[15]

These findings lead to a number of interpretations of our data, and these were discussed in depth at the 4th International Calreticulin Workshop (Oxford University, 2000) where a number of possibilities were tabled.

1. *Necator* truly secretes calreticulin, where it comes into contact with, and suppresses the host immune response.[16]
2. The host response to calreticulin is induced by calreticulin released by dead or dying worms in the tissues.
3. The response seen is autoimmune, induced by host calreticulin released following tissue damage during worm migration and feeding, or expressed on the surface of apoptotic cells. *Necator* secretions do in fact induce apoptosis in activated T cells.[17] Selection of calreticulin from *Necator* cDNA libraries by post infection IgE could be the result of antigenic cross reactivity between relatively conserved molecules.

On balance, we would like to hypothesize that *Necator* secretes calreticulin, as a component of its molecular armoury against immunological attack. This belief is based on the following facts:

1. We have no evidence using Western blot or SPR analysis that human IgE to *Necator* calreticulin cross reacts with human calreticulin.[1,16] Epitope mapping has revealed some cross-reactive epitopes, although human IgE recognizes the least cross-reactive regions backing up the Western blot data (see Fig. 21.5).
2. Monospecific antisera to *Necator* calreticulin recognize as yet undefined proteins in worm secretions.[16]
3. Recombinant *Necator* calreticulin binds human C1q (Fig. 21.1A and B) and inhibits IgG mediated haemolysis (Fig. 21.1C). It also binds in SPR to the cytoplasmic domains of integrins (Fig. 21.2) considered important to platelet activation.[16]

Furthermore, new data are presented below to support the view that hookworm calreticulin may well be secreted at the site of parasite attachment, and acts as a true allergen at this interface.

Affinity Purification of Native *N. americanus* Calreticulin

To ascertain whether native calreticulin could be retrieved from the excretory-secretory (ES) products of cultured parasites, a calreticulin-specific affinity matrix was prepared (Fig. 21.3), using a monospecific polyclonal antibody to hookworm calreticulin.

It can be seen from Figure 21.3 that columns loaded with hookworm somatic extracts (Fig. 21.3A) or secretions (Fig. 21.3B) retained protein, which eluted in fractions corresponding to an elution volume between 5 and 10 ml. Analysis of these factions on SDS-PAGE revealed protein bonds of appropriate mass (@ 56 kDa) from both somatic extracts, and worm secretions. This approximates to the molecular mass of *Necator* calreticulin on SDS-PAGE.

Although it may still be possible that the "calreticulin" protein band from secretions was released from dead or dying worms, the culture conditions employed were conducive to worm survival.[18] With regard to the immunogenicity of the eluted "calreticulin" (no protein sequence data has yet been obtained), column-retained protein from somatic extracts and secretions were predictably recognized by a monospecific polyclonal rabbit anti-*Necator* calreticulin (which was used as an IgG preparation on the affinity column). Only the somatic extract-derived material was recognized by IgE; this may be a matter of sensitivity given that the yield from somatic extracts was ten-fold greater, this is under investigation.

Figure 21.1. Recombinant *N. americanus* binds to human C1q and inhibits C1q dependent haemolysis of sheep red blood cells, as shown on the following page. A) The wells of a 96 well plate were coated overnight with 0-0.17 μg of recombinant *N. americanus* calreticulin. Following washing with 0.05 % PBS/Tween 20 the plate was blocked with 1% non-fat milk powder for 1 h at 37 °C. To assess C1q binding, the wells were incubated with 0.125 μg C1q in PBS for 2 h at 37 °C. The plate was washed again and 50 μl of goat anti-human C1q (1:100) serum added to each well and the plate incubated overnight at 4 °C. Following further washing C1q binding was detected by the addition of rabbit anti-goat IgG conjugated to alkaline phosphatase (1:5000) for 2 h at 37 °C. The plate was developed by the addition of *p*-nitrophenylphosphate (5 mg/ml) and the absorbance of each well measured at 405 nm. B) Similarly, the wells of a 96 well plate were coated with 0-0.3 μg of human C1q. Following washing and blocking as described above the wells were incubated with 0.15 μg of recombinant *N. americanus* calreticulin for 2 h at 37 °C. Following further washing 50 μl of rabbit anti *N. americanus* calreticulin (1:400) was added to each well and the plate incubated overnight at 4 °C. Binding of recombinant *N. americanus* calreticulin was detected by the addition of a goat anti-rabbit IgG conjugated to alkaline phosphatase (1:5000) for 2 h at 37 °C. The plate was developed as described above and the absorbance measured at 405 nm. C) C1q-deficient serum was diluted 1:20 in isotonic Veronal buffered saline containing 0.1 mM $CaCl_2$, 0.5 mM $MgCl_2$, 0.1 % (w/v) gelatin and 1 % (w/v) glucose. 100 μl of diluted serum was incubated with 0 to 4 μg of human C1q at 37 °C for 30 min. Following this period 100 μl (10^8 cells/ml) of sheep red blood cells (SRBC) previously sensitized with rabbit anti-SRBC IgG were added to each tube and incubated for 30 min at 37 °C. Intact cells were pelleted by centrifugation and the absorbance at 405 nm of 100 μl of the supernatant measured to monitor haemoglobin release. Under these conditions, 1 μg of C1q was found to cause approximately 40 % haemolysis. Using this assay 0-3.0 μg of recombinant *N. americanus* calreticulin was shown to inhibit C1q dependent haemolysis from 40% to 0% respectively. Recombinant *Schistosoma mansoni* calmodulin showed no inhibition of haemolysis under the same conditions.

The True Allergenicity of Hookworm Calreticulin?

Hookworm calreticulin is potentially a very interesting molecule in an allergenic sense. Its domain structure suggests that sites exist for the engagement of the basophil or mast cell bound IgE molecule through Fab antigen-binding domains and/or through Fc associated sugars, the latter property of the calreticulin lectin domain. It was therefore of interest to assess the ability of recombinant calreticulin to trigger histamine release from hookworm IgE sensitized basophils. In October 2001, 35 hookworm infected individuals living along the Bogia Coast Road of Madang Province were bled into heparin, and basophil degranulation tests conducted using worm secretions, recombinant calreticulin, and anti-human IgE as a positive control (Figure 21.4).

Early data indicate the competence of these basophils, to release histamine in the presence of the agonists described. This would suggest that CRT interacts directly with IgE which as a complex binds to surface IgE receptors and subsequently triggers an inflammatory response by the basophils. We have yet to determine the relative contributions of allergen sequences and the lectin domain to this process.

Antigenicity of Hookworm Calreticulin

The response of a pooled anti *Necator* plasma sample from infected individuals from Papua New Guinea was mapped onto 15 mer peptides covering the N-domain of human calreticulin. These peptides were chosen because earlier autoimmune studies indicated their antigenicity. This analysis could serve two purposes; (a) to determine the most antigenic sites of calreticulin, as recognized by infected patients, (b) to plot areas of potential cross reactivity. The results of this analysis are shown in Table 21.1.

The most antigenic region of this domain, as recognized by all antibody isotypes, was the segment ^{55}Ala to ^{90}Gly. Peptides from this region could be potentially useful for inclusion in candidate hookworm vaccines and in diagnostic kits, or for use in immune/epidemiological studies.

It may be significant to the host-parasite relationship that the Clq binding domains of calreticulin, ESKHKSDF, FNYKGKNV and KDIRCKDDE, which are considerably, or wholly, conserved sequences, do not fall within this highly antigenic region. This may allow the

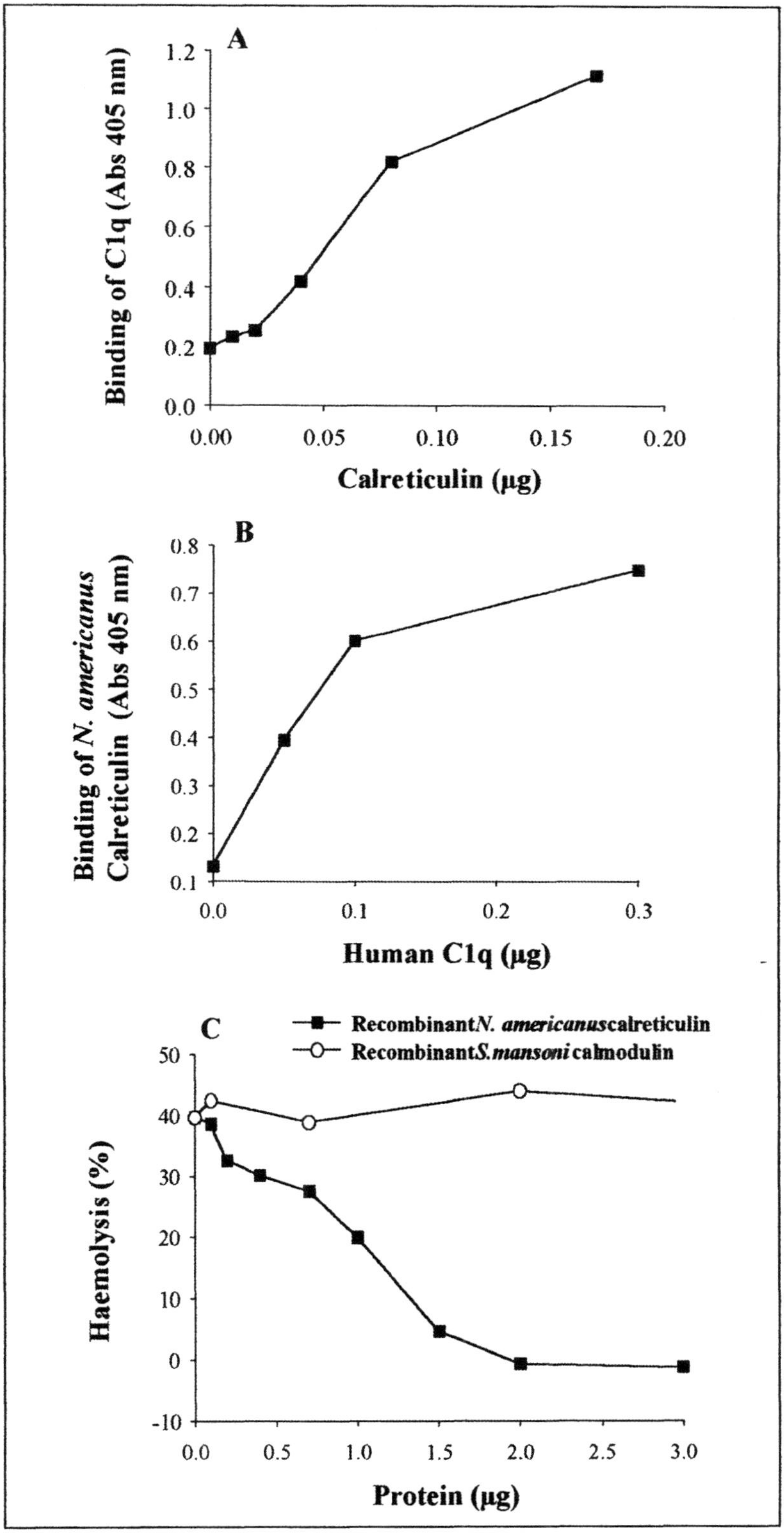
A
Binding of C1q (Abs 405 nm)
1.2
1.0
0.8
0.6
0.4
0.2
0.0
0.00
0.05
0.10
0.15
0.20
Calreticulin (µg)
B
Binding of N. americanus Calreticulin (Abs 405 nm)
0.8
0.7
0.6
0.5
0.4
0.3
0.2
0.1
0.0
0.1
0.2
0.3
Human C1q (µg)
C
Recombinant N. americanus calreticulin
Recombinant S. mansoni calmodulin
Haemolysis (%)
50
40
30
20
10
0
-10
0.0
0.5
1.0
1.5
2.0
2.5
3.0
Protein (µg)

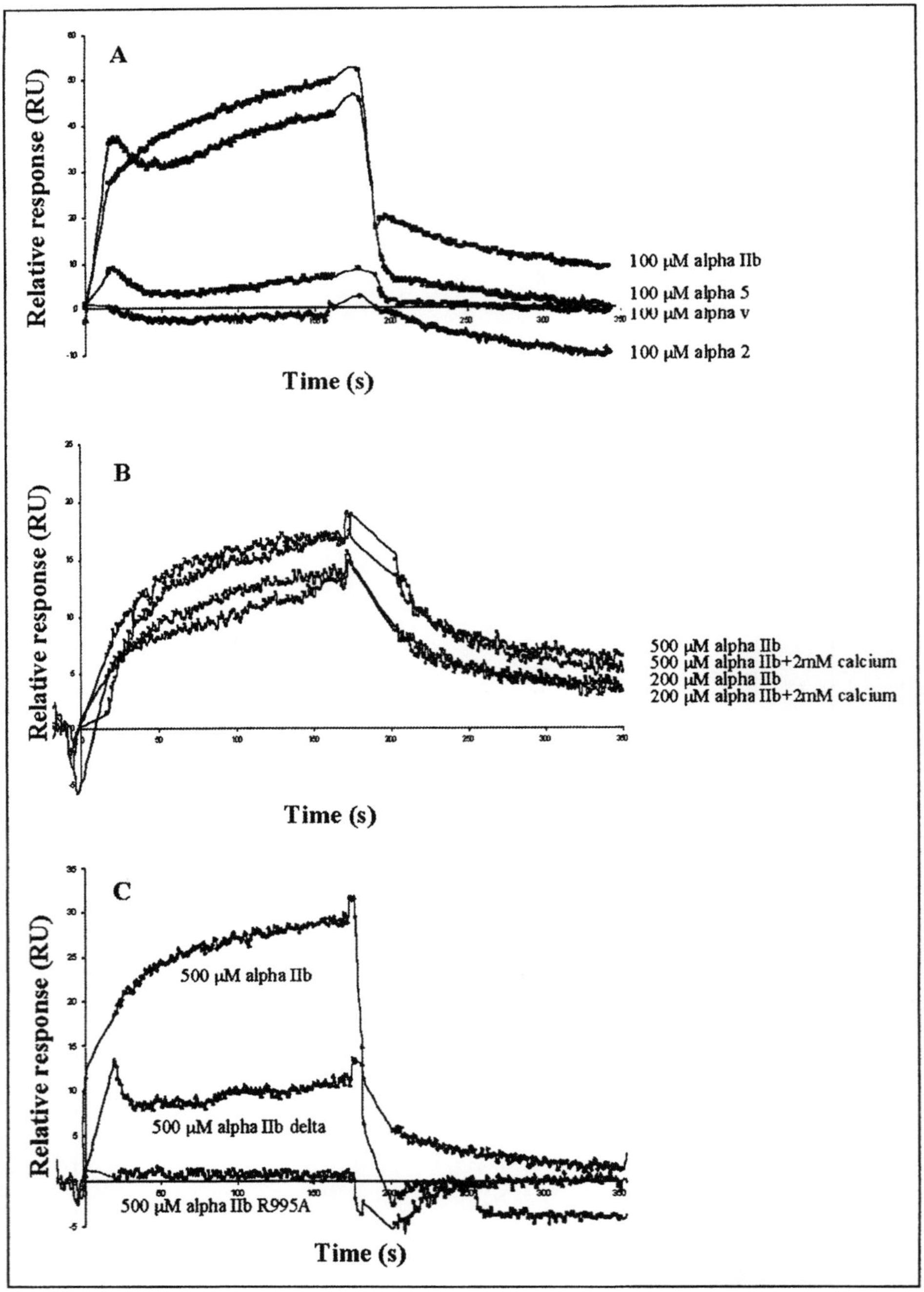

Figure 21.2. Surface plasmon resonance analysis of *N. americanus* calreticulin. A) Recombinant *N. americanus* calreticulin interacted weakly with αIIb and α5 cytoplasmic tail peptides of integrin α subunits but not with α2 or αv. B) This interaction was shown to be dose-dependent and in the case of *N. americanus* calreticulin Ca^{2+}-independent. C) Mutation or deletion within the highly conserved KVGFFKR membrane-proximal sequence of αIIb abolished the binding of *N. americanus* calreticulin.

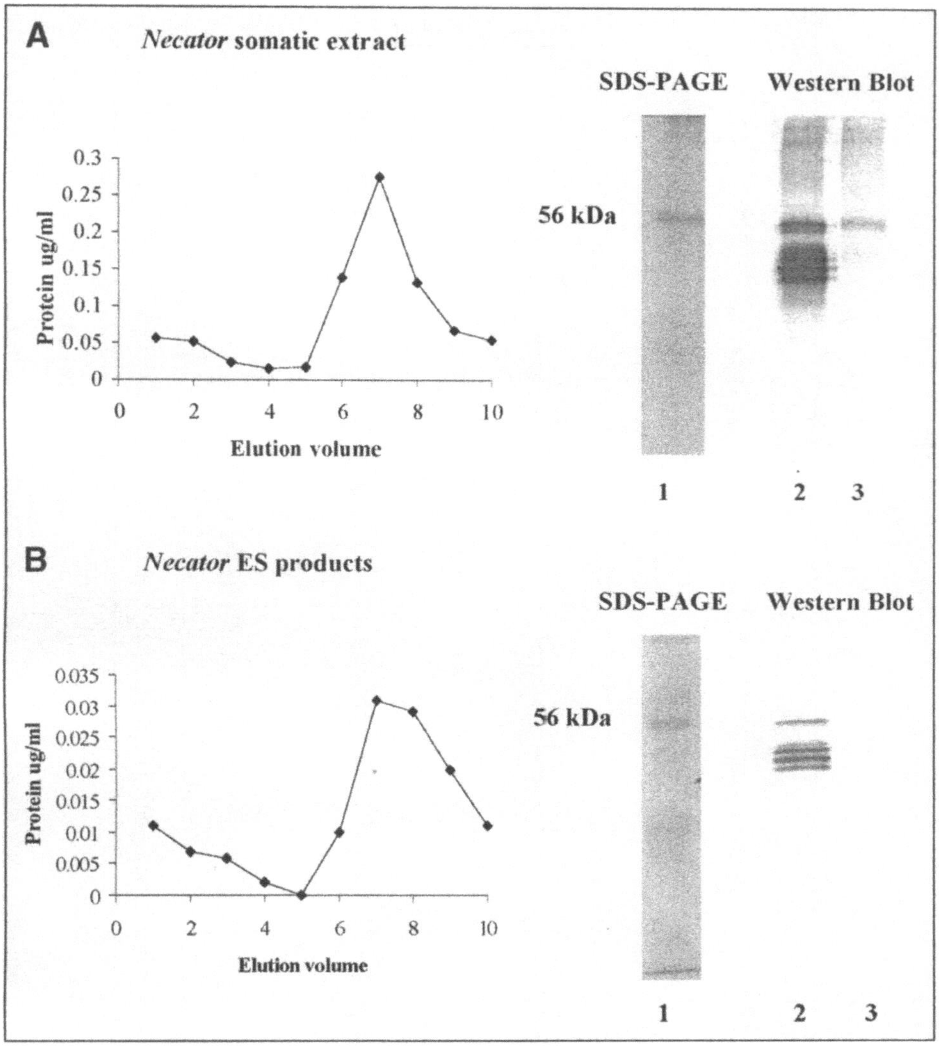

Figure 21.3. Antibody affinity purification of *N. americanus* calreticulin from somatic extracts and ES products. IgG prepared from polyclonal rabbit anti-calreticulin was bound to Affiprep Hz Hydrazide support (BioRad). *N. americanus* homogenate and excretory/secretory (ES) products were applied to the matrix (pre-equilibrated with 0.1 M sodium phosphate buffer, pH 7.4). The column was washed sequentially with 2 column volumes of 0.5 M NaCl, 0.1 M sodium phosphate buffer, pH 7.4 and 5 column volumes of 0.1 M NaCl, 0.1 M sodium phosphate buffer pH 7.4. Bound protein was eluted with 0.1 M glycine-HCl, pH 2.5. 900 μl of eluant were collected into 100 μl of 0.5 M Tris, pH 8.5, in order to neutralize the acidity of the elution buffer and preserve activity. 100 μl of fraction 7 from both somatic extracts and ES products was concentrated using StrataClean resin and resolved using 12 % SDS-PAGE, and either stained with Coomassie brilliant blue (lane 1) or tranfered onto nitrocellose and probed with either rabbit anti *Necator* calreticulin (lane 2) or human post-infection IgE (lane 3).

molecule to perform its anti-complementary activity without hindrance from the immune system. Furthermore, IgA and IgE recognisable exclusive peptide sequences were seen (IgE = peptides 8, 12, 17, 28, 44; IgA = 2, 13, 18, 34, 46, 47, 48, 54, 60, 61). This could be indicative of a mucosal response, and the presence of allergenic hot spots within the sequence.

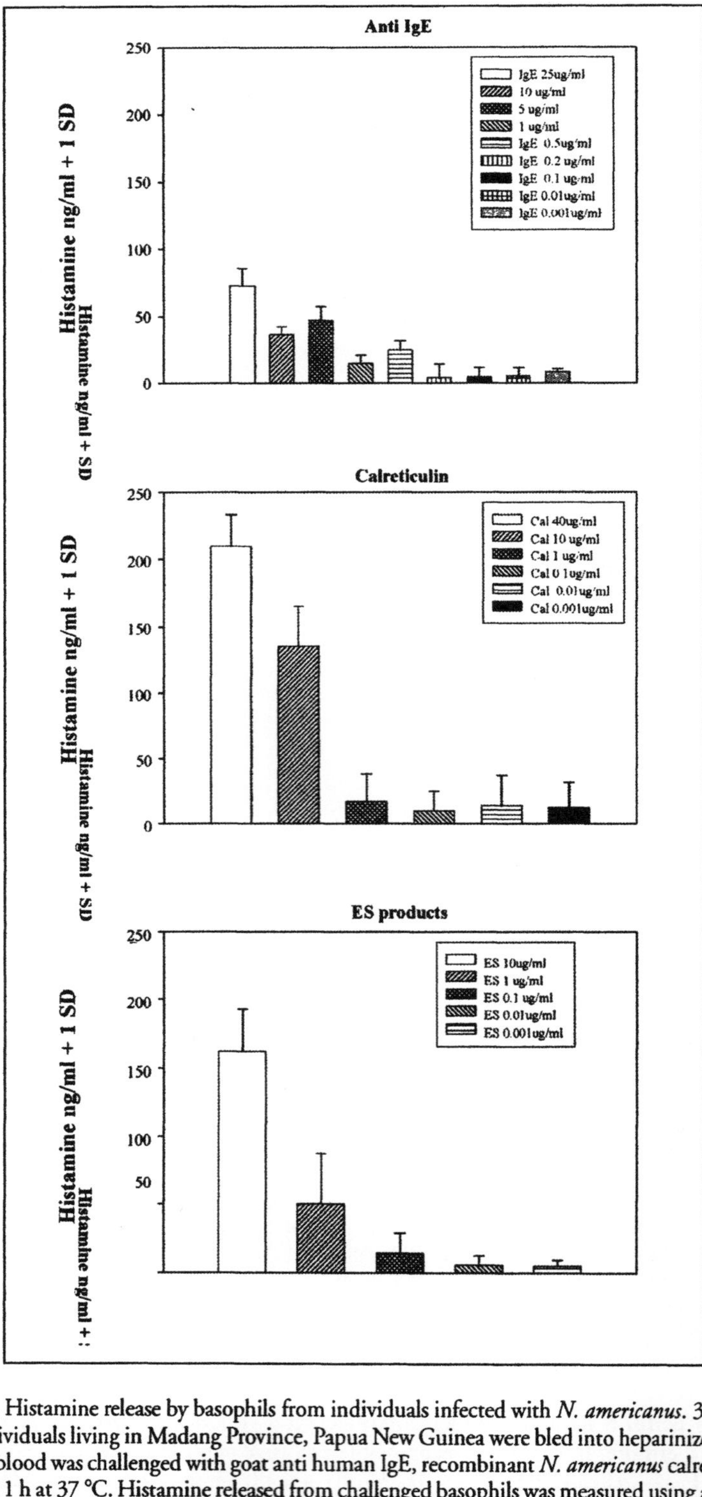

Figure 21.4. Histamine release by basophils from individuals infected with *N. americanus*. 35 hookworm infected individuals living in Madang Province, Papua New Guinea were bled into heparinized tubes. 100 µl of whole blood was challenged with goat anti human IgE, recombinant *N. americanus* calreticulin or ES products for 1 h at 37 °C. Histamine released from challenged basophils was measured using a competitive ELISA kit (Hycor Biomedical) as described by the manufacturer's instructions.

***Table 21.1. Prevalence of antibodies to multiple CRT synthetic peptides in sera from patients infected with* Necator americanus**

Peptide Number	Peptide Sequence	Immunoglobulin Class IgG	IgE	IgA
1	^{1}EPAVYFKEQFLDGDG		X	X
2	^{4}VYFKEQFLDGDGWTS			X
8	^{22}ESKHKSDFGKFVLSS		X	
10	^{28}DFGKFVLSSGKFYGD	X	X	X
12	^{34}LSSGKFYGDEEKDKG		X	
13	^{37}GKFYGDEEKDKGLQT			X
17	^{49}LQTSQDARFYALSAS		X	
18	^{52}SQDARFYALSASFEP			X
19	^{55}ARFYALSASFEPFSN	X	X	X
22	^{64}FEPFSNKGQTLVVQF	X	X	X
23	^{67}FSNKGQTLVVQFTVK	X	X	X
24	^{70}KGQTLVVQFTVKHEQ	X	X	X
25	^{73}TLVVQFTVKHEQNID	X	X	X
26	^{76}VQFTVKHEQNIDCGG	X	X	X
28	^{82}HEQNIDCGGGYVKLF		X	
34	^{100}LDQTDMHGDSEYNIM			X
36	^{106}HGDSEYNIMFGPDIC	X	X	X
37	^{109}SEYNIMFGPDICGPG		X	X
40	^{118}DICGPGTKKVHVIFN	X	X	
43	^{127}VHVIFNYKGKNVLIN	X	X	
44	^{130}IFNYKGKNVLINKDI		X	
46	^{136}KNVLINKDIRCKDDE			X
47	^{139}LINKDIRCKDDEFTH			X
48	^{142}KDIRCKDDEFTHLYT			X
51	^{151}FTHLYTLIVRPDNTY	X	X	X
54	^{160}RPDNTYEVKIDNSQV			X
60	^{178}SLEDDWDFLPPKKIK			X
61	^{181}DDWDFLPPKKIKDPD			X

Briefly, peptides were coated onto microtitre plates and 1/50 diluted pooled sera from hookworm infected individuals were incubated with the individual peptides shown below. Detection of specific binding of antibody isotype classes were performed by probing with anti-human IgG/E/A antisera conjugated to horse radish peroxidase. A 'X' denotes an OD reading at 450 nm of at least >2 SD than that observed against the control peptides. (Adapted from From S. Johnson's D.Phil thesis, University of Oxford).

We also sought to determine if the antigenic regions of the human form of calreticulin that leads to an immune response in autoimmune patients is also recognized by calreticulin antibodies from parasite infected individuals. Figure 21.5 demonstrates the cross-reactivity between antibodies with SLE and patients with *N. americanus* infection. There appears to be significant cross reactivity at the peptide level in four major segments; 34 Leu-Glu 42, 67 Phe-Gln 84, 109 Ser-Cys 120 and 151 Phe-Thr 156. The relevance of this finding has yet to be determined, given the lack of cross reactivity seen at the Western blot level in previous publications. Nevertheless, the subtle differences in the epitopes recognized by autoimmune sera and antibodies from hookworm patients might suggest that the mechanisms driving anti-CRT antibody production during the progression of the two diseases are different.

```
1   EPAVYFKEQF LDGDGWTSRW IESKHKSDFG  30
31  KFVLSSGKFY GDEEKDKGLQ TSQDARFYAL  60
61  SASFEPFSNK GQTLVVQFTV KHEQNIDCGG  90
91  GYVKLFPNSL DQTDMHGDSE YNIMFGPDIC 120
121 GPGTKKVHVI FNYKGKNVLI NKDIRCKDDE 150
151 FTHLYTLIVR PDNTYEVKID NSQVESGSLE 180
181 DDWDFLPPKK IKDPDASKPE DWDERAKIDD 210
211 PTDSKPEDWD KPEHIPDPDA KKPEDWDEEM 240
241 DGEWEPPVIQ NPEYKGEWKP RQIDNPDYKG 270
271 TWIHPEIDNP EYSPDPSIYA
```

Figure 21.5 Cross-reactivity between human and *N. americanus* calreticulin. Amino acid sequence of human CRT showing the antigenic sites recognized by autoantibodies from autoimmune disease patients and individuals infected with *N. americanus*. (- - - -) represents sequences found to be autoantigenic in 5 or more SLE patient sera. (——) represents sequences recognized by IgG, IgE and IgA antibodies from patients infected with *N. americanus*.

The Way Forward

Our experiments with hookworm calreticulin are far from complete:

- Preliminary data indicate that native calreticulin can be retrieved from worm conditioned media but we await detailed protein sequence analysis.
- Recombinant calreticulin does exhibit allergenicity against sensitized basophils, although the contributing true IgE-allergen epitopes and lectin interaction to this process needs investigating.
- The allergenicity of native calreticulin will be similarly explored.
- Should our studies conclude that calreticulin is indeed a hookworm secretory molecule, and given that hookworm sections induce apoptosis in activated T cells, it will be of considerable interest to investigate the interaction between a secreted calreticulin and CD91, the common receptor for heat shock proteins and calreticulin on leucocytes.[19]
- The observation that calreticulin-containing fractions of parasite extracts inhibit platelet glycoprotein Gp1b binding to von Willebrand factor therefore inhibiting platelet adhesion to damaged vessels also warrants further investigation.

Summary

The fortuitous discovery of calreticulin as a hookworm allergen in 1999 has led to a flurry of new research activity in our laboratory. The possibility that such an interesting multifunctional and multidomained molecule may be secreted opens up avenues for innovative research into the role of calreticulin in the host-parasite relationship, particularly its potential as an immune modulator and anti-haemostatic agent.

Acknowledgement

This work was initiated at the Centre for the Mechanisms of Human Toxicity at the University of Leicester during a research leave fellowship (DIP) funded by the Wellcome Trust.

References

1. Pritchard DI, Brown A, Kasper G et al. A hookworm allergen that strongly resembles calreticulin. Parasite Immunol 1999; 21:439-450.
2. Pritchard DI, Quinnell RJ, Walsh EA. Immunity in humans to N. americanus: IgE, parasite weight and fecundity. Parasite Immunol 1995; 71:71-75.
3. Scrivener S, Yemaneberhan H, Zebenigus M et al. Independent effects of intestinal parasite infection and domestic allergen exposure on the risk of wheeze in Ethiopia. Lancet 2001; 358:1493-1499.
4. Ostwald TJ, McLennan DH. Isolation of a high affinity calcium binding protein from sarcoplasmic reticulum. J Biol Chem 1974; 249:974-979.
5. Michalak M, Burns K, Andrin C et al. Endoplasmic reticulum form of calreticulin modulates glucocorticoid-sensitive gene expression. J Biol Chem Nov 15 1996; 271(46):29436-29445.
6. Sonnichsen B, Fullekrug J, Nguyen P, Diekmann W, Robinson DG, Mieskes G. Retention and retrieval: both mechanisms cooperate to maintain calreticulin in the endoplasmic reticulum. J Cell Sci 1994; 107:2705-2717.
7. Xiao G, Chung T-F, Pyun HY, Fine RE, Johnson RJ. KDEL proteins are found on the surface of NG108-15 cells. J Brain Res 1999; 72(121-128).
8. White TK, Zhu Q, Tanzer ML. Cell surface calreticulin is a putative mannoside lectin which triggers mouse melanoma cell spreading. J Biol Chem 1995; 270(27):15926-15929.
9. Gray AJ, Park PW, Broekelmann TJ, Laurent GJ, Stenmark KR, Mecham RP. The mitogenic effects of the β chain of fibrinogen are mediated through cell surface calreticulin. J Biol Chem 1995; 270(4):26602-26606.
10. Arosa FA, deJesus O, Porto G, Carmo AM, de Sousa M. Calreticulin is expressed on the cell surface of activated human peripheral blood T lymphocytes in association with major histocompatability complex class I molecules. J Biol Chem 1999; 274(24):16917-16922.
11. Andrin C, Tschopp J, Opas M, Bleackley RC, Michalak M. Interaction between a Ca^{+} binding protein calreticulin and perforin, a component of the cytotoxic T-cell granules. Biochemistry 1998; 37(29):10386-10394.
12. Jaworski DC, Higgins JA, Radulovic S, Vaughan JA, Azad AF. Presence of calreticulin in vector fleas (Siphonaptera). J Med Entomol 1996; 33(3):482-489.
13. Sanders ML, Jaworski DC, Sanchez JL et al. Antibody to a cDNA-derived calreticulin protein from Amblyomma americanum as a biomaker of tick exposure in humans. Am J Trop Med Hyg 1998; 59(2):279-285.
14. Sanders ML, Glass GE, Nadelman RB et al. Antibody levels to tick recombinant calreticulin increase in humans after exposure to Ixodes scapularis (Say) and are correlated with tick engorgement indices. Am J Epidemiol 1999; 149(8):777-784.
15. Kishore U, Sontheimer RD, Sastry KN et al. Release of calreticulin from neutrophils may alter C1q-mediated immune functions. Biochem J 1997; 322:543-550.
16. Kasper G, Brown A, Eberl M et al. A calreticulin-like molecule from the human hookworm Necator americanus interacts with C1q and the cytoplasmic signalling domains of some integrins. Parasite Immunol 2001; 23:141-152.
17. Chow SC, Brown A, Pritchard D. The human hookworm pathogen Necator americanus induces apoptosis in T lymphocytes. Parasite Immunol 2000; 22:21-29.
18. Carr A, Pritchard DI. Identification of hookworm (Necator americanus) antigens and their translation in vitro. Mol Biochem Parasitol 1986; 19:251-258.
19. Basu S, Binder RJ, Ramalingam T, Srivastava PK. CD91 is a common receptor for heat shock proteins gp96, hsp 90, hsp 70 and calreticulin. Immunity 2001;14(3):303-313.

Chapter 22

Calreticulin in *C. elegans*

Byung-Jae Park, Jin Il Lee and Joohong Ahnn

Abstract

The nematode *C. elegans* is an ideal organism to study the in vivo genetic and biochemical functions of calreticulin. In vitro studies show that the *C. elegans* CRT-1 protein, like other calreticulins, is a calcium-binding molecular chaperone. Mutants of *crt-1* in *C. elegans* mutants are viable and fertile, offering the opportunity for scientists to study in vivo functions of calreticulin at more depth. *crt-1* null mutants showed temperature-sensitive fertility defects, and transcription of *crt-1* was upregulated in stress conditions such as high temperature and ethanol treatment suggesting that calreticulin may be functioning in stress response. Mutants of the calreticulin gene were also shown to suppress necrotic cell death in neurons. The use of pharmological agents and the genetic application of mutants involved in ER calcium homeostasis showed that calreticulin was critical in the regulation of ER calcium levels during the neuronal degeneration process. Double mutants of *crt-1* mutants and *itr-1* IP3 receptor mutants displayed synergistic severity in defecation rhythm defects further suggesting the role of calreticulin in ER calcium homeostasis. Further genetic analysis in *C. elegans* between *crt-1* and other components involved in ER calcium regulation should deepen our understanding of calreticulin and calcium homeostasis at both the cellular and organism level.

Introduction

Calreticulin has been capturing the curiosity of many biochemists and cell biologists for over a quarter of a century. The popularity of this molecule among researchers stems from its diverse roles in protein folding, calcium homeostasis, and signaling while being primarily localized in the ER lumen.[1] Though it has been implicated in many biochemical and cellular functions, its in vivo role at the organismal level has only begun to be elucidated recently. The generation of calreticulin-deficient mice has made tremendous breakthroughs in learning about the role of calreticulin in modulating calcium signaling in the formation of mammalian cardiac tissue. But it has also made us realize that the loss of such a key protein to cellular processes is devastating to higher organisms that utilize numerous specialized and sensitive structures for its survival.

Caenorhabditis elegans As a Model Organism

Caenorhabditis elegans is a small (about 1 mm long) soil nematode found in temperate regions. *C. elegans* has been a model organism to study gene function during the past four decades since Sydney Brenner began using it to study the genetics of development and neurobiology in the 1960s,[2] nearly the same time calreticulin was first being purified in the laboratory of David MacLennan.[3] The size of the worm genome is about 100 million bases, and its whole-genome sequence was reported in 1998.[4] The entire sequence of the 19,000 or so genes that comprise the worm genome has been identified[4,5] and thousands of those genes have been and are being studied. *C. elegans*, with its simple 959 somatic cell body structure yet strikingly conserved genome compared to higher organisms, is a tremendous living tool for scientists to

Calreticulin, 2nd ed., edited by Paul Eggleton and Marek Michalak.

study in vivo roles of proteins.[6-8] It is in this model system that calreticulin, a protein whose functions at the biochemical and cellular level are well known, has been implicated in diverse functions at the organismal level.

crt-1 Gene and Protein

The calreticulin gene in *C. elegans* was identified and sequenced in 1991 and the gene consists of three exons.[9] Using a mouse calreticulin cDNA probe to screen a *C. elegans* genomic DNA library, the calreticulin gene was isolated and mapped toward the left-side end (mapped at -3.81 m.u.) of chromosome V and designated as *crt-1* although there is single calreticulin gene in *C. elegans*. The *crt-1* gene (Y38A10A.5) encodes the *C. elegans* homologue of calreticulin, which shows an overall 59% identity with human calreticulin and exhibits all the conserved functional domains including the cleavable N-terminal signal sequences, the proline-rich domain, a highly acidic C-terminal domain, and an ER-retention HDEL signal (Fig. 22.1A).

In Vitro Function

Though the calreticulin gene in *C. elegans* was highly conserved in all aspects, it remained to be known whether the biochemical functions of calreticulin were conserved in the nematode. Park et al purified *C. elegans* calreticulin (designated as CRT-1) and determined that CRT-1 could bind $^{45}Ca^{2+}$ in a calcium overlay experiment.[10] Not only was the role of CRT-1 as a calcium-binding molecule conserved but the role of calreticulin as a molecular chaperone was conserved as well. Saito et al previously demonstrated that calreticulin functions in vitro as a chaperone molecule by suppressing the thermal aggregation of the nonglycosylated protein citrate synthase (CS).[11] CS normally denatures and forms large aggregates upon heating at 45°C that can be suppressed by calreticulin. CRT-1 was also shown to suppress this thermal aggregation in a concentration-dependent manner demonstrating that the chaperoning function of calreticulin is conserved from the higher vertebrates all the way down to the lower invertebrates.

The Isolation of *C. elegans crt-1* Mutants

Since the conserved in vitro function of CRT-1 as a calcium-binding molecular chaperone now had been established, the question of whether these in vitro roles could be correlated to in vivo roles needed to be answered. Thus, target-selected mutagenesis to screen for a calreticulin mutant worm was the next essential step to elucidate this question. This powerful technique utilizes UV/trimethylpsoralen to induce random deletion mutations in the worm genome, followed by a PCR-based screening method that can identify possible deletions in individual genes.[12,13] Using this technique, Park et al isolated a viable mutant with a 1.1 kb deletion in the calreticulin gene which was designated as *crt-1(jh101)*.[13] Xu et al also isolated four alleles of calreticulin mutants independently (*bz29, bz30, bz31* and *bz50*).[14] Among these alleles the deletion mutant (*jh101*) and two point mutations (*bz29* and *bz30*) failed to produce any protein, while the other alleles encode proteins (Fig. 22.2). Surprisingly, this calreticulin mutant was not only viable but did not show any drastic or seriously debilitating phenotypes (possible reasons for this will be discussed at the end of this chapter).

In Vivo Functions of Calreticulin

Interestingly, though, the *crt-1(jh101)* mutants did display temperature-dependent reproductive defects. In normal culture conditions at 20°C *crt-1(jh101)* worms showed slightly lower numbers of progeny than the wild-type worms.[10,14] In a slightly stressful temperature of 25°C the fertility defect becomes much more severe. Upon closer examination calreticulin likely plays a role in both oocytes and sperm in *C. elegans*, which can account for the observed fertility defects. The severity of the defect was also temperature-dependent suggesting that calreticulin may have a function in stress response. The validity of this hypothesis was tested by observing transcript levels of *crt-1* before and after stress-induced treatments. At 25°C

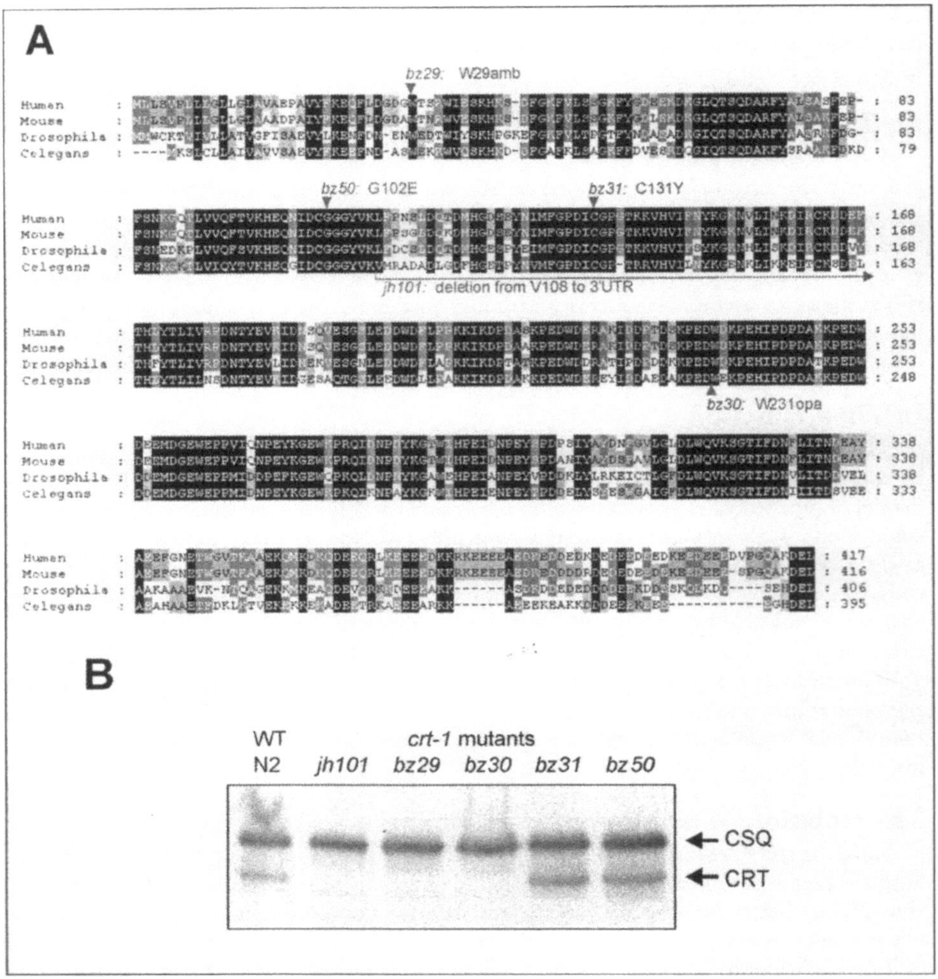

Figure 22.1. The conservation of calreticulin proteins and the western analysis of *crt-1* mutants in *C. elegans*. (A) Amino acid alignment of calreticulin proteins. Point mutations and deletion mutations of five alleles of *crt-1* are indicated. (B) Western analysis of alleles of *crt-1*. Three alleles (*jh101*, *bz29*, *bz30*) show no protein indicating null mutants while the two remaining alleles encode proteins. Calsequestrin (CSQ) polyclonal antibody was used as a control.

transcription of the *crt-1* gene measured by northern analysis increased by almost two-fold. Levels of the truncated *crt-1* transcript in *crt-1(jh101)* mutants also significantly increased. 7% ethanol treatment was also shown to induce stress in *C. elegans*, so transcription levels were also tested in ethanol-treated worms.[10] Wild-type animals treated in ethanol showed about a three-fold increase in *crt-1* transcription. Thus, a possible explanation for the function of calreticulin as a stress-response molecule likely involves its role as a chaperoning molecule. In stress-induced situations such as elevated temperatures or toxic ethanol levels, *crt-1* transcription is upregulated and an overall increase of CRT-1 protein occurs. CRT-1 chaperoning activity is then increased so the correct folding of proteins in the ER can faithfully occur even in unfavorable conditions. In *crt-1(jh101)* deletion mutants in which no CRT-1 protein is translated, problems in protein folding occur especially in stressful conditions when the need for

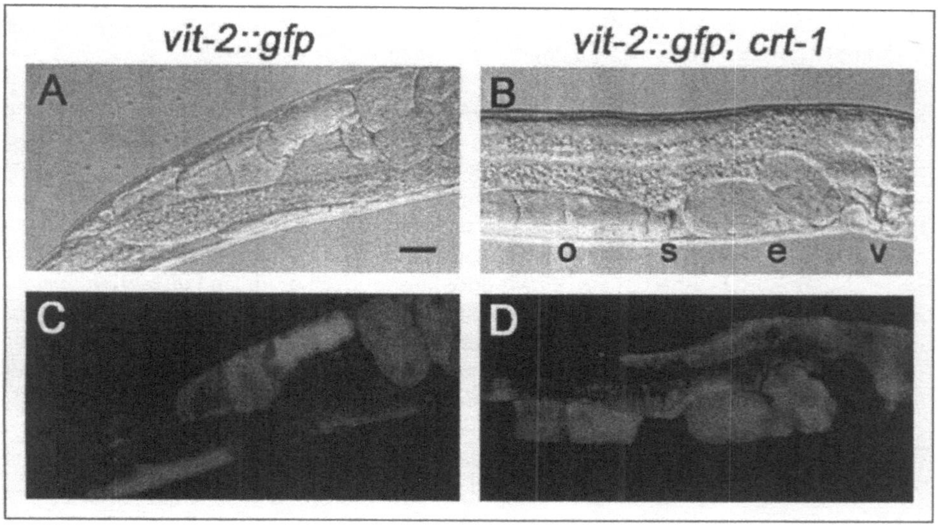

Figure 22.2. Calreticulin is not involved in receptor-mediated endocytosis. Nomarski images (A) and GFP images of the *vit-2::gfp* reporter (C) display normal receptor-mediated endocytosis in wild-type animals. In *crt-1(jh101)* mutants, receptor-mediated endocytosis appears to occur normally (B,D). Oocyte (o), spermatheca (s), embryo (e), and vulva (v) are marked as shown. Bar indicates 50 μm.

calreticulin as a molecular chaperone in the cells are most urgent. This results in calreticulin-deficient worms displaying temperature-sensitive defects such as decreased fertility.

ER-Mediated Calcium Homeostasis and Cell Death

As had been shown by Park et al, the role of calreticulin as a calcium-binding protein was conserved in *C. elegans*. Could calreticulin also play a role in calcium homeostasis and signaling in *C. elegans* as had been observed in studies in mice?[15,16] The answer to this question has been elucidated by the study of two different in vivo functions of calreticulin in the worm. The first involves necrosis, which is characterized by swelling of the dying cell usually as a result of cellular injury. Gain-of-function mutations in the degenerin Na^+ channel MEC-4 can also induce neurodegenerative necrotic cell death by allowing an excess flow of ions into the cell that is ultimately toxic.[17,18] To understand molecular mechanisms of necrosis in *C. elegans*, Xu et al used a complex classical genetic screen to identify genes that could suppress the necrotic cell death phenotype of *mec-4(d)* mutants.[14] Genetic mapping of four of these suppressor mutants showed that mutations in a single gene locus affected the *mec-4(d)*-induced necrotic cell death phenotype that was identified as the *crt-1* gene. *crt-1* was also shown to suppress necrosis induced by dominant channel-hyperactivating mutations in other members of the degenerin family.

Since calreticulin seemed to play an important role in *mec-4(d)*-induced necrosis it was thought that disruption of the calnexin gene, another calcium-binding molecular chaperone of the ER, may also suppress necrotic cell death. To test this hypothesis, Xu et al devised a scheme to introduce RNA interference effect into normally RNAi impenetrable neuronal tissues. Using this scheme they were able to knockout calnexin gene function in neurons of *mec-4(d)* mutants and observe the effects on necrosis. Indeed, necrosis was significantly suppressed by calnexin disruption, suggesting that either or both ER chaperoning activity and calcium regulation is necessary for necrotic cell death.

To address whether the role of calreticulin in calcium homeostasis might be a factor in necrosis, levels of calcium in the ER were manipulated both pharmacologically and genetically.

mec-4(d) mutants treated with the Ca^{2+}-specific chelator EGTA showed a decreased amount of neurodegeneration. When *mec-4(d)* mutants were treated with dantrolene, a chemical that specifically inhibits Ca^{2+} release from the ER stores,[19] necrosis was also partially reduced suggesting that Ca^{2+} stores in the ER are important for the progression of necrosis. To further verify this possibility, the effects of two Ca^{2+} release channels of the ER on necrotic cell death were also tested. Mutants of *mec-4(d)* carrying an additional mutation in either the *itr-1* gene, which encodes the IP_3 receptor in *C. elegans*, or the *unc-68* gene, which encodes the ryanodine receptor, both showed significantly reduced necrosis further supporting the hypothesis that calcium regulation in the ER is critical for neurodegeneration. Thus, if Ca^{2+} release from ER stores is an important aspect of *mec-4(d)*-induced necrotic cell death, and disruption of either calreticulin, calnexin, IP_3 receptor, or ryanodine receptor in *mec-4(d)* mutants can reduce neurodegeneration, then the release of remaining Ca^{2+} stores in the ER should restore the suppressed necrotic activity observed in *crt-1; mec-4(d)* double mutants. This was accomplished by treating the *crt-1; mec-4(d)* double mutants with thapsigargin, which both inhibits the SERCA $Ca2^{+}$ pump that reuptakes calcium into the ER and induces release of ER Ca^{2+}. Thapsigargin treatment restored necrosis in the progeny of the *crt-1; mec-4(d)* mutants indicating that elevated cytosolic Ca^{2+} levels regulated by calreticulin may be a cause of neurotoxity in *C. elegans*.

Defecation Cycle

Necrosis is not the only phenotype that calreticulin as an ER $Ca2^{+}$ regulator is involved in. The IP_3 receptor in *C. elegans* (*itr-1*) is predominantly expressed in the intestine, and mutations in the *itr-1* gene result in abnormal defecation rhythms.[20,21] As was previously mentioned both calreticulin and IP_3 receptor are involved in ER Ca^{2+} release during neurodegeneration and since both are expressed in the intestine it seemed reasonable to investigate whether calreticulin also may play a role in defecation cycles in conjunction with the IP3 receptor though calreticulin mutants alone show no defecation rhythm defects.[10,14] Double mutants between several different *itr-1* mutants and the *crt-1(jh101)* mutant resulted in worms with much more severe defects in defecation behavior suggesting a synergistic effect between the two mutations (Table 22.1).

crt-1 Is Not Essential for Receptor-Mediated Endocytosis

Observations that *crt-1*-deficient mutants had decreased fertility partially due to oocyte defects along with the fact that *crt-1* expressed in the intestine led Park et al to believe that there may be defects in receptor-mediated endocytosis. This intercellular process transports important oocyte proteins that are produced in the intestinal cells to the adjacent gonad and oocytes. Receptor-mediated endocytosis can be easily observed in vivo in *C. elegans* by constructing a fusion protein between green fluorescent protein (GFP) and a target protein in receptor-mediated endocytosis such as the yolk protein *vit-2*.[22] This construct is then transformed into *C. elegans* and the GFP pattern in the intestinal walls and the oocytes can be observed (Fig. 22.3A and C). Thus, mutants having defects in receptor-mediated endocytosis can be easily screened simply by observing the translocation of GFP-fusion from the intestinal cells to the oocytes. To determine whether disruption of calreticulin in *C. elegans* caused defects in this process, the *vit-2::gfp* expression was observed in *crt-1(jh101)* mutants (Fig. 22.3B and D). Loss of calreticulin did not result in any changes in expression of the *vit-2::gfp* reporter gene suggesting that *crt-1* does not function in receptor-mediated endocytosis.

Future Prospective

With the identification of genes coming to completion in the current post-genomic era, the relationships between proteins are becoming a major target of study for biologists. Since the full *C. elegans* genome has been known for almost four years, *C. elegans* researchers have just

***Table 22.1. The Effects of* crt-1 *and itr-1 mutants on the defecation cycles in* C. elegans**

Genotype	Defecation cycle (sec)
N2 (WT)	53.5 • 2.6
crt-1(jh101)	53.6 • 2.9
itr-1(sy328)[*1]	43.7 • 5.7
itr-1(sa73)[*2]	111 • 22
itr-1(n2559)[*3]	> 60 min
crt-1;itr-1(sy328)	60.0 • 2.8
crt-1;itr-1(sa73)[*4]	> 10 min

*1: _itr-1(sy328)_ is a gain-of-function mutant. (ref. 21)

*2: _itr-1(sa73)_ is a reduce-of-function mutant. (20. 10)

*3 : _itr-1(n2559)_ is a loss-of-function mutant. (20)

*4: This double mutant shows longer than 10 min or no defecation cycle. (10)

begun to tap into the field of proteomics to broaden the understanding of the worm at the molecular level. Newer and innovative computer techniques in two-dimensional gel electrophoresis analysis can aid scientists in elucidating relationships between proteins that were not previously known to be involved with each other and subsequently create new paths of study. Identifying proteins that are upregulated and downregulated in calreticulin-deficient mutants using this analysis (Fig. 22.3) will further widen our views for future research of this protein.

Finally, using classical methods of showing genetic interactions between calreticulin and other proteins may be the greatest usefulness that *C. elegans* has to offer in the future. Both Park et al and Xu et al have already demonstrated some genetic interaction between calreticulin and other genes that affect calcium homeostasis in the ER/SR in *C. elegans* (Table 22.2) using both double mutants and RNA interference experiments. Since processes of calcium regulation in the ER/SR are quite conserved in *C. elegans*, further analysis of genetic interaction between these genes by the generation of double mutants will deepen the knowledge of the roles of calreticulin in calcium homeostasis and calcium regulation in the ER/SR in general more conveniently and efficiently than studies in higher organisms.

An Evolutionary View of the Functions of Calreticulin

A viable calreticulin mutant was a fortunate necessity for *C. elegans* researchers to uncover some of the previously unknown in vivo functions of this fascinating molecule. Yet it may be a mystery to some why a complete knockout of calreticulin is viable in the first place. One obvious reason resides in the simplicity of *C. elegans* and the complexity of specialized structures in higher organisms. Nematodes lack sensitive specialized tissues such as the heart and brain where any malformation in these structures can be deadly to the organism. In a simple sense, the evolution of calreticulin's functions at the organism level diversified, specialized, and complexified at higher levels of the evolutionary ladder at the expense of viability in the case of mutation of the gene.

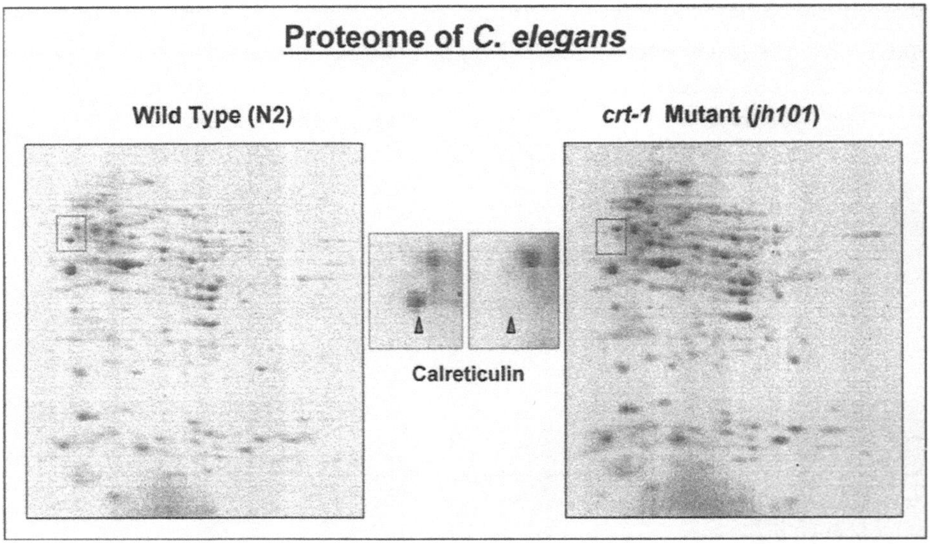

Figure 22.3. Proteomics of wild-type and *crt-1* mutant *C. elegans* using 2-dimensional gel electrophoresis. Inset indicates loss of calreticulin protein in *crt-1(jh101)* mutants.

More specifically, the ability of calreticulin to regulate intracellular calcium levels was likely exploited by newly developed calcium signaling pathways involving the evolutionarily novel transcription factor family NFATc. NFATc, which is essential for heart development and function,[28,29] is a mammalian transcription factor family that is not conserved in *C. elegans* or any other invertebrates. It has been thought, though, that evolutionary recombination of genes encoding conserved functional domains found in NFATc finally culminating with the recombination of a Ca^{2+}/calcineurin-sensing domain in vertebrates resulted in the formation of the NFATc transcription factor.[30] This helped fulfill Ca^{2+}-dependent signaling functions necessary for advanced neurological and cardiovascular development that did not exist in the lower invertebrates. The role of calreticulin as a regulator of intracellular Ca^{2+} concentrations was likely heightened by the development of Ca^{2+}/calcineurin/NFAT signaling mechanisms in vertebrate systems, evolving the role of calreticulin as an important in vivo chaperone and Ca^{2+} regulator in invertebrates to an essential molecule for the viability of advanced organisms. This has not only been observed in calreticulin but may also be true in the evolution of calcineurin function itself. Complete disruption of calcineurin activity in mice results in an inability to transclocate NFATc, causing abnormal vascular development and finally leading to embryonic lethality.[31] In *C. elegans*, though, complete loss of calcineurin function leads to several phenotypic defects, but the worm is viable and fertile.[32] This suggests that, as in the case of calreticulin, evolutionary functional diversification of protein domains of NFATc has led to more complex Ca^{2+} signaling pathways that may have augmented the role of calcineurin from an important protein to an essential molecule.

Acknowledgements

We would like to thank editors for letting us contribute a chapter in this book, Dr. M. Driscoll for sharing mutant strains, and also Dr. J. Bandyopadhyay for critical comments.

This was supported by BK21 (to B-J. Park) and Frontier 21(FG-3-8 to J. Ahnn).

Table 22.2. Genes affecting Ca^{2+} homeostasis in the ER/SR in C. elegans

Gene	Protein	Mutant phenotypes *1		References
		Lethality	Others	
itr-1	inositol triphosphate receptor	larva	defecation cycle and ovulation defects	Dal Santo, et al. (20) Clandinin, et al. (21)
ryr-1/ (unc-68)	ryanodyne receptor	subtle *1	muscle contraction and pumping defects	Maryon, et al. (23)
sca-1	SR/ER ATPase	embryo larva	pumping and movement defects	Zwaal, et al. (24) Cho, et al. (25)
crt-1	calreticulin	subtle *1	stress sensitivity suppression of cell death	Park, et al. (10) Xu, et al. (14)
cnx-1	calnexin	subtle *2	stress sensitivity suppression of cell death	Park, et al. *3 Xu, et al. (14)
csq-1	calsequestrin	normal	Ca^{2+} sensitivity	Cho. et al. (26) Cho, et al. *4

*1: All mutants show longer growth rates and reduced blood sizes, but *csq-1* mutant shows almost normal growth and development

*2: The mutant shows a little of embryonic and larva lethality

*3: Park, et al. unpublished data; Strains were kindly provided by the NemaPharm Group of Axys Pharmaceuticals. Liu, et al. (ref. 27)

*4: Cho, et al. unpublished data

References

1. Michalak M, Corbett EF, Mesaeli N et al. Calreticulin: one protein, one gene, many functions. Biochem J 1999; 344:281-292.
2. Brenner S. The genetics of Caenorhabditis elegans. Genetics 1974; 77:71-94.
3. Ostwald TJ, MacLennan DH. Isolation of a high affinity calcium-binding protein from sarcoplasmic reticulum. J Biol Chem 1974; 249:974-979.
4. The C. elegans Sequencing Consortium. (1998) Genome sequence of the nematode C. elegans: a platform for investigating biology. Science 1998; 282:2012-2018.
5. Wormbase home page: www.wormbase.org
6. Wood WB and the Community of C. elegans Researchers, eds. The Nematode Caenorhabditis elegans. Plainview: Cold Spring Harbor Laboratory Press, 1988.
7. Riddle DL, Blumenthal T, Meyer BJ et al, eds. C. elegans II. Plainview: Cold Spring Harbor Laboratory Press, 1997.
8. Epstein HE, Shakes DC, eds. Caenorhabditis elegans: Modern Biological Analysis of an Organisms. In: Methods in Cell Biology. Vol. 48. San Diego: Academic Press, 1995.
9. Smith MJ. A C. elegans gene encodes a protein homologous to mammalian calreticulin. DNA Seq 1992; 2:235-240.
10. Park B-J, Lee D-G, Yu J-R et al. Calreticulin, a Calcium-binding Molecular chaperone, Is Required for Stress Response and Fertility in Caenorhabditis elegans. Mol Biol Cell 2001; 12:2835-2845.
11. Saito Y, Ihara Y, Leqach MR, Cohen-Doyle MF et al. Calreticulin functions in vitro as a molecular chaperone for both glycosylated and non-glycosylated proteins. EMBO J 1999; 18:6718-6729.
12. Barstead RJ. Reverse Genetics. In: Hope IA, ed. C. elegans: A Practical Approach. Oxford: Oxford University Press, 1999:97-118.
13. Park B-J, Lee JI, Lee J et al. Isolation of deletion mutants by reverse genetics in Caenorhabditis elegans. Korean J Biol Sci 2001; 5:65-69.
14. Xu K, Tavernarakis N, Driscoll M. Necrotic Cell Death in C. elegans Requires the Function of Calreticulin and Regulators of Ca^{2+} Release from the Endoplasmic reticulum. Neuron 2001; 31:957-971.
15. Mesaeli N, Nakamura K, Zvaritch E et al. Calreticulin is essential for cardiac development. J Cell Biol 1999; 144:857-868.
16. Rauch F, Prud'homme J, Arabian A et al. Heart, brain, and body wall defects in mice lacking calreticulin. Exp Cell Res 2000; 256:105-111.
17. Hong K, Driscoll M. A transmembrane domain of the putative channel subunit MEC-4 influences mechanotransduction and neurodegeneration in C. elegans. Genetics 1994; 116:377-388.
18. Adams CM, Snyder PM, Price MP et al. Protons activate brain Na^+ channel 1 by inducing a conformational change that exposes a residue associated with neurodegeneration. J Biol Chem 1998; 273:30204-30207.
19. Song SK, Karl IE, Ackerman JJ et al. Increased intracellular Ca^{2+}: a critical link in the pathophysiology of sepsis? Proc Natl Acad Sci USA 1993; 90:3933-3937.
20. Dal Santo P, Logan MA, Chisholm AD et al. The inositol trisphosphate receptor regulates a 50-second behavioral rhythm in C. elegans. Cell 1999; 98:757-767.
21. Clandinin TR, DeModena JA, Sternberg PW. Inositol trisphospate mediates a Ras-independent response to LET-23 receptor tyrosine kinase activation in C. elegans. Cell 1998; 92:523-533.
22. Grant B, Hirsh D. Receptor-mediated endocytosis in the Caenorhabditis elegans oocyte. Mol Biol Cell 1999; 10:4311-4326.
23. Maryon EB, Coronado R, Anderson P. unc-68 encodes a ryanodine receptor involved in regulating C. elegans body-wall muscle contraction. J Cell Biol 1998; 134:885-893.
24. Zwaal RR, Baelen KV, Groenen JT et al. The Sarco-Endoplasmic reticulum Ca^{2+} ATPase Is Required for Development and Muscle Function in Caenorhabditis elegans. J Biol Chem 2001; 276:43557-43563.
25. Cho JH, Bandyopadhyay J, Lee J et al. Two isoforms of sarco/endoplasmic reticulum calcium ATPase (SERCA) are essential in Caenorhabditis elegans. Gene 2000; 261:211-219.
26. Cho JH, Oh YS, Park KW et al. Calsequestrin, a calcium sequestering protein localized at the sarcoplasmic reticulum, is not essential for body-wall muscle function in Caenorhabditis elegans. J Cell Sci 2000; 113:3947-3958.
27. Liu LX, Spoerke JM, Mulligan EL et al. High-throughput isolation of Caenorhabditis elegans deletion mutants. Genome Res 1999; 9:859-867.
28. Ranger AM, Grusby MJ, Hodge MR et al. The transcription factor NF-ATc is essential for cardiac valve formation. Nature 1998; 392:186-190.

29. Molkentin JD, Lu JR, Antos CL et al. A calcineurin-dependent transcriptional pathway for cardiac hypertrophy. Cell 1998; 93:215-228.
30. Graef IA, Chen F, Chen L et al. Signals transduced by Ca^{2+}/calcineurin and NFATc3/c4 pattern the developing vasculature. Cell 2001; 105:863-875.
31. Graef IA, Gastier JM, Francke U et al. Evolutionary relationships among Rem domains indicate functional diversification by recombination. PNAS 2001; 98:5740-5745.
32. Bandyopadhyay J, Lee J, Lee J et al. Calcineurin, a calcium.calmodulin dependent phosphatase, is involved in movement, fertility, egg laying, and growth in C. elegans. Mol Biol Cell 2002; 13:3281-3293.

CHAPTER 23

Calreticulin Deficient Mouse

Lei Guo

Calreticulin, from its initial discovery, has been considered a multifunctional protein.[1] Indeed, many diverse functions have been attributed to this protein, including roles in protein folding and quality control, in modulation of adhesion, in regulation of endoplasmic reticulum (ER) Ca^{2+} storage and SERCA function, and in modulation of gene expression and nuclear transport.[1-4]

To determine whether these functions do, in fact, involve calreticulin, investigators have created calreticulin-deficient mice. In a first step, embryonic stem cells that lack calreticulin were generated.[5] The phenotype of these cells indicates that calreticulin plays a role in both modulation of cell adhesiveness and in control of Ca^{2+} homeostasis.[5] Given the plethora of functions attributed to calreticulin, many scientists thought it would not be possible to generate calreticulin-deficient mice using these cells. To everyone's surprise, calreticulin-deficiency does not interfere with very early embryonic development. However, the deficiency is embryonic lethal at E14.5.[6,7] No live birth has been reported in mice with a calreticulin gene knockout.[6lRauch,2000#1569] The studies on calreticulin-deficient stem cells[5] indicate that calreticulin may not be essential for survival of single cells, but it appears that the protein is essential for proper organ development in the whole organism.[6,7] In this chapter I describe what we have learned from the calreticulin gene knockout mouse.

The Calreticulin Gene Knockout Mouse

Calreticulin gene knockout is embryonic lethal at approximately E14.5.[6,7] In contrast, heterozygous calreticulin gene knockout mice develop normally and show no specific phenotype, although they express 50% less calreticulin than wild-type animals.[6,7] This indicates that reduced expression of calreticulin is well tolerated. The embryonic lethality of calreticulin-deficiency is perhaps not surprising, since calreticulin is an important component of the ER membrane system. We expected that the absence of calreticulin would have devastating effects on ER function and, consequently, on many fundamental cellular pathways. Initially, the specific mechanism of the embryonic lethality was not obvious. Indeed, careful analysis of calreticulin-deficient embryos revealed few histological changes at the light microscope level, but these changes include failure to close the cranial neural tube, umbilical hernia and cardiac pathology.[6,7] It appears that the embryonic lethality of calreticulin-deficiency results from a lesion in cardiac development (see below).[6,7]

Cranial Neural Tube Closure and Umbilical Hernia in Calreticulin-Deficient Embryos

Histological analysis of brain tissue in calreticulin-deficient embryos indicates no major pathology. Rauch et al[7] reported that calreticulin gene knockout embryos developed exencephaly secondary to a failure to close the cranial neural tube. However, this was not a high frequency phenotype (they reported only 16%) and, likely because of its low frequency of occurrence, it has not been observed by others [6]. Although calreticulin-deficiency is embryonic lethal at E14.5,

Calreticulin, 2nd ed., edited by Paul Eggleton and Marek Michalak.

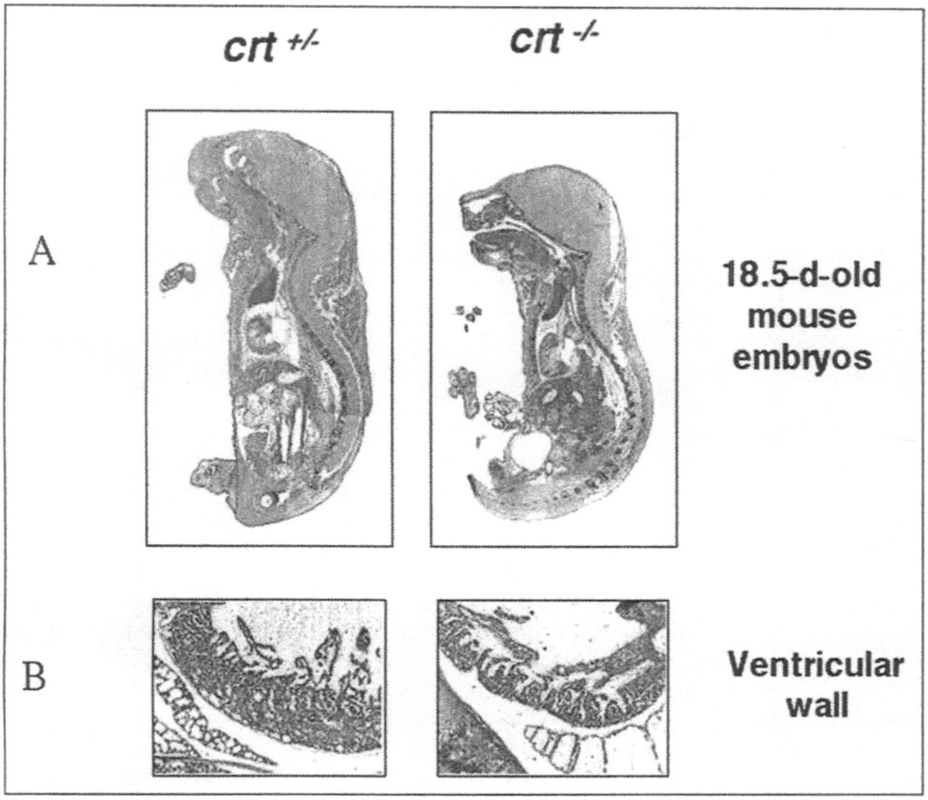

Figure 23.1. Histological analysis of calreticulin deficient mouse embryos. A) Low magnification of histology of 18-d-old *crt+/-* and *crt-/-* mouse embryos. *crt-/-* embryos showed the failure of absorption of the umbilical hernia. B) Histological analysis of the heart of *crt+/-* and *crt-/-* 18-d-old embryos. Sections were stained with the hematoxylin and eosin. Compared to the *crt+/-* embryos, *crt-/-* embryos showed increased fenestration, thinner ventricular wall, impaired growth of the compact layer of the ventricles. Reproduced with permission from.[30]

a significant number of embryos survive until E16.5 and E18.5.[6,7] A significant number of these surviving embryos fail to withdraw the physiological umbilical hernia[6,7] (Fig. 23.1). It is unlikely that the embryonic lethality observed in calreticulin-deficient mice results from this failure to absorb the umbilical hernia, since this pathology is not embryonic lethal in humans.

The omphalocele (umbilical hernia) and failure to close the cranial neural tube that have been observed in calreticulin-deficient mice may indicate that the protein does play a role in cell migration and adhesiveness, as proposed previously.[8,9] This is also in keeping with cell adhesion problems reported in calreticulin-deficient embryonic stem cells.[5] The molecular mechanisms behind calreticulin-dependent omphalocele and failure to close the cranial neural tube are not understood at present. Both are complex processes involving multiple cellular functions that might be influenced by calreticulin. For example, it has been suggested that calreticulin-dependent changes in cell sensitivity to apoptosis and/or differential expression of apoptotic molecules in calreticulin-deficient embryos may play a role.[7,10] Umbilical hernia is thought to be associated with the cyclin-dependent kinase inhibitory protein $p57^{KIP2}$, a regulator of cell proliferation and with disruption of Ca^{2+}-dependent protein kinase C pathways. Ca^{2+}-dependent transcriptional processes may also play a role in body wall development, since NF-AT3-deficient mice show signs of abdominal necrosis.[11] Calreticulin's role in modulation of Ca^{2+} homeostasis would then play a key role in this pathology. Cell adhesion and migration are of obvious importance for neural tube closure and umbilical hernia. Since cell migration and adhesion are essential processes in wound-healing, it is conceivable that calreticulin, Ca^{2+} and the ER may be important components in wound-healing mechanisms. However, this awaits further investigation.

Cardiac Pathology in Calreticulin-Deficient Embryos

Calreticulin-deficient hearts show a marked decrease in the thickness of the ventricular wall[6,7] (Fig. 23.1). In some cases, the ventricular wall comprises a single layer of cardiomyocytes.[6] There are deep intertrabecular recesses and increased fenestration, both of which are associated with the thinner ventricular wall and with impaired growth of the compact layer of the ventricles.[6] Hearts in calreticulin-deficient embryos also have thinner muscular walls in the ventricular septum.[6,7] Surprisingly, there are no severe pathological changes in the atrium, and there seem to be few pathological changes in developing atrial cardiomyocytes, indicating that calreticulin's role is more significant in development of the ventricular wall.[6]

Calreticulin is a minor component of cardiac cells[6,12-16] and, therefore, it is surprising that calreticulin-deficient mice die from impaired cardiac development [6]. However, recent studies looking at transactivation of the calreticulin promoter, and at the expression of calreticulin during mouse embryogenesis, show that the protein is highly expressed in cardiomyoctes during early stages of development.[6,15] We have inferred that calreticulin may be considered an embryonic cardiac gene. Interestingly, Grp94, another Ca^{2+}-binding chaperone located in the lumen of the ER, is also up-regulated during cardiac development.[17,18] This observation indicates that other ER lumenal proteins may be essential during cardiogenesis, and, in this case, mice deficient in ER luminal proteins (such as Grp94) might be expected to show a similar impaired cardiac development to that observed in calreticulin-deficiency.

Cardiac growth and development are extremely complex processes, which are kept under strict transcriptional control.[19-22] A number of transcription factors are critical for specific stages in vertebrate cardiac morphogenesis and hypertrophy, including the homeobox protein Nkx-2.5,[23-26] the basic helix-loop-helix proteins dHAND and eHAND,[20] MEF-2,[27-30] Sox-4, a member of the extended Sry family,[31] and GATA-4, a member of a family of zinc finger transcription factors.[29,32] Currently, we know that Nkx2.5 (tinman) activates transcription of the calreticulin gene and that COUP-TF1 binds to the Nkx2.5 site in the calreticulin promoter, repressing its transactivation.[33] This may explain why the expression of calreticulin is high in the embryonic heart (when Nkx2.5 expression is also high)[26] and down-regulated after birth.[6] Previously, the only cardiac-specific targets reported for Nkx2.5 were the ANF and the α-cardiac actin genes.[34-36] We have identified calreticulin as a new target for this transcription factor. Nkx2.5 is essential for cardiac development.[26] It is expressed at high levels during embryogenesis and disruption of the Nkx2.5 gene in mice leads to embryonic death resulting from cardiac morphogenetic defects.[37] This is very similar to the effects we see when the calreticulin-gene is disrupted.[6] The role of Ca^{2+} (if any) in the cardiac-specific transcriptional events that are regulated by Nkx2.5 and COUP-TF1 has not been investigated (Fig. 23.2).

Recent evidence indicates that intracellular, $InsP_3$-dependent Ca^{2+} stores play a role in cardiac development, growth and pathology, and that this role includes regulation and activation of the NF-AT/GATA-4/CaN transcriptional pathway.[6,38-41] Activation of this pathway depends on a sustained release of Ca^{2+} from ER stores,[6,38-41] which would have to be independent of sarcoplasmic reticulum (SR) Ca^{2+} stores. Consequently, we proposed that cardiac muscle may contain two functionally different Ca^{2+} storage compartments (which are not necessarily structurally distinct).[42] The SR is centrally involved in cardiac excitation-contraction coupling—contraction is initiated by the release of Ca^{2+} from the SR via ryanodine receptor (RyR) Ca^{2+} channels which are located in the cisternae of the SR, and relaxation results from active Ca^{2+} uptake by the SERCA Ca^{2+}-ATPase of the longitudinal SR.[43-47] Cardiac muscle must also contain ER, a membrane system that is continuous with the nuclear envelope and contains

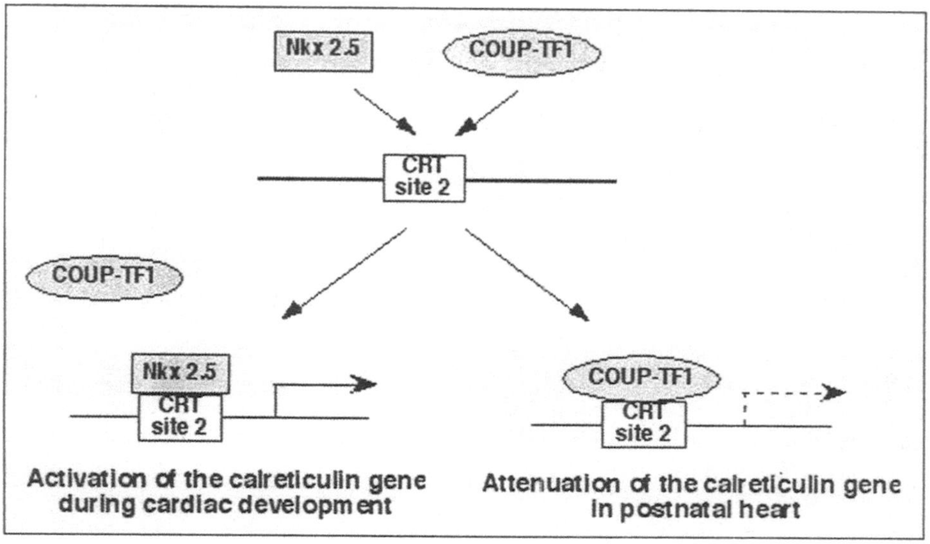

Figure 23.2. A model for transcriptional cooperation between Nkx2.5 and COUP-TF1. Figure shows a putative model for the role of Nkx.25 and COUP-TF1 in transcriptional activation of the calreticulin gene, although other transcription factors may also play a role in transactivation of the gene. During embryonic development Nkx2.5 plays a central role in activating the expression of calreticulin. Nkx2.5 and COUP-TF1 binds to the calreticuliun site 2 element in the calreticulin promoter.[20] In embryonic hearts Nkx2.5 expression is high and COUP-TF1 cannot bind to the calreticulin site 2 in the calreticulin promoter. Down-regulation of Nkx2.5 expression, which occurs as cardiac development progresses and in newborns, permits association between COUP-TF1 and the calreticulin promoter resulting in repression of the calreticulin gene. Reproduced with permission from.[20]

calreticulin and other ER proteins. Ca^{2+} fluxes from the ER may be involved in house-keeping functions in cardiomyocytes, including the regulation of gene expression, and protein and lipid turnover. This idea is supported by our work[6,33,48] and by other recent reports, which show that $InsP_3$ plays a regulatory role in cardiac autonomic activity.[42,49,50]

How Does Calreticulin-Deficiency Result in Impaired Cardiac Development?

Ca^{2+} handling via the $InsP_3$-dependent pathway of ER membranes and translocation of NF-AT from the cytoplasm to the nucleus are both impaired in calreticulin-deficient cells.[6] The simplest explanation for these findings is that calreticulin acts in the calcineurin/NF-AT/GATA-4 pathway described in cardiac hypertrophy,[41,51] and that this pathway is active during cardiac development.[6,52] Recently, we investigated whether calreticulin acts as an "upstream" molecule in the calcineurin/NF-AT pathway. Specifically, we wanted to determine whether activation of calcineurin by calreticulin is important in cardiac development. Importantly, we found that expression of Ca^{2+}/calmodulin-independent (constitutively active) calcineurin, in the heart, results in a rescue of the cardiac defects observed in calreticulin-deficient mice, such that these transgenic animals are born live despite total calreticulin-deficiency (Gou and Michalak, submitted). These findings clearly indicate that both calreticulin and calcineurin are essential in cardiac development, and that calreticulin exerts its effects in cardiac development through interaction with calcineurin, likely as a source of Ca^{2+}. This interaction implies a mechanism by which Ca^{2+} signaling might converge with regulation of cardiac gene expression during cardiac development. Clarification of this important point awaits further investigation.

The Calreticulin-Deficient Mouse Shows that Cardiac ER and SR Compartments are Functionally Distinct

One important lesson we have learned from the calreticulin-deficient mouse is that the SR and ER may be functionally distinct compartments in cardiomyocytes. The SR membrane contains a high density of SERCA2a, accumulates Ca^{2+} rapidly during muscle relaxation and stores it sequestered to calsequestrin. The ryanodine receptor/Ca^{2+} release channel is concentrated in terminal cisternae of the SR and plays a role in the Ca^{2+} release that triggers muscle contraction. It is well-established that the SR is formed from ER membranes during muscle development.[53,54] Calreticulin-deficient cardiomyocytes develop a functional SR and contract spontaneously.[6] However, these cells show impaired maintenance of Ca^{2+} homeostasis by the ER.[6,55] Study of calreticulin-deficient mice indicates that the Ca^{2+} pool that signals developmental responses, including activation of calcineurin-dependent transcriptional pathways, is distinct from the SR Ca^{2+} pool involved in excitation-contraction coupling. How cardiomyocytes distinguish these Ca^{2+} pools remains to be established. The suggestion that biologically distinct Ca^{2+} signals are generated by the ER and the SR is supported by recent reports which show that $InsP_3$ plays a role in regulation of cardiac autonomic activity[49] and that it may even modulate excitation-contraction coupling.[50]

The Effects of Calreticulin Over-Expression in Postnatal Heart and Its Role in Congenital Complete Heart Block

The expression of calreticulin in the embryonic heart is high, but it declines sharply after birth. This pattern likely results from activation of the calreticulin gene during development, by Nkx2.5, and from suppression of the calreticulin gene after birth, by COUP-TF1.[33] To determine the importance of this pattern of expression, we created transgenic mice which over-express calreticulin in the heart.[48] We found that these animals develop congenital arrhythmia (bradycardia), sinus node depression, and complete cardiac block, and that they die from heart failure. This indicates that calreticulin plays a role, which leads to pathology in the conductive system of the heart.[48] Histological analysis of the transgenic hearts revealed a dilated ventricular chamber, a relatively thinner ventricular wall, and ballooned atria with thrombosis and disarray of cardiomyocytes.[48] The enlargement of the heart is not due to a hypertrophic response. Using M-mode echocardiography we established that the transgenic animals exhibit a loss of systolic-function.[48] Electrocardiogram revealed that the P-R interval is significantly prolonged in mice that over-express calreticulin, and these mice develop a complete AV conduction block (Fig. 23.3). Prolongation of the QRS interval was not significant except in ECGs showing AV block.

Hearts that over-express calreticulin express very low levels of connexin43 (Cx43),[48] a major component of the gap junctions responsible for cell-cell communication.[56,57] Phosphorylation of Cx43 is an important modulator of intercellular communication via gap junctions, affecting conductance, metabolic coupling, growth and differentiation.[57] We found that the animals which over-express calreticulin also exhibit significantly reduced phosphorylation of connexin. It appears that the function of gap junctions is impaired when levels of calreticulin are high in postnatal hearts.

The current density of peak inward $I_{Ca,L}$ at 0 mV elicited in cardiomyocytes that over-express calreticulin was also significantly smaller than the $I_{Ca,L}$ measured in control cardiomyocytes.[48] Interestingly, the cardiac phenotype of mice, which over-express calreticulin is reminiscent of the phenotype seen in complete heart block in children.[48] The causes and molecular mechanism(s) behind complete heart block are not known, at present. However, it appears that calreticulin acts in a pathway responsible for the etiology of this disease.[48]

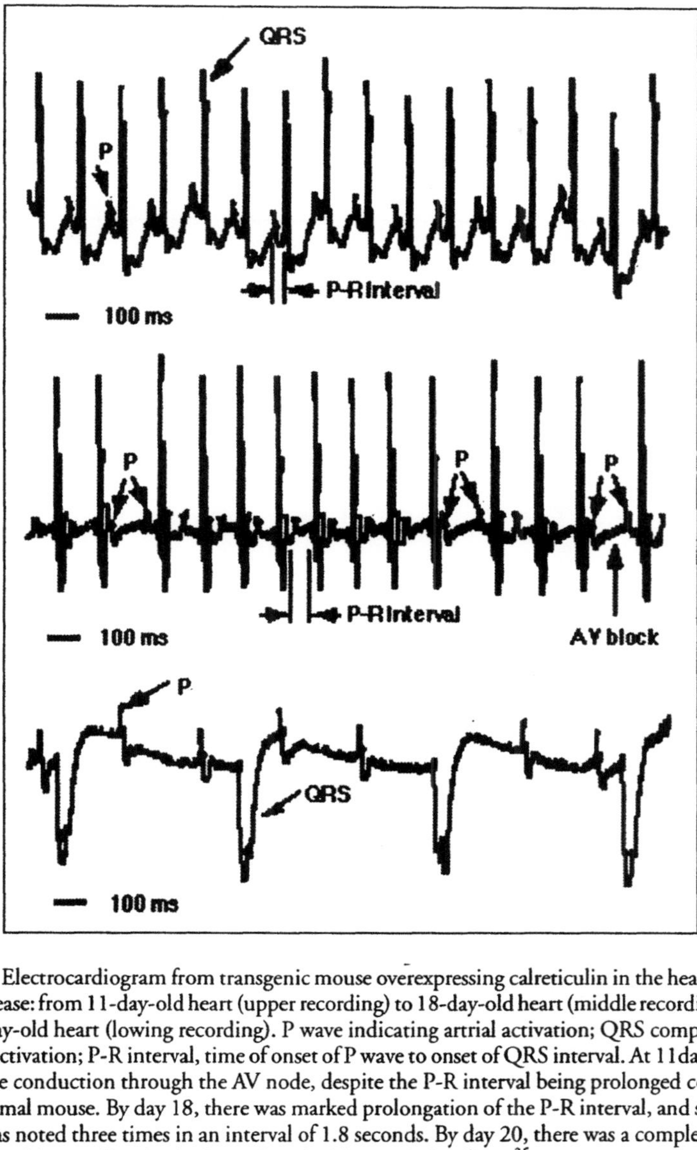

Figure 23.3. Electrocardiogram from transgenic mouse overexpressing calreticulin in the heart. Progressive AV node disease: from 11-day-old heart (upper recording) to 18-day-old heart (middle recording) to sudden death, 20-day-old heart (lowing recording). P wave indicating artrial activation; QRS complex indicating ventricular activation; P-R interval, time of onset of P wave to onset of QRS interval. At 11days of age there is one-to-one conduction through the AV node, despite the P-R interval being prolonged compared with that of a normal mouse. By day 18, there was marked prolongation of the P-R interval, and second-degree AV block was noted three times in an interval of 1.8 seconds. By day 20, there was a complete heart block followed by sudden cardiac death. Reproduced with permission from.[35]

Conclusions

Calreticulin is essential for proper cardiac development, and regulation of its expression is important in the development of the cardiac conductive system. These findings will add to our understanding of the molecular basis of congenital arrhythmias, and they may help us to identify targets for future therapies. To date, molecular studies of ER proteins have revealed that this membrane system plays a critical role in cardiac development. Most importantly, the ER is not only involved in the synthesis, modification and secretion of proteins, but is active in regulation of intracellular Ca^{2+} homeostasis in the developing heart. As we further our understanding of the role of ER membrane proteins in cardiac pathology, we will perhaps be able to

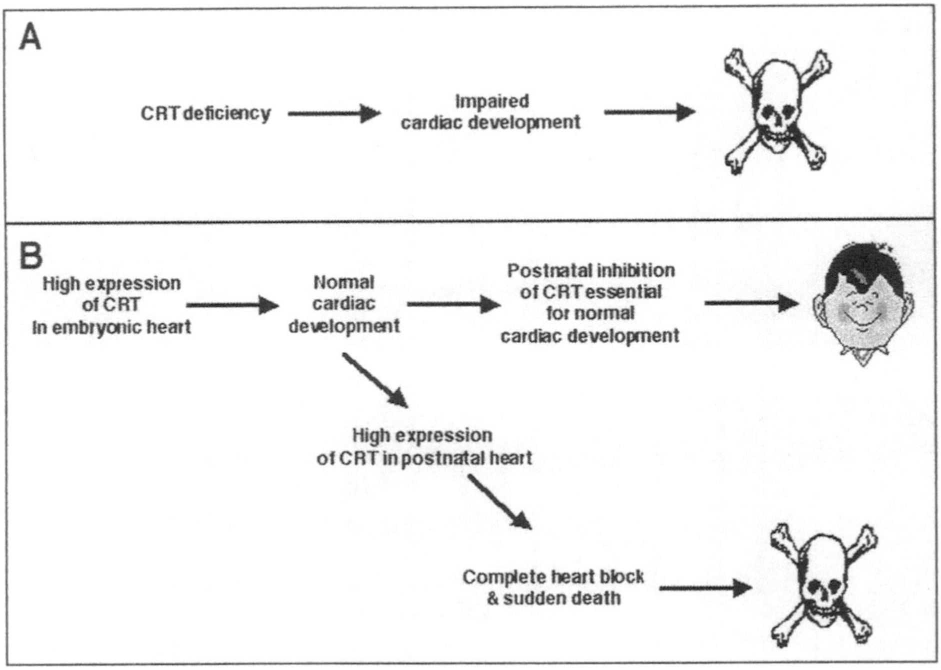

Figure 23.4. Calreticulin and cardiac development and pathology. A) Calreticulin plays an essential role in cardiac development during embryogenesis. Calreticulin deficiency is embryonic lethal because of a lesion in cardiac development.[30,44] B) Calreticulin is highly expressed in the embryonic heart but is down regulated in postnatal heart due to transcriptional inhibition.[20,30] Calreticulin level is maintained at low level in mature heart. If calreticulin highly expressed in the postnatal heart, it causes complete heart block followed by sudden death.[35]

address many clinical problems in the cardiovascular field. In summary, we have demonstrated that calreticulin plays an essential role in cardiac development during embryogenesis. Although calreticulin is not a transcription factor, it is a regulator of Ca^{2+} homeostasis and of the transcriptional pathways involved in proper cardiac development. Most importantly, expression of calreticulin is tightly regulated in the developed heart because high levels of the protein in developed heart are associated with complete heart block and sudden death and with some forms of hypertrophy (Fig. 23.4).

Acknowledgements

Research in our laboratory is supported by grants from the Canadian Institutes of Health Research, from the Heart and Stroke Foundations of Alberta. L.G. is a Heart and Stroke Foundation of Canada Postdoctoral Fellow.

References

1. Michalak M, Corbett EF, Mesaeli N et al. Calreticulin: one protein, one gene, many functions. Biochem J 1999; 344:281-92.
2. Crofts AJ, Denecke J. Calreticulin and calnexin in plants. Trends Plant Sci 1998; 3:396-99.
3. Nakhasi HL, Pogue GP, Duncan RC et al. Implications of calreticulin function in parasite biology. Parasitol Today 1998; 14:157-60.
4. Johnson S, Michalak M, Opas M et al. The ins and outs of calreticulin: from the ER lumen to the extracellular space. Trends Cell Biol 2001; 11(3):122-29.
5. Coppolino MG, Woodside MJ, Demaurex N et al. Calreticulin is essential for integrin-mediated calcium signalling and cell adhesion. Nature 1997; 386:843-47.

6. Mesaeli N, Nakamura K, Zvaritch E et al. Calreticulin is essential for cardiac development. J Cell Biol 1999; 144:857-68.
7. Rauch F, Prud'homme J, Arabian A et al. Heart, brain, and body wall defects in mice lacking calreticulin. Exp Cell Res 2000; 256:105-11.
8. Opas M, Szewczenko-Pawlikowski M, Jass GK et al. Calreticulin modulates cell adhesiveness via regulation of vinculin expression. J Cell Biol 1996; 135:1913-23.
9. Coppolino MG, Dedhar S. Calreticulin. Int J Biochem Cell Biol 1998; 30:553-58.
10. Nakamura K, Bossy-Wetzel E, Burns K et al. Changes in endoplasmic reticulum luminal environment affect cell sensitivity to apoptosis. J Cell Biol 2000; 150:731-40.
11. delaPompa JL, Timmerman LA, Takimoto H et al. Role of the NF-ATc transcription factor in morphogenesis of cardiac valves and septum. Nature 1998; 392:182-86.
12. Fliegel L, Burns K, MacLennan DH et al. Molecular cloning of the high affinity calcium-binding protein (calreticulin) of skeletal muscle sarcoplasmic reticulum. J Biol Chem 1989; 264:21522-28.
13. Milner RE, Baksh S, Shemanko C et al. Calreticulin, and not calsequestrin, is the major calcium binding protein of smooth muscle sarcoplasmic reticulum and liver endoplasmic reticulum. J Biol Chem 1991; 266:7155-65.
14. Tharin S, Dziak E, Michalak M et al. Widespread tissue distribution of rabbit calreticulin, a non-muscle functional analogue of calsequestrin. Cell Tissue Res 1992; 269:29-37.
15. Imanaka-Yoshida K, Amitani A, Ioshii SO et al. Alterations of expression and distribution of the Ca^{2+}-storing proteins in endo/sarcoplasmic reticulum during differentiation of rat cardiomyocytes. J Mol Cell Cardiol 1996; 28:553-62.
16. Tsutsui H, Ishibashi Y, Imanaka-Yoshida K et al. Alterations in sarcoplasmic reticulum calcium-storing proteins in pressure-overload cardiac hypertrophy. Am J Physiol 1997; 272:H168-H75.
17. Barnes JA, Smoak IW. Immunolocalization and heart levels of GRP94 in the mouse during post-implantation development. Anat Embryol 1997; 196:335-41.
18. Vitadello M, Colpo P, Gorza L. Rabbit cardiac and skeletal myocytes differ in constitutive and inducible expression of the glucose-regulated protein GRP94. Biochem J 1998; 332:351-59.
19. Chien KR, Zhu H, Knowlton KU et al. Transcriptional regulation during cardiac growth and development. Annu Rev Physiol 1993; 55:77-95.
20. Srivastava D, Cserjesi P, Olson EN. A subclass of bHLH proteins required for cardiac morphogenesis. Science 1995; 270:1995-99.
21. Olson EN, Srivastava D. Molecular pathways controlling heart development. Science 1996; 272:671-76.
22. Sucov HM. Molecular insights into cardiac development. Annu Rev Physiol 1998; 60:287-308.
23. Tonissen KF, Drysdale TA, Lints TJ et al. XNkx-2.5, a Xenopus gene related to Nkx-2.5 and tinman: evidence for a conserved role in cardiac development. Dev Biol 1994; 162:325-28.
24. Patterson KD, Cleaver O, Gerber WV et al. Homeobox genes in cardiovascular development. Curr Top Dev Biol 1998; 40:1-44.
25. Schwartz RJ, Olson EN. Building the heart piece by piece: modularity of cis-elements regulating Nkx2-5 transcription. Development 1999; 126(19):4187-92.
26. Evans SM. Vertebrate tinman homologues and cardiac differentiation. Semin Cell Dev Biol 1999; 10(1):73-83.
27. Edmondson DG, Lyons GE, Martin JF et al. Mef2 gene expression marks the cardiac and skeletal muscle lineages during mouse embryogenesis. Development 1994; 120:1251-63.
28. Ross RS, Navankasattusas S, Harvey RP et al. An HF-1a/HF-1b/MEF-2 combinatorial element confers cardiac ventricular specificity and established an anterior-posterior gradient of expression. Development 1996; 122(6):1799-809.
29. Fishman MC, Chien KR. Fashioning the vertebrate heart: earliest embryonic decisions. Development 1997; 124:2099-117.
30. Christensen TH, Kedes L. The myogenic regulatory circuit that controls cardiac/slow twitch troponin C gene transcription in skeletal muscle involves E-box, MEF-2, and MEF-3 motifs. Gene Expr 1999; 8(4):247-61.
31. Schilham MW, Oosterwegel MA, Moerer P et al. Defects in cardiac outflow tract formation and pro-B-lymphocyte expansion in mice lacking Sox-4. Nature 1996; 380:711-14.
32. Orkin SH. GATA-binding transcription factors in hematopoietic cells. Blood 1992; 80:575-81.
33. Guo L, Lynch J, Nakamura K et al. COUP-TF1 antagonizes Nkx2.5-mediated activation of the calreticulin gene during cardiac development. J Biol Chem 2001; 276:2797-801.
34. Grepin C, Dagnino L, Robitaille L et al. A hormone-encoding gene identifies a pathway for cardiac but not skeletal muscle gene transcription. Mol Cell Biol 1994; 14:3115-29.

35. Durocher D, Charron F, Warren R et al. The cardiac transcription factors Nkx2-5 and GATA-4 are mutual cofactors. EMBO J 1997; 16:5687-96.
36. Sepulveda JL, Belaguli N, Nigam V et al. GATA-4 and Nkx-2.5 coactivate Nkx-2 DNA binding targets: role for regulating early cardiac gene expression. Mol Cell Biol 1998; 18:3405-15.
37. Lyons I, Parsons LM, Hartley L et al. Myogenic and morphogenetic defects in the heart tubes of murine embryos lacking the homeo box gene Nkx2-5. Genes Dev 1995; 9(13):1654-66.
38. Timmerman LA, Clipstone NA, Ho SN et al. Rapid shuttling of NF-AT in discrimination of Ca^{2+} signals and immunosuppression. Nature 1996; 383:837-40.
39. Dolmetsch RE, Lewis RS, Goodnow CC et al. Differential activation of transcription factors induced by Ca^{2+} response amplitude and duration. Nature 1997; 386:855-58.
40. Rao A, Luo C, Hogan PG. Transcription factors of the NFAT family: regulation and function. Annu Rev Immunol 1997; 15:707-47.
41. Molkentin JD, Lu JR, Antos CL- et al. A calcineurin-dependent transcriptional pathway for cardiac hypertrophy. Cell 1998; 93:215-28.
42. Mesaeli N, Nakamura K, Opas M et al. Endoplasmic reticulum in the heart, a forgotten organelle? Mol Cell Biochem 2001; 224(1):1-6.
43. MacLennan DH. Molecular tools to elucidate problems in excitation-contraction coupling. Biophys J 1990; 58(6):1355-65.
44. Coronado R, Morrissette J, Sukhareva M et al. Structure and function of ryanodine receptors. Am J Physiol 1994; 266:C1485-C504.
45. Sutko JL, Airey JA. Ryanodine receptor Ca^{2+} release channels: does diversity in form equal diversity in function? Physiol Rev 1996; 76:1027-71.
46. MacLennan DH, Rice WJ, Green NM. The mechanism of Ca^{2+} transport by sarco(endo)plasmic reticulum Ca^{2+}-ATPases. J Biol Chem 1997; 272:28815-18.
47. Franzini-Armstrong C, Protasi F. Ryanodine receptors of striated muscles: a complex channel capable of multiple interactions. Physiol Rev 1997; 77:699-729.
48. Nakamura K, Robertson M, Liu G et al. Complete heart block and sudden death in mouse over-expressing calreticulin. J Clin Invest 2001; 107(10):1245-53.
49. Jaconi M, Bony C, Richards SM et al. Inositol 1,4,5-trisphosphate directs Ca^{2+} flow between mitochondria and the Endoplasmic/Sarcoplasmic reticulum: A role in regulating cardiac autonomic Ca^{2+} spiking. Mol Biol Cell 2000; 11(5):1845-58.
50. Lipp P, Laine M, Tovey SC et al. Functional $InsP_3$ receptors that may modulate excitation-contraction coupling in the heart. Curr Biol 2000; 10(15):939-42.
51. Sussman MA, Lim HW, Gude N et al. Prevention of cardiac hypertrophy in mice by calcineurin inhibition. Science 1998; 281:1690-93.
52. Nolan GP. Cardiac development. Transcription and the broken heart. Nature 1998; 392:129-30.
53. Franzini-Armstrong C, Jorgensen AO. Structure and development of E-C coupling units in skeletal muscle. Annu Rev Physiol 1994; 56:509-34.
54. Flucher BE, Franzini-Armstrong C. Formation of junctions involved in excitation-contraction coupling in skeletal and cardiac muscle. Proc Natl Acad Sci USA 1996; 93:8101-06.
55. Nakamura K, Zuppini A, Arnaudeau S et al. Functional specialization of calreticulin domains. J Cell Biol 2001; 154:961-72.
56. Gros DB, Jongsma HJ. Connexins in mammalian heart function. Bioessays 1996; 18(9):719-30.
57. Kardami E, Doble BW. Cardiomyocyte gap junctions: a target for growth factor signaling. Trends Cardiovasc Med 1998; 8:180-87.

APPENDIX I

Appendix I: Human Calreticulin Data Sheet

Paul Eggleton and Marek Michalak

Previous Names

High affinity Ca^{2+} binding protein, (HACBP). Calcium binding reticuloplasmin of molecular weight 55,000, (CRP55). Calregulin. Reticulin.

Physicochemical Properties

1. Calreticulin is a member of the endoplasmic reticulum Ca^{2+} binding lectin chaperones and shares amino acid sequence homology to calnexin.
2. The mature human calreticulin is approximately 400 amino acids long and has a cleaved 17 amino acid leader sequence.
3. The protein has been conveniently described as comprising of three domains namely, N-, P- and C-domain. The N-domain comprises of amino acid residues 1-180, P-domain 181-290 and C-domain, 291-400. 3D structure of the middle portion of the protein (the P-domain) has been solved by biophysical means and the NMR spectroscopy. The structure corresponding to amino acid residues 189–288 shows an extended hairpin fold comprising the entire polypeptide chain with amino and carboxyl termini in close spatial proximity. This unusual structure constitutes a new fold.

Mature Protein

pI	4.46
Extinction coefficient (1 mg/ml at 280 nm)	1.5
Sedimentation coefficient ($S^{\circ}_{20,w}$)	2.71 S
Predicted Stokes radius Rs (Å)	29.4 Å (for monomeric globular protein)
Calculated Stokes radius Rs (Å)	46.2 Å
Gel filtration elution size	158 Kd
Mr (Kd)	predicted 46.0 ± 1 Kd observed 55-60.0 Kd (on high pH SDS PAGE under reducing conditions)
Apparent length	29.8 nm
Apparent diameter	2.44 nm

N-Linked Glycosylation Sites (Species Specific)

1 in humans (unoccupied but in certain cell types may become occupied under stress) Asn^{327}. 2 in bovine sp. (residues 162 and 327).

Interchain Disulphide Bonds

1. Cys^{88}-Cys^{120}; free cysteine 162

Calreticulin, 2nd ed., edited by Paul Eggleton and Marek Michalak.

Phosphorylation

Several potential low homology consensus recognition sites for protein kinase C (residues 17-19; 36-38; 61-63; 68-70; 79-81), casein kinase II (residues 51-54; 172-175; 178-181, 196-200, 204-208, 307-311) and tyrosine kinase (residues 261-268).

Ion-Binding Characteristics

- Ca^{2+} (two binding regions) P-domain K_d = 1μM—1 mole/mole protein: C-domain K_d = 2mM—20-30 moles/mole protein
- Zn^{2+} (two binding regions) K_d = 0.8 μM -26 moles/mole protein; K_d = 47.6 μM—83 moles/mole protein
- Fe^{3+} K_d = 90μM—1 mole/mole protein

Gene Structure

The gene consists of nine exons and spans 3.6 kb and 4.6 kb in human and mouse genomic DNA respectively. The mRNA of the human form of calreticulin is 1.9 kb is size. The promoter of human and mouse genes for calreticulin has putative regulatory sites including: - AP-1/2; SpI, H4TF-1; Nkx2.5, COUP-TF1, MEF2c, 4 x CCAAT sequences.

Commercial Antibodies Raised Against Calreticulin

Company: Abcam Ltd.

31 Cambridge Science Park. Milton Road. Cambridge CB4 0FX. UK
Phone: +44 (0) 1223 472030
Fax: +44 (0) 1223 472038
Email: technical@abcam.com
www.abcam.com

Product Name (Number): Anti-Calreticulin (ab4109)

- Ab: Goat polyclonal against rabbit skeletal muscle calreticulin
- Cross reactivity: Human, mouse, rat, rabbit
- Applications: Western blotting (1/300 to 1/500); Immunochemistry (1/50 to 1/70) Flow cytometry—not described; Immunoprecipitation—does not work.

Company: Novus-Biological, Inc.

PO Box 802. Littleton, CO 80160. USA
Phone: 1-888-506-6887 or 303-730-1950
Fax: 303-730-1966
Email: novus@novus-biologicals.com
www.novus-biologicals.com/main.html

Product Name (Number): Anti-Calreticulin (600-101)

- Ab: Rabbit polyclonal against mouse calreticulin fusion protein containing amino acids 154-347.
- Cross reactivity: Human, mouse, rat, bovine and hamster.
- Applications: Western blotting (1/1000); Immunochemistry (1/250) Flow cytometry—(1/50 – 1/100); Immunoprecipitation—(1/300).

Company: Research Diagnostics Inc.

Pleasant Hill Road. Flanders NJ 07836. USA
Phone: (800) 631-9384 or (973) 584-7093
Fax: (973) 584-0210
Email: ResearchD@aol.com
www.researchd.com

Product Name (Number): Anti-Calreticulin (Cat No RDI-Calrtnabr)

- Ab: Rabbit polyclonal against recombinant human calreticulin produced in Baculovirus insect cell system.
- Cross reactivity: Human, rabbit, xenopus, non muscle ER and rat.
- Applications: Western blotting (1/1000); Immunochemistry (1/1000) Flow cytometry—(not stated) Immunoprecipitation—(1/100).

Company: Affinity Bioreagents

4620 Technology Drive, Suite 600. Golden, CO 80403, USA
Phone: 800-527-4535/303-278-4535
Fax: 303-278-2424
Email: affinity@bioreagents.com
www.bioreagents.com

Product Name (Number): Anti-Calreticulin (Cat No PA3-900)

- Ab: Rabbit polyclonal against recombinant human calreticulin produced in Baculovirus insect cell system.
- Cross reactivity: Human, rabbit and rat.
- Applications: Western blotting (1/1000); Flow cytometry—(1/100) Immunoprecipitation—(Assay dependent).

Product Name (Number): Anti-Calreticulin (Cat No PA1-902)

- Ab: Chicken polyclonal against synthetic peptide ($_{24}$KEQFLDGDAWTNRWVESKHK$_{43}$)
- Cross reactivity: human, mouse, canine hamster and rat.
- Applications: Western blotting (1/500); Immunohistochemistry—(1/100).

Product Name (Number): Anti-Calreticulin (Cat No PA1-903)

- Ab: Chicken polyclonal against synthetic peptide ($_{399}$DEKEEDEEESPGQAKD$_{414}$)
- Cross reactivity: human, mouse, canine hamster, rabbit and rat.
- Applications: Western blotting (1/2000).

Company: Upstate Biotechnology

199 Saranac Avenue. Lake Placid. NY. USA
Phone: 800-548-7853
Fax: 518-523-4513
Email: techserv@upstatebiotech.com
www.upstatebiotech.com

Product Name (Number): Anti-Calreticulin (Cat No 06-661)

- Ab: Rabbit polyclonal against KLH conjugated synthetic peptide ($_{394}$QAKDEL$_{400}$).
- Cross reactivity: Human, rabbit, Xenopus, rat and weakly with bovine and porcine.
- Applications: Western(0.5-2.0µg/ml); Immunohistochemistry (0.5µg/ml).

Company: Calbiochem

P.O. Box 12087. La Jolla, California 92039-2087, USA
Phone: (800) 628-8470 / (800) 854-3417
Fax: (800) 776-0999
Email: technical@calbiochem.com
www.calbiochem.com

Product Name (Number): Anti-Calreticulin Human (Mouse) (Cat No 208912)

- Ab: Protein G-affinity purified monoclonal antibody. Immunogen was a calreticulin-maltose binding protein fusion protein.

- Cross reactivity: Human, (not stated but detects calreticulin from diverse range of human cells e.g., epithelial cells, T-cells, and various leukemic cells.
- Applications: Western (0.5µg/ml); Immunohistochemistry (0.5µg/ml).

Company: Oxford Biomedical Research. (NH Biotech)

P.O. Box 522, Oxford M. 48371. USA
Phone: (810) 852-8815
Fax: 810 852 4466
Email: oxfordbio@aol.com
www.informagen.com

Product Name (Number): Sheep Antibody to Human Calreticulin (AP 62)

- Ab: Whole sera. Immunogen was carboxy terminal domain of human calreticulin fused to GST.
- Cross reactivity: Human, dog and rat.
- Applications: Western (1/5000 – 1/10000).

APPENDIX II

Amino Acid Sequence of Calreticulin

Comparison of amino acid sequence of selected calreticulins. Amino acid alignments were carried out using MacVector software. Residues numbered are for the human calreticulin starting with the N-terminal amino acid residue of the mature protein. Amino acid sequences with accession numbers in parenthesis Drosophila melanogaster (AB000718; X64461); Eptatretus burgeri (AB025323); Lampetra reissneri (AB25328); Hordeum vulgare barley 1a (L27348); Hordeum vulgare barley 2a (L27349); Bovine brain (L13462); C. annuum (X80756); C. elegans (X59589; S40213); Chlamydomonas reinhardtii (AJ000765); Rana rugosa (D78589); D. bioculata (Y09078); E. gracilis (Y09816); Human (X85727; X59053; S78537; M84739); Mouse (X14926); Necator americanum (AJ006790); N. plumbaginifolia (Z71395); Onchocerca volvulus (M20565); P. argentatum (X82578); Rabbit (J05138); Rat (D78308; X53363; X79327); Schistosoma mansoni (L24159); Schistosoma japanicum (M80524); Tritrichomonas suis (AJ011990); X. leavis (X67597; X67598); Z. mays (X89813; X78057; Z46772).

Calreticulin, 2nd ed., edited by Paul Eggleton and Marek Michalak.

		20	40	60
Human	MLLSVPLLLGLLGLAVA	EPAVYFKEQFLDGDGWTSRW	IESKHKSDFGK-FVLSSGKFY	GDEEKDKGLQTSQDARFYAL
Mouse	MLLSVPLLLGLLGLAAA	DPAIYFKEQFLDGDAWTNRW	VESKHKSDFGK-FVLSSGKFY	GDLEKDKGLQTSQDARFYAL
Rat	MLLSVPLLLGLLGLAAA	DPAIYFKEQFLDGDAWTNRW	VESKHKSDFGK-FVLSSGKFY	GDQEKDKGLQTSQDARFYAL
Rabbit	MLLPVPLLLGLLGLAAA	EPVVYFKEQFLDGDGWTERW	IESKHKSDFGK-FVLSSGKFY	GDQEKDKGLQTSQDARFYAL
X. leavis	LVLPLALGLCIA	EPAVYFKEEFTDGDGWTQRW	VESKHKTDYGK-FKLSAGKFY	GDSEKDKGLQTSQDARFYAM
Rana rugosa	IAV-FVLPLALALCLA	KPAVYFQEEFGDGDAWKERW	IQSKHKSDYGQ-FKLSAGKFY	GDEEKDKGLQTSQDAKFYAH
Bovine brain	LHHIVTF-LR	EETVFFSEQFLTLD—LKY	KASKL-SSI-R-EALSMSKV-	GIIE-N—F-CFSEISF—L
N. americanum	MRSLVALLPVLCIAVA-	E—VYFKEEFLD-DSWKERW	VQSKHKSDYGE-FVLSAGK-Y	FGDATRDQGMKTSQDAKFYS
Drosophila	TVTVLLATVGF-IS-A	E—VYLKENF-DNENWEDTW	IYSKHPKGEFGKFVLTPGTFY	NDAEADKGIQTSQDARFYA
Onchocerca	MQLSLLVGLVCFSAI	NAKIYFKEDF-SDDDWEKRW	IKSKHKDDFGK-WEISHGKFY	GDAVKDKGLKTTQDAKFYS
C. legans	LLAIVAV-VS-A	E—VYFKEEF-NDASWEKRW	VQSKHKDDFGA-FKLSAGKFF	DVESRDQGIQTSQDAKFYS
E. gracilis	MRKELWLLGLLSSQAVL	STIYYKETF-EPD-WETRW	THSTAKSDYG-KFKLTSGKFY	GDKAKDAGIQTSQDAKFYA
Z. mays	AVAALLALASVAAVAG	EVFFQEKF-E-DGWESRW	KSVEWKKDENAMGEWNHTSGKWN	GDAEDK GIQTSEDYRFYA
Barley 1a	LLRRLALLALASVAAVAA	DVFFQEKF-E-DGWESRW	KSVEWKKDENAMGEWNHTSGKWH	GDAEDK GIQTSEDYRFYA
N. plumbaginifolia	LHLITVFSLL-VAVVSA	EVFFEESF-N-DGWESRW	KSVEWKKDENAMGEWNHTSGKWN	GDANDK GIQTSEDYRFYA
C. reinhardtii	MKWGVVAVLATLVVA-ASAK	DYFKETF-DG-SWADRWT	KSSWVSDKGSAGEFKLTAGKWY	GDAEADKGIQTGPDSKFFA
Barley 2a	SVAAVAA	DVFFQEKF-E-DGWESRW	KSVEWKKDENAMGEWNHTSGKWH	GDAEDK GIQTSEDYRFYA
C. annuum			GEWNHTSGKWS	GDANDK GIQTSEDYRFYA
S. mansoni			FKIEAGKSPVNPIEDLGLKTTQDARFYG	
F. suis	TVPEG	NIFHFQSFGNSD-WSNTW	SVSKLDNYTGVWEVSESAQ	PQAIPGEKMIYMTKAMSYYG
D. biculata	LLAIVAV-VS-A	EVYFKEEF-NDASWEKRW	VQSKHKDDFGAFKLSAGKF	FDVESRDQGIQTSQDAKFYS
P. argentatum			EFGTRNYTSGKWN	GD—ANDKGIQTSEDYRFYA

	80	100	120		140
Human	SASFEPFSNKGQTLVVQFTV	VKHEQNIDCGGGYVKLFPNS	LDQTDMHGDSEYNIMFGPDI	CGPGTKKVHVIFNY KGKN	VL
Mouse	SAKFEPFSNKGQTLVVQFTV	VKHEQNIDCGGGYVKLFPSG	LDQKDMHGDSEYNIMFGPDI	CGPGTKKVHVIFNY KGKN	VL
Rat	SARFEPFSNKGQTLVVQFTV	VKHEQNIDCGGGYVKLFPGG	LDQKDMHGDSEYNIMFGPDI	CGPGTKKVHVIFNY KGKN	VL
Rabbit	SARFEPFSNKGQPLVVQFTV	VKHEQNIDCGGGYVKLFPAG	LDQKDMHGDSEYNIMFGPDI	CGPGTKKVHVIFNY KGKN	VL
X. leavis	SSRFDSFSNKDQTLVVQFSV	VKHEQNIDCGGGYVKLFPAA	LEQTEMHEESEYNIMFGPDI	CGPPTKKVHVIFQY KKKN	LQ
Rana rugosa	SAGFPAFSNKDKPLVVQFSV	VKHEQNIDCGGGYVKLFPST	LEQTDMHGESEYNIMFGPDI	CGPPTKKVHVIFNY KGKN	LQ
Bovine brain	QESIKSHGRRTLVGCSPWG	HEEQNIDCGGGYVNVFPAG	LDQTDMHGDSEYNIMFGPDI	CGPGTKKVHVIFNY KGKN	VL
N. americanum	RAAKPFKAFSNKGKTVVIQFT	VKHEQGIDCGGGYVKVMSSD	VDLSDFHGETPYNVMFGPDI	CGP-TKKVHDIFSY KGKN	HL
Drosophila	ASRKFDGFSNEDKPLVV-QFS	VKHEQNIDCGGGYVKLFDCS	LDQTDMHGESPYEIMFGPDI	CGPGTKKVHVIFSY KGKN	HL
Onchocerca	IGAKFKDSFSNKGKSLVIQFS	VKHEQDIDCGGGYVKLMASD	VNLEDSHGETPYHIMFGPDI	CGPGTKKVHVIFHY KDRN	HM
C. elegans	RAAKFKDDFSNKGKTLVIQYT	VKHEQGIDCGGGYVKVMRAD	ADLGDFHGETPYNVMFGPDI	CGP-TRRVHVILNY KGEN	KL
E. gracilis	ISSPAISSFSNEGKDLVLQFS	VKHEQDIDCGGGYLKLLP-S	VDAAKFTGDTPYHIMFGPDI	CG-ATKKIHFILTY KGKN	LL
Z. mays	ISAEYPEFSNKDKTLVL-QFS	VKHEQKLDCGGGYVKLLGGD	DQKKFGGDTSYSIMFGPDI	CGYSTKKVHTILTK DGKN	HL
Barley 1a	ISAEYPEFSNKDKTLVL-QFT	VKHEQKLDCGGGYVKLLGGD	VDQKKFGGDTPYGIMFGPDI	CGYSTKKVHTILTK NGKN	HL
N. plumbaginifolia	ISAEFPEFSNKGKNLVF-QFS	VKHEQKLDCGGGYMKLLSGD	VDQKKFGGDTPYSIMFGPDI	CGYSTKKVHAILTY NDTN	HL
C. reinhardtii	ISAPLAVTFDNTGKDTVVQFS	VKHEQDLDCGGGYIKVVPATS	KQEMGEFGGDTPYSIMFGPDI	CGYSTRKVHVILTY KGKN	YL
Barley 2a	ISAEYPEFSNKDKTLVL-QFT	VKHEQKLDCGGGYVKLLGGD	VDQKKFGGDTPYGIMFGPDI	CGYSTKKVHTILTK NGKN	HL
C. annuum	ISAEFPEFSNKGKNLVF-QFS	VKHEQKLDCGGGYMKLISGD	VDQKKFGGDTPYSIMFGPDI	CGYSTKKVHAILTY NDSN	HL
S. mansoni	IARKSIEPFSNRDKTLVLQFT	VKFDKTVTCGGAYIKLLGSD	IDPKTFHGETPYKIMFGPDI	CGMATKRIHVIFNY KGQN	HL
S. japanica		FT	VKFDKTVSCGGAYIKLLGSDI	DPKKFHGESPYKIF MGPDCGI-	
F. suis	LSTKFQDPLVLTDKTLVLQYE	LRLQDALNCGGAYIKLFKEN	FETPDLCNETRYVIMFGPDK	CG-STNKVHFIFRY KPNKTG—	
D. biculata	RAAKFKDDFSNKGKTLVIQYT	VKHEQGIDCGGGYVKV			
P. argentatum	ISAEYPEFSNKDKTLVF-QFS	VKHEQKLDCGGGYMKLLSGD	IDQKKFGGDTPYSIMFGPDI	CGYAQEVTHL——	K

	160	180	200	220
Human	INKDIRCKDDEFTHLYTLIV	RPDNTYEVKIDNSQVESGSL	EDDWDFLPPKKIKDPDASKP	EDWDERAKIDDPTDSKPEDW
Mouse	INKDIRCKDDEFTHLYTLIV	RPDNTYEVKIDNSQVESGSL	EDDWDFLPPKKIKDPDAAKP	EDWDERAKIDDPTDSKPEDW
Rat	INKDIRCKDDEFTHLYTLIV	RPDNTYEVKIDNSQVESGSL	EDDWDFLPPKKIKDPDAAKP	EDWDERAKIDDPTDSKPEDW
Rabbit	INKDIRCKDDEFTHLYTLIV	RPDNTYEVKIDNSQVESGSL	EDDWDFLPPKKIKDPDASKP	EDWDERAKIDDPTDSKPEDW
X. leavis	INKDIRCKDDSFTHLYTLIV	RPDNTYEVKIDNSKVESGSL	EDDWDFLPPKKIKDPEAKKP	DEWDERPKIDDPEDKKPEDW
Rana rugosa	INKDIRSKADVYSHLYTLIV	RPDNTYEVKIDNSKVESGNL	EDDWDFLPPKKVKDPEAKKP	DDWDERPKIDDPEDKKPEDW
Bovine brain	INKDIRCKDDEFTHLYTLIV	RPNNTYEVKIDNSQVESGSL	EDDWDFLPPKKIKDPDAAKP	EDWDDRAKIDDPTDSKPEDW
N. americanum	IKKDIRCKDDELTHLYTLIL	NPDNTYEVQIDGEKVESGEL	ESDWDLLPPKKIKDPDAKKP	EDWDEREYIDDADDKKPEDW
Drosophila	ISKDIRCKDDVYTHFYTLIV	RPDNTYEVLIDNEKVESGNL	EDDWDFLAPKKIKDPTATKP	EDWDDRATIPDPDDKKPEDW
Onchocerca	IKKDIRCKDDVFTHLYTLIV	NSDNTYEVQIDGEKAESGEL	EADWDFLPPKKIKDPDAKKP	EDWDEREFIDDEDDKKPEDW
C. elegans	IKKEITCKSDELTHLYTLIL	NSDNTYEVKIDGESAQTGSL	EEDWDLLPAKKIKDPDAKKP	EDWDEREYIDDAEDAKPEDW
E. gracilis	WKKEPRCETDTLSHTYTAVI	KADRTYEVLVDQVKKESGTL	EEDWEILKPKTIPDPEDKKP	ADWVDEPDMVDPEDKKPEDW
Z. mays	IKKDVPCETDQLTHVYTLII	RPDATYSILIDNEEKQTGSI	YEHWDILPPKKIKDPEAKKP	EDWDDKEYIPDPEDKKPEGY
Barley 1a	IKKDVPCETDQLSHVYTLII	RPDATYSILIDNEEKQTGSI	YEHWDILPPKEIKDPEAKKP	EDWDDKEYIPDPEDVKPEGY
N. plumbaginifolia	IKKEVPCETDQLTHVYTFIL	RPDATYSILIDNVEKQSGSL	YSDWDLLPPKTIKDPSAKKP	EDWDEKEFIDDPEDKKPEGY
C. reinhardtii	IKKDIKAETDQLTHVYTLVI	KPDNTYQVLIDLKEVASGSL	YEDWDMLPPKTIKDPKASKP	EDWDEREEIADPEDKKPEGW
Barley 2a	IKKDVPCETDQLSHVYTLII	RPDATYSILIDNEEKQTGSI	YEHWDILPPKEIKDPEAKKP	EDWDDKEYIPDPEDVKPEGY
C. annuum	IKKEVPCETDQLTHVYTFIL	RPDATYSILVDSVEFQSGSL	YSDWDILPPKTIKDPSAKKP	EDWDDK-N-SDPEDKKPEGY
S. mansoni	IKKDIPCKDDQKTHLYTLIV	RPDNSYEVLVDNEKVESGLL	EEDWNMLAPKMIDDPNDKKP	DDSQEEEYIDDPNDEKPLDW
S. japanica	MATKKVHVIFN-YKGKNHLI	KKEIPCKDDL-KTHLYTLIV	NPNNYEVLVDNADPN-DKKP	DDWVDEQFIDDPDDKKPDNW
F. suis	EEVEKHMEAPGQIKTDKLTHLYTLI	VRPDNSFEILIDAESVKQGNLLTDFSV	NPPKEIDDPSDTKP	ADWVDDEMMDDPDAIKPDDW

	240	260	280	300
Human	DKPEHIPDPDAKKPEDWDEE	MDGEWEPPVIQNPEYKGEWK	PRQIDNPDYKGTWIHPEIDN	PEYSPDPSIYAYDNFGVLGLD
Mouse	DKPEHIPDPDAKKPEDWDEE	MDGEWEPPVIQNPEYKGEWK	PRQIDNPDYKGTWIHPEIDN	PEYSPDANIYAYDSFAVLGLD
Rat	DKPEHIPDPDAKKPEDWDEE	MDGEWEPPVIQNPEYKGEWK	PRQIDNPDYKGTWIHPEIDN	PEYSPDANIYAYDSFAVLGLD
Rabbit	DKPEHIPDPDAKKPEDWDEE	MDGEWEPPVIQNPEYKGEWK	PRQIDNPDYKGTWIHPEIDN	PEYSPDANIYAYDSFAVLGLD
X. leavis	EKPEHIPDPDAVKPEDWDEE	MDGEWEPPVIQNPEYKGEWK	PRQIDNPDYKGKWIHPEIDN	PEYTPDDTLYSYDSFGVLGLD
Rana rugosa	DKPEHIPDPDAVKPEDWDEE	MDGEWEPPVITNPEYKGEWK	PRQIDNPDYKGKWVHPEIDN	PEYTPDPTLYSYADFGALVLD
Bovine brain	DKPEHIPDPDAKKPEDWDEE	MDGEWEPPLIQNPEYKGEWK	PRQIDNPEYKGIWIHPEIDN	PEYSPDSNIYAYENFAVLGLD
N. americanum	DKPEHIPDPDAKKPDDWDDE	MDGEWEPPMIDNPEYKGEWK	PKQIKNPAYKGKWIHPEIDN	PEYTPDDELYLYKDWGAIGFD
Drosophila	DKPEHIPDPDATKPEDWDDE	MDGEWEPPMIDDPEFKGEWQ	PKQLDNPNYKGAWEHPEIAN	PEYVPDDKLYLRKEICTLGFD
Onchocerca	DKPEHIPDPDAKKPEDWDDE	MDGEWEPPMVDNPEYKGEWK	PKQKKNPAYKGKWIHPEIEI	PDYTPDDNLYVYDDIGAIGFD
C. elegans	EKPEHIPDPDAKKPEDWDDE	MDGEWEPPMIDNPEYKGEWK	PKQIKNPAYKGKWIHPEIEN	PEYTPDDELYSYESWGAIGFD
E. gracilis	DEKPAQIPDPDATQPDDWDEE	EDGKWEAPMISNPKYKGEWK	AKKIPNPAYKGVWKPRDIPN	PEYEADDKVHIFDEIAAVGFD
Z. mays	DIDPKEIPDPDAKKPEDWDDE	EDGEWTAPTIPNPEYKGPWK	QKKIKNPNYQGKWKAPMIDN	PDFKDDPYIYAFDSLKYIGIE
Balrey 1a	DIDPKEVTDPDAKKPEDWDDE	EDGEWTAPTIPNPEYKGPWK	QKKIKNPNYQGKWKAPMIAN	PDFQDDPYIYAFDSLKYIGIE
N. plumbaginifolia	DIDPEEITDPDAKKPEDWDDE	EDGEWTAPTIPNPEYKGPWK	PKKIKNPNYKGKWKAPLIDN	PDFKDDPDLYVFPKLKYVGVE
C. reinhardtii	DIDPATIADKDAKKPEDWDDE	EDGTWEPPMIPNPEYKGEWK	AKMIKNPAYKGIWVAPDIDN	PDYVHDDKLYNFKDLKFVGFE
Barley 2a	DIDPKEVTDPDAKKPEDWDDE	EDGEWTAPTIPNPEYKGPWK	QKKIKNPNYQGKWKAPMIAN	PDFQDDPYIYAFDSLKYIGIE
C. annuum	DIDPEEITDPDAKKPEDWDDE	EDGEWTAPTIPNPEYKGPWK	A–KIKNPNYKGKWKAPMIDN	PEFKDDPDLYVFPKLKYVGVE
S. mansoni	DKPKTIPDMDAKKPDDWDDD	MDGEWKRPEKHNPEYKGEWS	PRRIENPKYKGQWKPAQIDN	PDYKPDPELYIQDDIGYVGFD
S. japanica	DQPKTIPDMDAKKPDDWDDA-	MDGEWERPQKDNPEYKGEWT	PRRIDNPKYKGEWKPVQIDN	PEY-KDPELYVLNDIGYVGFD
F. suis	ETDQPEFVKDPSIKDPPEGW	L-VD-EPKFITDPE—AAK	PEDWDD-DIHGEWEAPTIPN	PKCEVAPGCGEYEA-PLIRNE

	320	340	360		380
Human	DLWQVKSGTIFDNFLITNDE	AYAEEFGNETWGVTKAAEKQ	MKDKQDEEQRLKEEEEDKKR	KEEEEAEDKED	DEDKDEDEE
Mouse	DLWQVKSGTIFDNFLITNDE	AYAEEFGNETWGVTKAAEKQ	MKDKQDEEQRLKEEEEDKKR	KEEEEAEDKED	DDDRDEDED
Rat	DLWQVKSGTIFDNFLITNDE	AYAEEFGNETWGVTKAAEKQ	MKDKQDEEQRLKEEEEDKKR	KEEEEAEDKED	EDDRDEDED
Rabbit	DLWQVKSGTIFDNFLITNDE	AYAEEFGNETWGVTKTAEKQ	MKDKQDEEQRLKEEEEEKKR		KEEEEAEEDEEKDDDKEDEDE
X. leavis	DLWQVKSGTIFDNFLMTNDE	KHAEEYGNETWGVTKEAEKK	MKEQQDEEDRKKQEEEEKTR	K-EEEPQEEED	EDDDDEEEK
Rana rugosa	DLWQVKSGTIFDNFLITDDE	KFAEEHATKTWGVTKEGEKK	MKEQQDEEERKKQEEEEKKR	KEQEPAEEAED	DDDDDDDDD
Bovine brain	DLWQVKSGTIFDNFLITNDE	AYAEEFGNETWGVTKEGEKK	MKEQQDEEERKKQEEEEKKR	KEQEPAEEAED	DDDDDDDDD
N. americanum	DLWQVKSGTIFDNVLVSDSV	DEAKAHAAETFEKLKAAEKQ	MKDKQDEEQRLHEEEEEKKG	KEEEEAE-KDD	DEDKDEDEE
Drosophila	DLWQVKSGTIFDNVLITDDV	ELAAKAAAEVKN-TKPVEKE	LKEKADEENRKKMEEEAKK-	QEEEE-KKKKEK	EEKEEKED
Onchocerca	DLWQVKSGTIFDDVIVTDSV	EEAKKFGEKTLKITQAGEKK	MKEAQDEVQRKKDEEEAKKA	SDKDDEDEDDD	DEEKDDESK
C. elegans	DLWQVKSGTIFDNIIITDSV	EEAEAHAAETFDKLREGEKK	-KGKKTKKQK-KKEKNEKIK		KEKMKKRKRANRKKK
E. gracilis	DLWQVKSGTIFDNIIVTDSL	AEAKAFYDQTNGATKTVEKE	KKEKADEETR-KAEEEARKK	AEEEKEAKKDD	DEEEKEEEE
Z. mays	ELWQVKSGTLFDNIIITDDP	ALAKTFAEETWGKHKEAEKA	AFD-EAEKKK-EEEDAAKGG	DDEDDDLEDEE	DDEKADEDK
Barley 1a	ELWQVKSGTLFDNILITDDA	ALAKTFAEETWAKHKEAEKA	AFD-EAEKKK-EEEDAAKGG	DDEDDDLEDEE	DDEKADEDK
N. plumbaginifolia	ELWQVKSGTLFDNIVICDDP	EYAKAIAEETWGKQKDAEKA	AFE-EAEKKR-EEEESKAAP	ADSDAEEDDDA	DDDSDDADD
C. reinhardtii	ELWQVKSGSIFDNILVTDDL	EAAKKFAEDTWGKHKDEEKAFDMKVKKEEDEKKAKDAPPPP			VDAEAAEEEDDYEEDKEEPSG
Barley 2a	ELWQVKSGTLFDNILITDDA	ALAKTFAEETWAKHKDAEKA	AFD—EAEKK-KEEEDASK-	AGEDD-DDLDD	EDADDEDKD
C. annuum	ELWQVKSGTLFDNVVITDDP	EYAKSVAEETWGKQKDAEKA	AFQ—EAEKK-REEEESRLLP		AYSDAEDDDEADSDDAD
S. mansoni	DLWQVDSGSIFDNILITDSP	DFAKQEGERLW—RKRHDNEL	AEDQSATKSDSDKETDKAAE		EPTEEDEDVKPAENPSG
S. japanica	DLWQVDSGSIFDNILITDSP	DFAKEEGERLWRKRYDAEVA	KEQSSAKDDK-EEAEETKER		KELPD-DAKASDEPSGDHDE
F. suis	ELYKGK				

	401
Human	DEEDKEEDEEEDVPGQAKDEL
Mouse	EEDEKEEDEEES PGQAKDEL
Rat	EEDEKEED EEDATGQAKDEL
Rabbit	DEEDKDEEEEEAAAGQAKDEL
X. leavis	EEEKEEEEEEDEEETPLKDEL
Rana rugosa	EEIEEKEEEDEEESEAPQKDEL
Bovine brain	DEDEKEEEEEEDAAAAQAKDEL
N. americanum	EDEEKADEGHEEL
Drosophila	QDKDQSEHDE
C. elegans	GHDE
E. gracilis	EEDDKKDE
Z. mays	ADSDAEDSKDSDDEKQHDEL
Barley 1a	DKAGSDAEDDKDSDDEKHDEL
N. plumbaginifolia	KSESKDDEAHDEL
C. reinhardtii	MGSIKIPKEEEESGHDEL
Barley 2a	DKAGSDAEDDKDSDDEKHDEL
C. annuum	DADDKSESKDDDEAAHDEL
S. mansoni	DHDESDAEDGKDSDDEKHDEL

Index

N

O

P

Q

R

S

T

U

V

W

X

MIX
Papier aus verantwortungsvollen Quellen
Paper from responsible sources
FSC® C105338

If you have any concerns about our products,
you can contact us on
ProductSafety@springernature.com

In case Publisher is established outside the EU,
the EU authorized representative is:
Springer Nature Customer Service Center GmbH
Europaplatz 3, 69115 Heidelberg, Germany

Printed by Libri Plureos GmbH
in Hamburg, Germany